Advances in Intelligent Systems and Computing

Volume 452

Series editor

Janusz Kacprzyk, Polish Academy of Sciences, Warsaw, Poland
e-mail: kacprzyk@ibspan.waw.pl

About this Series

The series "Advances in Intelligent Systems and Computing" contains publications on theory, applications, and design methods of Intelligent Systems and Intelligent Computing. Virtually all disciplines such as engineering, natural sciences, computer and information science, ICT, economics, business, e-commerce, environment, healthcare, life science are covered. The list of topics spans all the areas of modern intelligent systems and computing.

The publications within "Advances in Intelligent Systems and Computing" are primarily textbooks and proceedings of important conferences, symposia and congresses. They cover significant recent developments in the field, both of a foundational and applicable character. An important characteristic feature of the series is the short publication time and world-wide distribution. This permits a rapid and broad dissemination of research results.

Advisory Board

More information about this series at http://www.springer.com/series/11156

Ramesh K. Choudhary · Jyotsna Kumar Mandal
Nitin Auluck · H.A. Nagarajaram
Editors

Advanced Computing and Communication Technologies

Proceedings of the 9th ICACCT, 2015

 Springer

Editors
Ramesh K. Choudhary
Asia Pacific Institute of Information
  Technology
Panipat, Haryana
India

Jyotsna Kumar Mandal
Department of Computer Science &
  Engineering
Kalyani University
Kalyani, West Bengal
India

Nitin Auluck
Indian Institute of Technology Ropar
Rupnagar, Punjab
India

H.A. Nagarajaram
Centre for DNA Fingerprinting
  and Diagnostics
Hyderabad, Telangana
India

ISSN 2194-5357   ISSN 2194-5365   (electronic)
Advances in Intelligent Systems and Computing
ISBN 978-981-10-1021-7   ISBN 978-981-10-1023-1   (eBook)
DOI 10.1007/978-981-10-1023-1

Library of Congress Control Number: 2016937965

Printed on acid-free paper

This Springer imprint is published by Springer Nature
The registered company is Springer Science+Business Media Singapore Pte Ltd.

Preface

This AISC volume contains selected papers presented at the 9th International Conference on Advanced Computing and Communication Technologies (9th ICACCT 2015), technically sponsored by Institution of Electronics and Telecommunication Engineers (India) and the Computer Society of India; held on 28 and 29 November 2015 at Asia Pacific Institute of Information Technology, Panipat, India.

The technical program at 9th ICACCT 2016 featured four keynote addresses, an industrial presentation, nine contributed paper sessions and a plenary session. Keynote speakers were Prof. J.K. Mandal, University of Kalyani; Dr. Dhananjay Bhattcharyya, Saha Institute of Nuclear Physics, Kolkata; Dr. Nitin Auluck, Indian Institute of Technology Ropar and Prof. Akhilesh Upadhyay, Sagar Institute of Research & Technology, Bhopal. Titles of their addresses were 'Security and Authentication in Public Domain Networks anchored with Chaos Dynamics'; 'Variation of Stacking Interactions in DNA and RNA: Dispersion Corrected Density Functional Theory Based Studies'; 'Scheduling Challenges for the Next Generation Heterogeneous Architectures'; 'Evolving Multimodal Framework for Human Behaviour Interpretation System' and the industrial presentation on 'Recent Trends in Mobile Networks' was made by Mr. Manish Vohra of Samsung Research Centre, New Delhi.

An overwhelming number of papers were received in response to the call for papers. The submitted papers were scanned for similarity index, and subsequently the contributions were peer reviewed by program committee members. After a rigorous peer-review process, the final selection of 58 presented and revised contributed papers, corresponding to an acceptance rate of 17 %, was determined by these evaluation processes with the recommendations of the editorial board.

The papers included in this volume cover a wide range of topics spanning theory, system and challenging applications pertaining to algorithms; soft computing algorithms, e.g. GA, SA, ACO and hybrid DS–fuzzy logic; system-on-chip design; real-time mitigation of DDoS; in silico drug design; epidemiological model; finite difference scheme for Stokes flows; simulation of Shor's algorithm; vertex bisection minimization problem; dynamic facility layout problem; 3D robot vision; Monte Carlo simulation; speech processing; data analytics; embedded systems testing; image compression and encryption; microstrip antenna design; RF-MEM systems; mesh-based networks; energy-efficient wireless sensor networks; etc.

We would like to thank the reviewers, the authors of all contributed papers and selected papers, the keynote speakers, and the session chairs, for their enormous contributions towards the success of this conference. We are thankful to the authors of the finally selected papers for suitably incorporating the changes suggested by the reviewers in the scheduled time frame. It is our pleasure to acknowledge the inspiring guidance of Conference General Chair, Prof. Lalit M. Patnaik, Adjunct Professor and INSA Senior Scientist, National Institute of Advanced Studies, Indian Institute of Science Campus, Bangalore, India in maintaining the high standards of the conference. The timely support of the Institution of Electronics and Telecommunication Engineers (India) and from Dr. Anirban Basu, Vice President and President Elect, Computer Society of India, in technically sponsoring the conference is gratefully acknowledged. We are indebted to Mr. Aninda Bose, Senior Editor, Springer India Pvt. Ltd. for his valuable advises on various issues related to the proceedings publication. The Organizing Committee at APIIT, Panipat did a wonderful job in coordinating various activities. Finally, we would like to thank the volunteers for their unforgettable, tireless and diligent efforts.

We hope the proceedings will inspire more research in computation, communication technologies and in solving complex and intricate real-world problems.

Ramesh K. Choudhary
Jyotsna Kumar Mandal
Nitin Auluck
H.A. Nagarajaram

Conference Organization

Conference General Chair

Lalit M. Patnaik, Adjunct Professor and INSA Senior Scientist, National Institute of Advanced Studies, Indian Institute of Science Campus, Bangalore, India

Advisory Committee

Sunanda Banerjee, Saha Institute of Nuclear Physics, Kolkata, India
Anirban Basu, Vice President cum President Elect, Computer Society of India
Pradeep K. Sinha, Centre for Development of Advanced Computing, Pune, India
Anthony Atkins, Staffordshire University, UK
Adrian Low, Staffordshire University, UK
Jyotsna K. Mandal, University of Kalyani, Kalyani, India
Raees Ahmed Khan, B.R. Ambedkar University, Lucknow, India
Manish Bali, NVIDIA Graphics Pvt. Ltd., Bangalore, India
Cai Wen, Guandong University of Technology, China
Yang Chunyan, Chinese Association of AI, China
Li Xingsen, Nigbo Institute of Technology, China
Adrian Olaru, University Politechnica of Bucharest, Romania
Nicole Pop, Universitatea Tehnica, Romania
Luige Vladareanu, Romanian Academy, Institute of Solid Mechanics, Romania
Andy Seddon, APIIT Malaysia, Kuala Lumpur, Malaysia
S. Venkatraman, University of Ballarat, Australia
Yashwant K. Malaiya, Colarado State University, USA
Ajay Kumar, Purnima University, Jaipur, India
B.S. Bodla, IMS, Kurukshetra University, India
Dinesh Khanduja, National Institute of Technology, Kurukshetra, India
Rachit Garg, Lovely Professional University, Phagwara, India

Sawtantar Singh, B.M. Singh College of Engineering, Muktsar, Punjab, India
Ekta Walia, South Asian University, New Delhi, India
V.P. Saxena, Jiwaji University, Gwalior, India
E.G. Rajan, Pentagram Research Centre, Hyderabad, India
Pradosh K. Roy, Asia Pacific Institute of Information Technology, Panipat, India

Technical Program Chair

Hongnian Yu, Bournemouth University, Bournemouth, UK

Program Committee

H.A. Nagarajaram, Centre for Computational Biology, CDFD Hyderabad, India
N. Seetha Rama Krishna, Intel India Pvt. Ltd. Bangalore, India
Dhananjay Bhattacharyya, Saha Institute of Nuclear Physics, Kolkata, India
Nitin Auluck, Indian Institute of Technology Ropar, Rupnagar, India
Vinay G. Vaidya, KPIT Technologies Ltd., Pune, India
Justin Champion, Staffordshire University, UK
P.N. Vinaychandran, Indian Institute of Science, Bangalore, India
E. Chadrasekhar, Indian Institute of Technology, Bombay, India
Peter Jun, Indian Institute of Technology Ropar, Rupnagar, India
Jagpreet Singh, Indian Institute of Information Technology, Allahabad, India
Sudarshan Iyengar, Indian Institute of Technology Ropar, Rupnagar, India
Apurva Mudgal, Indian Institute of Technology Ropar, Rupnagar, India
Gautam Garai, Saha Institute of Nuclear Physics, Kolkata, India
Shuang Cang, Bournemouth University, UK
Nikola Benin, Rusenski Universitet, Bulgaria
Yacine Ouzrout, Université Lumiere Lyon2, France
Ingrid Rugge, University of Bremen, Germany
M.N. Hoda, Chairman CSI Division I Hardware, India
Suresh Sathapathy, Chairman CSI Division IV Education & Research, India
Durgesh Misra, Chairman CSI Division V Communications, India
Akhilesh Upadhyay, Sagar Institute of Research, Technology & Science, Bhopal, India
Alok K. Rastogi, Institute of Excellence and Higher Education, Bhopal, India
Sunil Mishra, Institute of Excellence and Higher Education, Bhopal, India
Kanwalvir Singh Dhindsa, B.B.S.B. Engineering College, Punjab, India
Szabo Zoltan, Corvinus University of Budapest, Hungary
Matteo Savino, University of Sannio, Italy
Teresa Goncalves, Universidade de Évora, Portugal
Sergei Silvestrov, Mälardalen University, Sweden

Alamgir Hossain, University of Northumbria at Newcastle, UK
Sudip Roy, National Chemical Laboratory, Pune, India
Shilpi Gupta, S.V. National Institute of Technology, Surat, India
Kalyani Mali, University of Kalyani, Kalyani, India
Dhananjay Kulkarni, Asia Pacific Institute of Information Technology, Sri Lanka
Shankar Duraikannan, Asia Pacific Institute of Information Technology, Malaysia
Cheki Dorji, Royal University of Bhutan, Bhutan
Rameshwar Rijal, Kantipur Engineering College, Nepal
Pitipong Yodmongkon, Chiang Mai University, Thailand
V.K. Shrivastava, Asia Pacific Institute of Information Technology, Panipat, India
Chowdhury Mofizur Rahman, United International University, Dhaka 1209, Bangladesh
Trupti R. Lenka, National Institute of Technology, Silchar, India
Rajendra Sahu, Indian Institute of Information Technology, Hyderabad, India
K.R. Parpasani, M.A. National Institute of Technology, Bhopal, India
A.K. Mittal, Netaji Subhas Institute of Technology, Delhi, India
Akhilesh Swaroop, National Institute of Technology, Kurukshetra, India
Brahmjeet Singh, National Institute of Technology, Kurukshetra, India
B.B. Goyal, University Business School, Punjab University, Chandigarh, India
Akhilesh Tiwari, Madhav Institute of Technology & Science, Gwalior, India
J.S. Saini, DCR University of Science & Technology, Murthal, Haryana, India
A. Mondal, Burdwan University, Burdwan, India
A. Mukhopadhyay, Kalyani University, Kalyani, India
A.K Bhattacharya, National Institute of Technology, Durgapur, India
B. Basu, OPSIS System, Sector V, Salt Lake, Kolkata, India
B.K. Dey, Tripura University, Agartala, India
Bijay Baran Pal, Kalyani University, Kalyani, India
Balaram Bhattacharya, Vishva Bharati, Shantiniketan, India
C. Rosen, University of Derby, Derby, UK
D. Garg, Thapar University, Patiala, India
D.D. Sinha, University of Kolkata, Kolkata, India
J.P. Choudhury, Kalyani Government Engineering College, India
Jaya Sil, Bengal Engineering & Science University, Shibpur, India
K. Dasgupta, Kalyani Government Engineering College, India
M. Sandirigam, University of Peradeniya, Peradeniya, Sri Lanka
Md. U. Bokhari, Aligarh Muslim University, Aligarh, India
M.K. Bhowmik, Tripura University, Agartala, India
M.K. Naskar, Jadavpur University, Kolkata, India
N.R. Manna, North Bengal University, Siliguri, India
P. Dutta, Kalyani University, Kalyani, India
P.P. Sarkar, Purbanchal University, Koshi, Nepal
P. Jha, Indian School of Mines, Dhanbad, India
P.K. Jana, Indian School of Mines, Dhanbad, India
R.K. Samanta, Indian School of Mines, Dhanbad, India
R.K. Jena, Institute of Management Technology, Nagpur, India

S. Neogy, Jadavpur University, Kolkata, India
S. Mal, Kalyani Government Engineering College, India
S. Dutta, B.C. Roy Engineering College, Durgapur, India
S. Bhattacharyya, RCC Institute of Technology, Kolkata, India
S. Shakya, Tribhuvan University, Nepal
S. Bandyopadhyay, Calcutta University, Kolkata, India
S. Sarkar, Jadavpur University, Kolkata, India
S. Muttoo, Delhi University, Delhi, India
S. Changder, National Institute of Technology, Durgapur, India
S.K. Mondal, Kalyani Government Engineering College, India
S. Mukherjee, Burdwan University, Burdwan, India
Utpal Biswas, Kalyani University, Kalyani, India
U. Maulik, Jadavpur University, Kolkata, India
Tandra Pal, National Institute of Technology, Durgapur, India

Organizing Committee

Gaurav Gambhir, (Convenor), Department of Computer Science and Engineering, APIIT

Managing Committee

Virendra K. Srivastava, Department of Computer Science and Engineering, APIIT
Arun K. Choudhary, Department of Computer Science and Engineering, APIIT
Ravi Sachdeva, Department of Computer Science and Engineering, APIIT
Sachin Jasuja, Department of Mechatronics and Mechanical Engineering, APIIT
Rajesh Tiwari, Department of Management, APIIT
Prateek Mishra, Department of Computer Science and Engineering, APIIT
Geeta Nagpal, Department of Computer Science and Engineering, APIIT
Virendra Mehla, Department of Mechatronics and EE Engineering, APIIT
Vijayendra Sharma, Department of Mechatronics and EE Engineering, APIIT
Pradeep Singla, Department of Mechatronics and EE Engineering, APIIT
Sandeep Jain, Department of Applied Sciences, APIIT
Sipra Chaudhuri, Department of Management, APIIT
Sushil Kumar Sinha, Department of Applied Sciences, APIIT

General Chair's Message

The International Conference on Advanced Computing and Communication Technologies 2015 (ICACCT 2015) has witnessed several transformations over the years and was organized in Asia Pacific Institute of Information Technology, Panipat, in its present form during 28–29 November 2015. The papers submitted were subjected to similarity/plagiarism check and technical review before presentation. To ensure that the similarity percentage is consistently maintained within permissible limits, final versions of papers to appear in the proceedings volume were again subjected to similarity index check by Springer. The editors, Program Chair, Program Committee members, reviewers, keynote speakers, session chairs and volunteers have done their jobs excellently towards the success of the technical program.

We are grateful to the Computer Society of India and the Institution of Electronics and Telecommunication Engineers for being the technical co-sponsors.

As General Chair of Conference, I convey my profuse thanks to Prof. Ramesh K. Choudhary, Prof. J.K. Mandal, Prof. R.K. Sharma, Prof. Rakesh Kumar, Dr. Nitin Auluck, Dr. H.A. Nagarajaram, Dr. Dhananjay Bhattacharyya, Dr. Akhilesh Upadhyay, Mr. Manish Vohra and Prof. Pradosh K. Roy for their excellent technical support. Mr. Gaurav Gambhir and Mrs. Monika Gambhir provided continuous support to process the submitted papers. Professor Ramesh K. Choudhary has supported me in this task by extending the patronage of the management of APIIT.

Finally, my special appreciation to Dr. Anirban Basu, Vice President and President Elect, Computer Society of India, and to Mr. Aninda Bose, Senior Editor, Springer India Pvt Ltd, New Delhi for their timely support and advises.

I am sure this volume will find a proud place in the collection of books of researchers engaged in computing and communication technologies.

February 2016   Lalit M. Patnaik
General Chair, 9th ICACCT 2015
INSA Senior Scientist and Adjunct Professor
National Institute of Advanced Studies
Indian Institute of Science Campus
Bangalore, India

Contents

Part II   Communication Technologies

About the Editors

Ramesh K. Choudhary Director, Asia Pacific Institute of Information Technology, Panipat, obtained ME from Moscow Academy of Instrument Making and Information Science, Moscow and Ph.D. from VMU, Salem. Professor Choudhary has co-authored 'Testing of Fault Tolerance Techniques' published by Lambert Academic Publishing, Germany. He is the Indian Co-ordinator, Featured EUrope and South Asia MObility Network (FUSION), an ERUSMUS MUNDUS (EU) project to foster partnerships of emerging Asian countries with the EU countries to reinforce the existing collaborations developed through the EU funded projects. As a member of Extension Engineering Specialized Committee, Chinese Association for Artificial Intelligence, Guangdong University of Technology, China. Indian Coordinator of Project LEADER in collaboration with University of Sannio, Italy and Jao Tong University, China, he aims to stimulate EU-ASIA mutual recognition of studies in engineering, management and informatics. His current areas of research include software engineering and fault-tolerant computing. He is Fellow of Institution of Engineers (India), Institution of Electronics and Telecommunication Engineers, and Member of IEEE, ACM, Chinese Association for Artificial Intelligence.

Jyotsna Kumar Mandal former Dean, Faculty of Engineering, Technology and Management, University of Kalyani, obtained M.Tech. in Computer Science from Kolkata University and Ph.D. (Engg.) in Computer Science and Engineering from Jadavpur University. Professor Mandal has co-authored six books, viz. Algorithmic Design of Compression Schemes and Correction Techniques—A Practical Approach; Symmetric Encryption—Algorithm, Analysis and Applications: Low Cost based Security; Steganographic Techniques and Application in Document Authentication—An Algorithmic Approach; Optimization based Filtering of Random Valued Impulses—An Algorithmic Approach; and Artificial Neural Network Guided Secured Communication Techniques: A Practical Approach; all published by Lambert Academic Publishing, Germany. His total publication is more than 350 papers in international journals and proceedings. His current areas of research include coding theory, data and network security, remote sensing and

GIS-based applications, data compression, error correction, visual cryptography and steganography, distributed and shared memory parallel programming. Professor Mandal's biography is included in the 31st edition of Marque's World Who's Who published in 2013. He is Fellow of Institution of Electronics and Telecommunication Engineers, and Member of IEEE, ACM, and Computer Society of India.

Nitin Auluck is Associate Professor, Department of Computer Science and Engineering at the Indian Institute of Technology Ropar, Rupnagar, India. Prior to that, he was Assistant Professor, Department of Computer Science at Quincy University in Quincy, Illinois, USA. Dr. Auluck obtained Ph.D. in Computer Science and Engineering from the University of Cincinnati, Cincinnati, Ohio, USA in 2005 under the supervision of Prof. Dharma P. Agrawal. Dr. Auluck has published in the IEEE Transactions on Parallel and Distributed Systems, Transactions on Computers and Intelligent Systems, Proceedings of the IEEE International Symposium on Parallel Architectures, Algorithms and Programming 2012, Taipei, Taiwan, Proceedings of the 22nd ACM High Performance Computing Symposium 2014, Tampa, USA, Proceedings of the International Conference on Embedded and Ubiquitous Computing 2004, Aizu, Japan. His current areas of research include real-time systems, scheduling, parallel and distributed computing.

H.A. Nagarajaram is Group Leader, 'Laboratory of Computational Biology' at Centre for DNA Fingerprinting and Diagnostics (CDFD, Hyderabad) (Department of Biotechnology (DBT), Government of India). Dr. Nagarajaram obtained Ph.D. from Molecular Biophysics Unit (MBU), Indian Institute of Science, Bengaluru and did postdoctoral research with Sir Tom Blundell FRS, at the Department of Biochemistry, University of Cambridge, England. He joined CDFD in 2000 as one of its faculty members and established his research group. His current research interests mainly focus on proteins with regard to their structures, variations and interactions. He has guided several doctoral and graduate students. He has published more than 70 research articles in several international peer-reviewed journals of high impact, e.g. Proceedings of the National Academy of Sciences USA, Protein Science, Bioinformatics, Nucleic Acids Research, Journal of Proteome Research and Molecular Biosystems. Dr. Nagarajaram is member of The National Academy of Sciences, Allahabad, India.

About the Book

The book contains selected papers presented at the 9th International Conference on Advanced Computing and Communication Technologies 2015 (9th ICACCT 2015) held in Asia Pacific Institute of Information Technology, Panipat, India, during 28–29 November 2015. The volume has 58 papers, including the keynote address on computational biology, grouped into three parts, viz. Advanced Computing, Communication Technologies, Antenna Design and Power Systems. The book covers recent and significant research done in the fields of biocentric computing, programming systems and software engineering, data analytics and visualization, hybrid intelligent models, computer vision, robotics and automation, information theory and coding, mobile networks, wireless communications, antenna design and power systems. The volume is directed to researchers aspiring to solve complex and intricate real-world problems using advances in computation and communication technologies.

Part I
Advanced Computing

Variation of Stacking Interactions Along with Twist Parameter in DNA and RNA: DFT-D Studies

Sanchita Mukherjee, Manas Mondal and Dhananjay Bhattacharyya

Abstract In this study the structural features of dinucleotide sequences are quantitatively described in terms of the important orientation parameters of the bases and Watson-Crick base pairs using quantum chemical calculations. Variation of twist parameter in DNA and canonical RNA reveals the most probable value for each sequence. However, all the dinucleotide steps do not show the minima corresponding to specific twist value. All the quantum chemical calculations are done using DFT-D with ωB97X-D functional and 6-31G (2d,2p) basis set. In order to completely understand behavior of the base paired dinucleotide step, we have adopted Metropolis Monte Carlo simulation in 18 dimensional phase space using classical force-field. The results agree in general with the results obtained from DFT-D studies as well as experimental observations. We believe that Monte Carlo simulation with energy from DFT-D would better predict structural features of non-canonical base pairs, which appear frequently in RNA and lacks experimental details.

Keywords Non-canonical · Base pair · Dinucleotide · Monte carlo

S. Mukherjee
Biological Sciences, Indian Institute of Science Education and Research Kolkata, Mohanpur 741246, West Bengal, India

M. Mondal
Chemical, Biological and Macromolecular Science, S.N. Bose National Center for Basic Sciences, Sector III, Salt Lake, Kolkata 700098, India

D. Bhattacharyya (✉)
Computational Science Division, Saha Institute of Nuclear Physics, 1/AF Bidhannagar, Kolkata 700064, India
e-mail: dhananjay.bhattacharyya@saha.ac.in

© Springer Science+Business Media Singapore 2016
R.K. Choudhary et al. (eds.), *Advanced Computing and Communication Technologies*,
Advances in Intelligent Systems and Computing 452,
DOI 10.1007/978-981-10-1023-1_1

1 Introduction

Polymeric structures of nucleic acids are composed of five member sugars, phosphates groups and four varieties of nitrogenous planer aromatic bases, namely Adenine, Guanine, Thymine (or Uracil in RNA) and Cytosine. The bases form base pairs by specific hydrogen bonds between complementary bases and these base paired stacks give rise to the long anti-parallel double helical DNA. In the canonical Watson-Crick base pairing pattern in DNA, adenine (A) forms base pair with thymine (T), and guanine (G) with cytosine via hydrogen bonding interaction. RNA also has similar patterns of Watson-Crick base pairing between adenine (A) and uracil (U), and between guanine (G) and cytosine (C). However, RNA structures are collections of short helices interspersed by unpaired regions and packed together into compact structures, instead of monotonous long helices. Unlike DNA, RNA double helices do not have structural polymorphism. They can only adopt the A-form helical conformation with very narrow major groove and shallow and wide minor groove, in which minor groove sides of A:U and G:C Watson-Crick base pairs do not possess much sequence dependent variations of hydrogen bonding signatures required for specific recognition by proteins or ligands. Instead, RNA forms a range of structural patterns through base pairing and base stacking, giving rise to various motifs and folds as the building blocks of three-dimensional organization of double helices, hairpin loops, internal loops, kissing loops, pseudo-knots, coaxial stacks etc. [1]. The three dimensional structures are stabilized by long-range intra-molecular interactions between the secondary structural elements to yield complex motifs. It has been established through many experimental works that non-canonical base pairs are the key to tertiary structural organization of RNA. They are seen to be present in various types of RNA secondary structural motifs and help to maintain the helical architecture and reduce the conformational strain introduced in the double helix. The substantial stability of non-canonical base pairs is indicative of their importance as a seed or 'nucleation site' for folding of functional RNAs.

Along with hydrogen bonding interaction, structure of DNA and RNA is stabilized by the stacking interactions between the successive base pairs. This interaction between the base pairs arises from favorable van der Waals and hydrophobic contacts that can optimize the water-insoluble areas of contact. Experimental studies reveal that compared to the general Watson-Crick type of hydrogen-bonds, stacking interaction has larger effect on the structure and electronic properties of DNA [2, 3]. Stacking energy shows high sequence sensitivity and plays a key role in the stabilization of DNA and RNA secondary structure and specific types of DNA-protein or DNA-ligand interactions. The base-stacking interaction generally influences the physicochemical properties of base pair opening, formation of intercalating sites, bending propensity etc. Considering the nucleic acid bases as planer rigid moieties, their mutual orientations are routinely used to describe structure of base pairs formed by hydrogen bonding between the bases. These are the three translational (stagger, shear, stretch) and three rotational (buckle, open, propeller) parameters to describe orientation of one base with respect to its paired one (intra base pair parameters). Similarly there are three translational (shift, slide, rise) and three rotational (tilt, roll, twist) parameters to describe relative orientation of one base pair with respect to its neighboring stacked one, as recommended by IUPAC-IUB [4]. Hence a double helical fragment consisting of N base pairs can be completely characterized by 6 N intra-base pair parameters and 6(N-1) inter-base pair step parameters, all of them are considered to be mathematically orthogonal. These base pair parameters not only describe DNA or RNA double helical structural features but also can be used to construct molecular model of base pair coordinates with a given geometry [4–8]. Crystal and NMR structure database analyses reflect the sequence dependent preference of some of these parameters namely roll, slide and twist; and influence of them in base stacking and base pair step geometry [9–21]. Stability of non-canonical base pair and their base pairing pattern within a double helical region of RNA is also influenced by these parameters [8, 22–27]. However their preferred stacking geometry with other canonical and non-canonical base pairs are not understood well. The frequently observed G:U base pairs in RNA provide a complex array of hydrogen bond donors and acceptors creating a surface area for binding with proteins or metal ions. The structural transition of G:U base pairs is crucial to induce flexibility at the junction between canonical and non-canonical regions within RNA double helix [27]. A G:U base pair, within the acceptor stem of tRNA, is known to play a vital role in amino-acylation process and its bi-phasic stability at wobble position may also play an important role in the regulation of short-lived codon-anticodon recognition by acting as a conformational switch [28–32]. Due to shearing motion of the G:U base pairs the long axes directions of the G:U base pair changes significantly and the stacked conformation of the dinucleotide sequences containing this base pair becomes different from those of canonical base pair stacks. It was observed that difference in shear values (Δshear) between two consecutive base pairs in a dinucleotide sequence can influence the twist value [1, 21, 26].

There were a few attempts considering classical force field based methods and different quantum chemical methods to characterize preferred geometry of different base paired dinucleotide sequences in DNA and RNA in terms of stacking interaction with various values of few of the base pair parameters such as twist, rise, roll and slide keeping the others fixed and mixed results were obtained [11, 15, 18, 33–35]. For finding the stacking energy landscape in few of these sequence dependent base pair parameter hyperspace (e.g. roll and slide) ab initio quantum chemical methods are found to be accurate as it considered atomic details as well as their electron distribution and other non-covalent interactions explicitly [19, 34–44]. Dispersion corrected Density Functional Theory (DFT-D) is found to be most useful for such kind of calculations for these systems. This method can reproduce the preferred stacking geometry for few dinucleotide sequences of DNA and RNA as experimentally observed structure [34, 35]. However, role of different components of stacking energy in different zones of stability in terms of base pair and base pair step parameters for different dinucelotide steps are not quite clear from these studies. Consideration of all the eighteen base pair parameters of a dinucleotide sequence can be more realistic and informative to study the sequence dependence of preferred stacking geometry of a base paired dinucleotide step. However this is extremely difficult by using quantum chemical methods due to system size. On the other hand, from the previous studies the performance of classical force field based method is found to be quite accurate for predicting the preferred stacking geometry of the dinucleotide sequences [11, 16, 34, 45, 46].

In the present study we try to find out preferred stacking geometry and variation of stacking energy for the unique dinucleotide sequences of DNA and RNA in roll-slide parameters hyperspace with different twist values using DFT-D approach. Sequence dependent preferred stability zones of the dinucleotide sequences in roll-slide hyperspace with different twist values are observed. Consideration of classical force field based Monte Carlo simulation method using rigid base model is also found to be quite good for finding the proper stacking geometry of the dinucleotide sequences in multidimensional base pair parameters hyperspace. We have also used our methodology to find out the preferred stacking geometry of the non-canonical G:U base pair containing dinuclotide sequences and the study indicates their sequence dependent preference of twist value and stacking geometry. Overall the studies reproduce the sequence dependent motions between the base pair parameters to maintain different possible interactions between two base pairs, help to attain proper stacking geometry of the dinucleotide sequences.

2   Methodology

2.1   Modeling of Structure

NUCGEN software [47] has been used to generate coordinates of the atoms of the bases of the dinucleotide steps within significant domain of most sequence dependent parameters values [34]. The structures were generated for twist parameter variation in DNA and RNA, where the twist is varied from $0°$ and $45°$ in steps of $5°$. The DNA-like structures were generated with roll $0°$ and slide 0 Å values and different twist, whereas the RNA-like structures were generated with roll $+10°$ and slide -1. All the other parameters were kept fixed at their ideal values: buckle = $0.0°$, open = $0.0°$, propeller = $-10°$, stagger = 0.0 Å, stretch = 2.84 Å, tilt = $0.0°$, shift = 0.0 Å, rise = 3.2 Å.

For the roll-slide parameter hyperspace generation, the roll values were varied between $-20°$ and $+20°$ in steps of $4°$ and slide between -3.0 Å and 0 Å in steps of 0.3 Å (Fig. 1) for RNA [34].

RNAhelix program was used to generate coordinates of atoms for the dinucleotides having G:U base pairs. This program is quite similar to NUCGEN but is more versatile for non-canonical base pairs. The shear value was kept fixed at 0.0 Å for canonical A:T, A:U and G:C base pairs, at -2.25 Å for G:U base pair and 2.25 Å for U:G base pair. We have therefore generated 605 modeled dinucleotide step

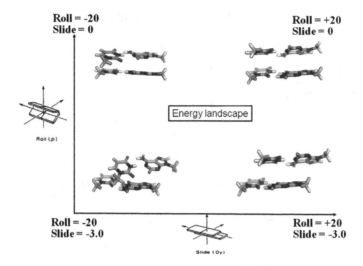

Roll = -20
Slide = 0

Roll = +20
Slide = 0

Energy landscape

Roll (ρ)

Roll = -20
Slide = -3.0

Roll = +20
Slide = -3.0

Slide (Dy)

Fig. 1 Energy map construction in roll-slide hyperspace with UU/AG base pair step (Roll in degree, slide in Å)

structures for each dinucleotide sequence using above parameters, where the sugar-phosphate backbone is replaced by methyl group as in normal practice [34, 35, 48].

2.2 Quantum Chemical Study

Single point energy calculations were carried out on the generated structures using empirical dispersion corrected DFT-D functional ωB97X-D [49] by Gaussian09 package [50] using 6-31G(2d,2p) basis set.

The total interaction energy for a base pair dinucleotide step (Fig. 1) is evaluated as,

$$E_{int} = E(UU/AG) - E(U{:}A) - E(U{:}G).$$

2.3 Classical Study

As an extensive grid search in 18 dimensional base pair parameters space is extremely difficult, statistical ensembles of dinucleotide geometries for different dinucleotide sequences have been generated by Monte Carlo Metropolis method at physiological temperature. In this study classical force field based van der Waals and Columbic potential energy is used between atoms of the four bases forming a

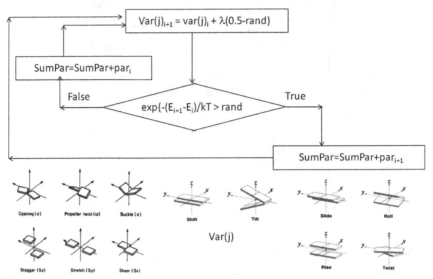

Fig. 2 Flow chart of Monte Carlo simulation method

base paired dinucleotide sequence. Considering distances between C1' atoms along the backbone chain remain within a well-defined range in the double helical structures of DNA or RNA, a harmonic potential corresponding to these distances to the above potential energy is incorporated. The harmonic force constant (k) corresponding to this harmonic potential across the C1'…C1' distance is considered as 5.0 kcal/mol/Å^2 [34]. In each Monte Carlo step we have randomly moved any one of the parameters (var(j)) among the 18 base pair parameters and accepted or rejected the corresponding move according to Metropolis criteria as shown in Fig. 2. Well established empirical AMBER force field (parm99bs0) [51] has been used for energy calculations.

Using this Monte Carlo simulation methodology we have generated ensembles of configurations for each dinucleotide sequences of DNA and RNA. There are eleven possible unique dinucleotide sequences formed by G:U non-canonical base pairs stacked on each other or on A:U or C:G canonical base pairs, i.e. (G:U::A:U) GA/UU, (G:U::G:C) GG/CU, (G:U::G:U) GG/UU, (U:G::C:G) UC/GG, (U:G::U: A) UU/AG, (G:U::C:G) GC/GU, (G:U::U:A) GU/AU, (G:U::U:G) GU/GU, (U:G:: G:C) UG/CG, (U:G::G:U) UG/UG, (U:G::A:U) UA/UG. From the ensembles of probable configurations for each base paired dinucleotide sequence their preferred stacking geometry has been analyzed. Stacking energy (E_{stack}) between any two base pairs as shown (Fig. 3), is defined as,

Fig. 3 A representative
model structure of
dinucleotide step (UU/AG;
represented as AC/DB)

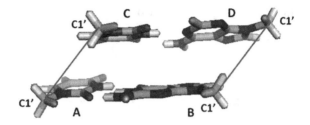

E_{stack} = Interaction Energy between (A and C + B and D + A and D + B and C) + Backbone penalty term.

The base pair and other structural parameters have been calculated using NUPARM [6, 8].

3　Results and Discussion

3.1　Stacking Energy Variations of the 10 Unique Dinucleotide Sequences of DNA and RNA with Twist

Stacking energy variations of the 10 unique dinucleotide steps along with twist (from 0° to 45°) show different favorable twist values for different steps (Fig. 4). We can broadly classify them as the type of bases involved, as purine-pyrimidine,

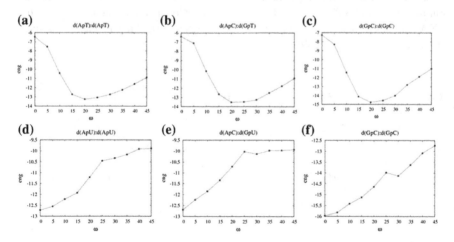

Fig. 4 Stacking energy variation with twist of the purine-pyriminide steps in DNA (a,b,c) and RNA (d,e,f). (X-axis is twist (ω) in degree and y-axis is stacking energy (eng) in kcal/mol.)

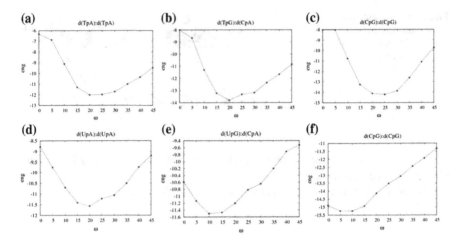

Fig. 5 Stacking energy variation of the pyriminide-purine steps in DNA (a,b,c) and RNA (d,e,f). (X-axis is twist (ω) in degree and y-axis is stacking energy (eng) in kcal/mol.)

pyrimidine-purine and purine-purine steps. Figure 4a, b, c show that for purine-pyrimidine steps in DNA-like structures, an energy minima is observed near 20–25° twist values. However no energy minima is observed in case of RNA-like structures (Fig. 4d, e, f). Pyrimidine-purine sequence dinucleotide steps show energy minima describing the favorable twists for the orientation of the step (Fig. 5). The DNA structures attend energy minima at 25° or above, whereas RNA steps show favorable twists at much lower values (Fig. 5d, e, f). For purine-purine steps the DNA base pair steps show much higher favorable twist values, where the minima is seen at 30° or larger (Fig. 6a, b, c, d). No such minima are seen for the RNA-line conformations. The stabilization energy for stacking in these structures are around −11 to −13 kcal/mol.

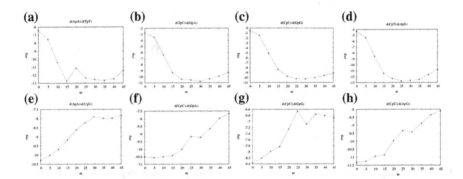

Fig. 6 Stacking energy variation of the purine-purine steps in DNA (a,b,c,d) and RNA (e,f,g,h). (X-axis is twist (ω) in degree and y-axis is stacking energy (eng) in kcal/mol.)

3.2   Stacking Geometry of the Non-canonical G:U Base Pair Containing Dinucleotide Sequences in Roll-Slide Hyperspace

The bases in G:U base pair are generally sheared with respect to each other (–2.42 Å for G:U and 2.42 Å for U:G). Due to the shearing motion of the G:U base pair the long axes directions will be different and the overall stacking geometry of the G:U base pair containing dinucleotide sequences are expected to be different from canonical ones. It was observed from crystallographic data that due to this shearing motion twist preference is different depending on the position of the G:U base pair in the dinucleotide stacks. Positive ΔShear values facilitate to acquire twist values smaller than 33°, whereas negative ΔShear values give the preference of higher twist values. Therefore twist is the most important parameter to maintain the overall stacking geometry of these dinucleotide sequences. Variations in average twist for different dinucleotide sequences containing G:U base pair is very prominent from the recent crystal and NMR structure database analysis [21]. Thus to analyze the twist preference of these dinucleotide sequences in roll-slide parameter hyperspace and in terms of stacking energy, we have generated the stacking energy contours for several different twist values ranging between 20° and 40°. Quantum chemical method is used for evaluation of stacking energy at different conformations in roll-slide hyperspace. Variation of minimum stacking energy along with twist indicates the preferred zones of stability and sequence dependent feature. The purine-pyrimidine sequences show stronger stacking interaction for smaller twist while the pyrimidine-purine sequences show the reverse trend. In the UA/UG, UG/CG, UU/AG and UG/UG sequences, where ΔShear values are negative, tendency of stronger total stacking energy for larger twist with larger negative stacking interaction energy have been observed. However these stacking energy contours are not well defined with respect to slide. The UC/GG sequence on the other hand, shows good agreement between experimental data and stacking energy at higher twist values (Fig. 7). The GG/CU, GC/GU, GU/AU, GA/UU and GU/GU sequences have positive ΔShear values and it is expected that these sequence will prefer smaller twist. For GC/GU, GU/AU and GU/GU sequences higher stacking energy is observed at smaller twist values and in the well defined stacking energy contour the slide values corresponding to minimum energy are also large negative as observed in crystal structures. In the GA/UU and GG/CU sequences the minimum total stacking energy is observed at higher twist. However at lower twist values negative slides become more preferred and we observe excellent agreement between experimental data clustering around the lowest total stacking energy regions (Fig. 7). Such large negative slide is almost a pre-requisite for A-form structure and hence these sequences may adopt lower twist values. The stacking energy values corresponding to these dinucleotide sequences indicate that stability is higher when G:U base pair stacks with one canonical base pair rather than the sequences where both the base pairs are G:U. The minor discrepancies between

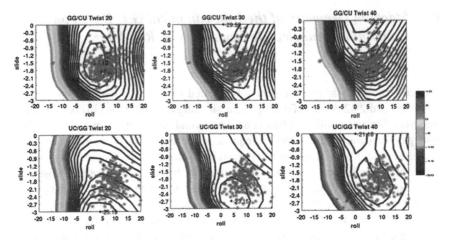

Fig. 7 Stacking energy contour for GG/CU and UC/GG dinucleotide sequences in roll-slide hyperspace at three different twist values

stacking energy based prediction and experimental values are possibly due to neglecting the variations of the other parameters such as propeller, open etc.

3.3 Sequence Dependent Stacking Geometry of the Unique Dinucleotide Sequences of DNA and RNA in 18 Dimensional Base Pair Parameters Hyperspace

The preferred geometry of the dinucleotide sequences are hence analyzed in terms of all the base pair parameters using Monte Carlo simulation by altering all 18 parameters. The average values of all the base pair parameters for each dinucleotide step of DNA and RNA are evaluated from the ensembles of geometries considering constant dielectric and distance dependent dielectric function. Among the base pair parameters roll, twist and slide are observed as most sequence dependent parameters. Sequence dependent preference of these parameters at different dielectric medium (Fig. 8) indicates their role to maintain the proper stacking geometry of the dinucleotide sequences. Generally preference of larger roll and slide values are observed corresponding to $\varepsilon = 1$, leading to poorer stacking. Positive roll and near zero slide values are mostly observed in case of pyrimidine-purine (YR/YR) (TA/TA, CA/TG and CG/CG) sequences. Negative slide and small positive roll values are observed for purine-pyrimidine (RY/RY) (AT/AT, AC/GT, GC/GC) sequences. Preferences of these base pair parameters are in agreement with Calladine's steric clash based rule [9, 13]. In case of RNA positive roll and negative slide values are preferred for GA/UC, AU/AU and UA/UA sequences. In case of RY/RY sequences positive roll and higher negative slide become more favorable in

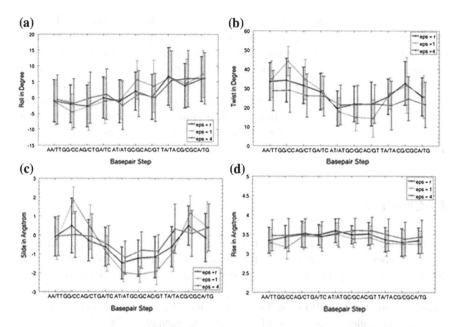

Fig. 8 Sequence dependent variation of average base pair step parameters (Roll, Twist, Slide and Rise) for unique dinucleotide sequences of DNA

lower dielectric medium ($\varepsilon = 1$) with lesser values of stacking overlap. In lower dielectric medium the rise values also reduces for most of the base pair sequences. Twist is found to be an important stacking parameter which controls the stacking geometry of the dinucleotide sequences in sequence dependent manner. Preference of lower twist values are observed for the AT/AT, AC/GT and GC/GC sequences. According to Calladine steric clash hypothesis these purine-pyrimidine steps may adopt smaller twist to avoid steric clash between purine residues in the major groove side [9, 11]. Depending on these base pair step geometry stacking energy is also found to be sequence dependent. Higher stacking energy at lower twist, small positive roll and negative slide value is observed for GC/GC and AT/AT sequences compared to CG/CG and TA/TA sequences. In case of RNA, AG/CU and CA/UG steps have higher stacking energy than the other sequences.

The G:U base pair containing dinucleotide sequences are also seen to follow the Calladine's rule. Negative or small roll values are observed for the purine-pyrimidine sequences, GC/GU, GU/AU and GU/GU. However positive roll values are not seen for few pyrimidine-purine sequences, UG/UG and UA/UG. Probably preferences of large positive slide for these dinucleotides help to reduce the purine-purine clash. In this case preference of smaller twist values are observed for GC/GU and GU/AU dinucleotides which have positive ΔShear values. However due to compensatory effect GU/GU, GA/UU dinucleotide appears to avoid smaller twist by adopting larger rise values while the GG/CU adopts larger twist associated with unusual positive slide. On the other hand, except UG/CG sequence, preference of large twist values are observed for UG/UG, UC/GG, UU/AG and UA/UG sequences, which have negative ΔShear values. Preference of large twist is observed for UG/UG sequences due to its large negative ΔShear value (–4.7). Thus, this study is able to predict the expected twist values for most of the dinucleotide sequences to maintain the favorable stacking geometry. A better method for calculation of energy, such as DFT-D method might improve prediction of the structural parameters. Like quantum chemical study here we have also observed lower stacking energy for the dinucleotide sequences having only G:U non-canonical base pairs.

4 Conclusion

We observed quite good agreement between DFT-D derived preferred geometrics and experimental data for most of the dinucleotide sequences. We presume the difference between predictions and observations are due to consideration of ideal values of most of the intra and inter base pair parameters. The average parameters obtained from Monte Carlo simulation using simplistic and fast energy calculation method also could explain some of the preferred geometrics. We believed that a Monte Carlo simulation using DFT-D derived energy might improve prediction method significantly. However, it is extremely difficult yet to perform such calculations using presently available computational power.

References

1. Halder, S., Bhattacharyya, D.: Prog. Biophys. Mol. Biol. **113**, 264 (2013)
2. Yakovchuk, P., Protozanova, E., Frank-Kamenetskii, M.D.: Nucleic Acids Res. **34**, 564 (2006)
3. Hagerman, K.R., Hagerman, P.J.: J. Mol. Biol. **260**, 207 (1996)
4. Olson, W.K., Bansal, M., Burley, S.K., Dickerson, R.E., Gerstein, M., Harvey, S.C., Heinemann, U., Lu, X.J., Neidle, S., Shakked, Z., Sklenar, H., Suzuki, M., Tung, C.S., Westhof, E., Wolberger, C., Berman, H.M.: J. Mol. Biol. **313**, 229 (2001)
5. Babcock, M.S., Olson, W.K.: J. Mol. Biol. **237**, 98 (1994)
6. Bansal, M., Bhattacharyya, D., Ravi, B.: Comput. Appl. Biosci. **11**, 281 (1995)
7. Lu, X.J., Olson, W.K.: Nucleic Acids Res. **31**, 5108 (2003)
8. Mukherjee, S., Bansal, M., Bhattacharyya, D.: J. Comput. Aided Mol. Des. **20**, 629 (2006)
9. Calladine, C.R.: J. Mol. Biol. **161**, 343 (1982)
10. Bhattacharyya, D., Bansal, M.: J. Biomol. Struct. Dyn. **8**, 539 (1990)
11. Hunter, C.A.: J. Mol. Biol. **230**, 1025 (1993)
12. Gorin, A.A., Zhurkin, V.B., Olson, W.K.: J. Mol. Biol. **247**, 34 (1995)
13. el Hassan, M.A., Calladine, C.R.: J. Mol. Biol. **259**, 95 (1996)
14. ElHassan, M.A., Calladine, C.R.: Philos. Trans. R. Soc. A-Math. Phys. Eng. Sci. **355**, 43 (1997)
15. Hunter, C.A., Lu, X.J.: J. Biomol. Struct. Dyn. **14**, 747 (1997)
16. Hunter, C.A., Lu, X.J.: J. Mol. Biol. **265**, 603 (1997)

17. Olson, W.K., Gorin, A.A., Lu, X.J., Hock, L.M., Zhurkin, V.B.: Proc. Natl. Acad. Sci. U.S.A. **95**, 11163 (1998)
18. Packer, M.J., Dauncey, M.P., Hunter, C.A.: J. Mol. Biol. **295**, 85 (2000)
19. Samanta, S., Kabir, M., Sanyal, B., Bhattacharyya, D.: Int. J. Quantum Chem. **108**, 1173 (2008)
20. Kailasam, B.S.: S, Bhattacharyya D. BMC Res. Notes **7**, 31 (2014)
21. Pingali, P.K., Halder, S., Mukherjee, D., Basu, S., Banerjee, R., Choudhury, D., Bhattacharyya, D.: J. Comput. Aided Mol. Des. **28**, 851 (2014)
22. Sponer, J.E., Spackova, N., Kulhanek, P., Leszczynski, J., Sponer, J.: J. Phys. Chem. A **109**, 2292 (2005)
23. Sponer, J.E., Spackova, N., Leszczynski, J., Sponer, J.: J. Phys. Chem. B **109**, 11399 (2005)
24. Roy, A., Panigrahi, S., Bhattacharyya, M., Bhattacharyya, D.: J. Phys. Chem. B **112**, 3786 (2008)
25. Sharma, P., Mitra, A., Sharma, S., Singh, H., Bhattacharyya, D.: J. Biomol. Struct. Dyn. **25**, 709 (2008)
26. Halder, S., Bhattacharyya, D.: J Phys Chem B **114**, 14028 (2010)
27. Halder, S., Bhattacharyya, D.: J Phys Chem B **116**, 11845 (2012)
28. Pley, H.W., Flaherty, K.M., Mckay, D.B.: Nature **372**, 111 (1994)
29. Scott, W.G., Finch, J.T., Klug, A.: Cell **81**, 991 (1995)
30. Baeyens, K.J., DeBondt, H.L., Pardi, A., Holbrook, S.R.: Proc. Natl. Acad. Sci. U.S.A. **93**, 12851 (1996)
31. Cate, J.H., Doudna, J.A.: Structure **4**, 1221 (1996)
32. Varani, G., McClain, W.H.: EMBO Rep. **1**, 18 (2000)
33. Gonzalez, O., Petkeviciute, D., Maddocks, J.H.: J. Chem. Phys. 138 (2013)
34. Mukherjee, S., Kailasam, S., Bansal, M., Bhattacharyya, D.: Biopolymers **101**, 107 (2014)
35. Mukherjee, S., Kailasam, S., Bansal, M., Bhattacharyya, D.: Biopolymers **103**, 134 (2015)
36. Sponer, J., Gabb, H.A., Leszczynski, J., Hobza, P.: Biophys. J. **73**, 76 (1997)
37. Hobza, P., Sponer, J.: J. Am. Chem. Soc. **124**, 11802 (2002)
38. Cooper, V.R., Thonhauser, T., Puzder, A., Schroder, E., Lundqvist, B.I., Langreth, D.C.: J. Am. Chem. Soc. **130**, 1304 (2008)
39. Morgado, C.A., Jurecka, P., Svozil, D., Hobza, P., Sponer, J.: Phys. Chem. Chem. Phys. **12**, 3522 (2010)
40. Svozil, D., Hobza, P., Sponer, J.: J. Phys. Chem. B **114**, 1191 (2010)
41. Banas, P., Mladek, A., Otyepka, M., Zgarbova, M., Jurecka, P., Svozil, D., Lankas, F., Sponer, J.: J. Chem. Theory Comput. **8**, 2448 (2012)
42. A.R. McDonald, E.J. Denning and A.D. MacKerell, Abstracts of Papers of the American Chemical Society, 245 (2013)
43. Parker, T.M., Hohenstein, E.G., Parrish, R.M., Hud, N.V., Sherrill, C.D.: J. Am. Chem. Soc. **135**, 1306 (2013)
44. Sponer, J., Sponer, J.E., Mladek, A., Jurecka, P., Banas, P., Otyepka, M.: Biopolymers **99**, 978 (2013)
45. Srinivasan, A.R., Torres, R., Clark, W., Olson, W.K.: J. Biomol. Struct. Dyn. **5**, 459 (1987)
46. Mohanty, D., Bansal, M.: J. Biomol. Struct. Dyn. **9**, 127 (1991)
47. Bhattacharyya, D., Bansal, M.: J. Biomol. Struct. Dyn. **6**, 635 (1989)
48. Panigrahi, S., Pal, R., Bhattacharyya, D.: J. Biomol. Struct. Dyn. **29**, 541 (2011)
49. Chai, J.D., Head-Gordon, M.: Phys. Chem. Chem. Phys. **10**, 6615 (2008)
50. Luo, R., Gilson, H.S., Potter, M.J., Gilson, M.K.: Biophys. J. **80**, 140 (2001)
51. Perez, A., Marchan, I., Svozil, D., Sponer, J., Cheatham, T.E., Laughton, C.A., Orozco, M.: Biophys. J. **92**, 3817 (2007)

Branch and Bound Algorithm for Vertex Bisection Minimization Problem

Pallavi Jain, Gur Saran and Kamal Srivastava

Abstract Vertex Bisection Minimization problem (VBMP) consists of partitioning the vertex set V of a graph $G = (V, E)$ into two sets B and B' where $|B| = \lfloor |V|/2 \rfloor$ such that its vertex width (VW) is minimized. Vertex width is defined as the number of vertices in B which are adjacent to at least one vertex in B'. It is an NP-complete problem in general but polynomially solvable for trees and hypercubes. VBMP has applications in fault tolerance and is related to the complexity of sending messages to processors in interconnection networks via vertex disjoint paths. In this paper, we propose a branch and bound algorithm for VBMP which uses a greedy heuristic to determine upper bound for the vertex width. We have devised a strategy to obtain lower bounds on the vertex width of partial solutions. A tree pruning procedure which reduces the size of search tree is also incorporated into the algorithm. This algorithm has been experimented on selected benchmark graphs. Results indicate that except for five of the selected graphs, the algorithm is able to, run through the search tree very fast.

Keywords Vertex bisection · Branch and bound · Vertex width · NP-complete · Graph layout

1 Introduction

Vertex Bisection Minimization problem (VBMP) consists of partitioning a vertex set of a connected graph $G = (V, E)$, $|V| = n$, into two sets B and B' where $|B| = \lfloor n/2 \rfloor$ such that vertex width (VW) is minimized where vertex width is

P. Jain (✉) · G. Saran · K. Srivastava
Department of Mathematics, Faculty of Science, Dayalbagh Educational Institute, Agra, India
e-mail: pallavijain.t.cms@gmail.com

G. Saran
e-mail: gursaran@dei.ac.in

K. Srivastava
e-mail: kamal.sri@dei.ac.in

© Springer Science+Business Media Singapore 2016
R.K. Choudhary et al. (eds.), *Advanced Computing and Communication Technologies*,
Advances in Intelligent Systems and Computing 452,
DOI 10.1007/978-981-10-1023-1_2

17

defined as the number of vertices in B which are adjacent to at least one vertex in B'. Formally, for a partition $P = (B,B')$, its vertex width is $VW(G,P) = |\{u \in B: \exists v \in B' \wedge (u,v) \in E(G)\}|$. VBMP is to find a partition P^* such that $VW(G,P^*) = \min_{\forall partitionP} VW(G,P)$. This problem has been treated as a graph layout problem by Diaz et al. [1]. VBMP is relevant to fault tolerance and is related to the complexity of sending messages to processors in interconnection networks via vertex disjoint paths [1]. NP-completeness of VBMP for general graphs is proved by Brandes and Fleischer [2] but it is shown to be polynomially solvable for trees and hypercubes. In spite of its practical applications, this problem remains almost unstudied so far. However, Fraire et al. [3] recently proposed two Integer Linear Programming (ILP) models and a branch and bound (B&B) algorithm for VBMP. They observed that ILP2 is the most promising method. In B&B, they have not used or provided any approach for determining the lower bound of the partial solutions. Their implementation also does not include any procedure for tree pruning. Therefore, it turns out to be an enumerative technique which is able to solve only 2 instances of graphs out of 108 instanced tested by them.

In this paper, we present a comprehensive B&B algorithm for VBMP. In order to generate an initial upper bound, a greedy heuristic has been designed (Sect. 4). A good heuristic which gives solution close to the optimal solution is always preferred as it is responsible for fathoming nodes at each level of the search tree in the B&B algorithm. We have also devised a strategy for finding lower bound of nodes representing partial solutions (defined in Sect. 3) at each level of the tree. Besides this, a procedure for tree pruning which helps to discard a large number of nodes in the search tree has also been designed (Sect. 5). The search tree is explored using depth first strategy. The proposed B&B algorithm (BBVBMP) is simulated on a large number of graphs including Small graphs, grids, trees and Harwell-Boeing graphs [3] and the algorithm is able to run through the search tree very fast except for Grid6×6, Grid7×7, bcspwr01, bcspwr02 and bcsstk1. We have also compared BBVBMP with ILP2 (Sect. 6). Conclusion is presented in Sect. 7.

2 Branch and Bound Algorithm

Branch and Bound Algorithm (B&B) is an exact combinatorial approach. It generates and explores the entire set of solutions to the problem by examining a search tree. It starts by generating an initial solution using some heuristic, or randomly, whose objective function value serves as upper bound of the optimal value. During the search process, it finds a lower bound at each node of the search tree. If this lower bound is greater than or equal to the upper bound then this node is fathomed because it guarantees that this node cannot result in a better solution. When exploration reaches a leaf node, it computes the objective function value of the corresponding solution and updates the upper bound if necessary. The B&B algorithm stops when all the nodes have been explored (either branched or fathomed), and returns the optimal solution to the problem [4]. B&B is an important class of algorithms and has been applied to a diversity of problems [4, 5, 6, 7].

In the context of VBMP, the search tree starts at the root which is an empty set. The tree branches into $\lceil n/2 \rceil + 1$ nodes from the root where each node is a singleton consisting of vertices from 1 to $\lceil n/2 \rceil + 1$ each. This forms level 1 of the search tree. At level 2, each node of level 1 is branched into nodes which now contain two vertices. Thus number of vertices in the nodes of level $i+1$ is one more than those at level i. Finally, level $\lfloor n/2 \rfloor$ consists of leaves each containing $\lfloor n/2 \rfloor$ vertices. In order to avoid duplicate nodes at each level, a pruning procedure is used which is described in Sect. 5. The tree is explored using depth first search strategy.

3   Lower Bound for Partial Solutions

Each node of the B&B tree represents a set $S \subseteq V$. This set will eventually be extended to a collection G of sets such that each B in G gives vertex bisection $(B, V \setminus B)$. We will refer to the set S itself as a partial solution hereafter. The set of adjacent vertices (neighbors) of $u \in V$ is denoted by $N(u) = \{v \in V: uv \in E\}$. The procedure *Compute_lowerbound(S)* (Fig. 1) is based on identifying those vertices of S whose neighbors cannot all be accommodated in B. In this strategy, number of vertices for which all the neighbors cannot be included in set B is identified.

Proposition 1 $|count \cup P|$ *in the procedure Compute_lowerbound(S) gives a lower bound for a node S.*

Proof In the procedure, *count* is the set of vertices whose all the neighbors cannot be included in the solution generated by S at the lower level of the search tree because the number of adjacent vertices is more than $\lfloor n/2 \rfloor - |S|$. Therefore, $|count|$ number of vertices will always contribute to the vertex width of the partition generated by S. It is clear that in order to minimize the contribution of a vertex to the vertex width, it is required to place all the adjacent vertices in the same partition

Fig. 1 Procedure for computing lower bound

```
Procedure Compute_lowerbound(S)
Step 1:  count = {v ∈ S : |N(v) \ S| > (⌊n/2⌋ - |S|)}
Step 2:  A = S\count
Step 3:  P = φ
Step 4:  for each vertex u ∈ A
Step 5:      P' = φ
Step 6:      C = S ∪ N(u)
Step 7:      P' = {v ∈ A : |N(v) \ C| > (⌊n/2⌋ - |C|)}
Step 8:      if |P·| < |P| or P = φ
Step 9:          P = P·
Step 10:     endif
Step 11: endfor
Step 12: lb = max(|count ∪ P|,1)
```

either in B or B'. Thus to accomplish this, for each vertex $v \in S \backslash count = A$, $N(v)$ is placed in S to give C and the number of vertices in S whose all the neighbors cannot be included in C are counted (Steps 6–7). Since P (Steps 8–9) records the smallest subset of $S \backslash count$ of which not all neighbors can be accommodated in B, $|count \cup P|$ represents the smallest number of elements of S whose neighbors cannot all be included in the complete set B. Hence, minimum number of vertices which will contribute to the vertex width have been considered which provides a lower bound guaranteeing that the solution generated by expanding the partial solution S will always give the vertex width greater than this lower bound.   □

4 Initial Upper Bound

We have designed a greedy heuristic to generate a solution which can give a close upper bound for the vertex bisection minimization problem. Pseudocode for this heuristic is presented in Fig. 2.

5 Tree Pruning

In this section, we describe the method of branching a node along with tree pruning. We have adopted two strategies for tree pruning.

1. The main idea for pruning is that at each level of tree, all those nodes are discarded which give duplicate nodes at this level or will give duplicate nodes in

```
Procedure H1
Step 1: B = ∅
Step 2: u = minimum degree vertex in G
Step 3: if |N(u)| < ⌊n/2⌋
Step 4:      B = {u}∪N(u)
Step 5: else
Step 6:      Randomly select X ⊂ N(u) s.t. |X|= ⌊n/2⌋-1
Step 7:      B = {u}∪X
Step 8: endif
Step 9: while |B| < ⌊n/2⌋
Step 10:     select a vertex v from B s.t. |N(v)\B|≠0 is
             minimum
Step 11:     Y = N(v)\B
Step 12:     if |Y|≤ ⌊n/2⌋ - |B|
Step 13:     B = B∪Y
Step 14:     else
Step 15:        B = B∪Z where Z is a randomly selected sub
                set of Y s.t. |Z| = ⌊n/2⌋ - |B|
Step 16:        endif
Step 17: endwhile
```

Fig. 2 Pseudocode of heuristic *H1*

Fig. 3 Procedure for branching a tree

```
Procedure Branch&Prune (Node)
Step 1:  R = { v ∈ V(G)\Node : v > max(Node) }
Step 2:  for i =1:|R|
Step 3:     P = {v ∈ V(G) : v > R[i]}
Step 4:       if |P| ≥ ⌊n/2⌋ - |Node ∪ {R[i]}|
Step 5:          Node_i = Node ∪ {R[i]}
Step 6:       endif
Step 7:  endfor
Step 8:  branches = {Node_i}
```

further branching at lower levels. Procedure *Branch&Prune(Node)* outlines the method for branching a node in the search tree without duplicate nodes (Fig. 3). Let the node to be branched be represented by an array *Node* representing the partial solution S. Now only those vertices $v \in V$ are considered for which $v > p$ where $p = \max\{w \in Node\}$. Let this set of vertices be represented by R. If the number of vertices having identifiers greater than $R[i]$ is more than $\lfloor n/2 \rfloor - |Node \cup \{R[i]\}|$, then *Node* is branched into $Node \cup \{R[i]\}$ otherwise it is not branched (Steps 3–6). In this manner the initial node is branched into $\lceil n/2 \rceil$ nodes $1, 2, \ldots, \lceil n/2 \rceil + 1$ each with a single different vertex. It may be noted that in VBMP maximum number of possible different leaves is $_{\lfloor n/2 \rfloor}^{n}C$.

2. A node is fathomed if lower bound (*lb*) ≥ upper bound (*UB*).

Proposition 2 *Procedure Branch&Prune(Node) (Fig. 3) eliminates the duplicate nodes of the same level of the search tree and the nodes which will give duplicate nodes in further branching.*

Proof In Step 1, only those vertices v are considered for which $v > \max\{w \in Node\}$ because the combination with the smaller ones is already present in their sibling nodes. Thus all the duplicate nodes at a level are eliminated. In Step 3, let v be the vertex considered for the branching. The set P consists of all those vertices w such that $w > v$. If $|P| \geq \lfloor n/2 \rfloor - |Node \cup \{R[i]\}|$ then it guarantees that this node will branch into at least one distinct node, otherwise, at any level *Node* will include a vertex whose identifier is less than $\max\{w \in Node\}$ which will result into a duplicate node at that level as this combination has already been taken. Hence the result follows.   □

6 Computational Experiments

This section describes the computational experiments performed to test the efficiency of our B&B algorithm. The algorithm has been implemented in C++ and the experiments are conducted on a Dual Xeon workstation with 12 GB RAM. The experiments have been conducted on four sets of instances: grids, trees, small graphs and harwell-boeing (HB) graphs [3]. The maximum time for the grid, tree and small instances was set to 5 min. while for HB instances time limit is 1 h as in

Table 1 Comparison between ILP2 and BBVBMP

Instances		Grids	Small	Trees	HB	Total
Number of instances		5	84	15	4	108
CPU (s)	ILP2	907.021	1598.902	73.198	14402.21	16981.33
	BBVBMP	120	16.63095	19.9	10802	10958.53
Number of optimal solutions	ILP2	2	69	15	0	86
	BBVBMP	3	84	15	1	103
Percentage of optimal solutions (%)	ILP2	40	82.1429	100	0	79.6296
	BBVBMP	60	100	100	25	95.37

[3]. Table 1 compares ILP2 and BBVBMP in terms of number of optimal solutions obtained and the CPU time.

For Grid6×6 and Grid7×7, tree was not explored completely. In HB graphs, the tree was explored completely only for can24. For the other graphs the tree was not explored completely in the specified time. Therefore, we cannot guarantee the optimality of result in these cases. Table 1 shows that BBVBMP outperforms ILP2 in terms of both the number of optimal solutions and CPU time.

7   Conclusion

An exact procedure based on B&B algorithm for vertex bisection minimization problem is developed in this paper. We have proposed a strategy for finding lower bound at each level of the tree and also a strategy for tree pruning. These strategies allow us to explore a smaller portion of the search tree. We have proposed a strategy for obtaining upper bound which is able to achieve optimal results for a large number of standard graphs for which experiments have been performed. It has been clearly shown that BBVBMP outperforms the state-of-art exact method for VBMP.

References

1. Diaz, J., Petit, J., Serna, M.: A survey of graph layout problems. ACM Comput. Surv. **34**, 313–356 (2002)
2. Brandes, U., Fleischer, D.: Vertex bisection is hard, too. J. Graph Algorithms Appl. **13**, 119–131 (2009)
3. Fraire, H., David, J., Villanueva, T., Garcia, N.C., Barbosa, J.J.G., Angel, E.R. del, Rojas, Y.G.: Exact methods for the vertex bisection problem. Recent Adv. Hybrid Approaches Des. Intell. Syst. Stud. Comput Intell. 547, 567–577 (2014)
4. Marti, R., Pantrigo, J.J., Duarte, A., Pardo, E.G.: Branch and bound for the cutwidth minimization problem. Comput. Oper. Res. **40**, 137–149 (2014)

5. McCreesh, C., Prosser, P.: A parallel branch and bound algorithm for the maximum labelled clique problem. Optim. Lett. **9**, 949–960 (2015)
6. Vlachou, A., Christos, D., Kjetil, N., Yannis, K.: Branch-and-bound algorithm for reverse top-k queries. In: Proceedings of the ACM SIGMOD International Conference on Management of Data, pp. 481–492. ACM Press, New York (2013)
7. Delling, D., Fleischman, D., Goldberg, A.V., Razenshteyn, I., Werneck, R.F.: An exact combinatorial algorithm for minimum graph bisection. Math. Program. **153**, 417–458 (2015)

Comparative Study of 7T, 8T, 9T and 10T SRAM with Conventional 6T SRAM Cell Using 180 nm Technology

Vinod Kumar Joshi and Haniel Craig Lobo

Abstract Data stability and power consumption have been reported two important issues with scaling of CMOS technology. In this paper, we have revisited these issues on 6T, 7T, 8T, 9T, 10T SRAM cells individually and a comparative analysis has been done based on different parameters like read delay, write delay, power consumption and static noise margin (SNM). The read/write delay and power consumption has been found 0.671/0.267 ns, 1.69 µW for 6T SRAM cell, 0.456/0.752 ns, 1.09 µW for 7T SRAM cell, 0.517/0.392 ns, 1.82 µW for 8T SRAM cell, 0.388/0.181 ns, 1.3 µW for 9T SRAM cell and 0.167/0.242 ns, 2.01 µW for 10T SRAM cell respectively. SNM has been calculated 0.4 V for 6T SRAM cell, 0.375 V for 7T SRAM cell, 0.65 V for 8T SRAM cell, 0.65 V for 9T SRAM cell and 0.6 V for 10T SRAM cell. All the circuit of SRAM cells and their layout has been designed using Cadence virtuoso ADE tool and Cadence virtuoso layout suite respectively using 180 nm CMOS technology. The post layout simulation results have been shown a good agreement with pre layout simulation results.

Keywords 6T SRAM cell · SNM · 7T SRAM cell · 8T SRAM cell · 9T SRAM cell · 10T SRAM cell · Write delay · Read delay

1 Introduction

The power dissipation has been become a major issue in SRAM design due to the downsizing of feature size with sub-micron technologies in recent years [1]. It has thus become very important to address this issue. Beside the low power dissipation, high speed and minimum die area are two other important VLSI cost function

V.K. Joshi (✉) · H.C. Lobo
Department of Electronics and Communication Engineering, Manipal Institute
of Technology, Manipal University, Manipal 576104, Karnataka, India
e-mail: vinodkumar.joshi@manipal.edu

H.C. Lobo
e-mail: lobo.hanielcraig@gmail.com

© Springer Science+Business Media Singapore 2016
R.K. Choudhary et al. (eds.), *Advanced Computing and Communication Technologies*,
Advances in Intelligent Systems and Computing 452,
DOI 10.1007/978-981-10-1023-1_3

during designing the memory. Here we have done the analysis of various SRAM cells made of 6, 7, 8, 9 and 10 transistors, at transistor level to highlight the above mentioned issues. To reduce power dissipation various approaches can be adopted like changing the threshold voltage or lowering the supply voltage [1]. But scaling the supply voltage effect the performance directly while scaling the threshold voltage shows highest impact in SNM [2]. In this paper, we have reviewed different SRAM cells like 7T SRAM [3, 4], 8T SRAM [5, 6], 9T SRAM [7, 8] and 10T SRAM [9, 10] based on read/write delay, power consumption and SNM compare to conventional 6T SRAM cell [1, 11–13].

2 6T, 7T, 8T, 9T and 10T SRAM Cells

2.1 6T SRAM Cell

6T SRAM cell is the most popular conventional cell made of 6 transistors. It consists of two access transistors M5 and M6 connected with common word line (WL). Each bit is stored in 4 transistors (M1, M2, M3, and M4) that form two cross coupled inverters [1, 11]. As long as the pass-transistors are turned off, the cell keeps the store data in the cell. The circuit of 6T SRAM cell and its layout has been shown in Fig. 1. Its circuit is symmetric and both bit lines BL and BLbar participate in read/write operations. While reading or writing, common WL controls accessibility to the cell nodes Q and Qbar through two pass transistor (or access transistors M5 and M6).

2.1.1 Write/Read Operation and SNM

During write operation WL is made high, data and its complement is fed through bit lines BL and BLbar respectively. The data gets stored at Q. To read a value from an SRAM cell, both bit lines (BL and BLbar) are precharged high (VDD) and the word line is raised turning on the pass transistors (M5 and M6). The bit line relative to the cell node that contains 0 begins discharging. Write and read delay of a 6T SRAM cell is found out to be 0.267 and 0.671 ns respectively (Fig. 2, Table 1).

SNM is defined as the minimum noise voltage present at each of the cell storage nodes necessary to flip the state of the cell. SNM is a critical metric for SRAM stability (Fig. 3). Higher the noise margin greater the stability of the cell. Read SNM of 6T SRAM cell is 0.4 V, which means the cell can handle a noise of 0.4 V during read mode (Table 1).

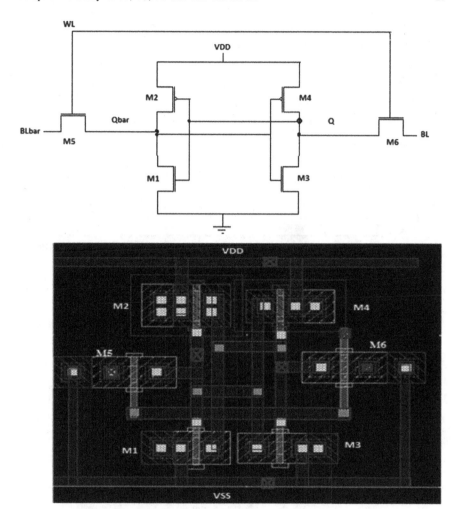

Fig. 1 Conventional 6T SRAM cell [11] and layout

2.2 7T SRAM Cell

7T SRAM cell has an extra transistor M7 when compared to a 6T SRAM cell (Fig. 4). The main difference in 7T SRAM cell when compared to 6T SRAM cell is that we set one of the bit lines at '1' and other at '0' during write operation while in case of 7T SRAM cell the data to be written is only given through one of the bit line (here BLbar). This ensures reduction in power dissipation [3, 4].

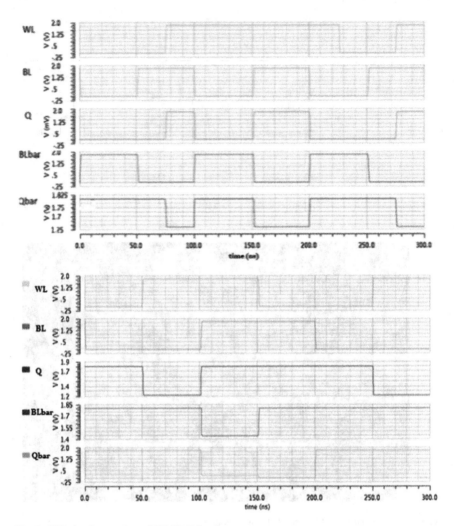

Fig. 2 Write/read operation of 6T SRAM cell

Table 1 Overall comparison of various SRAM cells

SRAM cells	Write delay (ns)	Read delay (ns)	SNM (V)	Power consumption (µW)
6T SRAM [11]	0.267	0.671	0.4	1.69
7T SRAM [3]	0.752	0.456	0.375	1.09
8T SRAM [5]	0.392	0.517	0.65	1.82
9T SRAM [8]	0.181	0.388	0.65	1.3
10T SRAM [9]	0.242	0.167	0.6	2.01

Fig. 3 SNM of 6T SRAM
cell during read mode

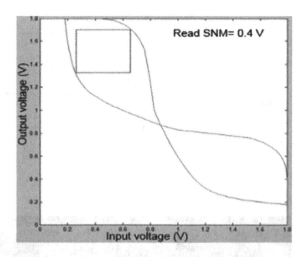

2.2.1 Write/Read Operation and SNM

The write operation starts by turning off transistors M7 (to cut off the feedback connection) and M6. To write in cell, the data is fed through BLbar and M5 is turned on which drives inverter (made up of M1 and M2) to develop the cell data at Q. While during read operation M5, M6 and M7 transistors are turned on which is same as that of reading using 6T SRAM cell. Write and read delay of a 7T SRAM cell is found out to be 0.752 and 0.456 ns respectively (Fig. 5, Table 1). Read SNM of 7T SRAM cell is 0.375 V, which means the cell can handle a noise of 0.375 V during read mode (Fig. 6, Table 1).

2.3 8T SRAM Cell

8T SRAM cell contains two extra transistors M7 and M8 to access read bit line (RBL) as compared to 6T SRAM cell (Fig. 7). These transistors also help to reduce the leakage current. The other architecture is same as that of 6T SRAM cell. During read operation when Qbar is '0', M5 tries to pull the node to a non-zero value and there is a chance of writing in the cell instead of reading which is known as "read upset". To overcome this problem, 8T SRAM cell has separate bit and word lines to have a better control over read and write operations (RBL, RWL for read and WL, BL, BLbar for write) (Fig. 7). The architecture of 8T SRAM cell improves the stability and provides a read upset free operation over 6T SRAM cell [5, 6].

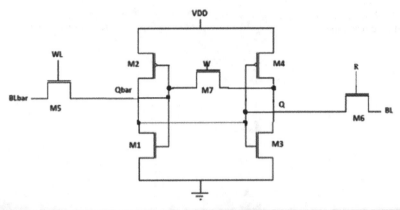

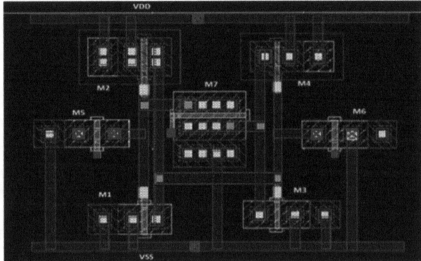

Fig. 4 7T SRAM cell [3] and layout

2.3.1 Write/Read Operation and SNM

During write operation, write word line (WL) is maintained at high whereas read word line (RWL) is made low. Data is written through bit line (BL) and complement of data through BLbar. During read operation, RBL is precharged to VDD whereas write word line (WL) is made low and read word line (RWL) is made high. Data is read at RBL. RBL gets discharged if the data read is 0, else it remains at precharged value. Write and read delay of 8T SRAM cell is found out to be 0.392 and 0.517 ns respectively (Fig. 8, Table 1). Read SNM of 8T SRAM cell is 0.65 V, which means that the cell can handle a noise of 0.65 V during read operation (Fig. 9, Table 1).

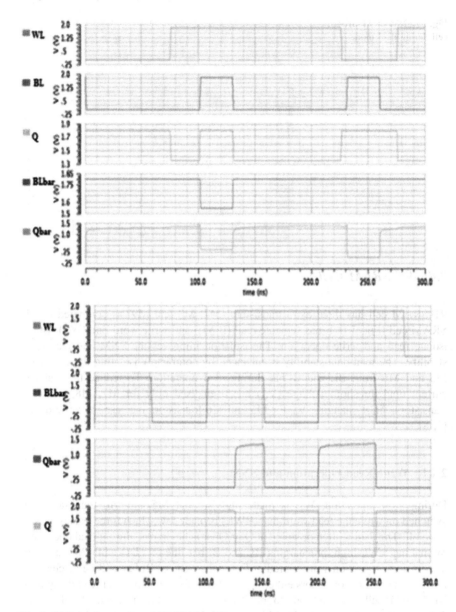

Fig. 5 Write/read operation of 7T SRAM cell

Fig. 6 SNM of 7T SRAM
cell during read mode

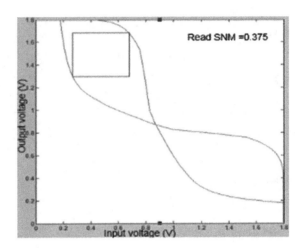

2.4   9T SRAM Cell

9T SRAM cell has three extra transistors (M7, M8 and M9) as compared to 6T
SRAM cell (Fig. 10). It has separate read (RWL) and write word line (WL) which
isolate read and write operation. The isolation improves the SNM and read data
stability. The read data stability is enhanced by completely isolating the transistor
M5 and M6 from BL and BLbar during read operation. To reduce the leakage
current dual threshold voltage (V_{th}) has been used only for M7, M8, and M9
transistors, since changing V_{th} for M5 and M6 transistors increase the write delay
[7, 8].

2.4.1   Write/Read Operation and SNM

During write operation, WL is made high so that M5 and M6 transistors are
switched on while the read word line (RWL) is made low which force transistor M9
to be off. The data written or stored at Q or Qbar is being read during read operation
by making RWL high, precharging BL and BLbar line to VDD while M5 and M6
are off now. Based on value stored at Q or Qbar, BL or BLbar will be discharged
through the pair of transistors either M7–M9 or M8–M9. Write and read delay of a
9T SRAM cell is found out to be 0.181 and 0.388 ns respectively (Fig. 11,
Table 1). Read SNM of 9T SRAM cell is 0.65 V, which means the cell can handle
a noise of 0.65 V during read operation (Fig. 12, Table 1).

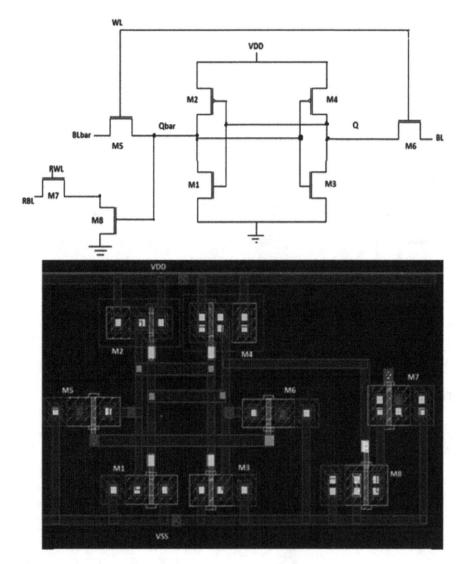

Fig. 7 8T SRAM cell [5] and layout

2.5 10T SRAM Cell

10T SRAM cell consists of 4 extra transistor (M7, M8, M9 and M10) as compared to 6T SRAM cell (Fig. 13). These extra transistor enhance the read SNM during the read operation by buffering the stored data. The hold SNM of 10T SRAM cell is same as that of a conventional 6T SRAM cell for same sized transistors M1 to M6 [9, 10].

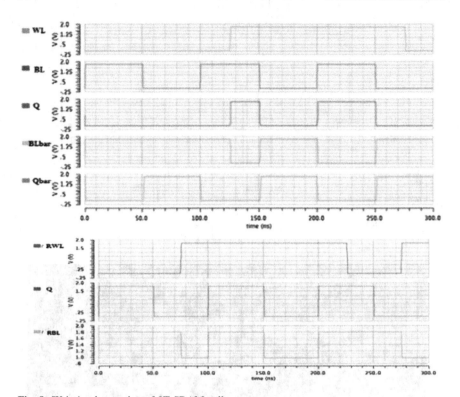

Fig. 8 Write/read operation of 8T SRAM cell

Fig. 9 SNM of 8T SRAM
during read mode

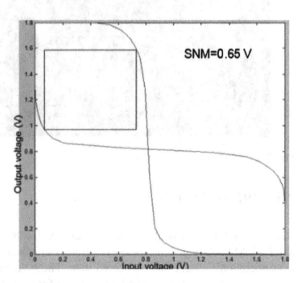

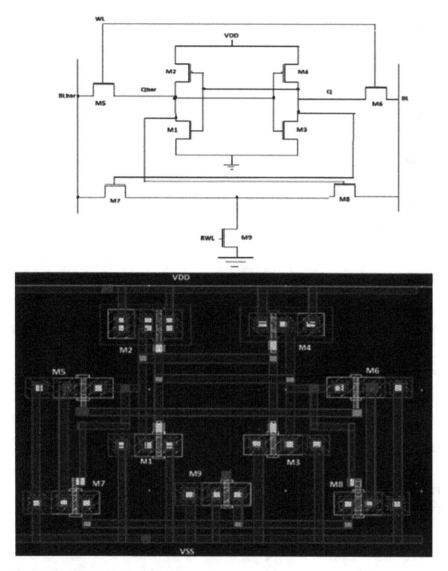

Fig. 10 9T SRAM cell [8] and layout

2.5.1 Write/Read Operation and SNM

During write operation, write word line (WL) is made high so that access transistors are on, while read word line (RWL) is made low. Data is fed through the bit lines and it is stored at Q. During read operation, WL is made low and RWL is made high which turns on transistor M9. The data to be read (Q), its inverse (Qbar) is fed to the inverter (comprising of transistors M7 and M8), and the data is read at read

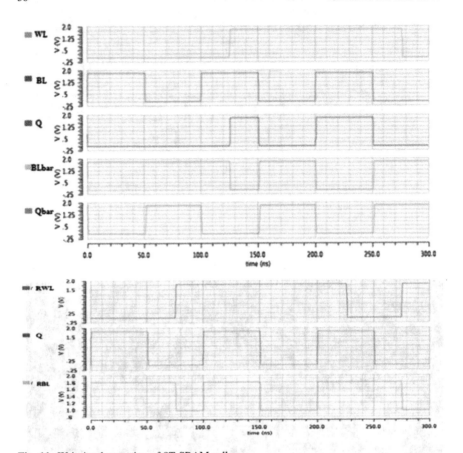

Fig. 11 Write/read operation of 9T SRAM cell

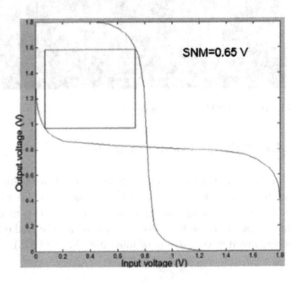

Fig. 12 SNM of 9T SRAM
during read mode

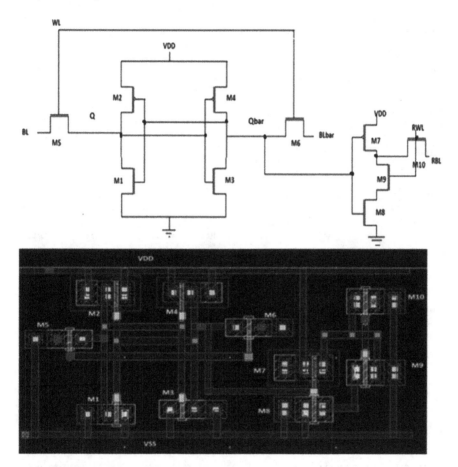

Fig. 13 10T SRAM cell [9] and layout

bit line (RBL). Write and read delay of a 10T SRAM cell is found out to be 0.242 and 0.167 ns respectively (Fig. 14, Table 1). Read SNM of 10T SRAM cell is 0.6 V, which means that the cell can handle a noise of 0.6 V during read operation (Fig. 15, Table 1).

3 Results Analysis

In 7T SRAM cell the data to be written is fed through one of the bit line (here BLbar) which makes the write delay 3 times more than 6T SRAM cell. There is no much change in SNM of both the cells, while ≈36 % reduction in power consumption of 7T SRAM cell as compared to 6T SRAM cell is observed (Table 1).

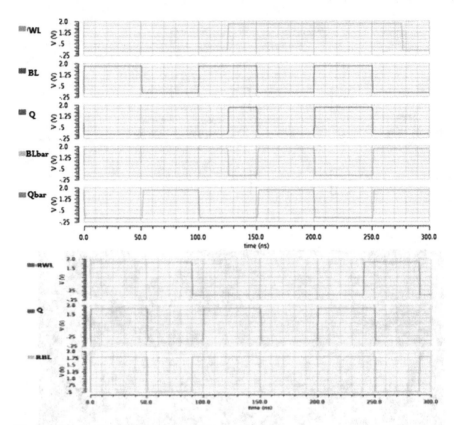

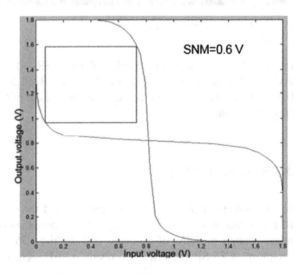

Fig. 14 Write/read operation of 10T SRAM cell

Fig. 15 SNM of 10T SRAM cell

In 8T SRAM cell the write delay is 1.5 times of 6T SRAM cell, while comparing SNM between 6T SRAM cell and 8T SRAM cell, SNM of 6T SRAM cell during read operation is 0.4 V but for 8T SRAM cell, SNM during read operation is 0.65 V, which is greater to 6T SRAM cell. Hence 8T SRAM cell is more stable compared to 6T SRAM cell during read operation. However the power consumption of 8T SRAM cell is increased by ≈8 % compared to 6T SRAM cell (Table 1).

The write delay for 9T SRAM cell is 1.5 times less than 6T SRAM cell. If we compare SNM between 6T SRAM cell and 9T SRAM cell, the SNM of 9T SRAM cell is ≈1.6 times of 6T SRAM cell. Hence 9T SRAM cell is more stable compared to 6T SRAM cell during read operation. Further comparing power consumption between 6T and 9T SRAM cells, there is 23 % decrease in power consumption in 9T SRAM cell compared to 6T SRAM cell (Table 1).

From Table 1, the write delay of 10T SRAM cell is almost same to 6T SRAM cell delay but there is an improvement in SNM. If we compare the SNM with 6T SRAM cell, the SNM of 10T SRAM cell is more than 6T SRAM cell. Hence 10T SRAM cell is more stable compared to 6T SRAM cell during read operation. There is also increase in power consumption of 10T SRAM cell by ≈19 % compared to 6T SRAM cell.

Appendix: Steps to Calculate SNM and MATLAB Code to Plot the Butterfly Curve

To calculate SNM the basic 6T SRAM cell and the process followed has been shown in Figs. 16 and 17 respectively. Below MATLAB code is given for it.

```
[a,T,aT] = xlsread('inverterA.xlsx')
t = a(:,1);y = a(:,2); plot(y,t,'red')
hold on
[b,bT,bT] = xlsread('inverterB.xlsx')
s = b(:,1);r = b(:,2)
axis square; plot(s,r,'red')
```

Fig. 16 Basic 6T SRAM cell

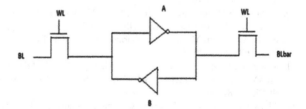

Fig. 17 Process to be
followed to calculate SNM

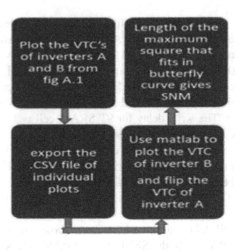

References

1. Rabaey, J.M., Chandrakasan, A.P., Nikolic, B.: Digital Integrated Circuits: A Design Perspective. PHI Learning, 2nd edn. ISBN-10: 8120322576 (2003)
2. Calhoun, B.H., Chandrakasan, A.P.: Static noise margin variation for sub-threshold SRAM in 65-nm CMOS. IEEE J. Solid State Circuits **41**(7), 1673–1679 (2006)
3. Ramy, E.A., Bayoumi, M.A.: Low-power cache design using 7T SRAM cell. IEEE Trans. Circuits Syst.-II. Express Briefs **54**(4), 318–322 (2007)
4. Kumar, V., Khanna, G.: A novel 7T SRAM cell design for reducing leakage power and improved stability. In: International Conference on Advanced Communication Control and Computing Technologies (ICACCCT-2014), pp. 56–59. Ramanathapuram (2014)
5. Chang, L., Fried, D.M., Hergenrother, J., Sleight, J.W., Dennard, R.H., Montoye, R. K., Sekaric, L., McNab, S.J., Topol, A.W., Adams, C.D., Guarini, K.W., Haensch, W.: Stable SRAM cell design for the 32 nm node and beyond. In: Symposium VLSI Technical Digest, pp. 292–293 (2005)
6. Moradi, F., Madsen, J.K.: Improved read and write margins using a novel 8T-SRAM cell. In: 22nd International Conference on Very Large Scale Integration (VLSI-SoC-2014), pp. 1–5 (2014)
7. Pasandi, G., Fakhraie, A.: 256-kb 9T near-threshold SRAM with 1 k cells per bit line and enhanced write and read operations. IEEE Trans. VLSI Syst. (2015)
8. Liu, Z., Kursun, V.: Characterization of a novel nine-transistor SRAM cell. IEEE Trans. VLSI Syst. **16**(4), 488–492 (2008)
9. Calhoun, B.H., Chandrakasan, A.P.: A 256 kb subthreshold SRAM in 65 nm CMOS. In: Solid State Circuits Conference (ISSCC-2006), session 34, pp. 2592–2601 (2006)
10. Islam, A., Hasan, M.: Leakage characterization of 10T SRAM cell. IEEE Trans. Electron Dev **59**(3), 631–638 (2012)
11. Khare, K., Khare, N., Kulhade, N., Deshpande, P.: VLSI design and analysis of low power 6T SRAM cell using cadence tool. Semiconductor electronics, IEEE international conference (ICSE-2008), pp. 117–121. (2008)
12. Akashe, S., Tiwari, N., Sharma, R.: Simulation and stability analysis of 6T and 9T SRAM cell in 45 nm. In: 2nd International Conference on Power, Control and Embedded Systems (ICPCES-2012), pp 1–6 (2012)
13. Chopade, S., Padole, D.: Stability analysis of 6T SRAM cell for nanoscale FD-SOI technology. In: Annual IEEE India Conference (INDICON-2014), pp. 1–6 (2014)

A Survey of Hardware and Software Co-design Issues for System on Chip Design

J. Kokila, N. Ramasubramanian and S. Indrajeet

Abstract The modern embedded system needs to be designed to meet the tremendous changes due to high speed and advancement in technologies. Encapsulating user needs into a small area and enhancing the specification of today's requirement is really a challenging task for System on chip designer. This review presents the practical view of Hardware/Software co-design and its main design issues tackled in recent years. The role of SOC cannot be limited to a single task, since it is an ocean of streams like Computer Architecture, VLSI and Embedded System, which has been joined together to deal with a current multi-tasking environment. The key motive of this survey is to cover the design issues faced by SOC designer and applying all these in Hardware Software co-design. A comparative study is made to show how the specific design is used efficiently and goals are reached in each proposal with its metrics and drawbacks.

Keywords SOC design · Driving factors · Methodologies · Hardware software co-design

1 Introduction

In Today's world 98 % of systems are using embedded systems. Hardware has to be cautiously designed and software has to be efficiently programmed in parallel in an embedded system so that overall system performance increases, reduction in power consumption and high Integrity has to be maintained. An embedded system

J. Kokila (✉) · N. Ramasubramanian · S. Indrajeet
Department of Computer Science & Engineering, National Institute of Technology,
Trichy, India
e-mail: jk.cse09@gmail.com

N. Ramasubramanian
e-mail: nrs@nitt.edu

S. Indrajeet
e-mail: indrajeet95@gmail.com

© Springer Science+Business Media Singapore 2016   41
R.K. Choudhary et al. (eds.), *Advanced Computing and Communication Technologies*,
Advances in Intelligent Systems and Computing 452,
DOI 10.1007/978-981-10-1023-1_4

is an electronic system that has a programming module which is embedded in computer hardware. It is used everywhere in modern life, starting from Consumer Electronics, Education, Telecommunication, Home Appliance, Transportation, Industry, Medical and Military. Software method with higher quality code, fixing errors and optimization is more important for the performance raise. In hardware building, more and more complex designs in RTL are becoming increasingly difficult. System-on-a-chip (SOC) is an Integrated Circuit, which houses all the critical elements of the electronic system on a single microchip. It is widely used across the embedded industry due to their small-form-factor, computational excellence and low power consumption. The hardware/Software co-design has some intention and purpose based on prefix 'co'. They are co-ordination, concurrency, correctness and complexity. The survey of historical details and future direction about this strategy has been discussed in detail from the scratch for advanced electronic systems [1]. The current and future direction is also very much related to the driving factors which has been examined in this work.

The remaining of this paper is organized as follows. Starting with the recent issues of the SOC design in Sect. 2 with latest comments, followed by the discussion of various proposals in Sect. 3 with different variants of hardware/software co-design, that have simultaneously emerged from 2013 to 2015. In Sect. 4 the designing strategies of SOC with its overview has been explained and highlighted the importance of core method. Finally, Sect. 5 remarks the main conclusion of this survey.

2 Design Issues

The design of the system will be affected by the choice of hardware and the type of software they use. The requirements of choice can be divided into two basic groups which may be used and achieve system goals. The users wish certain apparent properties in a system that is a system should be fast, convenient to use, reliable, easy to learn and to use, and safe. The system should be comfy to design, implement, and maintain and it should be precise for the below mentioned goals.

2.1 Performance

The demand for performance is increasing form data processing in documentation to high-speed computing and clustering, which in recent years plays main role in research and complex designing. The Performance is considered as the most important factor for design issues which may be due to double 'r' that is reconfiguration [2, 3] and reliability [3]. Multiprocessor SOC could provide immense computational capacity, achieving high performance in such systems is highly challenging.

2.2 Complexity

Progress in Embedded System guides us to grow a finite number of feature at low power integration and complexity into a single chip. This dispute requires a set of ways to integrate heterogeneous systems into the same chip to reduce area, power and delay. The five facets of complexity that range from functional complexity, architectural and verification challenges to design team communication and deep submicron (DSM) implementation.

2.3 Power Management

The SOC power issues have forced a reorganization of methodologies all over the design flow to explain power-related effects There are two types of power in SOC design one is static which is for memory and dynamic is for logic. Overall power in a chip may be dynamic power consumption, short-circuit current and leakage current will change dramatically. Increase in speed and decrease in size will lead to head dissipation and reliability issues. This may cause electro migration and IR drop for wide range of design applications.

2.4 Reliability

The parts of an SOC, Intellectual property and blocks related to memory, can go on to function while other parts do not. Some may work irregularly, or at lower speeds, others may increase in temperature to the point where they either shut down or slow down. This leads to some interesting issues across a variety of the electronics industry, ranging from legal liability, to what differentiates any functioning system especially critical systems from those that are non-functional or marginally functional. Just as systems are getting more complex, we need more and more metrics for analyzing it.

2.5 Security

If all the technology works as designed, there are yawning security holes in designs at every process node and in almost every IoT design even if it's only what a well-design piece of hardware or software is connected to. It is obvious that a device which is compromised is no longer reliable. But a device that can be flexible is not reliable, either. All of the processor companies have been active in securing their cores. ARM has its Trust Zone technology for compartmentalizing memory and processes. Software security alone is not needed for security but hardware plays a major role because hardware attacks are very much concerned With SOC.

2.6 Time-to-Market

The SOC designer can improve time-to-market and evade costly re-spins by using FPGA-based physical prototypes to speed up system validation. FPGA-based prototypes enable embedded software developers to use a real hardware earlier in the design cycle, allowing pre-silicon validation and hardware/software integration prior to chip fabrication. The use FPGA and simulation make the designing process easy and efficient.

3 Co-design: A Glance at Existing Approaches

The diagram (Fig. 1) targets the enhancement of SOC design as the effect of the design issue (cause) which in turn, become the effect of selected papers in this survey. A comparative study has been made in the Table 1 and each design issues are matched with some recent proposals.

3.1 Performance Increase Using Reconfiguration

The megablock is a basic repetitive block, the detection of which aids in finding the loops which can be mapped into a Reconfigurable Processing Unit (RPU), consisting of a specialized reconfigurable array of Functional Units. The synthesis of the RPU is done offline while the reconfiguration is done online. This work can be

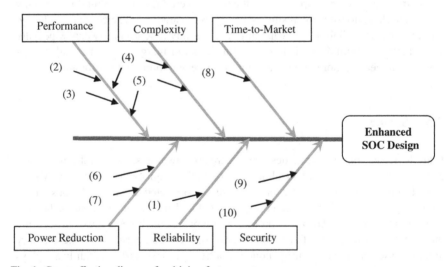

Fig. 1 Cause effective diagram for driving factors

Table 1 Comparative study of driving forces

S. No.	Design issues	Methods adapted	Hardware	Software	Metrics	Limitation
1	Performance	ZyCAP [4]	ZyCap-controller	Device configuration interface	Resource utilization performances	Need more resources
2	Performance	Reconfigurable HW/SW systems [3]	RPU	Megablocks	Execution time communication over-head	Blocks the processor during reconfiguration
3	Performance and complexity	Hardware software coherence protocols [5]	Local memory	Compiler	Speedup energy reduction	Uses of CPU cache
4	Performance and complexity	Accelerator coherency port using Xilinx Zynq [6]	Hardware accelerator and DMA	Application to transfer data between accelerator and memory	Energy efficiency, speed	Critical application will overloads cache space
5	Power reduction	Optimize battery life in SOC device [7]	Power management	Features and flexibility	Power variance	Lack of SOC IP block for power consumption
6	Power reduction	Device runtime power management [8].	Monitor to check bits	Central PM agent	Execution time and resources	Increase hardware complexity and not addressed the QOS
7	Reliability and complexity	Cardio [6]	Fault detection	System reconfiguration	Scalability performance	Overhead in broadcasting message at least one system should be active
8	Security	Authentications on FPGAs using PUF [7]	PUF	Authentication	Threshold error rate	Hardware resource occupied
9	Security	Locking phenomena using PUF [9]	TERO-PUF	Chip identification	Response time inter device variation	Architecture depend on clock input
10	Time-to-market	A metaprogrammed C++ framework [4]	Hw code Netlist (NOC)	SW code binary files	Memory FPA resource and latency	High through-put and cross-domain communication

carried out for cache memory, reduction in resource, floating point operation and for developing on line megablock for individual adaptable embedded systems.

In a system with multiple independent accelerators, a full reconfiguration forces reconfiguration of all of them even if not required. PR overcomes these limitations, and hence a key enabler for this paradigm of reconfigurable computing. Partial Reconfiguration in hybrid FPGA platforms like the Xilinx Zynq have definitely shown advantages and zycap, is a controller that significantly improves reconfiguration throughput in Zynq system over standard methods, while allowing overlapped executions, resulting in improved overall system performance. Zycap also plays a significant role in automating Partial Reconfiguration development in hybrid FPGAs.

3.2 Complexity Measure Using Memory and Accelerator

This paper focuses on coherence protocol for hybrid memory which may affect the power consumption and scalability. The two main concerns are the important quantity of power consumed in the cache hierarchy and the absence of scalability of current cache coherence protocols [5]. The optimized compiler are used mainly for generating machine language and it should be adaptable for any processor. The hybrid memory has incoherence between Local and cache hierarchy which is solved very efficiently by hardware-software protocols. The compiler generates an assembly code, even in memory aliasing problem. A comparative study has been made for cache-based architecture and hybrid memory. This work will increase average speedup of 38 % and energy reduction of 27 %.

Hardware accelerators can achieve significant performance for many applications. The motivation for enhanced execution speed and power efficiency for their specific computational workloads have been discussed in reference [10]. To perform a particular processing task using an accelerator there is a need to access memory. Xilinx Zynq provides three types of memory interface methods: GP (General Purpose), HP (High performance), and ACP (Accelerator Coherency Port). In the first approach hardware accelerator is connected to ARM CPU located on Zynq through GP interface for memory access through DMA. This brings down the performance. In the next approach accelerator is connected to ARM CPU through HP port, which increases the performance because HP ports have direct connection with DDR memory controllers on Zynq PS. In the third approach accelerator is connected through ACP port which further increases the performance because through ACP DMA can access caches of ARM cores which in turn gives low latency.

3.3 Power Management and Optimizing Battery Life

SOC contains many non CPU devices which are evaluated by drivers, so a device runtime Power management (PM) becoming very essential. PM is not used in drivers because to save chip area same power and clock source is used by different devices, if any one device PM performance is poor then the whole domain should be in enable which increases power loss. The solution for this is relieving drivers from PM responsibility by replacing the PM code in all drivers with a single kernel module called as Central PM Agent. Central PM Agent have two parts called monitor and controller. The monitor will check whether devices are in busy or idle mode using software and hardware approaches. Based on monitor information controller will keep the device in ENABLED or DISABLED mode. The software approach will monitoring whether a device register is accessed by CPU for threshold time. We can also simplify the driver source code by using a Power Advisor which suggests where to add PM calls in driver.

The surfacing of battery-constrained SOC device such as Mobile phones, laptops, modern electronic devices and tablets, requires hardware-software co-design to optimize battery life. The Offloading behavior from processor to dedicated engine on the SOC can help decrease power consumption. The Data Acquisition unit along with instrumental platform has been used to assemble the power consumption of various SOC IP blocks. In this work hardware component is designed in Hardware and adding features and flexibility was included through software module. This work can be extended by the software to design an energy aware application by collecting software based IP.

3.4 Reliability in CMP

The permanent failures in Chip Multiprocessor is a big issue in terms of reliability which have been rising above by cardio system. A Distribute Resource manager was developed to collect data about CMP Components which will test the hardware components at regular intervals and information's are collected and analyzed for future failure. If any core failure is detected then the core will be reconfigured and at least one core should be non-faulty to do the task at runtime. An experiment was conducted to measure performance and response time of Cardio used C++ simulator, gem5 and cacti 5.3. The benefit of this system is more reliable for distributed resources, fault—aware OS and scalable for CMP. This proposed work was compared with hardware only technique such as Immune in NOC. The Cardio can be embedded in microcontroller and heterogeneous SOC design for more efficient fault-tolerance systems.

3.5 Security Using PUF

The Hardware and software components communication using remote method Invocation in an object oriented programming concept. A Meta-programming framework for reducing the gap between hardware/software boundaries using

Remote method Invocation of object oriented approach [4]. The Proxy/Agent Mechanism have been implemented in software and hardware components to generate binary file and netlist. The communication framework has to focus on two important issues that is components mapping and framework portability. Several template-based implementations of serialize and reverse may be defined in order to provide the optimal marshalling for each data type. Synthesis constraints were set in both CatapultC and ISE with the goal of minimizing circuit area considering a target operating frequency of 100 MHz.

4 Comparisons of Methodologies Adapted for SOC Design

The best process has to be selected and using them appropriately in correct position for specific design is again a big overhead. The methods can be used in single or in combination to form a higher level of design strategy to support a modern SOC. The following design strategies have been widely used for designing SOC from early days to today. The First and selected method is Hardware/Software co-design which refers to parallel design of hardware components using Hardware description language and software module using any programming language of heavily complex electronic real time systems. The second method is programmable core debits the changes in functionality of post fabrication in SOC. The increase in complexity and time-to-market pressures makes the post-fabrication changes more essential for modern SOC design. One of the primary ways in which these cores will be used is to allow designers to leave certain aspects of the design unspecified until after fabrication [11].

The chip designer may not have the detailed IP knowledge for combining intellectual property (IP) blocks from various sources. Although much effort has gone into determining best practices for integrating IP blocks into an SOC design is the third method, the effort involved in integrating stand-alone block-level timing constraints in the chip-level timing environment is often overlooked. The quality of the configuration and integration of complex IP blocks can have an important impact on a SOC's development schedule and performance. The solutions comprise a component related language, component selection needs algorithm, type matching and inferencing algorithms, temporal property based behavioral typing, and finally a software system on top of an existing model which is created on the different platform.

This survey comprises of various journal papers like IEEE, ACM and elsiver which uses Hardware Software co-design method to explore the goals of system design and is comparatively analyzed based on the flavor of computer science and engineering. Various approaches have been compared and its limitations are listed in the Table 1. They are mainly based on resource management in CPU, cache and clock for performance and security, cache utilization and adoption of critical application for complexity, QOS and Lack of IP for power Management, depending on at least one active processor for Reliability.

5 Conclusion

System-on-chip has been an imprecise term that magically holds out a lot of enthusiasm, and has been gaining stamina in the computer industry. Though the possibility is enormous, complexities are numerous, and answering these to offer successful designs is a true engineering challenge. All these issues create a path for future research works. Finally, it is also very auspicious to see from researcher's point of view that many positive, technical and fundamental questions will arise and that these excite and keep our lives busy for the next decades of years.

References

1. Teich, J.: Hardware/software codesign: the past, the present, and predicting the future. In: Proceedings of the IEEE 100, Special centennial issue, pp. 1411–1430 (2012)
2. Vipin, K., Fahmy, S.: ZyCAP: efficient partial reconfiguration management on the Xilinx Zynq. Embed. Syst. Lett. IEEE **6**(3), 41–44 (2014)
3. Bispo, J., et al.: Transparent trace-based binary acceleration for reconfigurable HW/SW systems. IEEE Trans. Ind. Inform. **9**(3), 1625–1634 (2013)
4. Mück, T.R., Fröhlich, A.A.: A metaprogrammed C++ framework for hardware/software component integration and communication. J. Syst. Archit. **60**(10), 816–827 (2014)
5. Alvarez, L., et al.: Hardware-software coherence protocol for the coexistence of caches and local memories. IEEE Trans. Comput. **64**(1), 152–165 (2015)
6. Pellegrini, A., Bertacco, V.: Cardio: CMP adaptation for reliability through dynamic introspective operation. IEEE Trans. Comput. Aided Des. Integr. Circuits Syst. **33**(2), 265–278 (2014)
7. Aysu, A., Schaumont, P.: Hardware/software co-design of physical unclonable function based authentications on FPGAs. Microprocess. Microsyst. (2015)
8. Metri, G., et al.: Hardware/software codesign to optimize SoC device battery life. Computer **10**, 89–92 (2013)
9. Bossuet, L., et al.: A PUF based on a transient effect ring oscillator and insensitive to locking phenomenon. IEEE Trans. Emerg. Top. Comput **2**(1), 30–36 (2014)
10. Xu, C.: Automated OS-level Device Runtime Power Management. Rice University, Diss (2014)
11. Chen, Y.-K., Kung, S.-Y.: Trend and challenge on system-on-a-chip designs. J. Sig. Process. Syst. **53**(1–2), 217–229 (2008)
12. Sadri, M., et al.: Energy and performance exploration of accelerator coherency port using Xilinx ZYNQ. In: Proceedings of the 10th FPGAworld Conference. ACM (2013)
13. Rostami, M., et al.: Hardware security: threat models and metrics. In: Proceedings of the International Conference on Computer-Aided Design. IEEE Press (2013)

Safety Validation of an Embedded Real-Time System at Hardware-Software Integration Test Environment

Gracy Philip and Meenakshi D'Souza

Abstract As the complexity and functionality of embedded software is increasing steadily, ensuring that their behavior is safe is of primary concern. We propose a Safety Validation Method (SVM) that is used to monitor divergence in safe behavior during integration testing of an embedded system. The proposed monitor observes both application and system level parameters. It can capture effect of potential unsafe scenarios inadvertently created during testing that could recur during actual operation. We present a case study involving the safety validation of advanced fly-by-wire flight control system where the use of the SVM revealed safety critical failures during integration testing.

Keywords Safety validation · Embedded software · Verification · Testing · Monitors

1 Introduction

As the functionality of safety critical systems built these days is complex, safety validation has become a critical part of system development. Powerful cross development environments containing automatic code generators and testing tools are available these days. Commercial of the shelf hardware and software have also become part of systems being developed. In such a development scenario, visibility into the full system has decreased while complexity of the system has increased. Central focus of this paper is on safety assurance of an embedded real-time safety critical system.

G. Philip (✉)
CEMILAC, DRDO, Bangalore 560037, India
e-mail: gracy.philip@cemilac.drdo.in

M. D'Souza
IIIT, Hosur Road, Bangalore 560100, India
e-mail: meenakshi@iiitb.ac.in

© Springer Science+Business Media Singapore 2016   51
R.K. Choudhary et al. (eds.), *Advanced Computing and Communication Technologies*,
Advances in Intelligent Systems and Computing 452,
DOI 10.1007/978-981-10-1023-1_5

Most of the complex functionality of safety critical systems is implemented using software embedded in the system. However, safety is not considered to be a software property alone, it has to be defined at system level [1]. There are two aspects of safety: one is safety critical functionality, failure to achieve it will impact safety. Second aspect is platform-related, introduced by digital implementation in an embedded systems environment, which were not relevant in earlier electro mechanical environment.

Typically there are four main levels of testing: low level testing, software integration testing, hardware software integration testing, systems integration testing. In all these levels, requirement based testing is emphasized because this strategy has been found to be most effective for revealing errors [2]. Each level carries out both normal mode and robustness tests. Each level has its own granularity for fault simulation and result analysis to identify faults. A major lacuna with this compartmentalized testing strategy is that expected results are generated and results are analyzed at that level only, against test objectives at that level. Testing at these discrete levels fail to provide any concrete evidence about safety. Individual test results cannot be stitched together to get a clear picture on system safety. End result is that even after successful tests at all levels with adequate coverage; there can be faults left, hampering system safety.

We propose a safety validation method (SVM) that deploys a safety monitor to expose the vulnerabilities hampering safety. Such a monitor is developed using existing validation environments used for requirements level testing as per D0-178B [2]. There is no extra load on the system under test. We illustrate the working of such a safety validation using a case study involving safety assurance of a flight control system. As per [3], run time verification that deals with the application of formal verification techniques that allow checking whether a run of a system under scrutiny satisfies or violates a given correctness property. Our SVM combines run-time verification with testing in the integration testing environment, with safety criteria as the correctness property.

The paper is organized as follows: Sect. 2 describes the current safety assurance process followed by proposed SVM. Section 3 describes our case study in which the proposed method is applied towards safety validation of digital fly-by-wire flight control system of a modern fighter aircraft. Related work is presented in Sects. 4 and 5 concludes the paper.

2 Safety Assurance and Validation

In this section, we summarize the state-of-the-art processes followed during the development and life cycle of a safety critical system to ensure that it is safe. The term "safety" includes all aspects of software that generically mean that "nothing bad (in terms of behavior of the system) will happen". Current safety assurance methods are based on several stringent development processes and independent verification and validation methods as per DO-178B.

2.1 Safety Validation Method

Typically, hardware software integration testing is the phase in which requirements are validated. In this phase, software is installed in the platform and the integrated system is tested for meeting its requirements. Such a testing involves normal behavior of the system and fault mode behavior. In normal mode testing, a tester focuses on creating test cases to validate each requirement towards checking if the latter has been implemented properly. Fault mode testing is conducted by simulating the faults. This way, adequate coverage is achieved for safety critical software, as mandated by various standards [2].

While such an exhaustive integration testing validates individual requirements, it might fail to capture safety violations that occur at the system level. A tester who focuses on validating one particular requirement may ignore the system level behaviors, some of which could violate safety.

We propose a Safety Validation Method (SVM) to address this issue and capture unsafe behavior of the system during the integration testing phase. In our SVM, safety parameters are extracted through system safety and software safety analysis as shown in the Fig. 1. Figure 1 also describes the SVM, details of which are given in the subsequent sections. Behaviors violating safety are found automatically in real time by observing the values of the parameters as the application executes in the test environment.

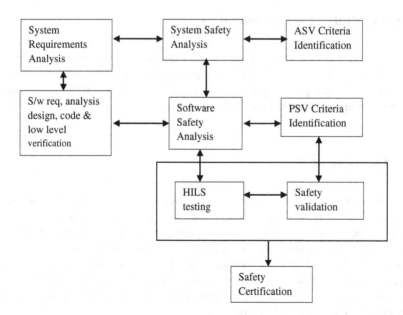

Fig. 1 Safety validation method

2.2 Safety Validation Types

Safety validation can be categorized into two types based on the aspects of safety that are being addressed: Application domain aware safety validation, and platform (system) aware safety validation.

Application-Aware Safety Validation

Application aware Safety Validation (ASV) takes care of safety captured as a part of the overall system functionality. Parameters to be monitored here include all application level data that capture the behavior of the system as running in the system simulated environment. System parameters indicating safety are to be arrived based on system specification analysis along with preliminary system safety analysis. System safety analysis brings out inter dependent sub-systems and their boundaries for achieving system functionality in a safe manner.

Platform-Aware Safety Validation

Platform aware Safety Validation (PSV) takes care of safety interpreted as a non-functional requirement. PSV validates additional hazards brought into the system due to digital implementation like channel failures, watchdog time out, unexpected exceptions and interrupts etc. Safety Monitor captures ASV and PSV parameters in real time, while HSI test scripts are run for normal and robustness test cases.

3 Case Study: Flight Control System

We now present a case study of safety validation of a digital fly-by-wire Flight Control System (FCS) using the SVM integrated with an existing test environment. Using the SVM, unexpected system failures could be captured, which would have occurred during flight operations if left undetected.

3.1 Overview of FCS

The FCS (refer to [4] for an example) is a quadruplex digital fly-by-wire system, it has four identical processing channels and quadruplex input sensors. All four channels process control laws for flight control with identical inputs. Channels work in frames-synchronous manner. FCS is designed to tolerate up to two critical failures. It is interfaced with one air data system per channel and has to control four primary actuators and two secondary actuators. It takes inputs from various sensors from pilot stick and other cockpit interfaces. It also gives out data for pilot displays, crash data recorder and get U home system.

The FCS has four identical computing channels; identical software is loaded in each channel. All the six actuators will continue to receive commands even in the case of failure of any two of the channels. Even when there is quadruplex redundancy available for hardware there is no redundancy in case of software, towards minimizing complexity in design and validation. So it is essential that the software is almost 100 % fault free, towards safe and reliable flight control. The FCS software is written in Ada programming language with SPARK Ada as the coding standard. Control laws were designed in Matlab [5] and Simulink [6] and converted into Ada programming language using Beacon code generator [7].

3.2 Platform-Aware Safety Validation of FCS

We now summarize the testing efforts that were undertaken for FCS. Testability was one of the main design features. Various levels of testing were planned with different test teams—unit testing, software-software integration testing, non-real time testing of control law packages, hardware-software integration testing, systems of system integration testing, pilot-in-loop testing and aircraft integration testing.

Each level of testing had its own objectives and coverage criteria and criterion for pass or fail of the test was limited to analysis at that level. A platform aware safety validation was designed at hardware software integration test level to validate system behaviors in terms of parameters that can be monitored. Criteria for passing a test were amended to include a no fault reply based on SVM along with the specified criteria.

3.3 Integration of PSV in Test Environment of FCS

During integration testing, the software is loaded as a part of the system that runs on an embedded platform containing all the required hardware. Sensor inputs and actuator outputs are simulated as in real-life conditions. Such a test environment is called Hardware Software Integration Test Environment (HSITE). HSITE is capable of fault simulation, and can access all processor and memory variables, along with sensors, actuators, flight control panel, air-data system and display system. It also has the capability to temporarily halt the embedded system processor and read any registers and memory variables.

Test script language provided by the HSITE enables writing of test cases and procedures. Test scripts are written corresponding to each requirement, test inputs generated and passed to the System under Test (SUT) and outputs are recorded. The following psuedocode provides details of how PSV was integrated with HSITE for FCS.

```
for requirements i = 1 to n do
        {
            power on rest of the system and initialize system parameters
            for test scripts j=1 to m do { // PSV monitoring
                initialize PSV & clear fault records
                run test scripts for each j
                compare results with expected output
        read PSV and fault records & compare results with expected output
            }
        }
```

Our platform aware SVM was implemented using the scripting language provided by the HSITE. Our SVM was able to configure, initialize and load the test scripts, record test outputs and simultaneously track all parameters to be monitored. The validation method was implemented once and designed to work as a stand-alone entity with several test scripts being run one after the other for validating their respective requirements. The monitor had features to clear all faults of the previous run and verify the integrity of each build.

The recorded data in the environment gives vital clues for analysis. The following embedded processor and system interrupts were monitored automatically for FCS: watch dog timer status, frame timer, arithmetic faults, floating point, constraint fault, events and machine fault, Mil bus interrupt, RS422 interrupt, real-time extensions, channels status, average and peak frame time.

3.3.1 Analysis of Results

Our safety validation revealed unexpected exceptions and interrupts which were not handled. We describe computing channel failures detection in detail as it is a safety-critical failure and later resulted in adding new requirements to the FCS.

Analysis of the data provided by SVM helped to observe a channel failure while conducting testing to prove some other requirement (whose test case passed) and this lead to a detailed analysis. The monitor recorded repeated occurrence of real time extension interrupt 74 leading to computing channel failure while running test scripts for proving requirements. As per design this interrupt was expected only when the frame time exceeds 12.5 ms causing the watch dog to time out. But on checking the frame recorded in the monitor it was found that none of the real time frames exceeded the 12.5 ms limit. Consequently, it was understood that triggering of real time overrun was spurious. This resulted in a new requirement to be added—the new one specified a method to distinguish between spurious and real interrupts, with the understanding that if an interrupt is real it will get repeated. For further analysis a derived requirements got added. The SVM could capture deviations in SUT configurations due to patches left behind by testers used for fault simulation. It also helped to capture peak execution time under multiple simulated failure conditions.

4 Related Work

Several researchers and practitioners working with safety critical systems have emphasized the need for exclusive practices, techniques and tools for safety analysis and assurance [8, 9, 1, 10]. In [11] the authors identify safety assurance parameters and techniques to be used throughout the development life cycle and elaborate on the need for independent verification and validation in all phases of the development life cycle. In [11], safety analysis of automated requirement models and their validation is discussed. The work presented here can be thought of as one concrete realization towards safety assurance at the integration testing phase. The proposed SVM not only provides safety assurance during requirements validation but can be continuously used in the later phases of development including safety assurance of incremental releases.

There is a large body of work on run-time verification [3] in the formal verification community. Our work is a case study in run-time verification but doesn't use any formal methods. We just do system level run-time testing, with specifications being given without using any formal notations. In [8] requirements based safety validation that takes an approach inclusive of both static and dynamic analyses for safety assurance is presented. Safety validation follows the requirements based dynamic analysis and testing. Here, we do automatic monitoring of all safety critical parameters, as a part of existing validation environment, combining run-time verification [3] and requirements based testing [8].

5 Conclusion and Future Work

The safety validation method is used for monitoring safety aspects of the system in the system integration and simulation environment, to analyze and perfect the system before its actual operational use. The case study involving application of safety validation to an FCS proved the concept of system aware safety validation, and its ability to catch safety critical errors in real time embedded environment. It was a partial implementation for platform aware SVM, where application aware safety parameters were not monitored. We are currently extending the safety validation to track all application aware parameters. We are also devising metrics to measure the efficiency of the safety validation.

References

1. Nancy, G.: Leveson. Systems Thinking Applied to Safety. MIT Press, Engineering a Safer World (2012)
2. RTCA Software Considerations in Airborne Systems and Equipment Certification DO-178B (1992)

3. Leucker, M., Schallhart, C.: A brief account of run time verification. J. Logic Algebraic Program. **78**, 293–303 (2009)
4. Official website of Tejas. http://www.tejas.gov.in/technology/fly_by_wire.html
5. Matlab. http://www.mathworks/in/products/matlab/
6. Simulink. http://www.mathworks.in/products/simulink/
7. Beacon coder. http://www.adi.com/products/b4s
8. Bhansali, P.V.: Software safety: current status and future directions. ACM SIGSOFT Softw. Eng. Notes **30**(1) (2005)
9. Hunter, B.: Assuring separation of safety and non-safety related systems. In: Proceedings of 11th Australian Workshop on Safety Critical Systems and Software (SCS) (2006)
10. Leveson, N., Alfaro, L., Alvarado, C., Brown, M., Hunt, E.B., Jaffe, M., Joslyn, S., Pinnel, D., Reese, J., Samarziya, J., Sandys, S., Shaw, A., Zabinsky, Z.: Demonstration of a Safety Analysis on a Complex System. Presented at the Software Engineering Laboratory Workshop, NASA Goddard (1997)
11. Modugno, F., Leveson, N., Reese, J.D., Partridge, K., Sandys, S.D.: Integrated Safety Analysis of Requirements Specifications, in Requirements Engineering (1997)

An Efficient Algorithm for Tracing Minimum Leakage Current Vector in Deep-Sub Micron Circuits

Philip C. Jose, S. Karthikeyan, K. Batri and S. Sivanantham

Abstract Leakage current has a large impact in the performance of a system. Dominant component of the leakage current is the subthreshold leakage. One of the most sophisticated techniques for reducing leakage current is the transistor stack. Leakage current primarily depends upon the input vectors applied to the circuit. It is possible to demote the leakage current further with the usage of 'IVC'. If it is possible to control this input vectors means leakage current can be reduced to a greater extent. A number of algorithms already exist to sort out this input vectors, but due to their exhaustive search nature they becomes ineffective. This paper propose an algorithm which focus on the best possible combination of the input vector that reduce the leakage current without trying for all the possible 2^k combinations. The problem can be treated as NP-Complete. Fan-out is not included in the algorithm since it is an independent factor of leakage current. The proposed algorithm precisely produces the input vector which gives the minimum leakage and shows a greater optimization in terms of time complexity and space complexity.

Keywords Gate replacement algorithm · Leakage current · Subthreshold leakage · Gate oxide leakage · SAT solver

P.C. Jose (✉) · S. Sivanantham
School of Electronics Engineering, VIT University, Vellore, Tamilnadu 632014, India
e-mail: philipc.jose2015@vit.ac.in

S. Sivanantham
e-mail: ssivanantham@vit.ac.in

S. Karthikeyan · K. Batri
Department of Electronics and Communication Engineering, PSNA College
of Engineering and Technology, Dindigul, Tamilnadu 624619, India
e-mail: skarthik@psnacet.edu.in

K. Batri
e-mail: krishnan.batri@gmail.com

© Springer Science+Business Media Singapore 2016   59
R.K. Choudhary et al. (eds.), *Advanced Computing and Communication Technologies*,
Advances in Intelligent Systems and Computing 452,
DOI 10.1007/978-981-10-1023-1_6

1 Introduction

Three critical parameters used to determine the performance of systems tends to be speed, area and power. Out of this power consumption has the upper hand since future devices will suffer from high battery pack up and very short battery life. Power consumption in CMOS circuits are mainly due to Standby current, Leakage current, Short circuit current and Capacitive loads. Leakage current tends to be the important factor of power consumption in VLSI. It is contributed by the following components which is applicable to both NMOS and PMOS [1]. Within the geometry of the devices many diodes are available because the way in which the MOS transistors are realized. In normal mode of operation these diodes are reverse biased and hence the name 'Reverse bias P-N junction diode leakage current'. In deep-sub micro meter technologies 'Band to Band Tunneling (BTBT) current' is becoming very large. High electric field across reverse-biased P-N junction causes significant current known as BTBT Current. When the electrons move from the substrate to the source or to the drain voltage drop across the electric barrier is lowered and electrons will move from valence band to the conduction band producing a flow of current. BTBT is predominant for electrons and not holes.

Subthreshold leakage current very significant since it passes from drain to the source through the channel. If the gate voltage is less than the threshold voltage means no channel will be present and when the gate voltage greater than the threshold voltage means drain current starts to flow. The channel current that flows with the application of gate voltage step by step from '0' volts leads to the significant amount of sub threshold current. 'Gate oxide tunneling current' mainly occurs when the thickness of gate oxide is reduced. On the other hand 'Gate current due to hot carrier injection' is because of the injection of hot electrons from substrate to oxide due to high electric field. 'Channel punch' through estimates the values V_{ds} for which the punch through occurs at $V_{gs} = 0$. It increases sub threshold current and degrades the sub threshold slope. Because of the overlap between gate and drain a phenomenon called 'Gate induced drain leakage current' and it is high when the electric field is in the drain junction. Out of this first 5 are off critical important and sub threshold leakage current is the dominant factor.

Leakage current reduction focused only during the standby mode. This is mainly because of the percentage difference in leakage reduction between active mode is very high compared to standby. The key parameter in the reduction of leakage current is the threshold voltage V_τ and with this leakage reduction classified as standby and runtime. Standby leakage reduction achieved by the well-known Transistor Stack, Variable Threshold CMOS (VTCMOS), Multi threshold CMOS (MTCMOS), Power Gating and combining power gating with Dynamic Voltage and Frequency Scaling (DVFS). Transistor stack tends to be the powerful tool mainly because of its added controllability with the addition of PMOS and NMOS hence a reduction in leakage. It relay on the principle that it is possible to demote the subthreshold leakage if more than one transistor in the stack and this finicky effect known to be as Stack effect. Because of the stack effect sub threshold leakage current depends on the input vector applied, so that further leakage reduction is possible if we are able to control the input vector applied to the transistors that are buried inside the CMOS gates.

2 Existing Algorithms

In this section a survey has been made to find what all are the works that been done in algorithmic area to reduce the leakage current. Genetic algorithm for bound leakage [2] explains the enumerate methods used to find minimum and maximum leakage. Since random search is used average power is also predicted. Based on the observation that leakage power dissipated by logic gates is on steady and is able to deduct the input vector by random sampling [3]. Leakage observability is used is used to find the minimum and maximum leakage in heuristics and exact algorithms for random combinatorial circuits [4]. It is based on the degree to which a particular input is observable in magnitude of leakage from power supply.

Robust SAT based search algorithm [5] use Incremental SAT solver to find the minimum and maximum leakages. For this purpose SAT problem is created which contains an objective function that has to be maximized or minimized. Leakage effects modelled using pseudo Boolean functions and coupled to VG-ILP and heuristic mixed integer linear programming [6]. VG-ILP Model: Miniature creates problem for finding input vectors with minimum leakage current with the aid of ILP formulation. MLP Model: Mainly oriented in reducing the run time. A new approach for finding minimum leakage vector is the 'Inspenum' and 'Mincutenum' algorithm [7]. Inspenum implicitly enumerates leakage with respect to the inputs whereas Mincutenum perform enumeration in mincut space of the circuit hyper graph with optimality. In Gate Replacement algorithm [8] internal gates subjected to worst case leakage are replaced with new gates which are of less leakage without modification of the circuit functionality.

Circuit modification and input vector approach reduces leakage current. Gates can be modified in such a way that its output helps to reduce the leakages associated with succeeding gates [9]. Here gate replacement is done to reduce the delay penalty in a slack-aware manner. Larger cells are created in the library to add much more flexibility to algorithm. Gate replacement is not reducing the leakage instead it is setting them to a low leakage state. A fast heuristic for tracing low-leakage is presented in [10]. A dynamic programming based algorithm is used for the fast evaluation of input vectors and replacement of gates which is of very high complexity. The main peculiarity of this work is the link-deletion-based decomposition which is better than tree decomposition. Fast Input Vector Algorithm (FIVA) is presented in [11]. Algorithm generates graph for each circuit and subthreshold leakage for each node of the graph is calculated. Boolean logic and probability used to trace the minimum leakage vector. Minimum leakage vector for test circuit can be produced using genetic algorithms [12]. Here algorithm is implemented using Verilog which saves run time and reduces leakage current in sleep mode.

3 Proposed Algorithm

This section put forwards the algorithm which can be used to reduce the leakage with proper selection of input vectors that must be applied to the input periphery of the circuits. It has been claimed that 'IVC' has been ineffective once the circuit is growing in depth. The problem was addressed with a modified algorithmic technique [13] of direct control of internal pins. The main principle is to insert multiplexers at the input of each pins of each gate and inputs in active mode are selected by *SLEEP* signals. This helps to select the input vectors with minimum leakage. With the insertion of multiplexers and *SLEEP* signals are very effective with CMOS circuits in leakage reduction. But this technique adds much more complexity since it is necessary to find some other techniques to reduce the leakage produced by the multiplexers.

This problem is addressed by Gate Replacement algorithm [8]. They are also internal point control but are of different from [13]. They have replaced the gates that are at Worst Case Leakage (WLS). With the assumption that for circuits with logic depth is higher the internal gates at WLS will be high and hence the technique will be effective. Exhaustive search nature of this algorithm makes it un-attractive since they are of very complex in terms of time and space. Research has been carried out to check whether fan-out has any significance with leakage. But the experiment clearly proves that leakage depends only on the input vectors and not on the circuit depth. So these complex algorithms with circuit modifications are not needed. From these results it is evident that an algorithm with less time and space complexity is needed rather than performing complex library modifications. The proposed algorithm for tracing minimum leakage current vector is depicted here. Some of the abbreviations used in the algorithm are as follows F, M, L, and SM. 'F' stands for the first value, 'M' for the middle value, 'L' for the last value and 'SM' for store middle.

The main objective of the proposed algorithm is to find the input vector which produces the minimum leakage current in less number of trials. For such a consideration it is necessary to cover the factors that affect leakage current in the algorithm. It has been proven that Fan-out an independent factor of leakage current. Leakage current can be claimed to be process dependent and it does purely dependent on input vectors applied at the input of the circuitry. Gate replacement algorithm focus with the replacement of the gates that are subjected for WLS and the algorithm proceed in an exhaustive manner. They check each input vector combination and hence the time complexity is very high. For further reduction of leakage current the replaced gates are fed with sleep signals, but for the production of this signals multiplexers will be required which are subjected for extra leakage. So additional circuitry needed increases the design complexity.

The pseudo code of the proposed algorithm is shown in Fig. 1. If a combinational circuit is of 'K' inputs then a total of $n = 2^k$ combinations. Searching an input vector in this space is a tedious process. So at the initial stage the whole search space is divided into 'K' equal groups. The intention of this work is to reduce the number of trails. So for the second step the leakage current associated with the middle term of each group is calculated and stored in an array. This array is sorted

If the combinational circuit is of k-inputs then the total number of input combinations is $n = 2^k$. Keep this assumption and follow the steps.

Step 1) Divide the whole search space, $n = 2^k$ into 'k' equal groups.

Step 2) Find the leakage currents associated with the middle term (M) of each group.

Step 3) Select the leakage currents associated with 'M' of each group, store it in an array, and sort the array in ascending order.

Step 4) Select the first leakage current value in the sorted array and store it in a new variable 'SM', and select the group in which this particular leakage current resides.

Step 5) Find the leakage current associated with the first (F), middle (M), and last (L) term of that group. Compare the leakage currents of this three and find minimum among F, M, and L.

Step 6) If (leakage current of 'F' is minimum & < SM)
{
a.   Store the leakage current in 'F' to 'SM'.
b.   Increment 'F' in position wise (first input position + next input position).
c.   Last leakage current value = Middle terms leakage current value. Find the middle term (M) of new sequence.
d.   Compare the leakage currents of 'F', 'M', 'L' of the new sequence and find the minimum leakage current among the three and compare it with leakage current stored in 'SM'.

   If (minimum < SM)
     {
     Replace the leakage current stored in 'SM' with   minimum and repeat from step 'b'.
   }
       Else
       {
         Stop search
         }}

Step 7) Else if (leakage current of 'L' is minimum & <SM)
{
a.   Store the leakage current in 'L' to 'SM'.
b.   Decrement 'L' in position wise (Last input-second last)
c.   Last leakage current value of new sequence = Middle terms leakage current of the new sequence. Find the middle term 'M' of new sequence.
d.   Compare the leakage currents of 'F', 'M', and 'L' of the new sequence and find the minimum leakage current among the three and compare the leakage current stored in 'SM'.

   If (minimum < SM)
     {
     Replace the leakage current in 'SM' with minimum and repeat from 'f'.
     }
       Else
       {
         Stop search
         }}

Step 8) Stop search

Fig. 1 Algorithm for tracing minimum leakage current vector

in ascending order according to the leakage current stored. It's up to the programmer to select which sorting algorithm must be used. Once the sorting in ascending order is over the first leakage current in the sorted array is selected and the corresponding group from which the leakage current resides is selected. This first leakage current is stored in a variable 'SM' which depicts the minimum leakage current in the sorted array.

Once the group is accessed then the first (F), middle (M) and last (L) terms leakage currents are compared and minimum among the three is selected. This minimum leakage current is compared with the minimum leakage stored in 'SM'. If the compared leakage current is minimum than the stored one in 'SM' then it will be replaced with the new minimum. After this a new search space is created. The first value of the new search space is the next value to the first and last will be the Leakage current corresponding to the middle term of the previous group. Next step is to find the new middle term of the new search space. Once this steps are over leakage current between 'F, M, and L of the new group are compared to find minimum among them and is again compared with the leakage current stored in 'SM' to check whether it is less than 'SM' or not. If it is less means again the above procedure will be repeated otherwise the search stops.

4   Analysis of Algorithm

This section focuses on the analysis of the proposed algorithm. Objective of the proposed algorithm is to find the input vector which produces the minimum leakage current. The algorithm has mainly 3 principle parts. For a circuit with 'K' inputs the whole search space constituted by $n = 2^k$ is divided into 'K' equal groups. The group elements can be made equal by adding dummy leakage values. Next is to find the leakage current associated with the middle term of each group and they are stored in an array. It's up to the programmer to select the type of sorting algorithm since elements in the array will vary according to the value 'K'. Here for to make the analysis simple a sorting algorithm of $O(n)$ is selected. Once the sorting is over the particular group in which the minimum leakage resides is traced off. Here the time and space complexity of the proposed algorithm is derived. First we will deal with the time complexity of the proposed algorithm.

4.1   Time and Space Complexity Analysis

4.1.1   Best Case

The first step is to create a recurrence relation. Once the recurrence relation is created then it can be solved with master's theorem. Here the analysis of the proposed algorithm is made w.r.t the C17 circuit and it is applicable to any circuits.

For C17 circuit $k = 5$. So $n = 2^k = 32$ combinations. So the recurrence relation can be modelled as follows

$$T(n) = 2T\left(\frac{n}{5}\right) + O(n). \tag{1}$$

The running time of the proposed algorithm equal to the running time of the linear time spend in splitting of groups and running time of the sorting process. For C17 'K' = 5, which means one $T\left(\frac{n}{5}\right)$ for splitting of groups. A sorting process with a worst case $O(n)$ is selected. Constant time is required for certain steps in the algorithm. C2, C4, C6.a, C6.b, C6.c, C6.d, C7.e, C7.f, C7.g, and C7.h and all this will constitute a constant total time of 'C'. Comparing 'C' with n, C << n. So the recurrence can be written as

$$T(n) = \begin{cases} 2T\left(\frac{n}{5}\right) + n; & n \geq 1 \\ 1 & ; n = 1 \end{cases}. \tag{2}$$

This recursion equation can be solved in many ways. By using master's theorem, creation of telescopic equation's or recursion tree method. Here the recurrence relation is solved by recursion tree method. In the proposed algorithm at each stage of search half of the elements are removed. At the start of the search a total of 'n' and once the algorithm proceeds one half of the elements will be removed. So it is possible to model the work done at each stage as $n, 2n, 4n, \ldots n = 2^k$. This means $\frac{n}{2^0}, \frac{n}{2^1}, \frac{n}{2^2}, \ldots \frac{n}{n}$. Therefore 1, 2, 3,… k, levels are possible. From this it's clear that for 2^k, $(K + 1)$ levels are mandatory. By solving this the best case for the proposed algorithm is $O(n\log n)$.

4.1.2 Worst Case

Analysis of worst case is somewhat complicated since the recurrence relation must have to be modified. The running time of the proposed algorithm is equal to the running time of two recursive calls plus the time taken for sorting. For worst case analysis the assumption made is that all leakage current values will be same and the whole 'n' must have to be searched at the same time because a minimum value of leakage is not possible. The simplest approach is to incorporate the case ($n = 0$). We assume $T(0) = 1$ and it has to be removed from the total 'n' in recurrence relation. Now the recurrence relation equals,

$$T(n) = \begin{cases} T(n-1) + n; & n \geq 1 \\ 1 & ; n = 1 \end{cases}. \tag{3}$$

By back substitution method worst case is $O(n)^2$.

4.1.3 Space Complexity

This part of the analysis is mainly focused on how much memory is required for the algorithm. For that purpose also a recurrence relation is needed. From (2) it is

$$T(n) = \begin{cases} 2\,T\left(\frac{n}{5}\right) + n; n \geq 1 \\ 1 \qquad\qquad ; n = 1 \end{cases} \tag{4}$$

The recurrence can be solved in the same manner in which best case was analysed. We have used the recursion tree to solve this particular problem. The amount of stack elements to be used depends on the number of levels present in the recursion tree. The partition starts from 'n' and at each time it will have equal portioned group. No need to consider the constants for space complexity analysis. Here we are only focused on the extra space required for the algorithm to perform the comparisons. For 'n' levels it requires $\lceil \log n \rceil + 1$ levels of stack. That is size of the stack is $(\lceil \log n \rceil + 1)k$. So the space complexity is $O(K(\log n)) = O(\log n)$.

This section intent was to do the complexity analysis of the proposed algorithm. The algorithm is said to have a Best Case of $O(n\log n)$. The Worst Case is $O(n^2)$ and takes a Space Complexity of $O(\log n)$.

5 Experimental Setup and Results

Leakage current for various logic gates is measured using Cadence Spectre with UMC 130 nm CMOS technology library. For a particular gate all the input combinations are applied separately and its leakage are measured. Here the leakage current measured for various logic gates contains only sub threshold and gate oxide leakage current. Leakage current depends on many factors. They are process dependent. Experiments have been executed to verify whether leakage current depends on fan-out. Tables 1, 2 and 3 represents the leakage measured for NOT gate, 2-input NAND gate, 2-input NOR gate.

Table 1 Leakage current measured for NOT gate

Input combination	Leakage current
0	1.278n
1	894.85p

Table 2 Leakage current measured for 2-input NAND gate

Input combination		Leakage current
0	0	234.06n
0	1	1.46092n
1	0	1.6764n
1	1	1.7897n

Table 3 Leakage current measured for 2-input NOR gate

Input combination		Leakage current
0	0	3.47671n
0	1	1.52719n
1	0	1.35257n
1	1	161.289n

It is evident that leakage current remains to be a constant irrespective of the number of fan-out stages [14]. Hence it can be claimed that fan-out is an independent factor of leakage current. So leakage current primarily depends on the input vectors applied at the input of the circuits and hence fan-out is not considered in the algorithm. Figure 2 shows the leakage measured for C17 circuit for an input combination '00010'. Measurement of leakage current for all the input combination is performed. The proposed algorithm is implemented using Turbo C. The computer is equipped with 4 GB internal RAM. The program takes 0.948060 μs to execute for C17 circuit. The program precisely finds the input vector, which produces the minimum leakage. Here for C17 circuit the input vector producing minimum leakage is '*00011*' producing a minimum leakage of '*5.95798*' namps.

Table 4 shows the comparisons of the various algorithms with the proposed algorithm in terms of time and space complexity. SAT Solver and Gate Replacement algorithms are compared with the proposed algorithm. The comparison table clearly depict that the proposed algorithm is having an upper hand with the existing SAT Solver and Gate Replacement algorithm. SAT Solver and Gate Replacement have a worst case of $O(n^2)$ which is same as that of the proposed one. Best case of proposed algorithm is $O(n\log n)$, whereas for SAT Solver and Gate Replacement it tends to be $O(n)$. The proposed algorithm needs a stack space of $O(\log n)$, whereas $O(n^2)$ for SAT Solver and Gate Replacement algorithm.

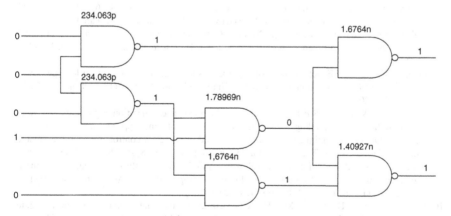

Fig. 2 C17 circuit with leakage current measured

Table 4 Comparison of complexity of algorithms

Algorithms	Worst case	Best case	Space complexity
SAT solver	$O(n^2)$	$O(n)$	$O(n^2)$
Gate replacement	$O(n^2)$	$O(n)$	$O(n^2)$
Proposed algorithm	$O(n^2)$	$O(n\log n)$	$O(\log n)$

6 Conclusion

The paper proposes an algorithm to find the input vector which produces the minimum leakage current without trying for all the 2^k combinations. The proposed algorithm has the upper hand in terms of space and time complexity and also no circuit modification is required. This means the algorithm can be applicable for practical cases. Experimental results and algorithm analysis proves that the algorithm is an optimized one.

References

1. Roy, K., Mukhopadhyay, S., Mahmoodi-Meimand, H.: Leakage current mechanisms and leakage reduction techniques in deep-submicrometer CMOS circuits. Proc. IEEE **91**(2), 305–327 (2003)
2. Chen, Z., Johnson, M.W.L., Roy, K.: Estimation of standby leakage power in CMOS circuits considering accurate modeling of transistor stacks. In: Proceedings of the International Symposium on Low Power Electronics and Design, Digest of Technical Papers, pp. 239–244 (1998)
3. Halter, Jonathan P., Najm, Farid N.: Gate-level leakage power reduction method for ultra-low-power CMOS circuits. In: Proceedings of the Custom Integrated Circuits Conference, pp. 475–478 (1997)
4. Johnson, M.C., Somasekhar, D., Roy, K.: Models and algorithms for bounds on leakage in CMOS circuits. IEEE Trans. Comput. Aided Des. Integr. Circuits Syst. **18**(6), 714–725 (1999)
5. Aloul, F.A., Hassoun, S., Sakallah, K.A., Blaauw, D.: Robust SAT-based search algorithm for leakage power reduction. In: Lecture Notes in Computer Science (including subseries Lecture Notes in Artificial Intelligence and Lecture Notes in Bioinformatics), vol. 2451, pp. 167–177 (2002)
6. Gao, F., Hayes, J.P.: Exact and heuristic approaches to input vector control for leakage power reduction. IEEE Trans. Comput. Aided Des. Integr. Circuits Syst. 25(11), art. no. 1715439, 2564–2571 (2006)
7. Chopra, K., Vrudhula, S.B.K.: Implicit pseudo boolean enumeration algorithms for input vector control. In: Proceedings—Design Automation Conference, pp. 767–772 (2004)
8. Yuan, L., Qu, G.: A combined gate replacement and input vector control approach for leakage current reduction. IEEE Trans. Very Large Scale Integr. VLSI Syst. **14**(2), 173–182 (2006)
9. Jayakumar, N., Khatri, S.P.: An algorithm to minimize leakage through simultaneous input vector control and circuit modification. In: Proceedings—Design, Automation and Test in Europe, DATE, art. no. 4211867, pp. 618–623 (2007)
10. Cheng, L., Chen, D., Wong, M.D.F.: A fast simultaneous input vector generation and gate replacement algorithm for leakage power reduction. ACM Trans. Design Autom. Electron. Syst. 13(2), art. no. 34 (2008)

11. Rjoub, A., Alajlouni, B.A., Almanasrah, H.: A fast input vector control approach for subthreshold leakage power reduction. In: 16th IEEE Mediterranean Electrotechnical Conference, MELECON, pp. 84–87 (2012)
12. Leelarani, V., Madhavilatha, M.: Verilog implementation of genetic algorithm for minimum leakage vector in input vector control approach. In: International Conference on Signal Processing and Communication Engineering Systems—Proceedings of SPACES 2015, in Association with IEEE, art. no. 7058231, pp. 132–136 (2015)
13. Abdollahi, A., Fallah, F., Pedram, M.: Leakage current reduction in CMOS VLSI circuits by input vector control. IEEE Trans. Very Large Scale Integr. VLSI Syst. **12**(2), 140–154 (2004)
14. Jose, P.C, Sivaprakasam, K., Krishnan, B.: Fan-out an independent factor of leakage current in deep-submicrometer circuits. Australian J. Basic Appl. Sci. (AJBAS), 718–723 (2015)

Age Invariant Face Recognition Using Minimal Geometrical Facial Features

Saroj Bijarnia and Preety Singh

Abstract This paper aims to determine a minimal set of geometrical facial features which can provide good accuracy in age invariant face recognition. The proposed method evaluates a geometrical model based on selected feature points. Feature selection is applied on computed features. These features are then used for face identification of a person across different ages. All experiments are performed on FGNET dataset. Results show that our feature subset shows enhanced recognition accuracy.

Keywords Age invariant face recognition · Geometrical features · Feature selection · Support vector machine · Classification

1 Introduction

Automation of face recognition is a rapidly growing research area with numerous applications such as crime investigation, biometrics, surveillance and image retrieval [14]. However, face recognition across aging is still a great challenge which has not been widely addressed.

Aging has a profound effect on an individual's face structure and causes variations in the appearance of the face in many different ways like wrinkles, weight gain etc. While the appearance of a person may change with age, there are certain features which are not affected much and can be employed for face recognition [9, 11] in an automated system. In our paper, we have used some of these geometrical features computed from the facial structure. We identify relevant features through feature selection techniques and use them for the task of face identification, matching a query image with database images of various persons.

S. Bijarnia (✉) · P. Singh
The LNM Institute of Information Technology, Jaipur, India
e-mail: saroj.bijarnia91@gmail.com

P. Singh
e-mail: prtysingh@gmail.com

© Springer Science+Business Media Singapore 2016   71
R.K. Choudhary et al. (eds.), *Advanced Computing and Communication Technologies*,
Advances in Intelligent Systems and Computing 452,
DOI 10.1007/978-981-10-1023-1_7

The rest of the paper is organized as follows: Section 2 presents related work. Section 3 discusses the proposed methodology. Section 4 analyzes the results of performed experiments. Section 5 concludes the paper.

2 Related Work

Face recognition across aging has been an active research area in recent years and several techniques have been proposed. A brief survey of age progression in human faces is discussed in [10]. Park et al. [8] proposed learning aging pattern based on Principal Component Analysis coefficients in separate 3D shape and texture space. They report an accuracy of 37.4 % for FGNET database and an accuracy of 66.4 and 79.8 % for Alb1 and Alb2 of MORPH database.

Lanitis et al. [6] used the active appearance model (AAM) and came up with a combined shape and intensity based model for representing face images. They have reported an accuracy of 68.30 %. Wang et al. [13] propose an age simulation model. For age simulation first they are using Active Shape Model. Eigen faces are computed by applying PCA on image for feature vector. Polynomial aging function is used for estimating age along with K-means classification. The accuracy reported is 63 % with a private database. Du et al. [4] propose a method using active appearance model and gradient orientation pyramid.

3 Proposed Methodology

Our proposed method shown in Fig. 1 is based on geometrical features derived from facial points. Firstly, localization of facial feature points is done to extract Euclidean distances and triangular areas. ReliefF Attribute Evaluation (RAE) and Correlation Feature Selection (CFS) are used to rank extracted features. Feature subsets are formed using ranked features and classified using Support Vector Machine [3].

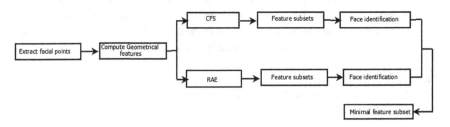

Fig. 1 Proposed methodology

3.1 Localization of Facial Feature Points

During age progression there are minor changes in craniofacial growth and most changes appears in the face shape [9, 11]. We consider 24 landmark points marking the main facial features like nose, mouth, eyebrow, eye. Aging related variations appear mostly in childhood. Since we are considering images above 17 years of age, it is an institution that geometric features will not change. Localization of these facial feature points is done by Active Appearance Model [2].

3.2 Computing Geometrical Features

After localization of facial feature points, we calculate 29 Euclidean distances and 8 triangular areas between selected points. These features are shown in Figs. 2 and 3 respectively. Thus, a total of 37 geometric features are computed.

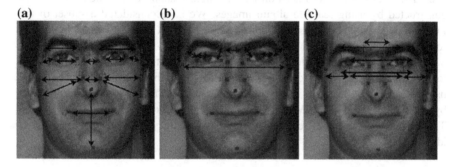

Fig. 2 Computed Euclidean distances

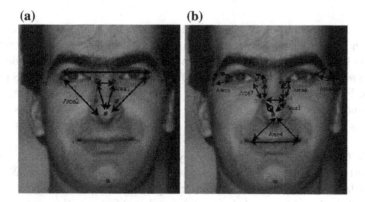

Fig. 3 Computed triangular areas

3.3 Feature Selection and Formation of Subsets

To determine relevant features, two feature selection techniques are used: ReliefF Attribute Evaluation [12] and Correlation Feature Selection [5]. These methods are used to rank the geometrical features. We start with the most prominent feature and keep adding features based on the next highest ranking.

This gives us 37 subsets for each feature selection method. We denote subset of ReliefF Attribute Evaluation by F_{ri} and subsets of Correlation Feature Selection using F_{ci}, where $i = 1, 2, 3… 37$.

We classify the subsets using Support Vector Machine [3]. We perform face identification using all possible generated subsets. Based on face identification results, a minimal set with good accuracy is chosen.

4 Experiments

We have conducted experiments on FGNET aging database [1]. The database was constructed by using real life album images. We have considered a subset of the FGNET database. This subset contains images of adult persons (age > 17) and is roughly frontal. There are a total of 269 such images.

From each image, we select 24 facial points. Using these points, we compute a total of 37 features. We rank features using RAE and CFS and form feature subsets using these ranked features. These subsets are used for the task of face identification by classifying them with SVM after parameter tuning of cost and gamma values. The subset with minimal features giving best accuracy for face identification is identified.

5 Results

For classification purpose, we divide the data into training and testing set in 70:30 ratio such that images of each person are in both the sets. Classification results are shown in Fig. 4 for subsets obtained from both RAE and CFS feature rankings.

From results, we can observe that with feature subsets formed using CFS ranked features, F_{c16} gives an accuracy of 76.60 %. Increase in number of features after that has no improvement in performance. In case of RAE ranked features, subset F_{r22} gives 77.66 % accuracy. Increase in number of features causes accuracy to decrease.

While CFS method gives best accuracy with a minimal subset size of 16 features, best accuracy for RAE is obtained with subset having 22 features. We want minimal facial features without compromising on accuracy. Accuracy of F_{c16} is comparable to that of F_{r22} while having less number of features. Thus, we can argue

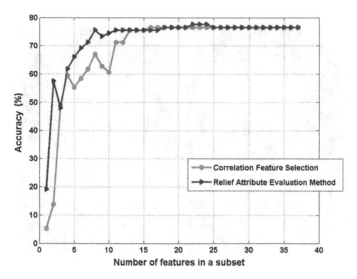

Fig. 4 Comparison of face identification results

that F_{c16} is a minimal set containing relevant features for the task of age invariant face recognition.

5.1 Comparison with State-of-Art

Lanitis et al. [6], have used a private database and accuracy reported is 68.30 %. Wang et al. [13] have also used a private database and report an accuracy of 63 %. Park et al. [8] have performed experiments using FGNET and MORPH (alb1) and accuracy reported are 37.4 and 66.4 % respectively. Nayak et al. [7], have used FGNET and accuracy reported is 70 %. D. Bouchaffra [15] have reported accuracy 48.96 % for FGNET. We see that our accuracy of 76.60 % results using the minimal set of geometrical features outperforms these results.

5.2 Analysis of Features in Minimal Subset

The geometric features contained in F_{c16} are shown in Fig. 5. Analyzing the features contained in F_{c16}, we observe that majority of the features are related to the periocular region. It is seen that aging has little effect on this region. It is also seen that the chin and face contour do not contribute significantly to the task of age invariant face recognition.

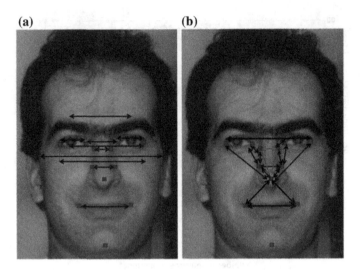

Fig. 5 Minimal facial features

6 Conclusions

This paper presents an age invariant face recognition system using geometric features extracted from the face. Feature selection is applied for determining prominent features across different ages. We determine a set of minimal features. We observe that our method provides good accuracy for face recognition across age progression.

References

1. FGNET Aging Database. http://www.fgnet.rsunit.com (2010)
2. Cootes, T., Edwards, G., Taylor, C.: Active appearance models. Pattern Anal. Mach. Intell., IEEE Trans. **23**(6), 681–685 (2001)
3. Cortes, C., Vapnik, V.: Support-vector networks. Mach. Learn. **20**(3), 273–297 (1995)
4. Du, Ji-XiangWu, X.C.M.: A method based on active appearance model and gradient orientation pyramid of face verification as people age. Mathematical Problems in Engineering pp. 11–16 (2014)
5. Hall, M.A.: Correlation-based feature selection for machine learning. Technical report, Department of Computer Science, Hamilton, New Zealand (1998)
6. Lanitis, A., Taylor, C., Cootes, T.: Toward automatic simulation of aging effects on face images. Pattern Anal. Mach. Intell., IEEE Trans. **24**, 442–455 (2002)
7. Nayak, J., Indiramma, M., Nagarathna, N.: Modeling self-principal component analysis for age invariant face recognition. In: 2012 IEEE International Conference on Computational Intelligence Computing Research (ICCIC), pp. 1–5 (2012)
8. Park, U., Tong, Y., Jain, A.K.: Age-invariant face recognition. IEEE Trans. Pattern Anal. Mach. Intell. **32**(5), 947–954 (2010)

9. Petra, K., Miroslav, B., Markus, S.: Analysis of craniofacial morphology changes during aging and their connection with facial age estimation. In: Proceedings of the ITI 2011 33rd International Conference on Information Technology Interfaces (ITI), pp. 481–486 (2011)
10. Ramanathan, N., Chellapa, R., Biswas, S.: Age progression in human faces: a survey. J.Vis. Lang. Comput. **15**, 3349–3361 (2009)
11. Ramanathan, N., Chellappa, R.: How would you look as you age? Image Processing
12. Robnik-Sikonja, M., Kononenko, I.: Theoretical and empirical analysis of Relief and RRelief. Mach. Learn. **53**(1–2), 23–69 (2003)
13. Wang, J., Shang, Y., Su, G., Lin, X.: Age simulation for face recognition. In: Pattern Recognition, 2006. ICPR 2006. 18th International Conference on. vol. 3, pp. 913–916 (2006)
14. Zhao, W., Chellappa, R., Phillips, P.J., Rosenfeld, A.: Face Recognition: A Literature Survey. ACM Comput. Surv. **35**(4), 399–458 (2003)
15. Bouchaffra, D.: Nonlinear Topological Component Analysis: Application to Age Invariant Face Recognition, IEEE Trans. On Neural Networks and Learning systems, 1375–1387 (2015)

Intelligent Prediction of Properties of Wheat Grains Using Soft Computing Algorithms

Manpreet Kaur, Paramita Guha and Sunita Mishra

Abstract In this paper, chemical properties of wheat dough are predicted using different soft computing tools. Here, back-propagation and genetic algorithm techniques are used to predict the parameters and a comparative study is made. Wheat grains are stored at controlled environmental conditions. The content of fat, moisture, ash, time and temperature are considered as inputs whereas protein and carbohydrate contents are chosen as outputs. The prediction algorithm is developed using back-propagation algorithm, number of layers are optimized and mean square errors are minimized. The errors are further reduced by optimizing the weights using Genetic Algorithm and again the outputs are obtained. The error between predicted and actual outputs is calculated. It has been observed that with back-propagation along GA model algorithm, errors are less compared to the simple back-propagation algorithm. Hence, the given network can be considered as beneficial as it predicts more accurately. Numerical results along with discussions are presented.

Keywords Proximate analysis · Back-propagation · Gradient descent · Genetic algorithm · Local minima

M. Kaur (✉)
Rayat and Bahra University, Mohali, India
e-mail: manpreetkaur7feb@gmail.com

P. Guha · S. Mishra
CSIR-CSIO, Chandigarh, India
e-mail: paramita.guha@csio.res.in

S. Mishra
e-mail: sunita_mishra@csio.res.in

© Springer Science+Business Media Singapore 2016   79
R.K. Choudhary et al. (eds.), *Advanced Computing and Communication Technologies*,
Advances in Intelligent Systems and Computing 452,
DOI 10.1007/978-981-10-1023-1_8

1 Introduction

In recent years modelling, simulation and control of internal parameters of physical systems or processes have become crucial areas of active research. Increasing demands for reliable and detailed analysis of various practical problems has made numerical modelling and simulation of physical systems an important area of research. This is true for both analyses of the systems themselves through simulations, as well as for design of associated controllers for system or process control. Moreover, most practical physical systems are governed by Partial Differential Equations (PDEs) and are of high dimensional in nature. They appear in various application areas such as thermal processes, chemical processes, agricultural and biological systems etc. These systems involve strong or weak interactions between different physical phenomena. Hence, lots of challenges are involved in modelling, simulation, prediction of the data from a physical system.

In current software-reliability research, the concern is how to develop general prediction models. Existing models typically rely on assumptions about development environments, the nature of software failures occurring. A possible solution is to develop models that don't require making assumptions about the development environment and external parameters. An interesting and difficult way to develop a model is using time-series prediction that predicts a complex sequential process. Recent advances in back-propagation (BP) networks [1, 2] show that they can be used in applications that involve predictions. This method has a significant advantage over analytical models as they do not require any assumptions. Using any input, BP can automatically develop its own internal model and predicts the future behaviour of the system. As it adjusts model complexity to match the complexity of the failure history, it can be more accurate than some commonly used analytical models. The main disadvantage of the BP method is that proper selection of weights is necessary to avoid slow convergence or local minima problem [3, 4]. To overcome these difficulties, in this work, genetic algorithm (GA) is used to optimize the weights of BP networks.

Here, wheat grains are stored at controlled temperature and humidity conditions and then dough is prepared from the grinded flour. Using AOAC standard methods [5], the internal parameters of wheat, viz., protein, carbohydrate, fat ash, moisture contents are determined. The main disadvantage involved here is that the chemical tests are time-consuming, laborious and costly. To overcome these difficulties, models are developed using different soft-computing tools like, BP and GA to predict these parameters. It has been observed from the literature that development of models based on GA-BP to predict the chemical parameters for stored grains has not been explored much. The paper is organized in the following way. In Sect. 2, Materials and methods, adopted to achieve the input data are discussed. Different soft-computing algorithms based on BP and GA are discussed in detail in Sect. 3. These algorithms are used to forecast the carbohydrate and protein values of wheat grains. Complete discussion along with numerical analysis is presented in Sect. 4. Finally the paper is concluded in Sect. 5.

2 Materials and Methods

2.1 Sample Storage

The wheat grains were collected from the markets of Punjab. The grains were placed in a storage chamber for duration of one year with temperature maintained at 20 °C and humidity around 25 %.The grains were packed in airtight zip-locked plastic bags so the humidity has least effect on grains properties.

2.2 Chemical Analysis

Sample of grains from the storage chamber was taken every month and immediate analysis was carried out using standard methods [5]. Moisture content in grains was checked using oven method [6] where grains were heated for one hour at 130 °C. Kjeldahl method [7] was applied for determining N_2 content in the grain sample. Soxhlet apparatus process [8] was used for finding the fat content where grain samples were refluxed using petroleum ether. After drying the sample for 6 h the percentage weight gives the fat content. For determining the carbohydrate contents Anthrone method [9] had been used. The grains were mixed with Anthrone solution and the absorption was measured in the UV region.

2.3 Dough Preparation and Rheological Analysis

For preparing dough, 5 gm of wheat sample was grinded using a blade grinder and water was mixed in ratio of 1 ml: 1 gm of wheat flour. The prepared dough was then wrapped in aluminium foil and was kept undisturbed for ten minutes to allow relaxation. Several essential rheological tests were carried out like, stress sweep (with small deformation); creep-recovery test (large deformation) [10, 11] temperature sweep dynamic oscillations etc.

3 Soft Computing Tools

3.1 Back Propagation

Back propagation (BP) is a technique of training multilayer ANNs which uses supervised learning technique [12]. Due to its flexibility it has been effectively employed in broad variety of applications like, modelling, control, prediction etc [13, 14]. In this algorithm, inputs are normalized and the network parameter such as number of hidden layers, number of neurons is selected. After that weights are selected for connecting input and hidden neurons and hidden and output neurons [15]. The values of weights are usually selected between −1 and 1. The weighted sum of inputs (x_j) is connected to the jth hidden layer and is given as

$$Net_j = w_{ij}x_j + \theta_j \tag{1}$$

with θ_j is taken as bias. Using the values of inputs, the output of hidden layer can be calculated as

$$O_j = \frac{1}{1 + e^{-Net_j}} \tag{2}$$

This function is termed as sigmoid function. Again, the output for the next layer can be calculated as

$$[I_o] = [W][O_j] \tag{3}$$

Finally, the outputs are evaluated and they are compared with the desired outputs and errors are evaluated. The error between the actual and desired outputs is given as

$$E_j = \sum_j [O_j - D_j]^2 \tag{4}$$

where D_j is the desired output. The weights are chosen in such a way that the errors are minimized [16, 17].

The main drawback of this method is that the solutions may stuck into a local minima. To overcome this difficulty, gradient decent algorithm may be used where the solutions do not stuck at local minima but it settles to global minima [3], [18]. In this paper, fat, moisture, ash, time and temperature are given as inputs to the BP network and protein and carbohydrate contents are chosen as outputs. Different layers can be placed between the inputs and outputs. Here we have chosen Elman network [19] to optimize the number of layers. It has been observed that only two layers with 20 units (12 units at first hidden layer and 6 units at second layer) are sufficient to efficiently determine and predict the output data.

3.2 Genetic Algorithm

It has been observed that the previous algorithm is slow and several times it got stuck at local minima. Hence, we have combined it with a new search algorithm known as Genetic Algorithm (GA) [20]. The GA is a searching technique which is based on natural selection process. The process is intended to repeat the process of natural systems [21, 22]. Randomly weights are selected, each weight is represented as gene and the string of genes is represented as one chromosome. The weight matrix can be given as

$$w_k = \left\{ \frac{+ \left(x_{kd+2} * 10^{d-2} + x_{kd+3} * 10^{d-3} \dots \right)}{10^{d-2}} \right\},$$

if $5 \le x_{kd+1} \le 9$, and

$$w_k = \left\{ \frac{- \left(x_{kd+2} * 10^{d-2} + x_{kd+3} * 10^{d-3} \dots \right)}{10^{d-2}} \right\}$$

if $0 \le x_{kd+1} < 5$,

where d is number of digits of a gene and $(x_{kd+1}, \dots)$ represent the genes of a chromosome. Now these weights are used in BP algorithm and the outputs are calculated using activation function. Population size is the number of chromosomes selected for the problem. The presentation of every chromosome is predicted using fitness function that measure how fine given data performs. The best performing chromosomes are selected for producing the next generation and the least performing chromosomes are removed from the previous population to generate new population. The errors are calculated using Eq. 4 and a fitness function is determined by reversing the error.

$$F = \frac{1}{E} \tag{5}$$

This algorithm is again used to predict the internal parameters of grains. In the next section the results obtained using both the algorithms are given and a comparative study is made.

4   Results and Discussions

Wheat grains were stored in a controlled atmosphere (with temperature at 20 °C and humidity at 20–25 %) for one year. The inputs provided are fat, ash, moisture, temperature and protein and carbohydrate values were predicted. Their range along with mean and standard deviation (SD) are summarized in Table 1 (Fig. 1 and Table 2).

The given data is divided into two portions. First about 80 % of data was used for training of the network while the left over 20 % data was retained for testing the algorithm. The input layer has five neurons, next the first hidden layer consists of twelve neurons, followed by the second hidden layer with six neurons and the at the output layer there are two neurons. So this network is stated as four layered network. Firstly random weights are selected and the input layer and hidden layers are trained using activation functions. The output of the layer becomes the input for the preceding layer and hence outputs are obtained and errors are evaluated using BP

Table 1 Input and output data

Parameters	Range	Mean	Standard deviation
Protein	10.78 to 19.5	10.98593	1.16002
Ash	1.54 to 1.96	1.7631	0.38886
Moisture	7.07 to 7.8	7.05603	1.12622
Carbohydrate	76.91 to 81.973	79.28084	1.77014
Fat	2.97 to 4.2	2.97973	0.79883
Temperature	20	20	0

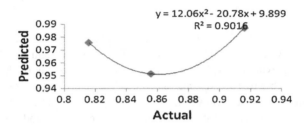

Fig. 1 Prediction of protein using BP

Table 2 Error occurring during the training

Nature of algorithm	Protein		Carbohydrate	
	MAE	% error	MAE	% error
BP	0.0762	18.56	0.0567	7.46
GA	0.0176	7.036	0.03456	2.087

Table 3 Correlation coefficient (R^2) during training and testing

Parameter	Training		Testing	
	BP	GA	BP	GA
Protein	0.970	0.969	0.9016	0.8782
Carbohydrate	0.969	0.984	`0.841	0.8912

and GA mechanism. At the end weights are adjusted and fed back to the network till the error gets reduced. For updating weights, the learning rate and momentum coefficient were chosen as 0.5 and 0.01 respectively. The correlation coefficients (R^2) given in Table 3 were evaluated by plotting the predicted data versus given data. Figures 2, 3, 4 show the relation between predicted and actual values of protein and carbohydrate content. It was observed that the average values of R^2 for the both models BP and GA were satisfactory.

The values of mean average error (MAE) and percentage error for both BP and GA models were calculated and given in Table 2. From that table we can conclude that GA based BP neural network provides very good results. The main advantages

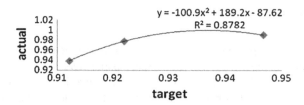

Fig. 2 Prediction of protein using GA-BP

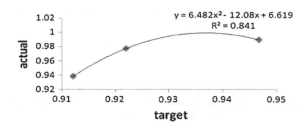

Fig. 3 Prediction of carbohydrate using BP

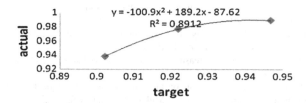

Fig. 4 Prediction of carbohydrate using BP GA-BP

of this algorithm as evaluated with Back Propagation neural network is that the number of iterations were lesser than simple BP and the network has never got stuck into global minima. With this network the protein and carbohydrate future data can be predicted without any chemical analysis.

From the above analysis, neural network model based on GA has predicted values were found to be very close to the observed values as compared to BP model. This shows that GA model show better potential in predicting data for future purposes. The better presentation of GA may be due to heuristic search for the optimal. Thus, this algorithm has a better possibility to attain the global minima. On the other hand, back propagation algorithm has many times fallen behind GA model might be due to local optima problem. Therefore, GA based NN model is considered to be more helpful for predicting data for future. The main disadvantage of the BP model is the proper selection of weights to avoid slow convergence and local minima problem. To overcome these difficulties, genetic algorithm based NNs can be used.

5 Conclusion

In this paper, internal parameters of wheat grains are predicted using a GA based BP model. As a comparative study, the same parameters are predicted using a simple BP based model. Advantages of GA based network over the conventional BP network are discussed. It has been studied that this model can be considered as better choice for achieving a complex relationship between inputs and outputs which is otherwise very difficult to obtain mathematically. Hence, it is better to apply GA algorithm for training the data, so that search space can be reduced to perform better and to avoid the difficulty of the convergence at local minima.

References

1. Zhixin, S., Bingqing, L.: Research of improved back-propagation neural network algorithm. In: 12th IEEE International Conference, China (2010)
2. Jung, I., Wang, G.:. Pattern classification of back-propagation algorithm using exclusive connecting network. World Acad. Sci. Eng. Technol. 1 (2007)
3. Toussaint, M.: Lecture Notes on Gradient Decent. Machine Learning & Robotics, Berlin (2012)
4. Bottou, L.: Large-scale machine learning with stochastic gradient descent. In: Proceedings of Compstat, pp 177–180 (2010)
5. AOAC, Official Methods of Analysis, 15th edition of Association of official Analytic alchemists, Arlington, VA. AOCS, vol. 1 (1998)
6. Can, K.S., Lim, S.C., Ong, C.E.: Measuring the moisture content of wood during processing. Forest Res. Inst. Malaysia 36 (2005)
7. Amin, M., Flowers, T.H.: Evaluation of kjeldahl digestion method. Journal of Research (Science), Bahauddin Zakariya University, Multan, Pakistan, vol. 15 (2004)
8. Ahmad, A., Alkarkhi, A.F.M., Hena, S.: Extraction, separation and identification of chemical ingredients of Elephantopus Scaber L. Int. J. Chem. 1 (2009)
9. Sadasivam, S., Manickam, A.: Biochemical methods, 2nd edn. New Age International (P) Ltd (2005)
10. Malkin, A., Isayev, A.: Rheology, Concepts, Methods and Applications. Chemtec Publishing (2006)
11. Zhai, H., Salomon, D., Miliron, E.: Using Rheological Properties to Evaluate Storage Stability and Setting Behavior of Emulsified Asphalts. Idaho Asphalt Supply, Inc. White Paper, USA, vol. 11 (2006)
12. Saduf, M.A.W.: Comparative study of back propagation learning algorithms for neural networks. Int. J. Adv. Res. Comput. Sci. Softw. Eng. 3 (2013)
13. Popescu, O., Popescu, D., Wilder, J., Karwe, M.: A new approach to modelling and control of a food extrusion process using artificial neural network and an expert system. J. Food Process Eng 24, 17–36 (2001)
14. Kashaninejad, M., Dehghani, A.A., Kashiri, M.: Modeling of wheat soaking using two artificial neural networks. J. Food Eng. 91, 602–607 (2009)
15. Llave, Y.A., Hagiwara, T., Sakiyama, T.: Artificial neural network model for prediction of cold spot temperature in retort sterilization of starch- based foods. J. Food Eng. 109, 553–560 (2012)
16. Sablani, S., Shaur Rahman, M.: Using neural networks to predict thermal conductivity of food as a function of moisture content. Temp. Apparent Porosity Food Res. Int. 36, 617–623 (2013)

17. Qiao, J., Wang, N., Ngadi, M., Kazemi, S.: Predicting mechanical properties of fried chicken nuggets using image processing and neural network techniques. J. Food Eng. **79**, 1065–1070 (2007)
18. Choi, B., Lee, J.H., Kim, D.H.: Solving local minima problem with large number of hidden nodes on two-layered feed-forward artificial neural networks. Neurocomputing **71**(16–18), 3640–3643 (2008)
19. Wagarachchi, N.M., Karunananda, A.S.: Mathematical modeling of hidden layer architecture in artificial neural networks. In: International Conference on Information Security and Artificial Intelligence, vol. 56 (2012)
20. Whitley, D.: Applying genetic algorithms to neural network problems. In: International Neural Network Society, pp. 1–6 (2008)
21. Koehn, P.: Combining Genetic Algorithms and Neural Networks, A Thesis Presented for the Master of Science Degree. The University of Tennessee, Knoxville, December (1994)
22. Mandal, S.N., Ghosh, A., Roy, S., Pal Choudhury, J., Chaudhuri, S.R.B.: A novel approach of genetic algorithm in prediction of time series data. Int. J. Comput. Appl. Adv. Comput. Commun. Technol. HPC Appl. ACCTHPCA(1) (2012)

An Approach to Improve the Classification Accuracy of Leaf Images with Dorsal and Ventral Sides by Adding Directionality Features with Statistical Feature Sets

Arun Kumar, Vinod Patidar, Deepak Khazanchi, G. Purohit and Poonam Saini

Abstract The basic purpose of this work is to study statistical feature set obtained from digital leaf image with dorsal and ventral sides and to find the degree of classification accuracy for each dorsal and ventral leaf image dataset. Moreover, the effect of adding directional features to statistical feature set on the overall classification accuracy, is also investigated. The work also studies whether the ventral side of the digital leaf image can be a suitable alternative for classification of leaf image data set or not.

Keywords Leaf images · Directionality · Statistical features · Dorsal and ventral sides

1 Introduction

The nature has given innumerable objects to view and to use them according to our requirements, but before starting to use them, our eyes must discern one object from the other. Once our brain has seen the characteristic features of the object of concern, we name it, and keep a copy of the object for future use. This copy of the image helps a person in segregating one object from the other. In computer based automatic classification of the images, several methods have been proposed in machine vision studies, which try to imitate the human visualization system. In a natural image, one portion of the scene has a boundary clearly demarcated from the

A. Kumar (✉) · V. Patidar · G. Purohit · P. Saini
Sir Padampat Singhania University, Udaipur, India
e-mail: arunkumarsai@gmail.com

D. Khazanchi
University of Nebraska, Omaha, USA

© Springer Science+Business Media Singapore 2016
R.K. Choudhary et al. (eds.), *Advanced Computing and Communication Technologies*,
Advances in Intelligent Systems and Computing 452,
DOI 10.1007/978-981-10-1023-1_9

other by its edges and ridges. The digital images are full of patterns inclined to angular positions. Therefore, getting appropriate information about the feature pattern helps in proper classification of digital images.

In case of leaves, nature provides two faces to the leaves i.e. dorsal or the front side and the ventral or the back side which is not the case for other types of objects. The classification on the basis of the dorsal side of the objects including leaves has been done by many researchers [1–3]. The role of ventral side of a leaf image in classification of data set remained untouched, therefore, there is a drastic need to study the role of ventral or the back side of the leaf in discriminating the leaf images because of the presence of prominent venation patterns.

The leaf image classification can be done by using leaf's geometrical features, texture and shape based features and color features [1–3] etc. A digital image is composed of pixels of different intensity values, the change in intensity values around the images leads to a scene with visible objects, if there is no change in the intensity, there is no image formation and this change in intensity values occur in a particular direction in the image. The concept of directionality histogram has been used for the characterization of brain micro-device interface using the device capture histology [4] and for 3D microstructure modeling of long fiber reinforced thermoplastics [5]. The concept of directionality also finds its use for finding texture features using the method of directionality histogram on geometric property of images [6].

In statistical jargon, a sample is a subset of values taken out from a digital image for understanding its characteristic properties through which detailed statistical analysis can be performed by utilizing first order statistics like mean, mode, median and standard deviation etc. for discriminating the images into various classes.

This work has tried to extract the various statistical features and directional features from the digital images and observing their effects in automated classification of digital images on the dorsal side, ventral side and combined dorsal-ventral sides. The statistical features like Mean Gray, Median, Integrated density, Skewness, Kurtosis, Standard Deviation, First order spatial moments (XM, YM) and Minimum Gray value have been computed for the dorsal, ventral and combined dorsal-ventral sides. We have computed the directional features like direction, dispersion and fitness etc. on the dorsal, ventral and combined dorsal-ventral sides of the leaf images. These directionality feature sets have been combined with the statistical feature sets for different sides. The classification algorithms like K-Nearest Neighbor (KNN), J48, Classification and Regression Tree (CART) and Random Forest (RF) have been used for classification of statistical features as well as combined statistical-directionality feature sets. The objective of this work is to find out whether the ventral side of the leaf image can be considered for leaf image classification or not and can the classification accuracy be improved by adding directionality features with the statistical features. The paper has been divided into different sections, where Sect. 2 describes the methodology adopted, Sect. 3 highlights the results and analysis and Sect. 4 states the conclusion.

2 Methodology Adopted

2.1 Database Creation

The leaf image data set with dorsal side is available from several sources including that of [7–9], but to study the objective of this work, i.e. to utilize both the dorsal and the ventral faces of the leaf images, the creation of a leaf image database was required with both the ventral as well as dorsal sides of the leaves. For the experimental research work, we have captured the images of dorsal and ventral sides of the leaves of 10 different plants which include: Helianthus annuus L., Psidium gujava, Alcia rosea, Jasminum sambac, Calotropis acia, Sarace indica, Cordia sebestena, Manilkara zapota, Hibiscus laevis, Ficus religiosa as shown in Fig. 1 using Sony Cybershot HX200V with 18.2MP "Exmor RTM" CMOS Sensor with extra high sensitivity technology, 30x optical zoom. The captured images include 25 dorsal side and 25 ventral side images for each of the above mentioned leaf categories totaling a sample size of 500 images with a pixel size of 1080 × 920.

All the colored images were converted into 8-bit gray scale and their size was reduced to 256 × 256 which reduced the size of the database and a stack was created using ImageJ (ver. 1.44) [10].

2.2 Preparation of Statistical Feature Set

Since, there are a variety of images available like X-ray images of the body parts, traffic scene etc., and all the images are different from one another but there are certain features which can be captured so that the images can be represented with minimum number of bits.

In an image, a pixel can take any value randomly from a set of values that appear in the same grid. Therefore a pixel becomes a random variable and the image becomes a random field.

The following statistical features have been extracted from the leaf image dataset for dorsal, ventral and combined dorsal-ventral sides [10]. The scale of calibration has been set in millimeter (mm).

Fig. 1 Colored sample of dorsal and ventral sides of the leaf images

- Mean Gray value: The mean gray value is calculated by making a selection of the region of the image and then mean value is calculated as mentioned in Eq. (1).

$$\mu = \frac{1}{N^2} \sum_{i,j=1}^{N} P_{i,j} \tag{1}$$

- Median value: The median value of the pixels in the selection region is the median of the data.
- Integrated Density (IntDen): The integrated density is a product of Area and Mean gray values.
- Skewness (Skew) and Kurtosis (Kurt): Spatial moments are a very simple and powerful way to describe the spatial distribution of values. Skewness measures the third order moment about mean, whereas the kurtosis measures the fourth order moment about the mean as mentioned in Eqs. (2) and (3) respectively.

$$\text{Skewness} = \mu_3 = \frac{1}{N^2 \sigma^3} \sum_{i,j}^{N} \left(P_{i,j} - \mu \right)^3 \tag{2}$$

$$\text{Kurtosis} = \mu_4 = \frac{1}{N^2 \sigma^4} \sum_{i,j=1}^{N} \left(P_{i,j} - \mu \right)^4 \tag{3}$$

- Minimum Gray Value (Min value): This represents the minimum gray value of the selection region.
- Standard Deviation (StdDev): Standard Deviation of the gray values finds the mean gray value of the data. Standard Deviation is the square root of variance (σ):

$$\text{Variance} = \sigma^2 = \frac{1}{N^2} \sum_{i,j=1}^{N} \left((P_{i,j} - \mu)^2 \right) \tag{4}$$

- First order spatial moments (XM and YM): This is the brightness-weighted average of the x and y coordinates of all pixels in the image or a selection.

2.3 Preparation of Directionality Feature Dataset

A digital image is made up of pixels of various intensity values which may vary over the entire region of the image space. This leads to the formation of texture structures in the images which can be studied with the help of directionality.

The basic concept of directionality in the digital images was given by Liu [11]. According to [11], when Gaussian filter as represented by Eq. (5) is applied to the image coordinates (x, y), with different scale (σ), it generates sets of images with different levels of smoothness. The next step is to find the edges in the images which can be obtained by finding the Laplacian of Gaussian (LOG) as mentioned in Eqs. (6) and (7).

$$g(x, y, \sigma) = \frac{1}{2\pi\sigma^2} \exp\left[\frac{1}{2}\left(\frac{r}{\sigma}\right)^2\right] \tag{5}$$

Here, $r^2 = (x^2 + y^2)$ and (x, y) represents the image coordinate values.

At the edges, the intensity of the pixels changes rapidly i.e. the zero-crossing detector looks for the places in the image where the Laplacian passes through zero. This results in the generation of binary image with single pixels thickness lines showing the position of zero crossing pixels, which is represented through Eq. (7). The Laplacian highlights the regions of rapid change which is used in edge detection.

$$L(x, y) = \frac{\partial^2 I}{\partial x^2} + \frac{\partial^2 I}{\partial y^2} \tag{6}$$

$$LOG(x, y) = \frac{-1}{\pi\sigma^4}\left[1 - \frac{x^2 + y^2}{2\sigma^2}\right] e\left(\frac{-(x^2 + y^2)}{2\sigma^2}\right) \tag{7}$$

According to Witkin [11], a concept of scale space can be applied to find information in images and it can be expressed as Eq. (8)

$$\psi(x, y; \sigma) = \left\{(x, y; \sigma)|z(x, y; \sigma) = 0; \left(\left(\frac{\partial z^2}{\partial x}\right) + \left(\frac{\partial z}{\partial y}\right)^2 \neq 0, \sigma > 0\right)\right\} \tag{8}$$

where $z(x, y; \sigma) = LOG(x, y) * I(x, y)$ and * here represents convolution operation, $I(x, y)$ is the image and Eq. (8) shows the scale space as applied to images to find the directional information.

The directionality plugin [12] used produces a histogram for the dorsal and ventral leaf image as shown in the Fig. 2. There are 90 bins for a total orientation of $180°$ that have been used for generating the histogram for the images. The plugin generates a comma separated values file (CSV) with direction column that reports the center of the Gaussian, the dispersion column that gives the standard deviation of the Gaussian calculated above. The amount column gives the sum of the histogram from the center minus standard deviation to the center plus standard deviation divided by the sum of the histogram. The goodness column reports the goodness of the fit and its value is 1 for good and 0 for bad. We have constructed a normalized directional histogram which shows the angles in the horizontal axis and

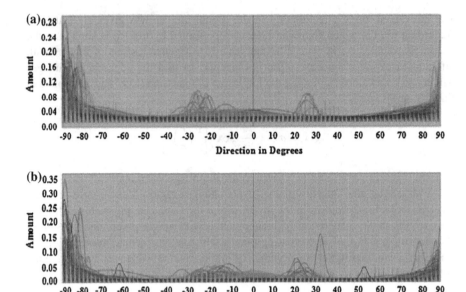

Fig. 2 Directionality histogram for, **a** dorsal, **b** ventral sides of the leaf images

vertical axis shows the percentage of pixels with different gradient angles as shown in Fig. 2. The major role of directionality is to find the amount of structure in a given direction in the digital image.

2.4 Application of Classification Algorithms

To find the classification accuracy, the KNN, J48, CART and RF classification algorithms have been used on the data sets obtained in Sects. 2.2 and 2.3 using "Caret" package under RStudio [13]. The datasets obtained from Sects. 2.2 and 2.3 are combined to find the effect of directionality on the statistical feature sets on the classification accuracy for the dorsal, ventral and combined dorsal-ventral sides of the leaf images [10, 12]. Each data set was split into two groups (Training and Testing sets) in the ratio 70:30. To obtain an accurate estimate to the accuracy of a classifier, k-fold cross validation is run several times, each with a different random arrangement of data sets. We have used 10-fold cross validation for the resampling process.

The accuracy for the analysis purpose have been calculated where the term accuracy refers to the number of times a correct match has been found for a particular leaf image in the dataset.

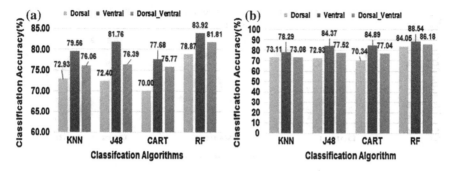

Fig. 3 Classification accuracy for, **a** statistical, **b** statistical-directionality feature sets

The classification accuracy for statistical features is shown in Fig. 3a and the combined statistical-directionality features is shown in the Fig. 3b for dorsal, ventral and combined dorsal-ventral sides of the leaf images.

3 Results and Analysis

3.1 Analysis of the Results on the Basis of Statistical Feature Set

As shown in Fig. 3a, the Random Forest (RF) classification algorithm is giving the highest accuracy values for dorsal, ventral and combined dorsal-ventral sides of the leaf images. Statistical feature set alone gives 83.92 % accuracy for ventral side using Random Forest which is more than the accuracy of dorsal (78.87 %) and combined dorsal-ventral (81.81 %) as represented in Fig. 3a.

3.2 Analysis of the Effect of Directionality Features on Combining Them with Statistical Features Sets

By comparing the Fig. 3a, b, it is analyzed that the overall accuracy for classification of leaf image data has been increased with the addition of directionality features into the statistical feature set. As in Fig. 3b, the classification accuracy is 88.54 % for ventral side, 86.18 % for combined dorsal-ventral side and 84.05 % for dorsal side of the leaf images as compared to accuracy shown in Fig. 3a i.e. 83.92, 78.87 and 81.81 % respectively. Therefore the overall accuracy has been increased for all the datasets by adding the directionality features with the statistical feature sets. The directionality features added additional information extracted from leaf images which resulted in improved accuracy results in all the sides of the leaf images.

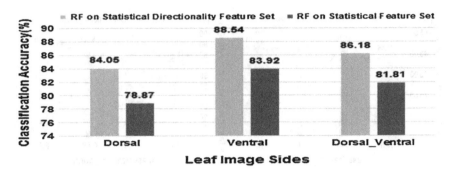

Fig. 4 Comparison chart for classification accuracy of different sides of leaf images using RF

3.3 Analysis on the Basis of Sides of the Leaf Images

The Random Forest algorithm is an ensemble learning technique used for classification and regression operations and based on the principle of constructing a multitude of decision trees at training time and it outputs the class that is the mode of the classes (in case of classification) or average prediction (in case of regression) of the individual trees. With the help of ensemble learning technique, it allows the algorithm to learn accurately both simple and complex classification features for the dataset.

As we are using a large number of variables in a dataset, and each variable due to its interaction with the other variable creates its own importance. The Random Forest algorithm estimates the importance of the variables by finding out by how much the prediction error increases when out of bag data is presented to the algorithm. The computations are done tree by tree as the random forest is constructed. This methodology of RF algorithm [14] makes it a better classifier as compared to other algorithms used.

As shown in Fig. 4, the ventral side is giving higher accuracy as compared to dorsal and combined dorsal-ventral sides of the leaf images for statistical feature set (83.92 %) and statistical-directionality combined feature set (88.54 %). This proves the assumption of this study, that ventral side can be considered for the classification purpose of the leaf images.

4 Conclusion

From Sect. 3.2, and Fig. 3, it has been observed that the directionality features with statistical feature sets improves the classification accuracy for dorsal, ventral and combined dorsal-ventral sides. On the basis of the results shown in Fig. 4 and Sect. 3.3, we propose to consider the ventral side of leaf images for the classification purpose which fulfills the objective of this research.

References

1. Kalyoncu, C., Toygar, O.: Geometric leaf classification. In: Computer Vision and Image Understanding, vol. 133, pp. 102–109 (2015)
2. Chaki, J., et al.: Plant leaf recognition using texture and shape features with neural classifier. In: Pattern Recognition Letter, vol. 58, pp. 61–68 (2015)
3. Perez, A.J., et al.: Color and shape analysis techniques for weed detection in cereal fields. Comput. Electron. Agric. **25**, 197–212 (2000)
4. Woolley, A.J., Desai, H.A., Steckbeck, M.A., Patel, N.K., Otto, K.J.: In situ characterization of the brain–microdevice interface using device capture histology. J. Neurosci. Methods **201**, 67–77 (2011)
5. Fliegener, S., Luke, M., Gumbsch, P.: 3D microstructure modeling of long fiber reinforced thermoplastics. Compos. Sci. Technol. **104**, 136–145(2014)
6. Islam, M.M., Zhang, D., Lu, G.: A geometric method to compute directionality features for texture images. www.users.monash.edu.au/∼dengs/resource/papers/icme08.pdf (2015). Accessed Aug 2015
7. Database from Caltech. http://imagej.nih.gov/ij/, http://www.vision.caltech.edu/Image_Datasets/leaves/leaves.tar (2013). (Web page last visited on Oct, 2013)
8. Database from Oxford. http://www.plant-phenotyping.org/CVPPP2014-dataset (2013). (Web page last visited on Oct, 2013)
9. Database from UCI. https://archive.ics.uci.edu/ml/machine-learning-databases/00288/leaf.zip (2013). (Web page last visited on Oct, 2013)
10. Rasband, W.S.: ImageJ, U.S. National Institutes of Health, Bethesda, Maryland, USA (1997–2014)
11. Liu, Z.Q.: Scale space approach to directional analysis of images. J. Appl. Opt. **30**(11), 1369–1373 (1991)
12. Directionality plugin. www.fiji.sc/Directionality (2015). Accessed Aug, 2015
13. R Development Core Team, R: A language and environment for statistical computing, R Foundation for Statistical Computing, Vienna, Austria (2008). ISBN 3-900051-07-0
14. http://www.bios.unc.edu/∼dzeng/BIOS740/randomforest.pdf (2015). Accessed Aug, 2015

Analyzing the Effect of Chaos Functions in Solving Stochastic Dynamic Facility Layout Problem Using CSA

Akash Tayal and Surya Prakash Singh

Abstract Chaotic Simulated Annealing (CSA) is a derivative meta-heuristic obtained by embedding chaos in Simulated Annealing (SA) meta-heuristic. Literature talks about various chaos maps and its analysis for efficient and convergent search for finding an optimal solution. In this article, CSA has been applied to Stochastic Dynamic Facility Layout Problem (SDFLP), which is a NP-hard combinatorial optimization problem. The goal is to evaluate the impact of seven known chaotic maps in solving the SDFLP. For optimizing SDFLP the best measure of its efficiency is by minimizing the total material handling cost (TMHC) of the layout. TMHC is an addition of material handling and rearrangement cost. The paper evaluates the solution on multiple aspects—material handling cost, rearrangement, cost and product demand distribution to prove the efficacy of chaos embedded algorithm (CSA) in solving SDFLP. The solutions obtained in some cases are better than the results reported in previous literature. Also, it was seen from the CSA simulation that '*sine*' and '*logistic*' chaotic maps has the best abilities in solving SDFLP.

Keywords Facility layout · Stochastic dynamic facility layout · Simulated annealing · Chaotic simulated annealing · Meta-heuristic · Chaotic functions

1 Introduction

Facility Layout Problem (FLP) deals with assignment of 'N' facilities to 'M' locations ($M \leq N$), to minimize associated material handling costs (MHC). The state of artwork on FLP can be found in Singh and Sharma [1] and Matai et al. [2].

A. Tayal (✉)
Indira Gandhi Delhi Technical University for Women, New Delhi, India
e-mail: akashtayal@yahoo.com

S.P. Singh
Department of Management Studies, Indian Institute of Technology Delhi, New Delhi, India
e-mail: surya.singh@gmail.com

© Springer Science+Business Media Singapore 2016   99
R.K. Choudhary et al. (eds.), *Advanced Computing and Communication Technologies*,
Advances in Intelligent Systems and Computing 452,
DOI 10.1007/978-981-10-1023-1_10

In the design of FLP, if the product flow is uncertain then it is referred to as stochastic facility layout problem (SFLP). SFLP could be single period (static) or multi period (dynamic). Single period SFLP is referred as Stochastic Static FLP (SSFLP) while multi period SFLP as Stochastic Dynamic FLP (SDFLP). Work on SFLP can be founded from Rosenblatt and Kropp [3], Palekar et al. [4], Braglia et al. [5], Krishnan et al. [6], Moslemipour and Lee [7], Moslemipour et al. [8] and Tayal and Singh [9–11].

MHC is the product of product demand and the distance between different locations. Tompkins et al. [12] advocated that MHC is a suitable and good measure for evaluating the efficiency of layout and it forms 20–50 % of the total manufacturing cost that can be further decreased by 10–30 % by an efficient and optimal design of facility layout. However, in SDFLP the product demand fluctuations leads to change in flow of materials thus same layout of facilities from one time period to another is inefficient. Hence, for SDFLP there is a need to rearrange the facilities for each time period, which is an additional cost to bear, known as rearrangement cost. To find optimal layout for SDFLP there is a need to minimize both material handling cost and rearrangement cost known as the total material handling cost (TMHC).

SDFLP is a class of NP-hard optimization problem having complexity of $(N!)^T$ where N and T stands for number of facility and time period, respectively. Thus, good meta-heuristic techniques are needed to solve SDFLP. In this paper, chaos embedded meta-heuristic i.e. Chaotic Simulated Annealing (CSA) is used for optimizing SDFLP. Chaos embedded meta-heuristics are classified into two general categories. First category contains the algorithms in which chaos is used instead of random number generators. On the other hand in the second category chaotic search that uses chaotic map is incorporated into meta-heuristics to enhance searching behavior of these algorithms and to skip local optima such as CSA. Work on analyzing various chaotic maps in solving SDFLP has not been carried out so far. In this paper seven chaotic map functions are discussed and their solution is evaluated on multiple aspects—material handling cost, rearrangement cost and product demand distribution to prove the efficacy of chaos embedded algorithm (CSA) and to find the most appropriate chaotic map for solving SDFLP. The results obtained are compared with existing literature on SDFLP.

The paper is organized as follows. In Sect. 2, the mathematical formulation of SDFLP is provided. In Sect. 3, the CSA along with the seven chaotic maps to solve SDFLP is explained. Section 4 gives the simulation results of problem set $N = 12$ facilities and time period, $T = 5$ taken from literature for two product density functions—Gaussian and Poisson. Also, inferences and comparison with existing literature is discussed. Section 5 concludes the paper.

2 Mathematical Model of SDFLP

The product flows between facilities are generally an expression of demand, which could be static, dynamic or uncertain. Stochastic facility layout (SFLP) has gained prominence because of uncertainty in product demand, which can be model as stochastic random variables. This random variable is expressed as Probability Distribution Function (PDF) with known mean and variance. Rosenblatt and Kropp [3] first proposed an analytical formulation of SFLP where the static product-demand flow was replaced by stochastic product-demand flow. Below are the notations used for the mathematical formulation (Table 1).

FLP was modeled as Quadratic Assignment Problem (QAP) by Koopman and Beckman [13], and is given below:

$$C(\pi) = \sum_{i=1}^{N} \sum_{j=1}^{N} \sum_{l=i}^{N} \sum_{q=1}^{N} f_{ij} d_{lq} x_{il} x_{jq} \tag{1}$$

Subject to:

$$\sum_{i=1}^{N} x_{il} = 1; \qquad \forall l \tag{2}$$

Table 1 Notations of SDFLP mathematical model

Notations	Description
i, j	Index for facilities (i, j = 1, 2, 3, ..., N); $i \neq j$
l, q	Index of locations (l, q = 1, 2, 3, ..., N); $l \neq q$
f_{ij}	Flow of material between facilities i and j
d_{lq}	Distance between locations l and q
N	Number of facilities
$C(\pi)$	Total MHC for layout π
$E(\pi)$	Expected value of the πth layout
$Var(\pi)$	Variance of the πth layout
$Pr(\pi)$	Probability of the πth layout
Z_p	Standard Z (random variable) value for percentile p
$U(\pi, p)$	Maximum value upper bound of $C(\pi)$ with confidence level p
K	Index for parts (k = 1, 2, ..., K)
M_{ki}	Operation number for the operation done on part k by facility i
D_{kt}	Demand for part k in period t
B_k	Transfer batch size for part k
C_{tk}	Cost of movements for part k in period t
Z	Random variable
a_{tilq}	Fixed cost of shifting facility i from location l to location q in period t

$$\sum_{l=1}^{N} x_{il} = 1; \qquad \forall i \tag{3}$$

$$x_{il} = \begin{cases} 1, & \textit{if facilities } i \textit{ is assigned to location } l \\ 0, & \textit{otherwise} \end{cases} \tag{4}$$

Equation (1) is modified for stochastic process then $C(\pi)$ becomes a function of random variables. Here, f_{ij} is changed to stochastic variable due to uncertainty of demand with mean μ_{ij} and variance σ_{ij}^2 respectively. If $U(\pi, p)$ is considered as the maximum value of $C(\pi)$ having confidence p then it can be minimized instead of minimizing $C(\pi)$. Following this basic assumption Kulturel-Konak et al. [14], Smith [15], Moslemipour and Lee [7] we can write,

$$\left\{ \begin{bmatrix} \sum_{t=1}^{T} \sum_{i=1}^{N} \sum_{j=1}^{N} \sum_{k=1}^{N} \frac{E(D_{tk})}{B_k} C_{tk} \sum_{l=1}^{N} \sum_{q=1}^{N} d_{lq} x_{til} x_{tjq} \\ + Z_p \sqrt{\sum_{t=1}^{T} \sum_{i=1}^{N} \sum_{j=1}^{N} \sum_{k=1}^{K} \frac{Var(D_{tk})}{B_k^2} C_{tk}^2 \left(\sum_{l=1}^{N} \sum_{q=1}^{N} d_{lq} x_{til} x_{tjq} \right)^2 } \\ + \begin{bmatrix} \sum_{t=2}^{T} \sum_{i=1}^{N} \sum_{l=1}^{N} \sum_{q=1}^{N} a_{tilq} x_{(t-1)il} x_{tiq} \end{bmatrix} \end{bmatrix} \right\} \tag{5}$$

Subject to:

$$\sum_{i=1}^{N} x_{til} = 1; \qquad \forall t, l \tag{6}$$

$$\sum_{l=1}^{N} x_{til} = 1; \qquad \forall t, i \tag{7}$$

$$x_{til} = \begin{cases} 1, & \textit{if facilities } i \textit{ is assigned to location } l \textit{ in period } t \\ 0, & \textit{otherwise} \end{cases} \tag{8}$$

$$|M_{ki} - M_{kj}| = 1 \tag{9}$$

In Eq. (5) the first term gives the Material Handling Cost (MHC) and second term describes the rearrangement cost RA_c.

3 Chaotic Simulated Annealing for Optimizing SDFLP

Mingjun and Huanwen [16] developed Chaotic Simulated Annealing (CSA) by introducing a chaotic system to SA. CSA Algorithm is similar SA on principles like cooling schedule, initial temperature and stopping criteria. However, there are basic two main differences between SA and CSA. Firstly, SA is stochastic phenomena and fully based on the Monte Carlo scheme where as CSA is a deterministic phenomenon. Secondly, the convergence of SA is done by thermal fluctuations using cooling schedules whereas in the case of CSA it is done by control of division structures. Figure 1 gives the illustration of CSA for solving SDFLP, Tayal and Singh [9]. The various chaotic functions to generate chaos are given in Table 2.

Initialize

    Start with a known or randomly generated initial solution, $s0$ and assign $s=s0$

    Generate the different chaotic variables, $H_{k_i}, i = 1, 2, \ldots, N$ by using the chaotic systems given in Table 2,

$$H_{k+1} = f(\mu, H_k) \qquad\qquad (a)$$

    where, $H_k \in [0,1]$. H_k is the value of the variable H at the kth iteration, k is a random integer min set $\{1, \ldots, 400\}$ and μ is called the bifurcation parameter of the system, in this paper μ is considered as 4.

    Initialize the temperature T_0

Generate new neighborhood solution, s'

    Compute the initial position value for the facility,

$$p_{0,i} = a_i + (b_i - a_i) \times H_{k_i} \qquad\qquad (b)$$

    where, a_i is the lower limit of the facility position and b_i is the upper limit of the facility position.

    Compute the neighborhood position value for the facility, using

$$y_{m,i} = p_{m,i} + \alpha \times (b_i - a_i) \times H_{k_m} \qquad\qquad (c)$$

    where, i is randomly chosen from the set $\{1,2, \ldots, N\}$, H_{k_m} is a chaotic variable produced by Equation (a), and k_m is a random integer in the set $\{1, \ldots, 400\}$

    Here, α is a variable which is decreased by the formula $\alpha = \alpha \times e^{-\beta}$ in each iteration. In this paper β is taken as 1.01

    Using above position vectors, s' is computed

Start Inner loop

Compute the TMHC i.e. $f(s)$ for $s0$ and s'

    Compute TMHC for $s0$ and s' given in Equation (5

Check,

    if $f(s) > f(s')$, assign $s = s'$

    else if $P((f(s')-f(s))/KT) < rand$, assign $s = s'$

Repeat until inner loop criteria

Decrease the temperature, using **exponential cooling**

Repeat until stopping criteria, reset inner loop criteria

Output the best solution's' it's material handling cost and rearrangement cost

---

Fig. 1 Chaotic simulated annealing algorithm to solve SDFLP

Table 2 Chaotic maps

Chaotic map	Chaotic function
Logistic	$H_{k+1} = \mu H_k(1 - H_k)$
Chebyshev	$H_{k+1} = \cos(k\cos^{-1}(H_k))$
Piecewise	$H_{k+1} \begin{cases} \frac{H_k}{P} & 0 \le H_k < P \\ \frac{H_k - P}{0.5 - P} & P \le H_k < \frac{1}{2} \\ \frac{1 - P - H_k}{0.5 - P} & \frac{1}{2} \le H_k < 1 - P \\ \frac{1 - H_k}{P} & 1 - P \le H_k < 1 \end{cases}$
Gaussian iterative	$H_{k+1} = \exp(-\alpha H_k^2) + \beta$
Sine	$H_{k+1} = \frac{a}{4}\sin(\pi H_k)$
Singer	$H_{k+1} = \mu(7.86H_k - 23.31H_k^2 + 28.75\ H_k^3 + 13.3028\ H_k^4)$
Tent	$H_{k+1} = \begin{cases} \dfrac{H_k}{0.7} & H_k < 0.7 \\ \dfrac{10}{3}(1 - H_k) & H_k \ge 0.7 \end{cases}$

4 Result and Discussion

The proposed methodology is illustrated in the flow chart given in Fig. 2. The author uses seven different chaotic functions in CSA meta-heuristic with exponential cooling schedule to solve SDFLP. The effectiveness of exponential cooling schedule has been studied and evaluated in, Tayal and Singh [11]. The CSA algorithm is programmed in Java 6 on a personal computer Intel core 2.53 GHz CPU and 4 GB RAM and is applied on two data set—one with Gaussian distribution product demand and second with Poisson distribution product demand for facility size, N = 12, and multiple time periods, T = 5. The first data set has been taken from Moslemipour and Lee [7] and the second data set has been hypothetically generated.

The numerical result, TMHC, along with its two components i.e. material handling and rearrangement cost is obtained for confidence level ($Z_p = 0.85$). The program is executed 10 times and the best value of the TMHC and computation time along with its layout is taken. Tables 3 and 4 displayed the results for two problem sets for Gaussian and Poisson product demand respectively. These results are statistically evaluated to show the worst, best, mean THMC and its standard deviation for the various chaotic maps. The following inferences can be drawn,

1. CSA with '*sine*' map has the best Minimum TMHC for Gaussian product demand and '*logistic*' map has best Minimum TMHC for Poisson product demand.
2. CSA perform better in terms of TMHC for the data set N = 12, time period T = 5 (Tayal and Singh [9]) as compared to Moslemipour and Lee [7].
3. Change of chaotic map in CSA does not impact the performance in terms of CPU time for solving. SDFLP.

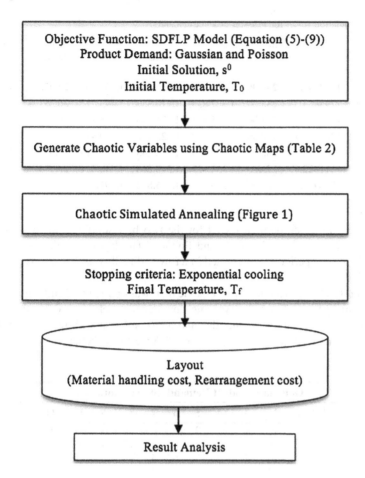

Fig. 2 Flow chart

Table 3 Simulation result for Gaussian product demand

	MHC	RA_c	Best TMHC	Mean TMHC	Worst TMHC	Standard deviation	CPU time (s)
Logistic	1213340.76	43,000	1,256,341	1,334,229	1,378,411	38,223	21.36
Chebyshev	1278358.00	24,000	1,302,359	1,365,520	1,410,778	36,994	19.77
Iterative	1483578.97	33,000	1,516,579	1,578,736	1,642,717	46,269	19.50
Sine	**1226019.12**	**28,000**	**1,254,019**	1,360,902	1,405,293	48,062	20.04
Singer	1320456.51	29,000	1,349,457	1,404,077	1,489,244	51,546	19.67
Tent	1273366.38	31,000	1,304,366	1,357,286	1,401,538	28,956	19.65
Piecewise	1279435.55	24,000	1,303,436	1,357,716	1,398,442	31,332	19.38

Table 4 Simulation result for Poisson product demand

	MHC	RA_c	Best TMHC	Mean TMHC	Worst TMHC	Standard deviation	CPU time (s)
Logistic	**1237264.57**	**29,000**	**1,266,265**	1,326,404	1,363,606	25,970	19.29
Chebyshev	1294833.04	28,000	1,322,833	1,370,357	1,445,417	36,257	19.41
Iterative	1417417.17	28,000	1,425,317	1,531,240	1,578,739	46,432	19.83
Sine	1234438.96	43,000	1,277,439	1,323,952	1,370,443	24,901	19.54
Singer	1346118.74	28,000	1,374,119	1,432,243	1,484,314	37,421	19.49
Tent	1311822.07	27,000	1,338,822	1,395,868	1,446,041	29,974	19.49
Piecewise	1260920.09	36,000	1,296,920	1,355,878	1,391,117	37,991	19.70

Figure 3 shows the optimal layout for the two best solutions for two different product demand case i.e. Gaussian and Poisson. The layout solution is provided below where time period is represented in a row and facility assignment is given for each period.

Further, CSA approach for '*sine*' chaotic map was applied to facility layout size, N = 12, time period, T = 5 and varying confidence level, Zp = 0.75, 0.85, 0.95 with Gaussian distribution product demand and compared with results from literature, as summarized in Table 5. It is seen CSA performs better.

Gaussian Product Demand: Sine Chaotic Map												
	N=1	2	3	4	5	6	7	8	9	10	11	12
T= 1	3	10	4	7	2	6	8	12	1	11	9	5
2	11	10	4	2	7	6	8	12	1	5	3	9
3	8	12	4	2	7	6	5	9	3	10	11	1
4	8	12	6	2	4	7	3	10	9	5	11	1
5	8	1	12	2	4	7	6	10	3	11	9	5

Poisson Product Demand: Logistic Chaotic Map												
	N=1	2	3	4	5	6	7	8	9	10	11	12
T= 1	3	10	4	7	2	6	8	1	12	11	9	5
2	12	8	4	7	2	6	5	9	3	10	11	1
3	12	8	1	7	2	6	5	4	3	10	9	11
4	1	12	4	2	7	6	5	9	3	10	11	8
5	1	2	4	7	6	5	9	3	10	11	12	8

Fig. 3 Optimal layouts

Table 5 Comparative results of THMC for N = 12, T = 5 at varying confidence level, Gaussian demand and '*sine*' chaotic map

	SA (Moslemipour and Lee [7])	CSA ('*sine*' chaotic map)
Confidence level (p)		
0.75	1,265,500	1,231,126
0.85	1,274,300	1,254,019
0.95	1,301,100	1,271,226

5 Conclusion

An important tool in optimization theory are meta-heuristic techniques, which explore the search space of the given data in both exploration and exploitation manner and provide a near-optimal solution within a reasonable time. In this paper, we turn the attention to chaos embedded Simulated Annealing meta-heuristic algorithms, known as Chaotic Simulated Annealing (CSA). Here, experimental studies of chaos and meta-heuristic were carried out. Seven different chaotic maps in CSA meta-heuristic with exponential cooling schedule are used to evaluate the impact in solving the SDFLP. Quality of solution is captured on both material handling cost and rearrangement cost aspect for confidence level i.e. $Z_p = 0.85$ for the data set N = 12, time period T = 5 with Gaussian and Poisson product demand. It can be inferred that CSA with '*sine*' map has the best minimum TMHC for Gaussian product demand and '*logistic*' has the best minimum TMHC for Poisson product demand. The main goal of this paper is to provide an experiment methodology and useful reference in the area of chaos embedded meta-heuristic optimization for solving SDFLP.

References

1. Singh, S.P., Sharma, R.R.K.: A review of different approaches to the facility layout problem. Int. J. Adv. Manuf. Technol. **30**(5–6), 425–433 (2006)
2. Matai, R., Singh, S.P., Mittal, M.L.: A non-greedy systematic neighbourhood search heuristic for solving facility layout problem. Int. J. Adv. Manuf. Technol. **68**(5–8), 1665–1675 (2013)
3. Rosenblatt, M.J., Kropp, D.H.: The single period stochastic plan layout problem. IIE Trans. **24**(2), 169–176 (1992)
4. Palekar, U.S., Batta, R., Bosch, R.M., Elhence, S.: Modeling uncertainties in plant layout problems. Eur. J. Oper. Res. **63**, 347–359 (1992)
5. Braglia, M., Zanoni, S., Zavanella, L.: Layout design in dynamic environments: analytical issues. Int. Transit. Oper. Res. **12**, 1–19 (2005)
6. Krishnan, K.K., Cheraghi, S.H., Chandan, N.N.: Facility layout design for multiple production scenarios in a dynamic environment. Int. J. Ind. Syst. Eng. **3**(2), 105–133 (2008)
7. Moslemipour, G., Lee, T.S.: Intelligent design of a dynamic machine layout in uncertain environment of flexible manufacturing systems. J. Intell. Manuf. **23**(5), 1849–1860 (2011)

8. Moslemipour, G., Lee, T.S., Rilling, D.: A review of intelligent approaches for designing dynamic and robust layout in flexible manufacturing systems. Int. J. Adv. Manuf. Technol. **60**(1–4), 11–27 (2012)
9. Tayal, A., Singh, S.P.: Chaotic simulated annealing for solving stochastic dynamic facility layout problem. J. Int. Manage. Stud. **14**(2), 67–74 (2014)
10. Tayal, A., Singh, S.P.: Integrated SA-DEA-TOPSIS based solution approach for multi objective stochastic dynamic facility layout problem. Int. J. Bus. Syst. Res. (2015a, in press)
11. Tayal, A., Singh, S.P.: Analysis of simulated annealing cooling schemas for design of optimal flexible layout under uncertain dynamic product demand. Int. J. Oper. Res. (2015b, in press)
12. Tompkins, J.A., White, J.A., Tanchoco, J.M.A.: Facilities Planning. Willey, New York (2003)
13. Koopmans, T.C.S., Beckman, M.: Assignment problem and the location of economic activities. Econometric **25**, 53–76 (1957)
14. Kulturel-Konak, S., Smith, A.E., Norman, B.A.: Layout optimization considering production uncertainty and routing flexibility. Int. J. Prod. Res. **42**(21), 4475–4493 (2004)
15. Smith, A.E., Norman, B.A.: Evolutionary design of facilities considering production uncertainty. In: Evolutionary Design and Manufacture: Selected Papers from ACDM 2000, pp. 175–186. Springer, London (2000)
16. Mingjun, J., Huanwen, T.: Application of chaos in simulated annealing. Chaos, Solitons Fractals **21**, 933–941 (2004)

Ranking of Stocks Using a Hybrid DS-Fuzzy System: An Application to Bombay Stock Exchange

Gour Sundar Mitra Thakur, Rupak Bhattacharyya,
Seema Sarkar (Mondal) and Partha Kar

Abstract The most crucial phase in the construction of any stock portfolio is proper selection of stocks before the assignment of investment ratio. Presence of various factors which directly or indirectly influence the stock market, have made this selection task uncertain and challenging. In this paper Dempster-Shafer (DS) evidence theory and fuzzy rule based system have been hybridized to rank the stocks under Bombay Stock Exchange based on their historical performance. This ranking proved to be very fruitful for the selection of stocks for short-term investment period.

Keywords Ranking of stocks · Fuzzy rule based system · Critical factors · DS theory

G.S. Mitra Thakur (✉)
Department of Computer Science and Engineering, Dr. B. C. Roy Engg. College,
Durgapur, West Bengal, India
e-mail: cse.gsmt@gmail.com

R. Bhattacharyya
Department of Mathematics, Bijoy Krishna Girls' College, Howrah,
West Bengal, India
e-mail: mathsrup@gmail.com

S. Sarkar (Mondal)
Department of Mathematics, National Institute of Technology,
Durgapur, West Bengal, India
e-mail: seemasarkarmondal@yahoo.co.in

P. Kar
Department of Computer Science, Proper Web Technologies Pvt. Ltd.,
Kolkata, West Bengal, India
e-mail: partha.kar.diatm@gmail.com

© Springer Science+Business Media Singapore 2016
R.K. Choudhary et al. (eds.), *Advanced Computing and Communication Technologies*,
Advances in Intelligent Systems and Computing 452,
DOI 10.1007/978-981-10-1023-1_11

1 Introduction

Portfolio selection is an optimal allocation of investors' asset among maximum number of stocks to provide good return with less risk. Predicting the performance and selecting the portfolio of stocks have been an important and complicated research area for researchers in modern stock exchanges because of uncertainties. Uncertainties come from various technical and fundamental factors which influence the stock market directly and indirectly.

In 1952, in a ground-breaking research by Markowitz [1], by representing portfolio return in the form of mean and risk as variance, he proposed a portfolio selection model to maximize return and minimize the risk. Then the relationship between risk and mean-variance were examined by many researcher like Zhou et al. [2], Yong et al. [3] etc. According to Modern Portfolio Theory (MPT), the investors always try to maximize their profit with minimum amount of risk by selecting a proper portfolio. Xidonas et al. [4], divided the portfolio selection process into two different stages. In the first stage some stocks are selected to construct the portfolio. In the second stage the percentage of the total value for each stock is identified.

In the last few decades Artificial Intelligence is playing a very vital role in the selection of stocks and in the construction of portfolio. Among various soft computing tools fuzzy set theory and fuzzy logic have been widely used in MPT. Details of few of these researches can be found in references like [5–8].

As the selection of proper stocks plays the most vital role in portfolio construction, main objective of this research is to identify top 10 stocks, which are expected to perform better for short-term investment period, out of 30 stocks registered in Bombay Stock Exchange (BSE) based on their historical performance. In this proposed work a new hybrid DS-fuzzy system has been developed for the ranking of stocks. Initially four critical factors namely Price to Earning Ratio (P/E), Price to Book Value (P/B), Price to Sales ratio (P/S) and Long Term Debt to Equity ratio (LTDER) have been identified which have direct or indirect influence in the performance of stocks. Then the historical data (2003–04 to 2011–12) of these four factors have been used to rank the stocks under BSE with the help of proposed DS-Fuzzy rule based system.

In Sect. 2 design of the proposed DS-Fuzzy model has been discussed and Sect. 3 draws the conclusion.

2 Design of the Proposed DS-Fuzzy Model for Stock Evaluation

Dempster-Shafer evidence theory and fuzzy rule based expert system have been hybridized in the proposed model to deal with the inherent uncertainty in stock selection problem. Dempster-Shafer evidence theory has been briefly introduced in this section before the discussion of the proposed model. For detailed explanation regrading expert system and its various components reader can consult [9].

2.1 Dempster-Shafer Evidence Theory

A research by A. P. Dempster in 1966 [10], proposed a multivalued mapping from one space to another space. It has been used for statistical inference, when we have multiple sample information and we need to identify a single hypothesis. The DS evidence theory of belief function was first proposed by A. P. Dempster in 1968 and then extended by Shafer in 1976 [11] which can deal with incomplete and uncertain information.

The basic idea of Dempster-Shafer Theory mainly deals with four components: frame of discernment, basic probability assignment (BPA), plausibility function (Pl) and belief function (Bel). Frame of discernment is defined as a finite set of mutually exhaustive and exclusive hypotheses. Let $\theta = \{\theta_1, \theta_2, \theta_3, ..., \theta_n\}$ is the frame of discernment, where n is the total number of hypotheses. X is assumed to be an arbitrary subset of θ. Then belief function over θ can be expressed as a mapping function m defined as:

$$m : 2^{\theta} \rightarrow [0, 1] \tag{1}$$

Such that $m(\phi) = 0$ and $\sum_{X \subseteq \theta} m(X) = 1$. $m(X)$ is the value of basic probability assignment (BPA) for a given set X.

Dempster's rule of combination for combining the degree of belief of two sets m_1 and m_2 can be defined as follows:

$$m_3(Z) = \frac{\sum_{X \cap Y = Z} m_1(X) m_2(Y)}{1 - \sum_{X \cap Y = \emptyset} m_1(X) m_2(Y)} \tag{2}$$

2.2 Proposed Model

The design of the proposed model is described through the following stages:

Collection of Historical Data and Fuzzification Total 30 stocks are registered in BSE. Last 9 years' (2003–04 to 2011–12) data of four factors (P/E ratio, P/B value, P/S ratio and LTDER) have been collected from different web sites like www. capitaline.com, www.bseindia.com, www.nseindia.com. For the simplicity of this system we have normalized the data within the range of [0, 10] by considering the maximum historical value of the last 9 years data to be 10.

Table 1 Membership functions for the linguistic values of the input variables

Factors in terms of linguistic variables	Linguistic values	Fuzzy trapezoidal membership
Price to earnings ratio (range 0–10)	Low	(0 0 1.8 2.8)
	Standard	(1.7 3.5 4.6 5.8)
	High	(5.3 7.5 10 10)

Fuzzification is a process of taking a crisp value as input and transforming it into the degree required by the terms. In the proposed model all four input factors have been converted to fuzzy variables with three linguistic values (Low, Standard and High) using fuzzy trapezoidal memberships in [0, 10]. Sample fuzzy membership for P/E is given in Table 1.

Fuzzy Rule Construction Dempster-Shafer theory has been used to construct fuzzy rules. Generally, the knowledge base of a Fuzzy Inference System is developed by using collection of fuzzy rules which determines how the output will be generated based on the given input. This proposed system consists of four input variables P/E, P/B, P/S and LTDER as discussed earlier. Based on these input parameters DS-fuzzy inference system determines whether the Selection of any stock will be *Highly Favorable, Moderately Favorable* or *Not Favorable* as an output.

As mentioned earlier in Table 1, as an example 'Price to Earnings Ratio' is having three linguistic values: *Low, Standard and High*. D-S theory has been applied to construct the fuzzy rules. The frame of discernment has been considered as $\theta = \{$*High Performance, Average Performance and Poor Performance*$\}$ to represent the three possible outcomes for each stocks.

Though there are various other ways to measure the performance of stocks, one of the most commonly used measure is Risk/Return ratio. Risk is generally measured by semivariance (S) of the stocks' previous returns and Return (R) is actually the mean of previous returns. Lower value of this ratio indicates better performance of stocks.

Now based on the linguistic value of each factor and comparing those with S/R value as a measure of their performance, basic probabilities have been assigned towards different hypotheses. The data of 2012–13 has been used to identify the stock. Initially a standard value for each of these four factors has been set for every stock with the help of expert opinions. From the last 9 years' historical data it has been found that if the value of P/E ratio is around its standard value, then 75 % of stocks under BSE perform better and S/R value is also found to be satisfactory. So 0.75 degree of belief has been assigned towards the hypothesis {*High Performance*}. If the value of P/E ratio is much lower than its standard value then it has been found that 60 % of stocks under BSE perform average and S/R value is average. So 0.6 degree of belief has been assigned towards the hypothesis {*Average Performance*}. Similarly, when the value of P/E ratio is much higher than the standard value then it has been found that 70 % of stocks under BSE perform poor and S/R value is also unsatisfactory. So 0.7 degree of belief has been assigned

Table 2 Membership functions for the linguistic values of the input variables

Factors	Linguistic values	High performance (H_P)	Average performance (A_P)	Poor performance (P_P)
P/E	Low		0.6	
	Standard	0.75		
	High			0.7
P/B	Low			0.8
	Standard		0.6	
	High	0.65		
P/S	Low			0.7
	Standard		0.65	
	High	0.75		
LTDER	Low	0.6		
	Standard		0.75	
	High			0.65

towards the hypothesis {*Poor Performance*}. In the same way initial believes have been assigned for all other factors (Table 2).

Let us consider a simple case

Rule 1. IF Price to Earnings Ratio is **Standard** THEN Performance will be **High** (m_1 (H_P) = 0.75).

Rule 2. IF Price to Book Value is **Low** THEN Performance will be **Poor** (m_2 (P_P) = 0.8).

Rule 3. IF Price to Sales Ratio is **High** THEN Performance will be **High** (m_4(H_P) = 0.75).

Rule 4. IF LTDER is **Low** THEN Performance will be **High** (m_6(H_P) = 0.6)

Now Dempster Rule of combination has been applied to calculate the final mass value of the final rule.

Let the IF part of Rule 1 and Rule 2 be the first two evidences and m_1, m_2 be two mass functions for them. Degree of belief has been assigned to the hypotheses based on the evidences. So from Rule 1, $m_1(H\_P) = 0.75$ and $m_1(\theta) = (1 - 0.75) = 0.25$, $m_1(\theta)$ represents the degree of belief in the rest of the hypotheses present in the hypothesis set. And from Rule 2, $m_2(P\_P) = 0.8$ and $m_2(\theta) = (1 - 0.8) = 0.2$, $m_2(\theta)$ represents the degree of belief in the rest of the hypotheses present in the hypothesis set.

Now these two evidences are combined and new mass m_3 is generated as mentioned in Table 3 and mass value for hypothesis m_3 can now be calculated with the help of Eq. (2).

Table 3 Combination of mass considering first two evidences

Combining m_1 and m_2	$m_2(P\_P) = 0.8$	$m_2(\theta) = 0.2$
$\mathbf{m_1(H\_P)} = 0.75$	$\emptyset = 0.6$	H_P = 0.15
$\mathbf{m_1(\theta)} = 0.25$	P_P = 0.6	$\theta = 0.05$

Table 4 Combination of mass considering first three evidences

Combining m_3 and m_4	$m_4(H\_P) = 0.75$	$m_4(\theta) = 0.25$
$m_3(H\_P) = 0.375$	H_P = 0.2813	H_P = 0.09
$m_3(P\_P) = 0.5$	$\emptyset = 0.375$	P_P = 0.125
$m_3(\theta) = 0.125$	H_P = 0.0934	$\theta = 0.03125$

Table 5 Combination of mass considering first three evidences

Combining m_5 and m_6	$m_6(H\_P) = 0.6$	$m_6(\theta) = 0.4$
$m_5(H\_P) = 0.7499$	H_P = 0.4499	H_P = 0.2999
$m_5(P\_P) = 0.2$	$\emptyset = 0.12$	P_P = 0.08
$m_5(\theta) = 0.05$	H_P = 0.03	$\theta = 0.02$

Now consider the IF part of the rule 3 to be the new evidence and m_4 be the mass function. A belief of 0.75 has been assigned towards the hypothesis {*High Performance*}.

So from Rule 3, $m_4(H\_P) = 0.75$ and $m_4(\theta) = (1 - 0.75) = 0.25$. Again m_3 and m_4 are combined to generate new mass m_5 as mentioned in Table 4 and new mass values for hypothesis m_5 can now be calculated with the help of Eq. (2).

Finally, the IF part of the Rule 4 is the last evidence and m_6 is the mass function. A belief of 0.6 has been assigned towards the hypothesis {*High Performance*}.

From Rule 4, $m_6(H\_P) = 0.6$ and $m_6(\theta) = (1 - 0.6) = 0.4$. Now, m_5 and m_6 are combined and final mass value m_7 is generated as mentioned in Table 5 and final mass value for hypothesis m_7 can be calculated using Eq. (2) as shown in Eq. (3).

$$m_7(H\_P) = \frac{0.4499 + 0.2999 + 0.03}{1 - 0.12} = 0.8861$$

$$m_7(P\_P) = \frac{0.08}{1 - 0.12} = 0.09 \tag{3}$$

$$m_7(\theta) = \frac{0.02}{1 - 0.12} = 0.023$$

In this way D-S theory is applied and final mass values for rest of the 80 rules are calculated. Maximum final mass value for *High Performance* (*H_P*) is found to be 0.9916 and minimum final mass value for *High Performance* (*H_P*) is found to be 0. The conclusion part of each rule is decided on the basis of their final mass values for *High Performance* (*H_P*). Now, favorability of the stocks is divided into three categories: *Highly Favorable* (Final mass value for H_P between 0.76 and 1), *Moderately Favorable* (Final mass value for H_P between 0.46 and 0.75) and *Not Favorable* (Final mass value for H_P less than 0.45). So the combined version of the above sample rules becomes:

"IF Price to Earnings Ratio is **Standard** AND Price to Book Value is **Low** AND Price to Sales Ratio is **High** AND LTDER is **Low** THEN the stock is **Highly Favorable**".

Because the final mass value for *High Performance* (*H_P*) was 0.8861 (between 0.76 and 1). Finally total of 81 (3^4) rules are formulated for the knowledge base of the proposed DS-fuzzy inference system and above three selection categories were converted into fuzzy linguistic variables with trapezoidal membership values for the output of the DS-fuzzy inference system as shown in Table 6.

Ranking of Different Stocks Under BSE Data of 2012–13 for all four input factors of 30 stocks are used as input to the DS-fuzzy system and then they are ranked based on their defuzzified values in descending order. This model has identified the top 10 stocks out of 30 stocks as short term investment preference. Table 7 shows the details of top 10 Stocks based on this ranking. As the consequent of every rule indicates about the favorability of stocks higher defuzzified value indicate higher favorability. Among all the 30 stocks, the highest defuzzified value derived as 0.8644 corresponds to *Hindustan Unilever Ltd.* and it has topped the ranking and the lowest defuzzified value is obtained for *Tata Steel Ltd.* as 0.1626.

As mentioned earlier S/R ratio is used as performance measure of stocks very often, a ranking of stocks is also done based on the S/R values of the stocks for the FY 2013–14. When we compared this ranking with our proposed ranking we found ten matches in top 15. It gives a clear indication about the efficiency of the model.

Table 6 Fuzzy trapezoidal membership of the output variable

Output variable	Linguistic values	Fuzzy trapezoidal membership
Selection (range 0–1)	Not favorable	(0 0 0.172 0.448)
	Moderately favorable	(0.34 0.46 0.57 0.75)
	Highly favorable	(0.64 0.88 1 1)

Table 7 Top 10 stocks

Rank	Name of the stocks	Defuzzified values (input data: FY 2012–13)
1	Hindustan Unilever Ltd.	0.8644
2	Sun Pharmaceutical Inds. Ltd	0.8457
3	I T C Ltd.	0.5369
4	Coal India Ltd.	0.5340
5	Tata Consultancy Services Ltd.	0.4936
6	Infosys Ltd.	0.3842
7	Dr. Reddy's Laboratories Ltd.	0.3553
8	Bajaj Auto Ltd.	0.3292
9	Hero Motocorp Ltd.	0.3002
10	Cipla Ltd.	0.2850

So it is obvious that any investor preparing a portfolio based on this proposed ranking for short-term investment period is expected to get better return in FY 2013–14.

3 Conclusion

Selection of suitable stocks from any stock market to maximize return is always a tough and challenging task for any kind of investor. This paper proposes a model to rank stocks under BSE based on their historical performance. A novel hybrid DS-Fuzzy model is proposed for this purpose. DS theory is used to assign beliefs towards different stocks and decide the consequents of fuzzy rule base and finally stocks are ranked based on the fuzzy rule based system. As the tested result shows a satisfactory comparison with the recent performance of the stocks this model can be implemented for any stock exchange. The ranking of this model can be easily used in any rank preference based portfolio selection model.

References

1. Markowitz, H.: Portfolio selection. J. Finance **7**, 77–91 (1952)
2. Zhou, X.Y., Li, D.: Continuous-time mean-variance portfolio selection: A stochastic lq framework. Appl. Math. Optim. **42**(1), 19–33 (2000)
3. Fang, Y., Wang, S.: An interval semi-absolute deviation model for portfolio selection. In: Fuzzy Systems and Knowledge Discovery, pp. 766–775. Springer (2006)
4. Xidonas, P., Ergazakis, E., Ergazakis, K., Metaxiotis, K., Askounis, D., Mavrotas, G., Psarras, J.: On the selection of equity securities: an expert systems methodology and an application on the athens stock exchange. Expert Syst. Appl. **36**(9), 11966–11980 (2009)
5. Hiemstra, Y.: A stock market forecasting support system based on fuzzy logic. In: Proceedings of the Twenty-Seventh Hawaii International Conference on System Sciences, 1994, vol. 3, pp. 281–287. IEEE (1994)
6. Bermudez, J.D., Segura, J.V., Vercher, E.: A fuzzy ranking strategy for portfolio selection applied to the Spanish stock market. In: Fuzzy Systems Conference, 2007. FUZZ-IEEE 2007, pp. 1–4 (2007)
7. Fasanghari, M., Montazer, G.A.: Design and implementation of fuzzy expert system for Tehran stock exchange portfolio recommendation. Expert Syst. Appl. **37**(9), 6138–6147 (2010)
8. Nair, B.B., Dharini, N.M., Mohandas, V.: A stock market trend prediction system using a hybrid decision tree-neuro-fuzzy system. In: 2010 International Conference on Advances in Recent Technologies in Communication and Computing (ARTCom), IEEE, pp. 381–385 (2010)
9. Giarratano, J.C., Riley, G.D.: Expert systems: principles and programming. CENGAGE Learning (2005)
10. Dempster, A.P.: Upper and lower probabilities induced by a multivalued mapping. The annals of mathematical statistics, pp. 325–339 (1967)
11. Shafer, G., et al.: A mathematical theory of evidence, vol. 1. Princeton University Press, Princeton (1976)

Empirical Assessment and Optimization of Software Cost Estimation Using Soft Computing Techniques

Gaurav Kumar and Pradeep Kumar Bhatia

Abstract Software Engineering especially project planning, scheduling, monitoring and control are based on accurate estimate of the cost and effort. In the initial stage of Software Development Life Cycle (SDLC), it is hard to accurately measure software effort that may lead to possibility of project failure. Here, an empirical comparison of existing software cost estimation models based on the techniques used in those models has been elaborated using statistical criteria. On the basis of findings of empirical evaluation of existing models, a Neuro-Fuzzy Software Cost Estimation model has been proposed to hold best practices found in other models and to optimize software cost estimation. Proposed model gives good result as compared to other considered software cost estimation methods for the defined parameters in overall but it is also dependent on type of project, data and technique used in implementation.

Keywords Back propagation neural network (BPNN) · Constructive cost model (COCOMO) · Function point (FP) · Fuzzy logic (FL) · Genetic algorithm (GA) · Particle swarm optimization (PSO) · Software cost estimation (SCE)

1 Introduction

Software Cost Estimation (SCE) of a software project starts from initial phase of software development which includes generating proposal requests, analysis, contract negotiations, planning, scheduling, designing, implementation, maintenance, monitoring and control. The estimation process includes size and effort estimation, initial project scheduling and finally estimation of overall cost of the project.

G. Kumar (✉) · P.K. Bhatia
Department of Computer Science and Engineering, Guru Jambheshwar University of Science & Technology, Hisar, Haryana, India
e-mail: er.gkgupta@gmail.com

P.K. Bhatia
e-mail: pkbhatia.gju@gmail.com

© Springer Science+Business Media Singapore 2016   117
R.K. Choudhary et al. (eds.), *Advanced Computing and Communication Technologies*,
Advances in Intelligent Systems and Computing 452,
DOI 10.1007/978-981-10-1023-1_12

Table 1 Features and limitations of techniques used in SCE

Technique used in SCE models	Features	Limitation
Analogy-Historical	Based on actual experience	Much information of historical projects is required
Expert judgment	Fast prediction, Easy to use	Dependency on experts
COCOMO	Common method	A large amount of data is required
ANN	Consistent, Ability of generalization	Training data dependency
FL	Flexible, Training not required	Hard to use
GA	Optimization	Initial values required
PSO	Optimization	Training data dependency

Accurate estimation of software cost is necessary to complete project within time and budget and to prevent failure of software project. If effort estimation is done too low, it may lead to problems in managing project, delay in delivery, overrun of budget and low software quality. If effort estimation is done too high, it may cause business loss and inefficient use of resources. Accuracy is important while software estimation for developers as well as customers as it determines what, where, and when resources will be used, analyzes impact of requirement change etc. Various SCE models have been developed to manage software project's budget and schedule. Estimation models developed so far has their own significance or importance and is applicable for specific type of projects. So the criteria to evaluate accuracy of software estimation model are much important to successfully complete a software project. Techniques used in estimating software cost have their own features as well as limitations. Some of which have been described in Table 1.

2 Review of SCE Models Based on Used Technique

A lot of research has been carried out for implementation of SCE models using various methodologies. A brief overview of research work done in the past for developing SCE model has been discussed here.

While developing SCE model, ANN acts as a proven practical way that reduces the model's input space (and thus computational complexity and human effort) while maintaining the same levels of effort prediction accuracy. An automated SCE applied on COCOMO data set using Feed forward BPNN tested on COCOMO NASA 2 dataset may help project manager for fast and realistic estimation of software cost for project effort and development time [1, 2]. Matlab Neural Network tool box with data from multiple projects can be used to validate, train and simulate the network with observations that neural network performs better than COCOMO; and Cascade correlation performs better than Neural Network [3]. BPNN model with COCOMO data works well for Small projects while neural network with Resilient Back Propagation is good for big projects [4]. Radial Basis Function Neural Network with K-means clustering algorithm can perform better in terms of accurate cost estimation [5]. A Neuro-Fuzzy Constructive Cost Model (COCOMO) proves that estimation accuracy can be improved as compared to COCOMO model using industry project data [6, 7].

FL solves the problems of vagueness, imprecise and incomplete data to make reliable and accurate effort estimates. FL can be used to develop SCE model by fuzzifying functional points, applying membership functions e.g. triangular, trapezoidal, Gaussian function etc. to represent the cost drivers and defuzzifying the results to get the resultant effort. SCE model developed using FL with membership functions gives better performance as compared to COCOMO model which was tested and evaluated on a dataset of software projects [8]. Triangular fuzzy logic on NASA software projects representing linguistic terms in Function Point Analysis (FPA) with complexity metrics estimates size in person hours [9]. FL with Gaussian Membership Function (GMF) applied on COCOMO cost drivers gives results close to the actual effort than the trapezoidal function [10]. FL with Takagi-Sugeno technique for estimation applied on COCOMO and SLOC using Function Point (FP) gives simple, better estimation capabilities and mathematical relationship between the effort and the inputs [11, 12, 13].

Genetic Programming provides a more advanced mathematical function to predict more accurate estimated effort. Data mining tool can be used to increase accuracy of effort estimation by selecting a subset of highly predictive attributes such as project size, development, and environment related attributes. GA can be used to assess software project in terms of effort computation that takes much less time and performs better than COCOMO model on NASA software project dataset. GA can provide better results as compared to COCOMO II as tested on Turkish and Industry data set [14–16].

PSO with clustering can perform efficient effort estimation with learning ability by providing an efficient, flexible and user friendly way to perform the task of effort estimation. More accuracy in SCE can be achieved than the standard COCOMO using PSO with K-means clustering applied on COCOMO model that enables learning from past project data and domain specific projection of future resource requirements. PSO with inertia weight applying on COCOMO data of NASA software project can be used to calculate MARE, VARE and VAF [17, 18, 19].

Any one of the Line of Code, Function Point and Cosmic FFP can be used to measure size of a software project. Cosmic FFP provides simple, easy to use, proven and practical solution for software size estimation and quality improvement. COSMIC FFP uses functional size unit for SCE where One Cosmic Functional Size Unit (CFSU) is assigned for each entry/exit of a data group and for each read/write operation by a data group [20].

3 Statistical Criteria to Analyze and Evaluate Performance of SCE Model

Statistical criteria to analyze and evaluate the efficiency of software cost estimation model have been shown in Table 2.

4 Proposed Model

Literature analysis reveals that the SCE models developed using Neural Network, Fuzzy Logic or combination of both provides good results as compared to other soft computing techniques. Neuro-Fuzzy model acts as a powerful tool to predict cost and quality by integrating numerical data and expert knowledge. Proposed neuro-fuzzy model has been derived from [2, 6, 7, 12]. Model has been validated through data got from PROMISE Software Engineering Repository of 93 NASA projects. For calculation of effort, COCOMO II model has been used:

$$Effort = A \times (KLOC)^{B + 0.01 \times \sum_{i=1}^{5} SF_i} \times \prod_{j=1}^{17} EM_j$$

$$Schedule(in months) = C \times Effort^D + 0.01 \times \sum_{i=1}^{5} SF_i$$

where A, B, C, D are domain specific parameters (By default A = 2.94, B = 0.91, C = 3.67, D = 0.28), SF is the scale factor and EM is the effort multiplier.

Cost drivers are used in calculation of development effort, such as analyst capability, application experience etc. Fuzzification converts the crisp data to linguistic variables which are passed to Inference Engine. A fuzzy set has been defined for six qualitative rating levels for every cost driver and expressed in linguistic terms as very low (VL), low (L), nominal (N), high (H), very high (VH) and extra high (XH). The membership functions used is triangular functions which is a three-point function, defined by minimum (α), maximum (β) and modal (m) values, i.e. T (α, m, β), where ($\alpha \leq m \leq \beta$). The rules can be on the basis of single parameter or combination of parameters e.g.

if (PREC is Very Low) then (EFFORT is Extra High)
if(PREC is Low) then (EFFORT is Very High)
if(FLEX is Very Low) then (EFFORT is Extra High) etc.

For defuzzification, Centeroide Method which calculates Centre of Gravity (COG) area under the curve has been used.

Table 2 Evaluation criteria for SCE models based on actual effort and predicted effort

S. No.	Evaluation term	Evaluation formula	Description
1	Relative Error (RE)	$RE(i) = \left\lvert \frac{E_P(i) - E_A(i)}{E_A(i)} \right\rvert \; where \; i = 1\ldots n$	RE is used to measure accuracy
2	Magnitude of Relative Errors (MRE)	$MRE = \frac{\lvert E_P - E_A \rvert}{\lvert E_A \rvert}$	SCE model with lower MRE is better as compared to higher MRE
3	Variance Absolute Relative Error (VARE)	$VARE = Var[MRE]$	SCE model with lower VARE is better as compared to higher VARE
4	Mean Magnitude of Relative Error (MMRE)	$MMRE = \frac{1}{n} \sum_{i=1}^{n} MRE_i$	MMRE assesses the performance of competing models to predict accuracy but has drawback of overestimation in case of many circumstances. An effort prediction models with MMRE $\leq$ 0.25 is considered as acceptable SCE model with lower MMRE is better as compared to higher MMRE
5	Median of Magnitude of Relative Error (MdMRE)	$MdMRE = Median(MRE_i)$	MdMRE is less sensitive to extreme values, while MMRE is sensitive to the outliers SCE model with lower MdMRE gives better accuracy
6	Magnitude of Error Relative to estimate (MER)	$MER = \frac{\lvert E_P - E_A \rvert}{\lvert E_P \rvert}$	SCE model with lower MER model is better as compared to higher MER

(continued)

Table 2 (continued)

S. No.	Evaluation term	Evaluation formula	Description
7	Mean Magnitude of Error Relative to estimate (MMER)	$MMER = \frac{1}{n} \sum_{i=1}^{n} MER_i$	SCE model with lower MMER model is better as compared to higher MMER Accuracy of an estimation technique is inversely proportional to the MMER/MMRE if (MMRE is large and MMER is small) then average actual effort < average estimated effort. else if (MMER is large) then average estimated effort < average actual effort
8	Percentage Relative Error Deviation (PRED)— Prediction of specific Level l	$pred(l) = \frac{k}{n}$	SCE model with higher PRED is better as compared to lower PRED as accuracy of SCE model is directly proportionally to pred(l). A prediction model is considered as acceptable when its accuracy level is 75% Pred(l) is the probability of the SCE model having relative error less than or equal to l i.e. MRE $\leq l\%$; where k is no. of observation and

(continued)

Table 2 (continued)

S. No.	Evaluation term	Evaluation formula	Description
			n is total no. of observations
9	Variance Account For (VAF)	$VAF(\%) = \left(1 - \frac{Var(E_A - E_P)}{Var(E_A)}\right)$	VAF measures future outcomes likely to be predicted by the SCE model. SCE model with *higher* VAF is better as compared to *lower* VAF
10	Pearson's Correlation Coefficient (CC)	$CC(n) = \dfrac{\sum_{i=1}^{n}\left[\left(E_A(i) - \bar{E_{A,n}}\right)\left(E_P(i) - \bar{E_{P,n}}\right)\right]}{\sqrt{\left[\sum_{i=1}^{n}\left(E_A(i) - \bar{E_{A,n}}\right)^2\right]\left[\sum_{i=1}^{n}\left(E_P(i) - \bar{E_{P,n}}\right)^2\right]}}$	CC between the actual and predicted estimation values indicates whether the actual and the predicted values move in the same direction. There are 3 conditions on behalf of CC: (a) $\lvert C\rvert \approx 1$ signifies a perfect estimation of the actual values by the predicted one (b) −ve CC signifies that the predicted values follow the same direction of the actual with negative mirroring i.e. with an 180° rotation about the time-axis (c) $C \approx 0$ signifies poor performance on the basis of predictions in capturing the evolution of actual values

(continued)

Table 2 (continued)

S. No.	Evaluation term	Evaluation formula	Description
11	Root Mean Squared Error (RMSE)	$RMSE(n) = \sqrt{\frac{1}{n}\sum_{i=1}^{n}[E_P(i) - E_A(i)]^2}$	RMSE is the square root of Mean Squared Error (MSE) that measures difference between predicted values by a SCE model and the actual values
12	Normalized Root Mean Squared Error (NRMSE)	$NRMSE(n) = \dfrac{RMSE(n)}{\sqrt{\frac{1}{n}\sum_{i=1}^{n}\left[E_A(i) - \bar{E}_{A,n}\right]^2}}$	NRMSE assess the quality of predictions using RMSE (a) NRMSE = 0 signifies predictions are perfect; (b) NRMSE = 1 signifies prediction is no better than taking E_P equal to the mean value of n samples
13	Logarithmic Standard Deviation (LSD)	$LSD = \sqrt{\dfrac{\sum_{i=1}^{n}\left[(\ln E_P - \ln E_A) + \frac{s^2}{2}\right]^2}{n-1}}$ where s^2 is an estimator of the variance of the residual	S.D. provides a measure of deviation that can be expected in the final number LSD should be minimized for a good model.

$$E = \frac{w_1\left(a\alpha^b\right) + w_2\left(am^b\right) + w_3\left(a\beta^b\right)}{w_1 + w_2 + w_3}$$

where w_1, w_2 and w_3 are weights of the optimistic, most likely and pessimistic estimate respectively. Here maximum weight is given for most expected estimate. $(a\alpha^b)$ denotes optimistic estimate, (am^b) denotes most likely estimate and $(a\beta^b)$ denotes pessimistic estimate.

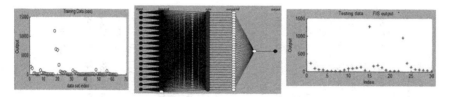

Fig. 1 ANFIS generation using clustering of training data, model structure and testing with error 0.0058664

Proposed model has been implemented in MatLab R2013 using ANFIS (Adaptive Neuro-Fuzzy Inference System) with a hybrid learning algorithm of least-squares method and back-propagation gradient descent (for small projects) and Resilient BPNN (for large project) are used to identify parameters of Sugeno-type fuzzy inference systems as shown in Fig. 1.

5 Empirical Analysis and Evaluation

Hereby we are going to evaluate some popular models with our proposed model based on the statistical criteria as defined in Sect. 3. The empirical evaluations have been derived from statistical analysis of predicted effort and actual effort given by model of that particular type of technique.

6 Result Analysis

From the data obtained by empirical calculation of selected SCE models, it can be verified that models based on Neural Network, Fuzzy Logic or their combination performs better than other methods i.e. GA and PSO.

Based on Table 3 to find the optimized model, Table 4 and Fig. 2 reveal that for all statistical criteria, no model is at 1st rank while proposed model gives at least 2nd rank. Although the proposed Neuro-Fuzzy model does not give best results for all statistical parameters, still if we compare in overall, we can say that the proposed model provides optimized result for SCE.

Table 3 Evaluation of techniques used in SCE model using statistical parameters

Technique	Author/Parameter	Concept used	VARE	MMRE	MdMRE	PRED (25 %)	MMER	LSD	VAF	Corr. Coeff.	NRMSE
SCE model evaluation criteria →			(Min.)	(Min.)	(Min.)	≈1	(Min.)	(Min.)	(Max.)	≈1	≈0
FUZZY LOGIC	H. Mittal (2007)	Triangular	0.008984	0.121702	0.094704	0.9	0.104432	0.153540	0.985009	0.993011	0.131985
		Triangular	0.025013	0.108036	0.052932	0.9	0.089959	0.196215	0.991506	0.995744	0.094367
	S. Reddy (2009)	Gauss. MF	0.026421	0.170208	0.147061	0.8	0.190517	0.241454	0.900426	0.951539	0.328191
	A. Mittal (2010)	Triangular	0.129992	0.392957	0.260556	0.4615	0.496572	0.641570	0.305086	0.657381	0.874154
	S. Kumar (2011)	FL Poly. Reg.	0.035704	0.421978	0.408259	0.1538	0.742063	0.543016	0.344225	0.757476	0.892070
	Ziauddin (2013)	Triangular	0.003763	0.075120	0.074931	0.9666	0.078539	0.097462	0.987720	0.998464	0.128870
	A. F. Sheta (2013)	Takagi-Sugeno	0.180757	0.433817	0.335163	0.4166	0.347397	0.769945	0.961584	0.980603	0.192640
		FP	**0.003580**	**0.049489**	**0.021706**	**1**	**0.048550**	**0.077305**	**0.997413**	**0.998705**	**0.050862**
NEURAL NETWORK	G. Kumar (2014)	BPNN	0.044214	0.276810	0.2710459	0.45	0.364313	0.392395	0.930055	0.967339	0.267792
	S. Reddy (2008, 2010)	RBFN	0.023156	0.240936	0.262962	0.4	0.330535	0.293232	0.793182	0.929154	0.532589
		BPNN	0.077821	0.327361	0.219964	0.5384	0.373670	0.495869	0.940268	0.990797	0.250661
		RBPNN	0.046815	0.302310	0.250113	0.4615	0.616836	0.445945	0.960052	0.984979	0.217702
	A. Kaushik (2013)	BPNN	0.004498	0.098403	0.089	1	0.114903	0.119214	0.985564	0.996473	0.155138
		BPNN	**0.003328**	**0.062864**	**0.04625**	**1**	**0.060725**	**0.084652**	**0.993602**	**0.999942**	**0.092258**
	A. Bawa (2012)	BPNN	1.384785	1.021585	0.517221	0.3	0.651527	2.935350	-0.274271	0.206923	1.132091
		Cascade Corr.	0.700430	0.622814	0.374323	0.5	0.388678	1.650447	0.436012	0.703326	0.878329
NEURO—FUZZY	X. Huang (2003)	all data	0.002424	0.077057	0.058932	1	0.075707	0.089723	0.966624	0.995148	0.196549
		partial data	0.001684	0.084438	0.068816	1	0.084578	0.092836	0.966193	0.994410	0.192589
		large weight	0.003004	0.060044	**0.036102**	1	0.059734	0.078563	0.975723	0.995815	0.166063
		w/o mono. const.	**0.000723**	**0.043597**	0.040457	1	**0.042495**	**0.050128**	**0.990973**	**0.999256**	**0.082849**
	Proposed Model	ANFIS with BPNN, RBPNN	**0.000736**	0.046238	**0.030176**	1	0.042359	0.061820	0.993437	0.999442	0.071682

(continued)

Table 3 (continued)

Technique	Author/Parameter	Concept used	VARE	MMRE	MdMRE	PRED (25 %)	MMER	LSD	VAF	Corr. Coeff.	NRMSE
SCE model evaluation criteria →			(Min.)	(Min.)	(Min.)	≈1	(Min.)	(Min.)	(Max.)	≈1	≈0
GENETIC ALGORITHMS	N. Sharma (2013)	LOC	0.020419	0.232549	0.210933	0.6111	0.279118	0.286695	0.927928	0.974623	0.270758
		LOC + ME	0.018985	0.205639	0.225520	0.6666	0.262881	0.257503	0.933351	0.966105	0.286109
		LOC + ME + d	**0.014125**	**0.203292**	0.194742	**0.7222**	**0.247370**	**0.243233**	0.919930	0.964684	0.287741
	A. F. Sheta (2006)	LOC	0.073210	0.237887	**0.145028**	0.6111	0.288359	0.414983	0.970821	0.985310	0.173369
		LOC + ME	0.329407	0.636396	0.492722	0.3888	0.322478	1.127379	0.975648	0.987779	0.272828
	A. Dhiman (2013)	LOC	0.372690	0.491585	0.185454	0.6	0.268257	0.960162	**0.989669**	**0.996946**	**0.103586**
PSO	Hari (2011)	PSO Clustering	0.009694	0.126253	0.105285	0.7777	0.119264	0.159424	**0.983950**	**0.994124**	**0.126812**
	P. Reddy (2011)	PSO	**0.002565**	**0.049020**	**0.031160**	1	**0.049372**	**0.068906**	0.981987	0.991859	0.134442
	Sweta (2013)	Multi. Obj. PSO	2.930059	0.801342	0.263577	0.4761	0.295084	7.570433	-0.129914	0.585991	1.101915
		Sup. Vect. Reg.	1.828605	0.870634	0.33775	0.3809	0.502089	3.723324	0.657335	0.831674	0.626645

Table 4 Evaluation of SCE techniques w.r.t. statistical parameters

Technique/Parameter	Neural N/w	Fuzzy logic	Neuro-fuzzy	Neuro-fuzzy
	A. Kaushik (2013)	A.F. Sheta (2013)	X. Huang (2003, 2007)	Proposed model
NRMSE (≈0)	0.089353	**0.051238**	0.083758	0.071682
VARE (Min)	0.002347	0.004136	**0.000613**	0.000736
MMRE (Min)	0.073145	0.052658	**0.042164**	0.046238
MdMRE (Min)	0.043192	**0.023515**	0.041347	0.030176
MMER (Min)	0.071634	0.049124	**0.043373**	**0.042359**
LSD (Min)	0.096564	0.076632	**0.061282**	0.06182
VAF (Max)	0.991533	**0.996134**	0.991763	0.993437
Corr. Coeff. (≈1)	**0.999856**	0.998615	0.999467	0.999442
PRED (25 %) (≈1)	**1**	**1**	**1**	**1**

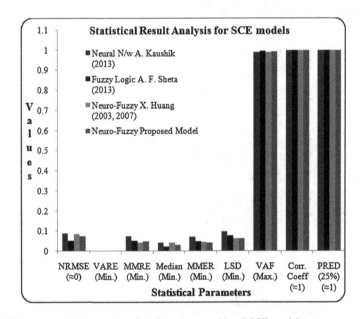

Fig. 2 Performance of Proposed model with other considered SCE models

7 Conclusion and Future Works

In this paper, a detailed empirical analysis and evaluation of SCE models developed through soft computing techniques (e.g. ANN, FL, GA, PSO etc.) have been done using an in-depth review and statistical criteria. Final results indicates that none of the models shows perfect behavior as in terms of certain measure, certain model qualifies better than another, but for other measures it may be worse. Analytical review of considered models shows that SCE models based on NN, FL or combination of NN and FL can give better results as compared to SCE models based on other techniques. Keeping this view, an optimized Neuro-Fuzzy SCE model has been proposed that provides optimum results for considered statistical parameters as compared to other considered SCE models. Due to limitations of NN and FL, proposed model is dependent on size and type of project and data used for training/learning. As per the empirical analysis, it seems that while developing SCE model there is still some scope of improvement in Neuro-Fuzzy techniques and can be accommodated in near future. In future, some improvements may be done by developing SCE models using other optimization techniques like Ant Colony Optimization, Bee Colony Optimization etc. to overtake the performance given by proposed Neuro-Fuzzy SCE model.

References

1. Kumar, G., Bhatia, P.K.: Automation of software cost estimation using neural network technique. Int. J. Comput. Appl. **98**(20), 11–17 (2014)
2. Kaushik, A., Soni, A.K., Soni, R.: A simple neural network approach to software cost estimation. Global J. Comput. Sci. Technol. **13**(1), Version 1, 23–30 (2013)
3. Bawa, A., Chawla, R.: Experimental analysis of effort estimation using artificial neural network. Int. J. Electron. Comput. Sci. Eng. **1**(3), 1817–1824 (2012)
4. Reddy, C.S., Raju, K.: An optimal neural network model for software effort estimation. Int. J. Softw. Eng. **3**(1), 63–78 (2010)
5. Reddy, C.S., Sankara Rao, P., Raju, K., Valli Kumari, V.: A new approach for estimating software effort using RBFN network. Int. J. Comput. Sci. Netw. Secur. **8**(7), 237–241 (2008)
6. Huang, X., Ho, D., Ren, J., Capretz, L.F.: Improving the COCOMO model using a neuro-fuzzy approach. Elsevier J. Appl. Soft Comput. **7**, 29–40 (2007)
7. Huang, X., Capretz, L.F., Ren, J., Ho, D.: A neuro-fuzzy model for software cost estimation. In: Proceedings of the IEEE 3rd International Conference on Quality Software, 126–133, 6–7 Nov 2003
8. Mittal, A., Parkash, K., Mittal, H.: Software cost estimation using fuzzy logic. ACM SIGSOFT Softw. Eng. Notes **35**(1), 1–7 (2010)
9. Mittal, H., Bhatia, P., Optimization criteria for effort estimation using fuzzy technique. CLEI Electron. J. **10**(1), Paper 2, 1–11 (2007)
10. Reddy, C.S., Raju, K., An improved fuzzy approach for COCOMO's effort estimation using gaussian membership function. J. Softw. **4**(5), 452–459 (2009)
11. Ziauddin, K.S., Khan, S., Nasir, A.J.: A fuzzy logic based software cost estimation model. Int. J. Softw. Eng. Appl. **7**(2), 7–17 (2013)
12. Sheta, A.F., Aljahdali, S.: Software effort estimation inspired by COCOMO and FP models: a fuzzy logic approach. Int. J. Adv. Comput. Sci. Appl. **4**(11), 192–197 (2013)
13. Swarup Kumar, J.N.V.R., Mandala, A., Vishnu Chaitanya, M., Prasad, G.V.S.N.R.V., Fuzzy logic for software effort estimation using polynomial regression as firing interval. Int. J. Comput. Technol. Appl. **2**(6), 1843–1847 (2011)
14. Sharma, N., Sinhal, A., Verma, B.: Software assessment parameter optimization using genetic algorithm. Int. J. Comput. Appl. **72**(7), 8–13 (2013)
15. Sheta, A.F.: Estimation of the COCOMO model parameters using genetic algorithms for NASA software projects. J. Comput. Sci. **2**(2), 118–123 (2006)

16. Dhiman, A., Diwaker, C.: Optimization of COCOMO II effort estimation using genetic algorithm. Am. Int. J. Res. Sci. Technol. Eng. Math. 208–212 (2013)
17. Hari, C.V.M.K., Sethi, T.S., Jagadeesh, M.: SEEPC: A toolbox for software effort estimation using soft computing techniques. Int. J. Comput. Appl. **31**(4), 12–19 (2011)
18. Prasad Reddy P.V.G.D., Hari, C.V.M.K.: Software effort estimation using particle swarm optimization with inertia weight. Int. J. Softw. Eng. (IJSE) **2**(4), 87–96 (2011)
19. Kumari, S., Pushkar, S.: Comparison and analysis of different software cost estimation methods. Int. J. Adv. Comput. Sci. Appl. **4**(1), 153–157 (2013)
20. Kumar, G., Bhatia, P.K.: A detailed analysis of software cost estimation using COSMIC-FFP. PAK Publishing Group J. Rev Comput. Eng. Res. **2**(2), 39–46 (2015)
21. Kaushik, A., Chauhan, A., Mittal, D., Gupta, S.: COCOMO estimates using neural networks. Int. J. Intell. Syst. Appl. **9**, 22–28 (2012)

An Innovative 'Cluster-then-Predict' Approach for Improved Sentiment Prediction

Rishabh Soni and K. James Mathai

Abstract Sentiment analysis is a field related to data mining in which subjective information is extracted from source materials such as Twitter, blogs, newspaper articles, etc. Twitter data presents an opportunity for companies to analyze the sentiment the customers or potential users have towards its products. This paper presents an innovative sentiment prediction approach in which the data is first clustered using K-means clustering and then CART algorithm is applied to each cluster to classify the tweets as positive or negative. The results of innovative 'cluster-then-predict' approach directs towards an improved overall prediction accuracy with an increased collected sample data size leading to better clustering and improved classification of each cluster. Also, clustering of data provides useful insights, which helps the companies to gauge consumer sentiment. For the purpose of this paper, tweets related to 'Windows 10' which was launched by Microsoft on July 29, 2015; have been extracted.

Keywords Twitter · Clustering · Classification · Social network · Sentiment analysis

1 Introduction

Opinions which are expressed by users in social networks such as Twitter, plays an important role in determining public sentiment across diverse areas, such as buying products, stock market sentiment or election results prediction. The companies regularly monitors these social networks for mining user sentiments towards their products. Sentiment analysis uses Natural Language Processing and Computational Linguistics to identify text sentiment, typically as positive or negative. One of the challenges in this field of research is to improve the sentiment prediction accuracy. In this paper, the data from twitter is collected to analyze and predict the sentiment

R. Soni (✉) · K. James Mathai
National Institute of Technical Teachers' Training and Research, Bhopal, MP, India
e-mail: soni_rishabh@yahoo.com

© Springer Science+Business Media Singapore 2016
R.K. Choudhary et al. (eds.), *Advanced Computing and Communication Technologies*,
Advances in Intelligent Systems and Computing 452,
DOI 10.1007/978-981-10-1023-1_13

of the users towards the 'Windows 10' operating system, which was recently launched by Microsoft on July 29, 2015. The tweets showing some sentiment towards 'Windows 10' were collected using Twitter API. The main task is a binary classification to classify tweets as positive or negative.

To perform such a classification, various researchers have used algorithms such as CART, Naïve Bayes, RandomForest, Support Vector Machines (SVM), etc. In this paper, a hybrid methodology has been proposed in which k-means clustering is performed on the data, to cluster the tweets based on the words they contain. After this, each of the subset of data is trained separately by using CART algorithm. The final result is combined to find out the prediction accuracy on the test set, using confusion matrix. This approach can also be called as 'cluster-then-predict' approach. The use of unsupervised learning, such as clustering, enhances the interpretability of the solution; enabling analysis of the clusters and related patterns in them. Also, using CART algorithm, trees are drawn to find out which keywords related to 'Windows 10' affects the public perception of the product.

2 Related Work

Twitter has been used extensively for performing sentiment prediction on a variety of topics, ranging from stock market prediction [6–9], crime prediction [10], movie sentiment or box-office prediction [11, 12] and election forecasting [13, 14]. Authors Alexander Pak et al. [12], build a corpus using automated collection of Twitter data. Using this corpus, they have built a sentiment classifier, which is able to determine positive, negative and neutral sentiments for a document. Agarwal et al. [15] investigated two kind of models viz. tree kernel and feature based, and demonstrated that both these models outperform the unigram baseline.

Several authors use different supervised machine learning techniques such as CART, SVM and Naïve Bayes. In this research [16], authors perform classification of sentiment as positive or negative using algorithms such as MaxEnt (Maximum Entropy), SVM and Naïve Bayes. In the study [7], author Linhao Zhang examines the effectiveness of various machine learning techniques on providing a positive or negative sentiment on a twitter corpus.

Author Westling in his paper [1] realize that clustering can be done to divide the texts into groups such that the text in the same group are more similar to each other than to text in other groups. When combined with other methods, clustering can be useful when no classified data is available. Author Dodd in his research [2] recognizes that using k-means clustering may provide more valuable insights when combined with sentiment analysis. A clustering algorithm may discover groups of tweets about a particular feature of the product, or other information related to its release. Authors Jose et al. in their project [3] also propose the idea of using data clustering for making the classification domain-specific and thus to improve accuracy. In a research [4], authors Akhavan Rahnama et al. recognizes clustering of social streams in real-time as an improvement in sentiment analysis. In his book

[5], author Zhao extracts text from Twitter to build a document-term matrix. He also performs clustering to find groups of words and also groups of tweets. But he does not partition his tweet data according to the cluster to perform prediction on it.

From the studies and analysis of papers published before, it was observed that classification algorithms like SVM and RandomForest gives better classification accuracy, especially when compared with CART algorithm. But CART has advantage over others in that trees can be visualized to observe which keywords are significant. The 'cluster-then-predict' approach improves the accuracy of CART to a level comparable with other algorithms, while significantly enhancing the interpretability of the result.

3 Methodology for Implementation

Authors proposes a hybrid approach named 'cluster-then-predict' which comprises of both unsupervised and supervised learning for predicting the sentiment. After obtaining the data and pre-processing it (Sect. 3.1), K-means clustering is performed on form clusters of tweets data points, such that similar tweets (based on the words they contain) gets clustered into one cluster. Then, CART algorithm is applied to each cluster of data and prediction on test data is made. Finally, result is analyzed and compared with other classification algorithms. The steps are shown in Fig. 1.

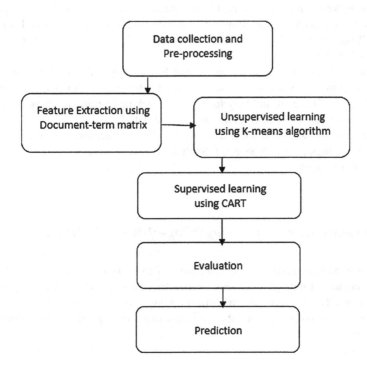

Fig. 1 Flowchart of methodology

3.1 Data Collection and Preprocessing

Twitter data containing the term 'Windows 10' was collected using Twitter API. The tweets showing some sentiment towards the product are kept as a CSV (Comma Separated Values) file. The sentiment rating of each tweet had been entered in a separate column. The sentiment rating was taken by finding average of rating given by five people who are familiar with Windows Operating System.

A database of 1522 tweets was collected for analysis. It contained 855 tweets which showed positive sentiment and 667 tweets which showed negative sentiment towards 'Windows 10'. The CSV data file is imported to 'R' statistical programming language GUI environment [17]. 'R' is a very popular language for data analytics and contains many packages for text mining and analysis.

It is hard for a machine to understand raw twitter data because of the heavy use of homonyms and metaphors. Also, sarcasm is widely used in tweets which are hard for computers to interpret. A process of data cleaning was also done in this regard. The aim of the data cleaning process is to remove any unwanted content from the tweets. Data cleaning and preprocessing can not only simplify the classification task for the machine learning model but it also serves to greatly decrease processing cost in the training phase.

The pre-processing steps done to ensure algorithm works well with data are:

(a) Change the text case—so that all the words are either lower-case or upper-case.
(b) Punctuations can also cause problems—basic approach is to remove everything that is not alphabet or numeric. All punctuations were removed so that "@Windows10", "#Windows10", "Windows10!" may be treated as just 'Windows10'.
(c) Removing unhelpful terms—many words are frequently used but are only meaningful in a sentence, these are called "stopwords". Such as: is, the, at, which, etc. These are unlikely to improve machine learning prediction quality. Removing them will also reduce size of the data.
(d) Stemming—this step is motivated by the desire to represent words with different endings as the same word. Such as argument, argued, arguing is simply changed to argue.

3.2 Feature Extraction Using "Bag-of-Words" Method

The feature extraction process is derived from "Bag-of-Words" technique [18]. This builds a document-term matrix, which generates a matrix with rows corresponding to documents (tweets) and columns corresponding to the words in those tweets. The values in the matrix are the number of times that word appears in corresponding document.

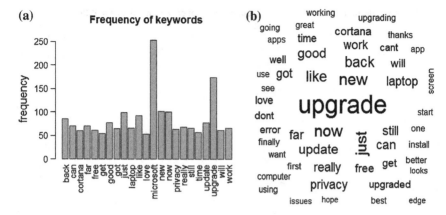

Fig. 2 **a** Showing bar graph of 'frequency of keywords' and **b** showing 'word cloud'

The data we get is called sparse, means that there are many zeroes in our matrix. The number of terms (or columns) in the matrix is problematic for two reasons, first one is computational, more terms means more independent variables, which usually means that it takes longer to build our models. Secondly, in building the models the ratio of independent variables to observations will affect how good the model will generalize. So, the terms which does not appear very often were removed by setting sparsity threshold. A new matrix is produced which contains the terms which appear in at least 1 % of the tweets.

Using this document-term matrix the keywords appearing most can be visualized using bar graph and word cloud (b) in Fig. 2.

3.3 Performing K-Means Clustering

K-means clustering had been performed, which takes the values in the matrix as numeric. The tweets are then clustered with *kmeans()* function with the number of clusters set to three. The K-means algorithm use the document-term matrix (Sect. 3.2) as data. A silhouette plot Fig. 3 shows how well each object lies within its clusters.

This plot shows for each cluster:

- the number of plots per cluster = number of horizontal lines,
- the mean similarity of each plot to its own cluster minus the mean similarity to the next most similar cluster (given by the length of the lines) with the mean in the right hand column, and
- the average silhouette width.

Fig. 3 Silhouette plot

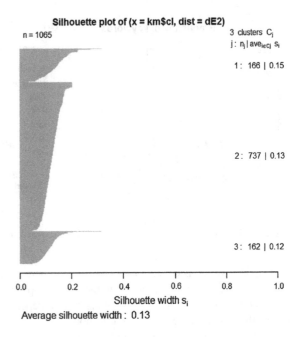

Silhouette plot of (x = km$cl, dist = dE2)

n = 1065

3 clusters C_j
j : n_j | ave$_{i \in C_j}$ s_i

1 : 166 | 0.15

2 : 737 | 0.13

3 : 162 | 0.12

0.0   0.2   0.4   0.6   0.8   1.0

Silhouette width s_i

Average silhouette width : 0.13

The contents of the each cluster shows the following popular words in each of the four clusters:

cluster 1: start menu back
cluster 2: install microsoft error
cluster 3: microsoft upgrade thank

From the above top words, generate from R code, it can be observed that the clusters are of different topics. The cluster 1 focuses on the 'start menu' on Windows 10, which was brought back after Windows 7. Similarly, remaining two clusters focus on different topics. This cluster analysis is very helpful for the companies to analyze consumer sentiment towards their products. Relation between various keywords within a cluster can give useful insights.

It must be noticed that the k-means clustering works only on numerical data. Since, the document-term matrix consists of numerical values, k-means clustering algorithm can be used.

3.4 Performing Supervised Learning Using CART

Decision trees such as CART are one of the most widely used machine learning algorithms. Much of their popularity is due to the fact that they can be adapted to almost any type of data. They are a supervised machine learning algorithm that divides the training data into smaller and smaller parts in order to identify patterns

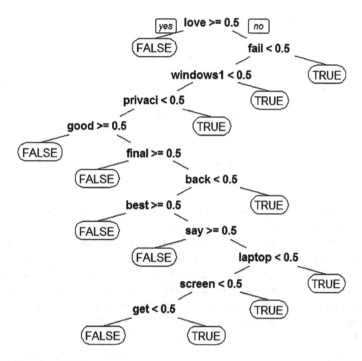

Fig. 4 Plotting the CART tree

which can be used for classification. The knowledge is then presented in the form of logical structure similar to a flow chart that can be easily understood without any statistical knowledge.

In supervised learning, to provide training to the data, the data was divided into training and testing sets, each containing 70 and 30 % of the data, respectively. After that, CART algorithm is applied to each of the three (see Sect. 3.3) training datasets.

Here, the CART algorithm is used to classify the tweets as positive or negative, By plotting the model, prominent keywords are generated can be depicted which sway public perception.

Using decision trees such as CART makes the solution much more interpretable. For example, from the above Fig. 4, it can be inferred that if the keywords such as 'privacy' and 'fail' are likely to appear in the tweet, then its sentiment is likely to be negative.

4 Result Analysis

The models which were built using CART algorithm were applied to the corresponding testing set, to predict the outcome variable. This outcome variable showed for each tweet, a binary value with '1' indicating negative sentiment and '0'

Table 1 Performance of 'cluster-then-predict' approach

Technique	Parameters					
	Accuracy (in %)	TPR/sensitivity	FPR/specificity	Precision	F score	AUC
K-means + CART (proposed)	69.73	0.7937	0.5728	0.7058	0.7472	0.7587
SVM	70.67	0.80	0.585	0.7128	0.754	0.7634
CART	62.80	0.8538	0.325	0.6220	0.723	0.6675
Naïve Bayes	65.64	0.712	0.585	0.688	0.70	0.6945
Random forest	72.86	0.7665	0.68	0.7547	0.76	0.7815

indicating positive. For analysis, all the results were combined. A confusion matrix was then created using the 'table' command in R.

The performance of the proposed 'cluster-then-predict' approach was compared with other classification algorithms, such as Naïve Bayes, SVM and RandomForest. They were evaluated based on the parameters such as Accuracy of Prediction, True Positive Rate (TPR)/Sensitivity, False Positive Rate (FPR)/Specificity, Precision, F1 score, area under ROC curve (AUC) and interpretability of the solution.

Accuracy gives the measure of how often the classifier is correct. It can be observed in the Table 1; that the prediction 'accuracy' of the innovative 'cluster-then-predict' as well as 'AUC' is almost similar to traditionally more accurate algorithms like SVM and RandomForest, without compromising on interpretability. The increase in accuracy is significant as compared to the base CART algorithm. Additionally, the clustering provides useful insight into the group of keywords which are significant, so the interpretability quotient of proposed algorithm is high, as compared to others.

Sensitivity (or True Positive Rate or Recall) measures the proportion of positives that are correctly identified as such (in this case, the percentage of tweets showing negative emotion which are correctly identified as having negative emotion). Similarly, Specificity measures the proportion of negatives that are correctly identified as such. Precision means if '1' or negative sentiment is predicted, how often is it correct. F Score is the weighted average of the true positive rate (recall) and precision. Lastly, AUC is the area under ROC (Receiver Operator Characteristic) graph. AUC is a good way to summarize a binary classifier's performance in a single number.

It is believed that if the data size is increased, the 'cluster-then-predict' approach should perform even better. This is because, processing with more dataset would lead to better clustering and improved classification on each cluster. This should improve the performance and accuracy significantly. Clustering can also be performed in a distributed environment where subsets of data are distributed to different computing machines.

5 Conclusion

This paper presents a hybrid approach—'cluster-then-predict'—to improve accuracy of predicting twitter sentiment. The possibility of combining both the supervised and unsupervised learning, in the form of K-means clustering and CART, have been investigated. The benefit of CART is that solution is more interpretable than other methods such as SVM or Naïve Bayes. But, it has traditionally less accuracy. The use of k-means clustering in the innovative 'cluster-then-predict' approach, before making predictions using CART, improves the accuracy of the classification significantly. Using clustering, cluster analysis can also be done to further gain useful insights from the data.

In future work, author(s) will try to include sentiment from emoticons, make the classification multi-class i.e. showing strongly positive, positive, neutral, negative or strongly negative sentiments.

References

1. Westling, A.: Sentiment analysis of microblog posts from a crisis event using machine learning. Master of Science Thesis, KTH CSC, Stockholm, Sweden
2. Dodd, J.: Twitter sentiment analysis. Final Project Report. National College of Ireland (2015, May)
3. Jose, A.K., Bhatia, N., Krishna, S.: Twitter sentiment analysis. Major Project Report, NIT Calicut (2010)
4. Akhavan Rahnama, A.: Real-time sentiment analysis of Twitter public stream, 84pp. University of Jyväskylä, Jyväskylä (2015)
5. Zhao, Y.: R and Data Mining: Examples and Case Studies. Academic Press. (2012)
6. Mittal, A., Arpit, G.: Stock Prediction Using Twitter Sentiment Analysis. Stanford University, CS229 (2012)
7. Zhang, L.: Sentiment analysis on Twitter with stock price and significant keyword correlation. Diss. 2013, The University of Texas at Austin (2013)
8. Ohmura, M., Kakusho, K., Okadome, T.: Stock market prediction by regression model with social moods. Int. J. Soc. Behav. Educ. Econ. Manage. Eng. 8(10) (2014)
9. Vu, T.-T., et al.: An experiment in integrating sentiment features for tech stock prediction in twitter. In: Workshop on Information Extraction and Entity Analytics on Social Media Data, 9 Dec 2012, Mumbai, The COLING 2012 Organizing Committee, pp. 23–38 (2012)
10. Wang, X., Gerber, M.S., Brown, D.E.: Automatic crime prediction using events extracted from twitter posts. In: Social Computing, Behavioral-Cultural Modeling and Prediction, pp. 231–238. Springer Berlin Heidelberg (2012)
11. Lu, Y., et al.: Integrating predictive analytics and social media. In: 2014 IEEE Conference on Visual Analytics Science and Technology (VAST). IEEE, Paris, France (2014)
12. Pak, A., Patrick, P.: Twitter as a corpus for sentiment analysis and opinion mining. LREC 10 (2010)
13. Kagan, V., Andrew, S., Subrahmanian, V.S.: Using twitter sentiment to forecast the 2013 Pakistani election and the 2014 Indian election. In: IEEE Intelligent Systems, vol. 1, pp. 2–5 (2015)

14. Gayo-Avello, D.: I wanted to predict elections with twitter and all I got was this Lousy Paper. A Balanced Survey on Election Prediction using Twitter Data. arXiv preprint arXiv:1204.6441. University of Oviedo, Spain (2012)
15. Agarwal, A., Xie, B., Vovsha, I., Rambow, O., Passonneau, R.: Sentiment analysis of twitter data. In: Proceedings of the Workshop on Languages in Social Media, pp. 30–38. Association for Computational Linguistics, June 2011
16. Selmer, O., Brevik, M.: Classification and visualisation of Twitter sentiment data. Master's Thesis, NTNU-Trondheim (2013)
17. The R project for statistical computing. https://www.r-project.org/
18. "Bag-of-Words" feature extraction technique. https://en.wikipedia.org/wiki/Bag-of-words_model

A Hybrid Algorithm with Modified Inver-Over Operator and Genetic Algorithm Search for Traveling Salesman Problem

Dharm Raj Singh, Manoj Kumar Singh and Tarkeshwar Singh

Abstract In this article, we develop a novel hybrid approach to solve the traveling salesman problem (TSP). In this approach, we first initialize suboptimal solution using Nearest Neighbor (NN) tour construction method, followed modified Inver-over operator and then proposed crossover with 2-opt mutation applied to improve for optimal solution. We use 14 TSP data sets from TSPLIB to evaluate the performance of proposed hybrid method. The proposed hybrid method gives better results in terms of best and average error. In experimental results of the tests we show that the proposed hybrid method is superior to available algorithm in literature.

Keywords Traveling salesman problem · Basic inver-over operator · Modified inver-over operator · 2-opt mutation · Crossover operator

1 Introduction

1.1 Traveling Salesman Problem

Combinatorial optimization problem contains a broad study of the traveling salesman problem (TSP) [1]. Given a set of n cities and distance between each pair of cities, the traveling salesman visit all cities only once and return back to starting city with minimum distances. This type problem is modeled in graph theory with

D.R. Singh (✉) · M.K. Singh
DST-Centre for Interdisciplinary Mathematical Sciences (DST-CIMS),
Banaras Hindu University, Varanasi 221005, India
e-mail: dharmrajsingh67@yahoo.com

M.K. Singh
e-mail: manoj.dstcims@bhu.ac.in

T. Singh
Department of Mathematics, Birla Institute of Technology and Science Pilani,
K K Birla Goa Campus, Zuarinagar 403726, Goa, India
e-mail: tksingh@goa.bits-pilani.ac.in

© Springer Science+Business Media Singapore 2016   141
R.K. Choudhary et al. (eds.), *Advanced Computing and Communication Technologies*,
Advances in Intelligent Systems and Computing 452,
DOI 10.1007/978-981-10-1023-1_14

help of vertices and edges. Vertices represent cities and edges represent roads between the two cities and the weight of an edge represents distance between two adjacent cities. Thus, we have a weighted graph $G = (V, E, w)$, where w: $E \rightarrow Z$ and Z is a set of nonnegative integer. A closed $C = (u = u_0, u_1, u_2, \ldots, u_n = u)$ tour in which all the vertices are distinct which is known as Hamiltonian cycle. Finding Hamiltonian cycle with minimum travel cost $w(C) = \sum w(e_i)$, where summation is taken over all edges in C in the weighted graph is the desired solution. The matrix representation of weighted graph is known as cost matrix which is denoted by $C = (c_{ij})$, $i, j = 1, 2, \ldots, n$. An entry of cost matrix c_{ij} represents the cost between ith vertex to jth vertex. The total cost of Hamiltonian cycle is the sum of cost of each edge in Hamiltonian cycle. In general; this problem is NP-hard.

The Euclidean distance (cost) between two adjacent vertices is calculated as fallow:

$$c_{ij} = \sqrt{\left(x_i - x_j\right)^2 + \left(y_i - y_j\right)^2}, \tag{1}$$

where (x_i, y_i) and (x_j, y_j) are coordinates of two adjacent vertices.

1.2 Literature Review

According to Arora [2], TSP is NP-hard combinational optimization problem. The most commonly used heuristic and meta-heuristic algorithms such as Greedy algorithms, 2-opt algorithms, Simulated Annealing (cf., [3, 4]), Tabu Search, Ant Colony Optimization [5, 6], Genetic Algorithms (cf., [7–10]), Neural Networks [11], Weed optimization [12] and Memetic Algorithm (MA) (cf., [11, 13]) etc.

TSP algorithms are categorized into two classes such as exact and approximate algorithms. The exact algorithm is always used for optimal solution. Cutting plane or branch and bound method is one of the most effective exact algorithms, which have been solved large TSP instances (see [14]). But the exact algorithms have taken high time complexity. However, an approximate algorithm is used for the solution of TSP instances in reducing computation, while the obtained solution using approximate algorithm is nearest to the optimal solution. In [15], we have used nearest neighbor tour construction method for generating a tour and tour improvement methods used to get better quality of tour by applying the various exchanges. The 2-opt optimal heuristic is generally considered as one of the most effective methods, which is generating approximate optimal solutions for the traveling salesman problem (TSP). Guo Tao proposed an algorithm based on the Inver-over operator and memetic algorithm with improved Inver-over operator, which is presented in [9, 16], respectively.

The remaining part of the paper is constructed as follows. Section 2 describes the overview of Basic Inver-over operator. Proposed hybrid method is described in Sect. 3. Experimental results are presented in Sect. 4. In the last Sect. 5 provides the concluding part of the article.

$$(2,1,4,6,8,7,3,5)\xrightarrow[inverse(8,7,3)]{c'=3}(2,1,4,6,3,7,8,5)\xrightarrow[inverse(2,1,4,6)]{c'=2}(6,4,1,2,3,7,8,5)\xrightarrow{c'=3}end$$

$$c=6 \qquad\qquad\qquad\qquad\qquad c=3 \qquad\qquad\qquad\qquad c=2$$

Fig. 1 Single iteration of basic inver-over operator

2 Basic Inver-Over Operator

The Inver-over operator [9] performs both operation crossover and mutation. In this method, first we randomly select a city from the individuals, after then we generate a random number (rnd). If the random number (rnd) < certain threshold value (prd) then select the second city from the rest of individuals for inversion. In this case, inversion performs mutation. If the random number (rnd) > prd then randomly select another new individual from the population. After then select the second city just after to the selected first city from the new individual. In this case, the inversion operation performs crossover. The whole procedure of Basic Inver-over operator is given in [16, 17].

Let us assume that starts with selected chromosome S' is (2, 1, 4, 6, 8, 7, 3, 5). Figure 1 shows a single iteration of this operator and the inversion function which is responsible for the inversion.

3 Proposed Hybrid Method

Proposed hybrid method provides the approximate solution which depend heuristics with exploration and exploitation. The overall procedures of our algorithm that combines the nearest-neighbor tour construction, modified Inver-over operator and genetic algorithms with 2-opt mutation heuristics.

In the proposed hybrid method, the first step in algorithm is using Nearest Neighbor tour construction heuristic (see, [18]) to generate initial population of population size. In the second step, we find fitness value of each chromosome in population. In third step, we select best chromosome S' from population. In the fourth step, we applied the modified Inver-over operator on the selected chromosome S'. After the fourth step, we generate new chromosome. If cost of new generated chromosome is less than old chromosome then replace the old chromosome by new chromosome. In the fifth step randomly select two chromosomes from population and then we applied the proposed crossover operator on selected chromosomes with crossover probability rate (pc). In the last step, we applied 2-opt optimal mutation operator on selected two parents or new individuals that generated after crossover, with mutation probability rate (pm) and update the population. The whole process repeated until termination condition is satisfied. Figure 2 describes a detail description of the proposed hybrid algorithm.

```
Using Nearest Neighbor tour construction heuristic to generate initial population of size P;
while (terminate condition are not satisfied) do
  Find the fitness value f for each chromosome in P;
  Select best chromosome S' = Sᵢ from P;
  Randomly select a city c from S';
  Generate a random number (rnd1) between (0, 1);
  while (true)
    Generate a random number (Rnd2) between (0, 1);
    if (Rnd2< prd) then
      Apply 2-opt optimal mutation to best selected chromosome S';
      Randomly select city c from S';
      Continue;
    else if (Rand2<prd + pcs)
      Randomly select city c' from a set of kα- nearest neighbors of the city c;
    else
      Randomly select a new chromosome of P;
      Assign to city c' the subsequently to the city c in the newly selected chromosome;
    end if
    if ( city c' is the previous or next city of c in S' ) then
      Break;
    else
      Apply Inverse operation from city c to the city c' in S' (excluding c);
    end
    if (rnd1< puc (gn)) then
      Remain city c unchanged;
    else
      c = c'
    end
  end while
    if (cost of new generated chromosome is lesser than old chromosome) then
      Replace the old chromosome by new chromosome;
    end
  Random select two parents for crossover (S₁, S₂ ∈ P );
  Apply proposed crossover on selected parents with probability rate pc;
  Apply 2-opt optimal mutation on selected two parents or new child with probability rate pm;
  Update population;
end while
```

Fig. 2 Proposed hybrid algorithm with modified inver-over operator

3.1 Modified Inver-Over Operator

Figure 1 depicts Basic Inver-over operator process, in which first inversion operator to city (8, 7, 3), then the new age (6, 3) is added to the current solution and again apply inversion operator to city (2, 1, 4, 6), then the edge (3, 2) is added. After the second inversion edge (6, 3) is deleted from solution. Therefore, Basic Inver-over operator ignores direction of path. The direction of path involved in Modified Inver-over operator. In this case, the newly added edge, by applying first inversion operator will not be removed till the next inversion. Figure 3 shows Modified

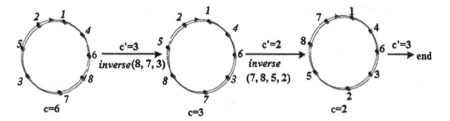

Fig. 3 A single iteration of modified inver-over operator (involving direction of path)

Inver-over operator that considers direction of path. Therefore, effect of each inversion will be sure.

It has verified by experimental result. The speed of convergence will be increased by receiving instructor of candidate set to select the city c'. In Modified Inver-over operator, randomly we select city c' from a set of α-nearest neighbors of the city c with certain threshold value [19].

Moreover, if the city c is unchanged with city c' then the population diversity will be kept until the possibility of traffic at local optimum will be reduced. Then, both hybrid methodologies have developed to kept population changes with effectiveness. If the random number (rnd1) < puc then the city c unchanged, otherwise city c replace by c'. In the algorithm, the possible adaptive value of puc increases with iteration of the algorithm, the city c will be unchanged with the higher possibility in a later period of the algorithm, while the population changes will be managed suitably. The relationship between iteration of algorithm and puc is given as follows:

$$puc(gn) = \exp\left(\log\left(\frac{puc\_max - puc\_min}{N}\right)\right) \times gn, \qquad (2)$$

where, *puc* denotes the possibility unchanged city c, *puc_min* and *puc_max* are minimum and maximum possibilities, *gn* indicates iteration number while *N* is total number of iteration.

3.2 2-Opt Mutation

Croes [20] proposed the 2-Opt method. Basically, 2-opt mutation deletes two edges from path, and add two new edges that are not in solution in such way that the cost of new path is lesser. The process is continued till no further improvements are possible [19]. The resulting path is referred to as a 2-opt optimal (see [21]). For example, inverting the subsequence $(x_{i+1}, \ldots, x_j)$ of sub path $(x_i, x_{i+1}, \ldots, x_j, x_{j+1})$ is replaced with $(x_i, x_j, \ldots, x_{i+1}, x_{j+1})$.

Finally, performing the 2-exchange mutation, the resulting cost is expressed as follow:

$$\Delta_{ij} = c(x_i, x_j) + c(x_{i+1}, x_{j+1}) - c(x_i, x_{i+1}) - c(x_j, x_{j+1}). \tag{3}$$

The resulting cost is obtained by iteratively applying 2–exchange mutation till impossible move yields a negative Δ value.

3.3 Proposed Crossover Operator

In the proposed crossover, the first city of chromosome s_1 is copied to first position in the child c_1 and first city of chromosome s_2 is copied to first position in the child c_2. The remaining cities changed accordingly in Fig. 5. For example, the first city of parent s_1, s_2 are copied at the first position in the child c_1 and c_2 respectively in Fig. 4, remaining positions 2, 3, 4, 5, 6, 7, 8 and 9 cities are swapped as procedure given in Fig. 5.

Fig. 4 Proposed crossover

Parent s_1: | 2 | 4 | 7 | 1 | 8 | 9 | 5 | 6 | 3 |

Parent s_2: | 7 | 4 | 3 | 5 | 8 | 9 | 2 | 6 | 1 |

Child c_1 : | 2 | | | | | | | | |

Child c_2 : | 7 | | | | | | | | |

Child c_1 : | 2 | 4 | 3 | 5 | 8 | 9 | 7 | 6 | 1 |

Child c_2 : | 7 | 4 | 2 | 1 | 8 | 9 | 5 | 6 | 3 |

Fig. 5 Algorithm for proposed crossover

```
for i = 1: n
    for j = 2:n
        if (s_2 (i) = = s_1 (j))
            c_1(j) = s_2(j);
        end
        if (s_1 (i) = = s_2(j))
            c_2(j) = s_1(j);
        end
    end
end
for i = 2:n
    if (c_1 (1) = = c_1 (i))
        c_1 (i) = s_2 (1);
    end
    if (c_2 (1) = = c_2 (i))
        c_2 (i) = s_1 (1);
    end
end
```

4 Experimental Results

4.1 Experimental Setup

For evaluating the performance of experimental results, Intel (R) Core (i5) 3.20 GHz processor, 2GB RAM on MATLAB is used. In this experiment, we used 14 different TSP benchmark instances taken from the TSPLIB. The value of parameters used in experiment are population size (P) = 40, minimum probability of city (c) unchanged is $puc\_min = 0.2$, maximum probability of city (c) unchanged is $puc\_max = 0.5$ and 5α-nearest neighbour. The value of prd = 0.02, pcs = 0.05, pc = 0.8, pm = 0.25 with N = 2000 iteration.

The efficiency of proposed hybrid method is based on Percentage Best Error (% Best Err.) and Percentage Average Error (% Ave. Err.). The Percentage Best Error and Percentage Average Error are given as follows:

$$\% \text{ Best Err.} = \frac{(\text{best solution from } n - \text{trail}) - (\text{best known solution from TSPLIB})}{\text{best known solution from TSPLIB}} * 100$$

$$\% \text{ Ave. Err.} = \frac{(\text{average solution from } n - \text{trail}) - (\text{best known solution from TSPLIB})}{\text{best known solution from TSPLIB}} * 100$$

4.2 Experimental Results and Analysis

In this section, we compared the proposed hybrid algorithm with the recent algorithms as presented in [7, 15, 16]. We performed $n = 10$ trails for our present method. The comparative results are presented in Table 1, the best results of the method for particular instances are in bold. From last row of the table, it is clear that our proposed method gives better result for all instances with respect to all parameters, except berlin52 for % Best Err. and % Ave. Err., kroA100 for % Best Err. and % Ave. Err., pr144 for % Ave. Err., ch150 for % Best Err. and % Ave. Err., pr152 for % Best, rat195 Best Err. and % Ave. Err. and ts225 for % Best Err. Moreover, obtained result by proposed method is better for TSP instances pr144 and kroB150, the best known solutions 58,537 and 26,130 reported by TSPLIB are replaced by the new best obtained solutions 58535.2 and 26127.36.

The last row of Table 1 has shown the average performance of each method over the data set. Figure 6 indicates % Ave. Err. with respect to instances size. From Fig. 6, we conclude that the proposed hybrid method is best among Inver-over and LK MA without local search, NN+2-opt mutation and GSTM algorithm for the traveling salesman problem.

Table 1 Performance comparison of different algorithm

Instance	Modified inver-over operator with genetic algorithm			Inver-over and LK MA (remove local search)			NN+2-opt mutation			GSTM algorithm		
	Best Err. (%)	Ave. Err. (%)	Ave. time(s)	Best Err. (%)	Ave. Err. (%)	Ave. time (s)	Best Err. (%)	Ave. Err. (%)	Ave. time (s)	Best Err. (%)	Ave. Err. (%)	Ave. time (s)
berlin52	0.0314	0.0314	1.2775	**0.0000**	**0.0000**	**0.4867**	0.0318	0.0318	3.042	**0.0000**	**0.0000**	0.836
kroA100	0.0162	0.0204	2.9336	**0.0000**	**0.0000**	**0.6193**	0.0141	0.1142	7.394	**0.0000**	1.1836	6.987
pr144	**-0.003**	0.0439	3.9659	0.0564	0.1350	**0.6878**	**-0.003**	**0.0174**	14.780	0.0000	1.0809	13.598
ch150	0.2778	0.6062	5.1288	**0.0000**	0.3585	**0.8566**	0.7077	1.2467	13.213	0.4596	0.6357	11.240
kroB150	**-0.0101**	**0.6271**	5.7666	0.0421	0.6456	**0.7829**	0.0421	0.9885	14.148	0.9644	1.7616	11.684
pr152	0.0022	**0.0769**	5.8002	**0.0000**	0.1259	**0.7144**	0.1886	0.4072	18.364	0.7695	1.6202	7.937
rat195	0.6518	0.8275	6.9373	**0.4305**	**0.6586**	**0.8503**	1.2613	1.9513	20.248	0.6027	1.8425	15.050
d198	**0.2050**	**0.4456**	9.3652	0.3359	0.6800	**0.9391**	0.5830	0.9157	22.302	0.3866	1.2193	12.096
kroA200	**0.0889**	**0.4293**	7.9241	0.4052	0.5816	**0.9047**	0.5925	1.0440	21.719	0.8683	1.5432	13.292
ts225	0.0023	**0.0209**	8.7002	**0.0000**	0.4850	**0.9815**	0.0055	0.3040	37.201	0.2527	0.4994	11.559
pr226	**0.0016**	**0.3257**	8.9554	0.1344	0.4270	**0.9206**	0.0535	0.3831	31.887	0.7242	1.5287	13.843
pr299	**0.2103**	**0.6718**	15.9916	0.6661	2.3401	**1.1949**	1.2139	2.1958	51.794	1.2326	2.9169	17.424
lin318	**0.5389**	**0.8847**	20.8173	1.3610	2.3058	**1.2731**	1.0027	2.2865	77.049	0.9827	3.3099	14.643
pcb442	0.9038	1.3331	37.5992	1.4494	2.1127	**1.6723**	1.8177	2.9822	122.998	2.0501	2.7758	19.132
Average	**0.2084**	**0.4532**	10.0831	0.3486	0.7754	**0.9203**	0.5368	1.0621	31.637	4.0423	1.5655	12.094

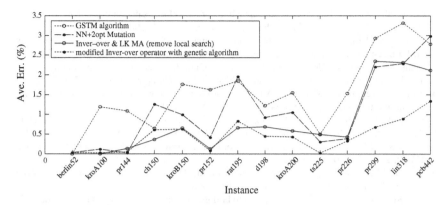

Fig. 6 % Ave. Err. (over 10 runs) for the 14 TSPLIB instances ordered by size

Our proposed method gets speed up the convergence of optimal solution because in this hybrid proposed method. Firstly, we generate initial global solution by using NN tour construction algorithm, then modified Inver-over operator is being used, after then proposed crossover with a powerful local improvement method (2-opt mutation) that refines the solution for global optimality.

5 Conclusion

In this article, we have discussed a novel proposed hybrid method for the Euclidean traveling salesman problem. It is combination of NN tour construction, Modified Improved Inver-over operator and genetic algorithm. In proposed method we considered the direction of path, randomly select a city from a set of α-nearest neighbors and proposed crossover with a powerful local improvement method (2-opt mutation). The combination affects the quality of solution. Therefore, our proposed hybrid method gives better performance than Greedy Sub Tour Mutation (GSTM), NN+2-opt mutation and Inver-over and LK MA (remove local search). Thus, the proposed hybrid method i.e. Modified Inver-over operator with genetic algorithm remarkably increases the performance of genetic algorithm in TSP solutions.

References

1. Lawler, E.L.: The traveling salesman problem: a guided tour of combinatorial optimization. Wiley-Interscience Series Discrete Mathematics (1985)
2. Arora, S.: Polynomial time approximation schemes for Euclidean traveling salesman and other geometric problems. J. ACM (JACM) **45**(5), 753–782 (1998)

3. Chen, Y., Zhang, P.: Optimized annealing of traveling salesman problem from the nth-nearest-neighbor distribution. Physica A **371**(2), 627–632 (2006)
4. Jeong, C.S., Kim, M.H.: Fast parallel simulated annealing for traveling salesman problem on SIMD machines with linear interconnections. Parallel Comput. **17**(2), 221–228 (1991)
5. Deng, W., Chen, R., He, B., Liu, Y., Yin, L., Guo, J.: A novel two-stage hybrid swarm intelligence optimization algorithm and application. Soft. Comput. **16**(10), 1707–1722 (2012)
6. Yun, H.Y., Jeong, S.J., Kim, K.S.: Advanced harmony search with ant colony optimization for solving the traveling salesman problem. J. Appl. Math. (2013)
7. Albayrak, M., Allahverdi, N.: Development a new mutation operator to solve the traveling salesman problem by aid of genetic algorithms. Expert Syst. Appl. **38**(3), 1313–1320 (2011)
8. Louis, S.J., Li, G.: Case injected genetic algorithms for traveling salesman problems. Inf. Sci. **122**(2), 201–225 (2000)
9. Tao, G., Michalewicz, Z.: Inver-over operator for the TSP. In: Parallel Problem Solving from Nature—PPSN, pp. 803–812. Springer, Berlin (1998)
10. Ahmed, Z.H.: An experimental study of a hybrid genetic algorithm for the maximum traveling salesman problem. Math. Sci. **7**(1), 1–7 (2013)
11. Créput, J.C., Koukam, A.: A memetic neural network for the Euclidean traveling salesman problem. Neurocomputing **72**(4), 1250–1264 (2009)
12. Zhou, Y., Luo, Q., Chen, H., He, A., Wu, J.: A discrete invasive weed optimization algorithm for solving traveling salesman problem. Neurocomputing **151**, 1227–1236 (2015)
13. Merz, P., Freisleben, B.: Memetic algorithms for the traveling salesman problem. Complex Syst. **13**(4), 297–346 (2001)
14. Laporte, G.: La petite et la grande histoire du probleme du voyageur de commerce (2006)
15. Singh, D.R., Singh, M.K., Singh, T.: A hybrid heuristic algorithm for the Euclidean traveling salesman problem. In: 2015 International Conference on Computing, Communication and Automation (ICCCA), pp. 773–778. IEEE (2015)
16. Wang, Y.T., Li, J.Q., Gao, K.Z., Pan, Q.K.: Memetic algorithm based on improved inver–over operator and Lin-Kernighan local search for the Euclidean traveling salesman problem. Comput. Math Appl. **62**(7), 2743–2754 (2011)
17. Wang, Y., Sun, J., Li, J., Gao, K.: A modified inver-over operator for the traveling salesman problem. In: Advanced Intelligent Computing Theories and Applications. With Aspects of Artificial Intelligence, pp. 17–23. Springer, Berlin (2012)
18. Johnson, D.S., McGeoch, L.A.: The traveling salesman problem: A case study in local optimization. Local Search Comb. Optim. **1**, 215–310 (1997)
19. Helsgaun, K.: An effective implementation of the Lin-Kernighan traveling salesman heuristic. Eur. J. Oper. Res. **126**(1), 106–130 (2000)
20. Croes, G.A.: A method for solving traveling-salesman problems. Oper. Res. **6**(6), 791–812 (1958)
21. Gutin, G., Punnen, A.P. (eds.): The traveling salesman problem and its variations, vol. 12. Springer Science & Business Media (2002)

Exploration of Feature Reduction of MFCC Spectral Features in Speaker Recognition

Mohit Kumar, Sachin Katti and Pradip K. Das

Abstract Most recognition systems heavily depend on the features used for representation of speech information. Over the years, there has been a continuous effort to generate features that can represent speech as best as possible. This has led to the use of larger feature sets in speech and speaker recognition systems. However, with the increasing size of the feature set, it is not necessary that all features are equally important for speech representation. This paper investigates the relevance of individual features in one of popular feature sets, MFCCs. The objective of the study is to identify features which are more important from speech information representation perspective. Experiments were conducted for the task of speaker recognition. Results indicate that it is possible to reduce the feature set size by more than 60 % without significant losses in accuracy.

Keywords Feature reduction · Speaker recognition · PCA · LDA · Factor analysis

1 Introduction

Speech recognition deals with the recognition of the content that is embedded in the speech signal. Speaker recognition deals with determining the identity of the person who generated the given speech signal.

Features are needed to represent the speech signal so that it can be processed by machines. Over the years, Mel Frequency Cepstrum Coefficients (MFCCs) [1, 2]

M. Kumar (✉) · S. Katti · P.K. Das
Department of Computer Science and Engineering,
Indian Institute of Technology Guwahati, Guwahati 781039, Assam, India
e-mail: mohit.2013@iitg.ernet.in

S. Katti
e-mail: s.katti@iitg.ernet.in

P.K. Das
e-mail: pkdas@iitg.ernet.in

© Springer Science+Business Media Singapore 2016   151
R.K. Choudhary et al. (eds.), *Advanced Computing and Communication Technologies*,
Advances in Intelligent Systems and Computing 452,
DOI 10.1007/978-981-10-1023-1_15

have been considered as the most popular feature set. It has also been observed that the feature set sizes have become larger over time. The goal behind the idea has been to represent as much information as possible with a large sized feature vector. But it may not be the case that all the features in the feature set are representing similar amount of information. Some of the features may be less important than others in conveying speech information. We explore this direction in the current study. The motivation behind this work is to understand the importance of individual features in the feature sets. With this information, we can achieve a reduction in the feature set size without significant loss in the accuracy of the recognition system.

Several studies have been conducted exploring dimensionality reduction and speaker modeling. Hu and Zahorian [3] has tried non-linear reduction techniques for HMM based phonetic recognition. Quan et al. [4] has used the Principal Component Analysis reduction technique to develop a speech emotion engine. Fodor [5] provides a lucid survey of major dimensionality reduction techniques that can be applied to various problems. Reynolds and Rose [6] discussed the use of Gaussian Mixture Models to represent the speakers. This was a novel approach at that time. Further, [7] proposed modifications to existing GMM models. More advanced models were also introduced such as Joint Factor Analysis (JFA) [8], i-vector modeling [9], deep learning and neural networks based methods [10].

The organisation of the rest of the paper is as follows. The next section discusses the concepts and modeling techniques used in speaker recognition systems. Section 3 deals with various feature reduction techniques. Section 4 describes the experimental process and layout. Section 5 presents and discusses the results obtained. Section 6 presents the conclusions of the study. The paper ends with acknowledgements and references.

2   Speaker Modeling Approaches

Gaussian Mixture Models (GMMs) have formed the backbone of the speaker recognition and verification tasks for decades [6]. Using a finite number of Gaussian components, we can represent a speaker's characteristics. A GMM model can be represented as a weighted sum of M component Gaussian densities as follows:

$$p(x|\lambda) = \sum_{i=1}^{M} w_i g(x|\mu_i, \Sigma_i) \tag{1}$$

where x is a D-dimensional vector, w_i, $i = 1, 2, ..., M$ are the mixture weights, and $g(x|\mu_i, \Sigma_i)$, $i = 1, 2, ..., M$ are the component densities of the Gaussian. Each individual component density is a D-variate Gaussian function as follows:

$$g(x|\mu_i, \Sigma_i) = \frac{1}{(2\pi)^{\frac{D}{2}}|\Sigma_i|^{\frac{1}{2}}} \exp\left\{-\frac{1}{2}(x - \mu_i)' \Sigma_i^{-1}(x - \mu_i)\right\} \tag{2}$$

where μ_i is the mean vector for component i and Σ_i is the covariance matrix. The sum of all the weights must add to 1.

Further research led to the emergence of Universal Background Models (UBMs) [7]. A UBM is speaker independent Gaussian mixture model which is trained from a large set of speakers to represent the general speech characteristics. The major use of the UBM is the adaptation of the individual speaker models with UBM parameters as priors. It was observed that this approach led to performance improvements in recognition experiments. The models obtained using UBMs are known as Adapted GMMs.

3 Feature Reduction Approaches

Feature reduction techniques [5] can be classified in two categories: Feature selection methods and Feature re-extraction methods.

3.1 Feature Selection Methods

Feature Selection is one way of achieving feature reduction where we use the merit of each feature in the feature vector and decide whether to retain it or not. The critical part of this approach is deciding the merit of the features. In this study, we have used two methods to rank the features: Naïve dropping and F-Ratio.

In Naïve dropping method, we consider the order of the feature in the feature set as their rank. This is not a sophisticated method for the ranking of features. Another method which uses inter class and intra class variance to decide the merit of the features is the F-Ratio [11, 12] method. F-Ratio for ith feature can be calculated as follows:

$$F_i = \frac{B_i}{W_i} \tag{3}$$

where B_i is the interclass variance and W_i is the intra class variance.

3.2 Feature Re-extraction

Feature re-extraction is a way to transform features according to some given criteria. This approach can be used to obtain reduced features from existing features. One of the most popular methods is Principal Component Analysis.

Principal Component Analysis Principal Component Analysis (PCA) [4] is a statistical technique that uses orthogonal transformations to transform a set of observations of possibly correlated variables into a set of linearly uncorrelated variables. These are called principal components. The formulation of PCA ensures that the first principal component has the largest variance. Each successive component has the next largest variance while being orthogonal to each of the previous components. Using this transformation the feature space is projected onto a smaller subspace which best preserves the variance of the input data [12].

Linear Discriminant Analysis Linear Discriminant Analysis (LDA) [13] is a method used for dimensionality reduction that aims to retain as much class discriminatory information as possible. The idea in LDA is to project the feature space onto a smaller subspace that contains most of the class variability. This is achieved by maximizing the F-Ratio of the training data in the transformed space. In this way, it can be considered as a generalized F-Ratio method.

Factor Analysis Factor Analysis (FA) [14] is a linear generative model in which the latent variables are assumed to be Gaussian. The core idea of the techniques is to model the observed variables in terms of a smaller number of 'factors' which are believed to cause the observations. The transformation of features into a smaller subspace is achieved by a loading matrix.

4 Experimental Layout

4.1 Data

For our experiments, we have used our own recorded data. It consists of the ten English digits spoken by 10 male speakers in the age group from 23 to 27 years. Each digit was recorded 20 times at 16,000 samples/s with 16 bit representation using Cool Edit Pro 2.1 recording studio software. A total of 2000 utterances were used. For training, 1500 utterances were used. The rest 500 utterances were used as test data.

4.2 Feature Extraction

For our experiments, we have used the Hidden Markov Model Toolkit (HTK) [15]. We have used the HCOPY tool from HTK to extract Mel Frequency Cepstrum Coefficients for each frame of the utterance. Each frame consisted of 320 samples with the step size of 80 samples.

4.3 Feature Transformation and Reduction

In case of feature selection techniques, the feature vector was reduced by leaving out a particular feature as determined by the rank of that particular feature. At each reduction stage, one feature was dropped. In case of feature re-extraction techniques, each feature vector was multiplied with the transformation matrix to obtain the reduced feature vector. For both the cases, the reduced feature vectors were used for training and testing.

4.4 Experiments

We performed 5 experiments in this study. First, we used the Naïve dropping technique to reduce the features. We started by reducing the data to 11 dimensional data by leaving out the last feature. It was continued till only two features were left in the feature vector. For each reduction, we performed speaker recognition tests to determine the accuracy of the reduced feature set. This process was followed for F-Ratio as well with the exception that the merit criterion was now changed to F-Ratio of the features. The feature with the least F-Ratio was dropped first.

In case of feature re-extraction techniques, we obtain a transformation matrix. This transformation matrix is then multiplied with feature vectors to reduce the feature size. This process was followed in case of PCA, LDA and Factor Analysis. For each of the speaker recognition experiments, we used the mixture count of 32 to train speaker models.

5 Results and Discussion

Figures 1, 2, 3, 4, 5 depict the trends of accuracy with reduced features. In Naïve dropping method, we observe that as we start dropping coefficients from the feature vector, reducing the number for coefficients from 12 to 7, the accuracy oscillates between 78.2 and 83.4 %. But as we continue dropping further, the accuracy drops significantly to 66.2 % with 4 coefficients, 47.2 % with 3 coefficients and 34 %

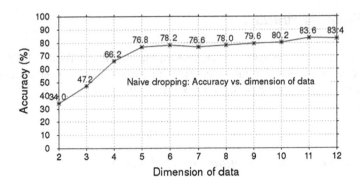

Fig. 1 System performance as dimension is reduced with Naïve dropping

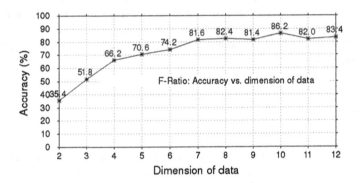

Fig. 2 System performance as dimension is reduced with F-Ratio

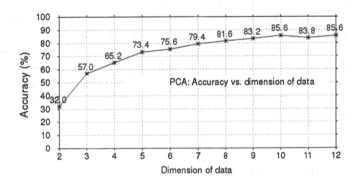

Fig. 3 System performance as dimension is reduced with PCA

with 2 coefficients. This indicates that the first 5 coefficients are very important to
the recognition process. The rest of the features help increase the performance only
slightly as compared to the first 5 coefficients. Taking only first 5 coefficients, we
can reduce the feature set by more than 60 %.

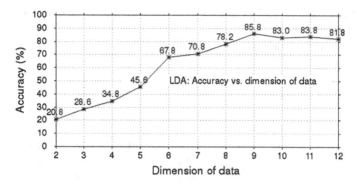

Fig. 4 System performance as dimension is reduced with LDA

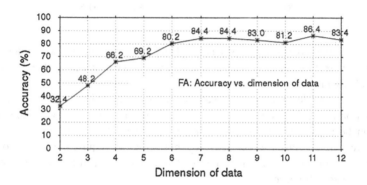

Fig. 5 System performance as dimension is reduced with FA

In case of F-Ratio method, again we observe that dropping the last 5 coefficients do not affect the recognition performance significantly, with accuracy ranging between 81.4 and 86.2 %. However, on further drop of coefficients, performance starts to degrade rapidly. This also indicates the importance of first 7 features as ranked by their F-Ratios. Retaining only the first 7 features, we can reduce the size of our feature set by more than 41 %.

Similarly, we observe that for PCA, the jump in the performance is significant for the first 6 features in the feature set. The rest of the features do not contribute as much to the performance. The reason for this is that the features are ranked in order of decreasing variance. Thus, the first 6 features explain most of the variance.

In case of LDA, we note that first 9 features in the feature set are contributing heavily to the performance as we observe sharp jumps in accuracy with addition of each feature. But the improvements are not seen after adding more features. On the contrary, performance degrades a bit. Thus, we can reduce the feature set by 25 % along with achieving a better performance.

In case of Factor Analysis, we observe that performance stays between 84.4 and 86.4 % when feature set size is reduced from 12 to 7. On further reduction, sudden

Table 1 Results showing accuracy versus different dimension of data for different dimensionality reduction techniques

Dimension of data	Naïve dropping	F-Ratio	PCA	LDA	Factor analysis
2	33.8	35.4	32	20.8	32.4
3	47.2	51.8	57	28.6	48.2
4	66.2	66.2	63.8	34.8	66.2
5	76.8	70.6	73.4	45.6	69.2
6	78.2	74.2	75.6	67.8	80.2
7	76.6	81.6	79.4	70.8	84.4
8	78	82.4	81.6	78.2	84.4
9	79.6	81.4	83.2	85.8	83
10	80.2	86.2	85.6	83	81.2
11	83.6	82	83.8	83.8	86.4
12	83.4	83.4	85.6	81.8	83.4

performance degradations are observed. Thus, we can obtain more than 41 % reduction in feature set size with a small drop in performance. Table 1 lists the results obtained.

We can conclude from the observations that we can achieve significant amount of reduction in the feature set size with a small drop in the accuracy of the system. This can be highly useful for cases where we have a two-layer system or cases where we can compromise on accuracy but speed is required. For a two pass system, this can serve as the first layer of the system that can be used for quickly pruning the search space.

6 Conclusion

In this work, we studied the performance degradation with respect to reduction in size of the feature set. We first used the Naïve dropping technique. It produced promising results but there was no proper way to determine the rank of the features in this method. We then used the F-Ratio of the individual features to decide on the merit of each feature. This is a better approach to reduce the feature set size. Next, feature re-extraction methods were applied on the data. First in this category is the PCA. It transformed the data and then projected it to a lower subspace. Significant reductions in feature set size were observed with this method. LDA, another method to re-extract the features, was also used for reduction. It reduced the feature set but the behaviour of PCA was much better as it reduced the feature set size to much more than LDA with similar degradations in performance. In case of Factor Analysis, we again observed that we could reduce the feature set size to more than 40 % with minor performance degradations. However, all re-extraction methods incur an overhead of transformation of the test feature vector as each test vector has to be multiplied by a loading matrix. But both selection as well as re-extraction techniques gave significant reductions which can be used to improve the response time of the system.

Acknowledgments The authors gratefully acknowledge the funding support provided by DEITY, New Delhi to carry out this research work.

References

1. Liu, Q., Sung, A.H., Qiao, M.: Temporal Derivative-Based Spectrum and Mel-Cepstrum Audio Steganalysis. IEEE Trans. Inf. Forensics Secur. **4**(3), 359–368 (2009)
2. Sharma, S., Kumar, M., Das, P.K.: A technique for dimension reduction of MFCC spectral features for speech recognition. In: International Conference on Industrial Instrumentation and Control (ICIC), pp. 99–104 (2015)
3. Hu, H., Zahorian, S.A.: Dimensionality reduction methods for HMM phonetic recognition. In: IEEE International Conference on Acoustics Speech and Signal Processing (ICASSP), pp. 4854–4857 (2010)
4. Quan, C., Wan, D., Zhang, B., Ren, F.: Reduce the dimensions of emotional features by principal component analysis for speech emotion recognition. In: 2013 IEEE/SICE International Symposium on System Integration (SII), pp. 222–226 (2013)
5. Fodor, I.K.: A survey of dimension reduction techniques (2002)
6. Reynolds, D.A., Rose, R.C.: Robust text-independent speaker identification using Gaussian mixture speaker models. IEEE Trans. Speech Audio Process. **3**(1), 72–83 (1995)
7. Reynolds, D.A., Quatieri, T.F., Dunn, R.B.: Speaker verification using adapted Gaussian mixture models. Digit. Signal Proc. **10**(1), 19–41 (2000)
8. Kenny, P., Dumouchel, P.: Disentangling speaker and channel effects in speaker verification. In: IEEE International Conference on Acoustics, Speech, and Signal Processing. pp. I–37. IEEE (2004)
9. Dehak, N., Kenny, P., Dehak, R., Dumouchel, P., Ouellet, P.: Front-end factor analysis for speaker verification. IEEE Trans. Audio Speech Lang. Process. **19**(4), 788–798 (2011)
10. Hannun, A., Case, C., Casper, J., Catanzaro, B., Diamos, G., Elsen, E., Ng, A.Y., et al.: DeepSpeech: Scaling up end-to-end speech recognition. arXiv preprint arXiv:1412.5567 (2014)
11. Chow, D., Abdulla, W.H.: Speaker identification based on log area ratio and gaussian mixture models in narrow-band speech. In: PRICAI 2004: Trends in Artificial Intelligence. LNCS, vol. 3157, pp. 901–908. Springer Berlin Heidelberg (2004)
12. Paliwal, K.K.: Dimensionality reduction of the enhanced feature set for the HMM-based speech recognizer. Digit. Signal Proc. **2**, 157–173 (1992)
13. Sheela, K.A., Prasad, K.S.: Linear discriminant analysis F-Ratio for optimization of TESPAR & MFCC features for speaker recognition. J. Multimedia **2**(6), 34–43 (2007)
14. Suhr, D.D.: Principal component analysis vs. exploratory factor analysis. SUGI 30 Proceedings, pp. 203–230 (2005)
15. Young, S.: The HTK Book, 1st edn. Entropic Cambridge Research Laboratory, Cambridge (1997)

MFAST Processing Model for Occlusion and Illumination Invariant Facial Recognition

Kapil Juneja

Abstract Illumination Variation and wearable objects loses the partial facial information that it degrades the accuracy of recognition process. In this paper, a high performance driven accurate method is provided for facial recognition. The proposed MFAST (Multi-Featured Analog Signal Transformed) Model genuinely transmute the substantial facial information in analog featured conformation. This analog featured structured is formed using segmented featured elicitation. These features include center difference evaluation as moment, the asymmetric structure analysis as Skewness and Outlier Prone Measure as Kurtosis. These analogous features are shaped to justified form and generate a compound signal form. Mapping of these distillates signal points over facial dataset with specification of threshold window. The decomposed form recognition method enhanced the accuracy and performance. The experimentation on FERET, LFW and Indian Databases signify that the model outperformed than existing algorithms.

Keywords MFAST · Analog form · Segmented featured · Asymmetric structure

1 Introduction

The natural face image is taken in diverse environment condition which cause sundry illumination [1, 2] problem. Illumination affects the image entirely or moderately. Partial featured change is difficult to analyze and repair which increases the recognition time criticalities. Some illumination adjustment methods exist including gamma correction, Contrast adjustment approaches. But if the database image also suffers the illumination variations, it will be difficult and performance degrading approach to repair each DB image with each recognition process. Even the repair must be relative to the contrast observations of input image. Natural capturing images also face the problem of improper or discriminative [3–6] image

K. Juneja (✉)
307, Sector 14, Rohtak 124001, Haryana, India
e-mail: Kapil.juneja81@gmail.com; kapil.juneja.1981@ieee.org

© Springer Science+Business Media Singapore 2016
R.K. Choudhary et al. (eds.), *Advanced Computing and Communication Technologies*,
Advances in Intelligent Systems and Computing 452,
DOI 10.1007/978-981-10-1023-1_16

161

capturing. Such capturing results partial capturing [7], pose variation, facial information loss [8] etc. Subject specific variations and data loss requires a series of filter [9, 10] operation to repair image even then it will impact similarity score. The classified facial mapping process can be applied to represent the entire face recognition process. Based on the captured facial segments and features, the accuracy of the method depends. Different kind of deficiencies that can be identified in natural captured image are shown in Fig. 1. This paper has provided the solution against all these problems.

If the face image is captured in normalized environment and without any capturing mistake, even then it can hide some of the facial features. Such high level feature loss occurs, if the features are covered by some wearable object such as glasses, scarf, cap, etc. Pose change of the person, head position also affects the facial features. Such kind of feature variation and feature loss is not recoverable through any filter. Some of researchers provide some sparse methods with multiple instance based recognition [11, 12] process. The dynamic feature [13] selection and transformation methods are also applied to recognize [14] image under these abruptions.

In this paper, the image transformation model is provided based on multiple feature aspects. This method is significant to covert the image information to analog curve by applying the segmented feature derivation method. This analog form based frequency point mapping will be able to map the selected feature points over the facial image that will provide a solution to most of the captured image

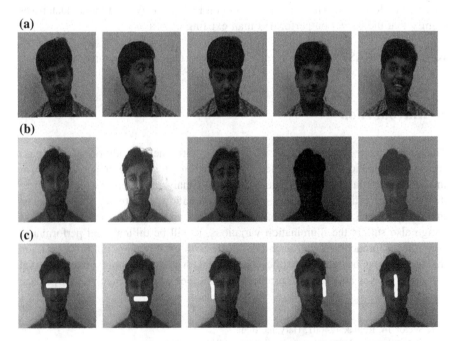

Fig. 1 Spotted deficiencies in natural images **a** head position deficiencies, **b** illumination variation deficiencies, **c** missing information

irregularities. In this section, all the deficiencies identified in facial image capturing and recognition process are discussed. In Sect. 2, the proposed work model is presented with a description of each of inclusive stage. In Sect. 3, the comparative experimentation is provided to evaluate the accuracy of this model.

2   MFAST Model

In this section, the proposed MFAST model is described to identify the solution of improper facial capturing and hidden partial facial regions. The model is robust against pose variation, illumination variation and wearable object problems. As discussed earlier, the model will transform each of DB images and input image in the form of analog feature curves. Instead of comparing high resolution face image, the analog curve obtained from MFAST model will be compared. To apply accurate recognition process, varying window ratio specific weight ratio method is applied to frequency points of analog curve. The section describes the integrated process stages.

2.1   MFAST: Feature Extraction

The natural image taken either as input or the DB image is in raw form, which is required to transform into analog featured form. The MFAST model for featured conversion to analog form is shown in Fig. 2. At first stage of this model, the

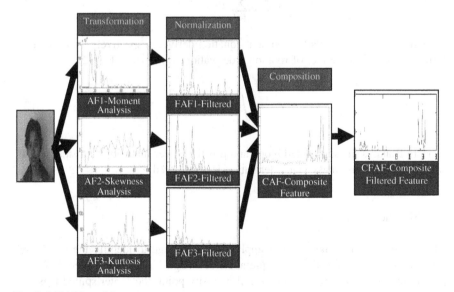

Fig. 2 MFAST model

transformation is applied to acquire the segmented features and represent as analog form. Normalization process is applied to analog curve to obtain key featured points. Finally the feature curves are combined in composing featured form. Second level filtration is applied to normalize the composite featured signal.

2.1.1 Transformation

To apply the transformation, the complete image is divided into smaller blocks of equal size. For each block, single feature points will be obtained with feature specification. The block size identifies the number of feature points which are presented as the feature curve. The experimentation signifies that the lower block size, improved the accuracy of recognition process.

AF1-Moment Analysis

AF (Analong Feature) Moment is quantified form of mass value of segmented region. This quantized vector can be described in different forms based on the order specification. Zeroth order moment itself represents mass value, 1st order moment is the center of mass; second order moment represents rotational inertia. In the same way, each order generates the structural and quantified features of the image. In this work, the mass based rotational inertia is considered as featured data. The statistical derivation of this feature is shown here below

$$M_k = \frac{1}{N} \sum_{i=1}^{N} (Blk_i - \mu)^k \tag{1}$$

Here Eq. (1) represents the moment equation derived respectively to mean. The formulation with the effect of rotational derivation is shown in Eq. (2)

$$M_k = \int_{-\infty}^{\infty} (Blk_i - \mu)^k \tag{2}$$

Here Blake is the segmented block and μ is the center intensity poi respective to which the rotational mass value derivation is considered.

AF2-Skewness Analysis

AF (Analog Feature) Skewness is applied on the image to identify the degree of asymmetric as the quantized interpretation of image block is considered. While presenting the probability distribution of intensity point, the center spitted legs are observed. Stretched left leg where represents the negatively skewed image and right stretched leg represents positively skewed. The quantized image block based skewness formulation is shown in Eq. (3)

$$Sk = \frac{\frac{1}{n}\sum_{i=1}^{n}(Blk_i - \mu)^3}{\left(\sqrt{\frac{1}{n}\sum_{i=1}^{N}(Blk_i - \mu)^2}\right)^3} \tag{3}$$

This equation represents mean specific coefficient derivation is considered for symmetric feature identification.

AF3-Kertosis Analysis

Third block segmented Analog Feature (AF) parameter taken here is Kurtosis which generates the outlier prone block evaluation. This evaluation is based on normal distribution applied to intensity regulation of image block. A comparative significant decision value considered is 3. The kurtosis value higher than 3 resembles higher outlier proneness in block, whereas the lower value signify lesser error proneness. The formulation of this feature vector is shown in Eq. (4)

$$Kr = \frac{E(Blk - \mu)^4}{\sigma^4} \tag{4}$$

Here E(x) is expected to quantify the image along with population pixel evaluation.

2.1.2 Normalization

Each of the individual evaluated block featured value is quantized and transformed into a curve form. This generated analog curve is irregular and requires to dissolve the inclusive impurities and disturbance. Band pass filtration along with dynamic thresholding is applied here for probabilistic featured evaluation. This feature vector independent normalization process is shown in Table 1.

Here Table 1 described the method to normalized analog featured curve. This model includes successive processes including low pass filtration and high pass filtration. The derivative coefficient signal is obtained using derivative evaluation, which successively following by squared transformation to obtain the absolute signal value considered as a processed signal form.

Table 1 AF normalization

Input :   AF : Featured transformed Analog Signal
              L : Block Length Specification
Process :
              n=Length(AF)
              AF(nL) = 2AF(nL - L) - AF(nL – 2L) + AF(nL)- 2AF(nL- 6L)+AF(nL- 12L)
              /*Low Pass Fileration*/
              AF(nL) = 64AF(nL – 32L) - [AF(nL - L)+ AF(nL) - AF(nL – 63L)]
              /*High Pass Fileration*/
              AF(nL) = (1/16L)[-AF(nL–2L)–2AF(nL - L) + 2AF(nL + L) +AF(nL+ 2L)].
              /*Derivative Evaluation*/
              AF(nL) = [AF(nL)]^2
              /*Absolute Signal Form Tranformation*/
              Th=Mean(AF)/3
              ResAF=AF>Th
              /*Thresholding*/
Output : ResAF

2.1.3 Composition

The final step of MFAST model is to combine these AF features curves and formulate a combined feature form which is implied in later stage for facial recognition. The composted curve formulation is shown in Eq. (5)

$$CAF = AF1 \,|AF2\,|AF3. \qquad (5)$$

The composed curve is finally applied by feature normalization phase to remove the aggregative impurities over the signal curve. This aggregative normalized curve is obtained by Eq. (6) which describes the algorithmic process of Table 1.

$$CFAF = \text{Normalization (CAF)}$$

This CFAF form feature is obtained from the input image and the database. The recognition process applied to composition is defined in the next subsection.

2.2 MFAST: Recognition

This composed CFAF curve is finally compared with each of DB images under curve point analysis. The matching is here identified in terms of ratio between mapped points and total curve points. If input and target CFAF curves are same, each point of the input CFAF curve will overlap the target CFAF curve. If the difference is in illumination values, then some of the curve points of input CFAF will be above or below the target curve with little difference. Applied threshold window based matching is able to perform recognition on such curves. If the input image is having hidden features or obscure features, then some of the points can exactly map, some can map with window difference and some completely differ points will not map under window formulation. To cover all aspects of recognition, the educational formulation of the recognition process is shown in Eq. (6) and its visualization is shown in Fig. 4 (Fig. 3).

$$RRatio = EM + FWM * 0.9 + SWM * 0.8 \qquad (6)$$

Here

EM   Exactly Matched Featured Ratio

FWM First Window (3 Points) Specified Matched Feature Ratio

SWM Second Window (5 Points) Specified Matched Feature Ratio

Here Fig. 4 shown the input CFAF mapping applied on each DB CFAF curve. Here blue line is DB image curve points and the red line represents the input image curve points. The result shows the clear disqualification of this mapping. The experimentation applied on different databases is shown in Sect. 3.

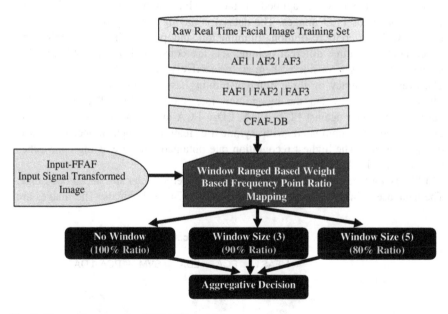

Fig. 3 Recognition process of MFAST model

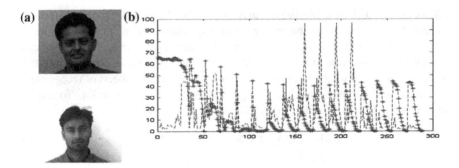

Fig. 4 MFAST recognition process **a** example of input and target images, **b** CFAF curve map on input and target image

3 Experimentation

MFAST Model is designed in this paper to improve the recognition accuracy against the illumination problem, pose variation, partial occlusion problems. To obtain the natural images the experimentation is applied on FERET (Face Recognition Technology), LFW (Labeled Faces in the Wild) and Indian Database. At first experimentation is applied on the FERET database which consist about 14,126 images of 1199 images. The dataset is having multi aged and mix gender images with different resolution, illumination and pose. While performing the experimentation, three different training sets are considered of 200, 500 and 700 images. The test set considered here are respectively of 100, 200 and 300 images. The average accuracy on three different training-testing pairs is shown in Fig. 5. MFAST model provided the recognition rate about 92.5 %. The comparative observations taken against PCA, LDA, LDA–PCA and SVM approaches. As the existing methods compared with symmetric features which reflected in low recognition rate. The highest recognition rate obtained among existing approaches is 80.1 % for PCA–LDA approach.

LFW is another online acquired database considered under size and scope range. The database is having about 13,233 images for 5749 individuals. The images are

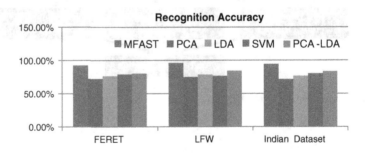

Fig. 5 Recognition accuracy analysis

captured in a natural environment and relative variables. All the real time problems and impurities are the part of the database. These highly diverse features based database is applied for accuracy improvement for MFAST model. To perform the analysis, two training and test set pairs are taken. The first set consists 200 training images and 50 test images, whereas the second considered dataset pair is wider with 500 training images and 150 test images. The proposed model has provided the evaluation respective to the structural and asymmetric formulation of facial image. Based on these constraint specific observations, the model has provided the average accuracy of 96 %. The comparative observations show that the existing models provided maximum upto 84 % average accuracy for the PCA–LDA method.

Final experimentation is applied to more complex Indian face database. The dataset is having images of 61 individuals with 11 instances of each. The variation in database images is in terms of illumination, head position, pose and expression. To apply experimentation, two training sets of 300 and 400 images is taken with 50 and 200 test images. MFAST model provided 94 % accuracy rate, which is much higher than the highest recognition rate of existing approaches as shown in Fig. 5.

4   Conclusion

The paper has presented an image to segmented analog featured transformation based MFAST model. The model applied a segmented featured derivation for three main vectors called moment analysis, skewness analysis and kurtosis feature analysis. These features are applied to obtain the symmetric observation, fault prone strength analysis and structural feature extraction. After extracting the features, these quantized features are transformed to analog curve form. A band pass filtration based constraint derivation is applied to normalize the feature curve. The composed CFAF feature is formed by collecting all three feature curves. Finally, feature points of input and target DB CFAF curves are mapped under window threshold specification. The experimentation is here implied for three FERET, LFW and Indian Face Databases. The comparative observations obtained against PCA, SVM, LDA and PCA–LDA methods show that the model has enhanced the recognition accuracy because of asymmetric feature based evaluation. The average accuracy against multiple variations provided by the model is up to 96 %, whereas the existing methods provided the accuracy maximum upto 83 %.

References

1. Lee, M., Park, C.H.: An efficient image normalization method for face recognition under varying illuminations. In: Proceedings of the 1st ACM International Conference on Multimedia Information Retrieval, pp. 128–133 (2008)
2. Kim, W., Suh, S., Hwang, W., Han, J.-J.: SVD face: illumination-invariant face representation. IEEE Signal Process. Lett. **21**(11), 1336–1340 (2014)

3. Han, Z., Fang, C., Ding, X.: A discriminated correlation classifier for face recognition. In: Proceedings of the 2010 ACM Symposium on Applied Computing, pp. 1485–1490 (2010)
4. Xu, Y., Fang, X., Yang, J., You, J., Liu, H., Teng, S.: Data uncertainty in face recognition. IEEE Trans. Cybern. **44**(10), 1950–1961 (2014)
5. Hu, H.: Face recognition with image sets using locally grassmannian discriminant analysis. IEEE Trans. Circuits Syst. Video Technol. **24**(9), 1461–1474 (2014)
6. Wan, J., Yang, M., Gao, Y., Chen, Y.: Pairwise costs in semisupervised discriminant analysis for face recognition. IEEE Trans. Inf. Forensics Secur. **9**(10), 1569–1580 (2014)
7. Kekre, H.B., Sarode, T.K., Natu, P.J., Natu, S.J.: Transform based face recognition with partial and full feature vector using DCT and walsh transform. In: International Conference and Workshop on Emerging Trends in Technology, pp. 1295–1300 (2011)
8. De Marsico, M., Nappi, M., Riccio, D.: Measuring sample distortions in face recognition. In: Proceedings of the 2nd ACM Workshop on Multimedia in Forensics, Security and Intelligence, pp. 83–88 (2010)
9. Rong, J., Gritti, T., Shan, C.: Upscaling faces for recognition systems using trained filters. In: Proceedings of the 1st International Workshop on Interactive Multimedia for Consumer Electronics, pp. 105–112 (2009)
10. Divya, R., Rath, A., Manikantan, K., Ramachandran, S.: Astroid shaped DCT feature extraction for enhanced face recognition. In: Proceedings of the CUBE International Information Technology Conference, pp. 95–101 (2012)
11. Wang, J., Lu, C., Wang, M., Yan, S., Hu, X.: Robust face recognition via adaptive sparse representation. IEEE Trans. Cybern. **44**(12), 2368–2378 (2014)
12. Xu, Y., Li, X., Yang, J., Lai, Z., Zhang, D.: Integrating conventional and inverse representation for face recognition. IEEE Trans. Cybern. **44**(10), 1738–1746 (2014)
13. Girgensohn, A., Adcock, J., Wilcox, L.: Leveraging face recognition technology to find and organize photos. In: Proceedings of the 6th ACM SIGMM International Workshop on Multimedia Information Retrieval, pp. 99–106 (2004)
14. Hu, S., Maschal, R., Young, S.S., Hong, T.H., Phillips, J.P.: Performance assessment of face recognition using super-resolution. In: Proceedings of the 10th Performance Metrics for Intelligent Systems Workshop, pp. 195–200 (2010)

Fuzzy Soft Set Theory and Its Application in Group Decision Making

T.R. Sooraj, R.K. Mohanty and B.K. Tripathy

Abstract Soft set theory was introduced by Molodtsov to handle uncertainty. It uses a family of subsets associated with each parameter. Hybrid models have been found to be more useful than the individual components. Earlier fuzzy set and soft set were combined to form fuzzy soft sets (FSS). Soft sets were defined from a different point of view in Tripathy et al. (Int J Reasoning-Based Intell Syst 7(3/4), 224–253, 2015) where they used the notion of characteristic functions. Hence, many related concepts were also redefined. In Tripathy et al. (Proceedings of ICCIDM-2015, 2015) membership function for FSSs was defined. We propose a new algorithm by following this approach which provides an application of FSSs in group decision making. The performance of this algorithm is substantially improved than that of the earlier algorithm.

Keywords Soft sets · Fuzzy sets · Fuzzy soft sets · Group decision making

1 Introduction

The Fuzzy set introduced by Zadeh [1] in 1965 has been found to be a better model of uncertainty and has been extensively used in real life applications. In order to bring topological flavour into the models of uncertainty and associate family of subsets of a universe to parameters, soft sets were introduced by Molodtsov [2] in 1999. The study on soft sets was carried forward by Maji et al. [3, 4]. As mentioned

T.R. Sooraj (✉) · R.K. Mohanty · B.K. Tripathy
School of Computing Science, VIT University, Vellore, Tamil Nadu, India
e-mail: soorajtr19@gmail.com

R.K. Mohanty
e-mail: rknmohanty@gmail.com

B.K. Tripathy
e-mail: tripathybk@vit.ac.in

© Springer Science+Business Media Singapore 2016   171
R.K. Choudhary et al. (eds.), *Advanced Computing and Communication Technologies*,
Advances in Intelligent Systems and Computing 452,
DOI 10.1007/978-981-10-1023-1_17

in the abstract, hybrid models obtained by suitably combining individual models of uncertainty have been found to be more efficient than their components. Several such hybrid models exist in the literature. Following this trend Maji et al. [5] put forward the concept of FSSs will systematize many operations defined upon them as done in [5]. Extending this approach further, we introduced the membership functions for FSS in [6]. Maji et al. discussed an application of soft sets in decision making problems [3]. Some applications of various hybrid soft set models are discussed in [1, 7–12]. This study was further extended to the context of FSSs [6] where they identified some drawbacks in [3] and took care of these drawbacks while introducing an algorithm for decision making [6]. In this paper, we have carried this study further by using FSS in handling the problem of multi-criteria group decision making.

2   Definitions and Notions

A soft universe (U, E) is a combination of a universe U and a set of parameters E

Definition 2.1 (*Soft Set*) We denote a soft set over (U, E) by (F, E), where

$$F : E \rightarrow P(U) \tag{2.1}$$

Here, P(U) denotes the power set of U.

Definition 2.2 (*Fuzzy soft set*) We denote a FSS over (U, E) by (F, E) where

$$F : E \rightarrow I(U) \tag{2.2}$$

3   Fuzzy Soft Sets (FSS)

Here, we discuss some definitions and operations of FSSs. Let (F, E) be a FSS. In [13] the set of parametric membership functions was defined as $\mu_{(F,E)} = \left\{ \mu_{(F,E)}^a | a \in E \right\}$ of (F, E).

Definition 3.1 For any $\forall a \in E$, the membership function is defined as follows.

$$\mu_{(F,E)}^a(x) = \alpha, \alpha \in [0, 1] \tag{3.1}$$

For any two FSSs (F, E) and (G, E) the following operations are defined.

Definition 3.2 $\forall a \in E$ and $\forall x \in U$, the union of (F, E) and (G, E) is the fuzzy soft set (H, E), is given by

$$\mu^a_{(H,E)}(x) = \max\left\{\mu^a_{(F,E)}(x), \mu^a_{(G,E)}(x)\right\} \qquad (3.2)$$

Definition 3.3 $\forall a \in E$ and $\forall x \in U$, the intersection of (F, E) and (G, E) is the FSS (H, E), is given by

$$\mu^a_{(H,E)}(x) = \min\left\{\mu^a_{(F,E)}(x), \mu^a_{(G,E)}(x)\right\} \qquad (3.3)$$

Definition 3.4 Given (F, E) is said to be fuzzy soft subset of (G, E), $(F, E) \subseteq (G, E)$ and $\forall a \in E$ and $\forall x \in U$,

$$\mu^a_{(F,E)}(x) \le \mu^a_{(G,E)}(x) \qquad (3.4)$$

Definition 3.5 (F, E) is said to be equal to (G, E) written as $(F, E) = (G, E)$ if $\forall x \in U$,

$$\mu^a_{(F,E)}(x) = \mu^a_{(G,E)}(x) \qquad (3.5)$$

Definition 3.6 The complement (H, E) of (G, E) in (F, E) is defined $\forall a \in E$ and $\forall x \in U$.

$$\mu^a_{(H,E)}(x) = \max\left\{0, \mu^a_{(F,E)}(x) - \mu^a_{(G,E)}(x)\right\} \qquad (3.6)$$

4 Application of FSS in Group Decision Making

Several applications of soft sets theory are given in [2]. In [5] Maji et al. provided an application of FSSs in a decision making system. But the algorithm given in that paper has some issues and those issues are discussed in [6]. Tripathy et al. rectified the issues and provided suitable solution for the problems addressed in [3] and also introduced the concept of negative and positive parameters in [6].

Most of the real-life problems cannot be effectively resolved by a single decision maker. Depends on the uncertainty and the amount of knowledge available, it is not easy to take a suitable decision for a single decision maker. So, it is needed to gather multiple decision makers with different knowledge structures and experience to conduct a group decision making (GDM). Here we discuss an application of group decision making in FSSs.

Algorithm

1. Input the priority given by the panel $(J_1, J_2, J_3, \ldots J_n)$ for each parameter, where 'n' is the number of judges.
2. For each judge J_i $(i = 1, 2, 3, \ldots, n)$ repeat the following steps.

a. Input the fuzzy soft set (F, E) provided by Judge J_i and arranges it in tabular form.

b. Construct the priority table (PT). This table can be obtained by multiplying priority values with the corresponding parameter values. Also, calculate the row-sum of each row in the PT.

c. Construct comparison tables (CT). This can be achieved by finding the entries as differences of each row sum in PTs with those of all other rows.

d. Find the row sum for each row in the CT to obtain the score.

e. Construct the decision table by taking the row sums in the CT. Assign rankings to each candidate based upon the row sum obtained.

3. 3. Create a rank table based on the results obtained from the above step which contains rankings provided by all the judges.

4. 4. Calculate the row-sum of each candidate in the rank table to find the rank-sum of each candidate. The candidate with lesser row sum value is the best choice. If more than one candidate is having the same rank-sum, then the candidate having higher value in highest absolute priority column will be selected. This process is continued till final ranking list is obtained.

Assume that 'n' candidates are applying for a job in an organization. From these n candidates, the organization filters out many candidates based on some criteria (For e.g.: Those who got more than 60 % marks are eligible to attend the interview). The candidates, who passed the elimination criteria, will be eligible to attend the interview. The interview performance of each selected candidate is analyzed by a panel of different judges. Here, the panel assigns some parameters to evaluate the performance of each candidate. Some parameters are communication skills, personality, reactivity etc.

Let U be a set of candidates $\{c_1, c_2, c_3, c_4, c_5, c_6\}$. The parameter set E be {knowledge, communication, reaction, presentation, extracurricular activities}. Consider a FSS (U, E) describing the 'performance of candidates'. Consider J_1, J_2 and J_3 are the judges who analyze the performance of the candidates and each judge is assigning a rank to each candidate according to his/her performance.

The panel of judges assigns priority values to the parameters and based upon the impact of the parameters, they assign rankings to each parameter. This is shown in the following Table 1. The parameters knowledge, communication, behaviour, presentation, extracurricular activities are represented by e_1, e_2, e_3, e_4 and e_5.

The parameter values assigned by each judge to the candidate depend upon the performance of the candidate in the interview. The FSS for the candidates from the judge J_1 is shown in Table 2.

The priority for parameters e_1, e_2, e_3, e_4 and e_5 is given by the judge panel as 0.4, 0.3, −0.15, 0.05 and 0.1. Here, the parameter 'e_3' is a negative parameter.

Table 1 Priority rank table

	e_1	e_2	e_3	e_4	e_5
Priority	0.4	0.3	−0.15	0.05	0.1
Parameter rank	1	2	3	5	4

Table 2 FSS (F, E) by judge J_1

U	e_1	e_2	e_3	e_4	e_5
c_1	0.2	0.3	0.8	0.8	0.6
c_2	0.4	0.6	0.2	1	0.5
c_3	0.8	0.9	0.7	0.9	0.7
c_4	0.8	0.9	0.8	0.9	0.7
c_5	0.4	0.9	0.6	0.1	0.8
c_6	0.9	1	0.3	0.2	0.3

Table 3 Priority table for J_1

U	e_1	e_2	e_3	e_4	e_5	Row-sum
c_1	0.08	0.09	−0.12	0.04	0.06	0.15
c_2	0.16	0.18	−0.3	0.05	0.05	0.14
c_3	0.32	0.27	−0.105	0.045	0.07	0.6
c_4	0.32	0.27	−0.12	0.045	0.07	0.585
c_5	0.16	0.27	−0.09	0.005	0.08	0.425
c_6	0.36	0.3	−0.045	0.01	0.03	0.655

Parameters are classified into positive parameter and negative parameter. We use the notion of negative parameter as in [6]. The priority table is as follows (Table 3).

The comparison table obtained for the candidates by the judge J_1 is obtained is shown in the Table 4.

Comparison table (CT) shows the ranking of each candidate by the judge J_2. Here, the candidate c_6 is the best choice. Since the selection of the best candidate is governed by a panel of 3 members, we cannot take this as the best choice. So, we have to find the comparison table of the judges J_2 and J_3 to decide the optimum choice. Representation of FSS of candidates by judge J_2 is shown below (Table 5).

After applying the algorithm in the above FSS, we will get the comparison table as shown in the Table 6.

The FSS of candidates by Judge J_3 is given as follows (Table 7).

After applying the algorithm in the above FSS, we will get the comparison table for the judge J_3 as follows (Table 8).

Table 4 Comparison table

U	c_1	c_2	c_3	c_4	c_5	c_6	Score	Rank
c_1	0	0.01	−0.45	−0.435	−0.275	−0.505	−1.655	5
c_2	−0.01	0	−0.46	−0.445	−0.285	−0.515	−1.715	6
c_3	0.45	0.46	0	0.015	0.175	−0.055	1.045	2
c_4	0.435	0.445	−0.015	0	0.16	−0.07	0.955	3
c_5	0.275	0.285	−0.175	−0.16	0	−0.23	−0.005	4
c_6	0.505	0.515	0.055	0.07	0.23	0	1.375	1

Table 5 FSS (F, E) by judge J_2

U	e_1	e_2	e_3	e_4	e_5
c_1	0.3	0.2	0.6	0.7	0.7
c_2	0.5	0.5	0	0.9	0.6
c_3	0.9	0.8	0.5	0.8	0.8
c_4	0.9	0.8	0.6	0.8	0.8
c_5	0.5	0.8	0.4	0	0.9
c_6	1	0.9	0.1	0.1	0.4

Table 6 Comparison table of judge J_2

c_i	c_j							
	c_1	c_2	c_3	c_4	c_5	c_6	Score	Rank
c_1	0	−0.26	−0.45	−0.435	−0.275	−0.505	−1.925	6
c_2	0.26	0	−0.19	−0.175	−0.015	−0.245	−0.365	5
c_3	0.45	0.19	0	0.015	0.175	−0.055	0.775	2
c_4	0.435	0.175	−0.015	0	0.16	−0.07	0.685	3
c_5	0.275	0.015	−0.175	−0.16	0	−0.23	−0.275	4
c_6	0.505	0.245	0.055	0.07	0.23	0	1.105	1

Table 7 FSS (F, E) by judge J_3

U	e_1	e_2	e_3	e_4	e_5
c_1	0.5	0.4	0.8	0.9	0.9
c_2	0.7	0.7	0.2	1	0.8
c_3	1	1	1	1	1
c_4	1	1	0.8	1	1
c_5	0.7	1	0.6	0.2	1
c_6	1	1	0.3	0.3	0.6

Table 8 Comparison table for judge J_3

c_i	c_j							
	c_1	c_2	c_3	c_4	c_5	c_6	Score	Rank
c_1	0	−0.255	−0.365	−0.395	−0.265	−0.395	−1.675	6
c_2	0.255	0	−0.11	−0.14	−0.01	−0.14	−0.145	5
c_3	0.365	0.11	0	−0.03	0.1	−0.03	0.515	3
c_4	0.395	0.14	0.03	0	0.13	0	0.695	2
c_5	0.265	0.01	−0.1	−0.13	0	−0.13	−0.085	4
c_6	0.395	0.14	0.03	0	0.13	0	0.695	1

Table 9 Rank table

	J_1	J_2	J_3	Rank-sum	Final-rank
c_1	5	6	6	17	6
c_2	6	5	5	16	5
c_3	2	2	3	7	2
c_4	3	3	2	8	3
c_5	4	4	4	12	4
c_6	1	1	1	3	1

Rank of all candidates given by judges J_1, J_2 and J_3 are shown in the rank table (Table 9). From this rank, we can find the final rank of the candidates.

From the above table, we can see that the panel has selected the candidate c_6 as the best choice.

5 Conclusions

The definition of soft set using the characteristic function approach was provided in [12], which besides being able to take care of several definitions of operations on soft sets could make the proofs of properties very elegant. Earlier FSSs were used for decision making in [5]. Some flaws in the approach were pointed out in [6] and rectifications were made. Due to the lack of information and uncertainty in real life scenarios, a single decision maker cannot able to take proper decision. So, a new algorithm is introduced in this work with respect to decision making by a group of decision makers.

References

1. Zadeh, L.A.: Fuzzy sets. Inf. Control **8**, 338–353 (1965)
2. Molodtsov, D.: Soft set theory—first results. Comput. Math Appl. **37**, 19–31 (1999)
3. Maji, P.K., Biswas, R., Roy, A.R.: An application of soft sets in a decision making problem. Comput. Math Appl. **44**, 1007–1083 (2002)
4. Maji, P.K., Biswas, R., Roy, A.R.: Soft set theory. Comput. Math Appl. **45**, 555–562 (2003)
5. Maji, P.K., Biswas, R., Roy, A.R.: Fuzzy Soft Sets. J. Fuzzy Math. **9**(3), 589–602 (2001)
6. Tripathy, B.K., Sooraj, T.R, Mohanty, R.K.: A new approach to fuzzy soft set and its application in decision making. In: Proceedings of ICCIDM 2015, Dec 5–6, Bhubaneswar
7. Tripathy, B.K., Sooraj, T.R, Mohanty, R.K.: A new approach to interval-valued fuzzy soft sets and its application in decision making. Accepted in ICCI-2015, Ranchi
8. Tripathy, B.K., Sooraj, T.R., Mohanty, R.K.: A new approach to interval-valued fuzzy soft sets and its application in group decision making. Accepted in CDCS-2015, Kochi
9. Tripathy, B.K., Mohanty, R.K., Sooraj, T.R.: On intuitionistic fuzzy soft sets and their application in decision making. Accepted in ICSNCS-2016, New Delhi
10. Tripathy, B.K., Mohanty, R.K., Sooraj, T.R.: On intuitionistic fuzzy soft set and its application in group decision making. Accepted for presentation ICETETS-2016, Thanjavur

11. Tripathy, B.K., Mohanty, R.K., Sooraj, T.R., Tripathy, A.: A new approach to intuitionistic fuzzy soft set theory and its application in group decision making. Presented at ICTIS-2015, Ahmedabad, Springer publications, (2015)
12. Tripathy, B.K., Arun, K.R.: A new approach to soft sets, soft multisets and their properties. Int. J. Reasoning-Based Intell. Syst. 7(3/4), 244–253 (2015)
13. Tripathy, B.K., Mohanty, R.K., Sooraj, T.R., Arun, K.R.: A new approach to intuitionistic fuzzy soft sets and its application in decision making. In: proceedings of ICICT-2015, Oct 9–10, Udaipur, (2015)

Discriminating Confident and Non-confidant Speaker Based on Acoustic Speech Features

Shrikant Malviya, Vivek Kr. Singh and Jigyasa Umrao

Abstract Nowadays affective computing is one of the most interesting and challenging research areas among the human computer interaction (HCI) researchers. One of the potential applications of emotion detection is to analyze confidence level of a speaker. In human-computer or human-human interaction systems, speech based confidence level check can provide users with improved services. Confidence level checking would be useful in various applications e.g. job recruitment process. In this paper a solution to similar kind of applications is proposed to differentiate between a confident person and non-confident person. A set of speech features have been selected empirically to support the concept that speech conveys not only the linguistic messages, but also emotional content. A protruding result has been sought on using SVM and KNN classifiers in determining confidence level of a human. The accuracy and efficiency of result is found good and reliable.

Keywords Human-centered computing · Speech analysis · Feature extraction · Feature analysis · Affective computing

1 Introduction

With the advancement in the field of Human Computer Interaction, various solutions have been proposed to automate the communication between human and machine. Among various researches in this field, speech based HCI solutions are quite

S. Malviya (✉) · V.Kr.Singh · J. Umrao
Department of Information Technology, Indian Institute of Information
Technology Allahabad, Allahabad, India
e-mail: shrikant.iet6153@gmail.com

V.Kr.Singh
e-mail: vivekkr.singh@gmail.com

J. Umrao
e-mail: jigyasaumrao@gmail.com

© Springer Science+Business Media Singapore 2016   179
R.K. Choudhary et al. (eds.), *Advanced Computing and Communication Technologies*,
Advances in Intelligent Systems and Computing 452,
DOI 10.1007/978-981-10-1023-1_18

popular nowadays. In present scenario speech is being used widely in biometric devices. It is obvious that every human has some distinct voice features in his voice, based on which, applications perspective to human identification can be developed [1]. Another application of speech is to give a voice for a written material which would be beneficial for the blind people in their daily life. In such a magnanimous research field, lots of researchers are trying to get insights of affective computing through speech mode [1, 2].

In presented work, an analysis is done to distinguish between a confident and non-confident person on the basis of their speech. This may help to find out confident candidates from a group of people involved in debate during interview selection process. For a string of phonemes, different sound signal is generated for different people due to differences in dialect and vocal tract length and shape based on reader's emotion [3–5]. This phenomenon is the key point to discriminate between the confident and non-confident people. This discrimination technique could be found very useful in education system e.g. in automatic grading of students based on their presentations on the subject topic. Alternatively, automatic grades could be awarded automatically to the students based on the performance in telephonic interview and group discussion as required for company's recruitment. In case of any such applications, a large database has to be maintained. And analysis of each recorded data manually is a tedious and tiring task. Therefore, there is an urgent need to provide some solution to extract confident people among a large database in order to save man power and time. Hence the objective of current research is confidentiality check of a person based on speech [6, 7]. Here, we are proposing a technique which is in its pilot stage.

No direct researches have been found in the field of detecting confidence level of a speaker. However, emotion based features can be best utilized to obtain the desired objective. Ververidis et al. [2] has proposed emotional states, language, speakers and kind of speech that were addressed and also presented all acoustic features which affects the emotion such as pitch, formants, the vocal tract cross-section area, the Mel-frequency cepstral coefficient, speech rate, intensity of speech signal, Teager energy operator-based features on the dataset of emotional speech. El Ayadi et al. [1] has compared various emotion detection features of speech signal. They discussed important issues in the design of emotional speech databases and possible techniques to select as well as extract the suitable features from the speech.

Based on the above study, features are chosen on the basis of experiments done for confidentiality check. Several processes such feature selection, labeling and classification are discussed briefly in further sections. Section 2 gives a detailed description about the proposed approach with providing proper reasoning of in the selection of feature. Further, the next Sect. 3, a detailed comparative description of SVM and KNN has given to analyze the obtained result. Finally conclusive remarks are pointed to summarize the work done and future possibilities.

2 Proposed Approach

To differentiate personalities on the basis of confidence level through voice analysis, several speech features have to be selected empirically by observing its effect on human spoken system. It could easily be observed that a non-confident people used to stammer while delivering a speech. This stammering voice could be identified by the number of pauses, repetitions of words, filled pauses and speech rate. Uniform number of pauses in the entire speech is required for listeners to understand the spoken utterances. If the number of pauses is more frequent and if they are non-uniformly distributed, speaker could be categorized in non-confident class. Similarly, various features like short term energy (STE), speech rate, articulation rate, mean pause duration (MPD), mean syllable duration (MSD), phonetic time ratio (PTR) might be used to classify a speaker as a confident or non-confident speaker [6, 7].

Further in this section, a discussion is given for the selection of proper speech features suitable for distinguishing confident and non-confident spoken utterances. Some features suitable in detecting emotions could also be used for confidentially check. For example a person in stress emotion and the same person in excited emotion would express the two different states of confidence. Excited emotion exhibits properties more towards confident people whereas stress emotion more related to non-confident people. Emotions like stress, hesitation and nervousness are useful in the identification of non-confident speakers similarly emotions like calm, happy and neutral will be suitable for the confident speakers. Acoustic features e.g. prosodic features have given more preference than other linguistic features e.g. syllables, diphthong. Work is divided into two parts. First, is the selection of suitable affective features and second, is to establish robust mathematical classification model.

The architecture of proposed system is shown in Fig. 1. Firstly, the audio files are collected in.wav format. In pre-processing, audio files are made noise free and clean by using BRIL noise reduction algorithm [6]. Suitable features are extracted from these audio files as discussed in Sect. 2.3. Further, the extracted feature vector is labeled into confident or non-confident class. Some classifiers are applied on these labeled data to retrieve the result. A detailed study is given in further sections.

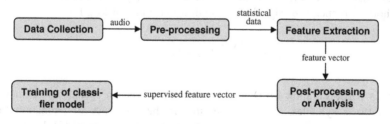

Fig. 1 Architecture of the proposed approach

2.1 Data Collection

Proposed approach has been applied on two different datasets. One obtained from LDC (Linguistic Data Consortium) [8], a dataset of emotional speech. Second dataset is locally developed through performing some vivas, presentations and interviews. A comparison of results on both has shown in the result and discussion section. 116 emotion wave files have considered. Files belong to anxious and panic emotions, assigned to non-confident class likewise neutral and happiness data to confident class. Hence finally 58 wave file of confident and 58 wave file of non-confident class are collected. 76 data files have been taken for training and 40 files for testing in equal partition of groups. The proposed approach is also checked on a locally developed dataset of 10–12 persons. Obtained speech wav files are labeled manually into confident and non-confident categories. In total, 74 wav files are selected divided equally in confident and non-confident class. 56 files, used to train the model, which comprised of 28 files of confident and 28 files of non-confident speeches. Rest 18 files which are used for testing are also equally divided in confident and non-confident speeches.

2.2 Pre-processing

Speech produced from human vocal tract system is time varying and glottal value remain same for 10–30 ms. Hence, short time analysis seems suitable to process the speech signal. Short term processing is helpful to see the changes in its features e.g. short-term energy, short term Zero-crossing rate. Speech is divided in the sequence of equal size frames (10–30 ms) to get the required discerning speech characteristics.

2.3 Feature Selection and Extraction

This section elicits the speech feature selection criteria about their relevance in the classification of confident and non-confident speakers. Short term energy (E) shows time varying property of speech signal [6]. High variation of energy throughout the speech characterizes the non-confident nature of the speaker. Formula for finding the short term energy of signal x of length n via frame of length m, is:

$$E = \sum_{n=-\infty}^{\infty} (x[n]w[m-n])^2 \qquad (1)$$

Mean pause duration MPD is a time duration when a speaker is silent with respect to its syllables. Total silence time is the difference of phonation time duration of speech signal and total time duration of speech signal [7]. It would be higher for non-confident rather than the confident. It can be assumed that a non-confident person's total pause duration would be more than confident speaker. Method of calculation is:

$$MPD = \frac{Total\ speech\ time - phonation\ time}{all\ possible\ no\ of\ syllables} \qquad (2)$$

Speech rate is defined as the ratio of number of syllables and total speech duration time [9]. Higher speech rate indicates more syllables had occurred in respective time duration which represents that the related speaker is more confident than the speaker having less speech rate.

$$speech\ rate = \frac{total\ no\ of\ syllables}{total\ time\ duration} \qquad (3)$$

Articulation Rate is a prosodic feature of speech which is similar to speech rate in which all silent portions had been removed from the speech signal [10, 11]. This also works in same way as speech rate works because for an expert person duration of silence will be less and phonation time will be more and also total no of syllable will be more. Therefore, the articulation rate of expert speaker will be more as compared to non-expert.

$$Articulation\ rate = \frac{total\ possible\ no\ of\ syllables}{phonetic\ time\ duration} \qquad (4)$$

Phonetic Time ratio (PTR) is ratio of the time between phonation time and total time. This will give higher value for expert speaker because number of syllables and phonation time will be higher over here. The period of silence will be less in case of expert so maximum total time will be phonation time. Maximum value of ratio can be 1.

2.4 Classification and Decision

Speech signal are processed frame by frame. Standard size of frames should be between 20–40 ms (millisecond). After feature calculation, each feature is passed through classification models. 65 % of dataset is used to train the classification models. Hence, a supervised classification method is used which during its training it takes the known class data. Artificial Neural Network and Support Vector Machine classifiers are used for said purpose. A comparative analysis of result on the basis of these two classifiers is reported in next section.

3 Result and Analysis

Performance of the model has been evaluated by sensitivity and specificity value which are calculated on the basis of true positive (TP), true negative (TN), false positive (FP), false negative (FN).

Sensitivity and Specificity is calculated by:

$$Sensitivity = \frac{TP}{TP + FN} \quad Specificity = \frac{TN}{TN + FP}$$

3.1 Result of LDC Dataset

LDC dataset is labeled manually into confident and non-confident. 20 wav files of confident speakers and 20 wav files of non-confident speakers are chosen. After applying KNN classifier, 17 confident files are classified positively whereas 3 miss classified. For non-confident wav files, system shows 100 % accurate classification. SVM is shows comparatively poor results as shown in Table 1 (Fig. 2).

KNN classifier computes an approximate 92.5 % accuracy for the confident and non-confident class based on the statistics of speech rate, articulation rate, mean pause duration, phonation ratio. In the current two class problem the Support Vector Classifier (SVM) is showing the classification accuracy of 75 % on same number of features as used with KNN approach. In clean data result of K-nearest neighbor classifier is good as compare to support vector machine due to definite margin and linear boundary.

Table 1 Confusion matrix of KNN and SVM on LDC dataset

Class	Total data	KNN		SVM	
		Classified	Miss-classified	Classified	Miss-classified
Confident	20	17	3	16	4
Non-confident	20	20	0	14	6

Fig. 2 Comparison of KNN and SVM on LDC dataset

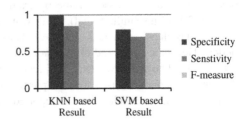

Table 2 Confusion matrix of KNN and SVM on manual dataset

Class	Total Data	KNN		SVM	
		Classified	Miss-classified	Classified	Miss-classified
Confident	9	8	1	7	2
Non-confident	9	7	2	6	3

Fig. 3 Comparison of KNN and SVM on our dataset

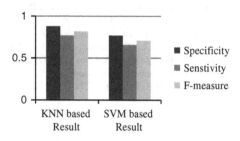

Result of manually created dataset

A data is collected locally during viva, presentation of students. The collected sound wav files are labeled as confident and non-confident manually. After applying both the classifiers KNN and SVM, it is found that KNN is showing good result as shown in Table 2 and Fig. 3.

4 Conclusion

In this paper, a new application area of speech has been scratched. Some experiments, using the KNN and SVM classifier, have been done for the identification of spoken utterances as confident or non-confident using feature e.g. speech rate, syllable count, total no of pause, total time duration and phonation time. The proposed approach has considered acoustic speech features to extract the statistical information. Further these statistical features passed to trained classification model. We got result as 91 and 83 % in K-nearest neighbor and 75 and 72 % in support vector machine on LDC (Standard Speech Dataset on emotion) and manually created dataset respectively.

References

1. El Ayadi, M., Kamel, M.S., Karray, F.: Survey on speech emotion recognition: features, classification schemes, and databases. Pattern Recognit. **44.3**, 572–587 (2011)
2. Ververidis, D., Kotropoulos, C.: Emotional speech recognition: resources, features, and methods. Speech Commun. **48**(9), 1162–1181 (2006)

3. Picard, R.W.: Affective computing: challenges. Int. J. Hum. Comput. Stud. **59**(1), 55–64 (2003)
4. Tao, J., Tan, T.: Affective Information Processing. Springer, London (2009)
5. Chen, L., et al.: Speech emotion recognition: features and classification models. Digital Signal Process. **22.6**, 1154–1160 (2012)
6. Rabiner, L.R., Schafer, R.W.: Digital processing of speech signals. Prentice Hall (1978)
7. Rabiner, L., Biing-Hwang, J.: Fundamentals of speech recognition (1993)
8. Liberman, M., et al.: Emotional prosody speech and transcripts LDC2002S28. Web Download. Linguistic Data Consortium, Philadelphia (2002)
9. Cairns, D.A., Hansen, J.H.L.: Nonlinear analysis and classification of speech under stressed conditions. J. Acoust. Soc. America **96**(6), 3392–3400 (1994)
10. Arrabothu, A.R., Chennupati, N., Yegnanarayana, B.: Syllable nuclei detection using perceptually significant features. In: INTERSPEECH, pp. 963–967 (2013)
11. Davis, S.B.: Acoustic characteristics of normal and pathological voices. Speech Language: Adv. Basic Res. Practice **1**, 271–335 (1979)

Multi-deterministic Prioritization of Regression Test Suite Compared: ACO and BCO

Shweta Singhal, Shivangi Gupta, Bharti Suri and Supriya Panda

Abstract. Regression Test Suite Prioritization has become a very prominent area of research in software engineering due to the advancements in the field of technology. Software development budget generally keeps very little room for the software maintenance phase. Hence instead of developing new test cases for any version of the software, it is intelligent to prioritize the available test suite to check the correctness of the available code. Researchers have come across many actual natural systems that are remarkable examples of solving any problem efficiently. In this paper we have compared the work of two nature inspired systems: Ant Colony Optimization (ACO), Bee Colony Optimization (BCO). The comparison has been analyzed using eight examples used to solve the regression test prioritization problem. The effectiveness of the two techniques discussed here have been compared using several metrics namely Average Efficiency (AE) and Average Percentage of Test Suite Size Reduction (ASR), Percent Average Execution Time Reduction (AETR).

Keywords. Regression testing · Ant colony optimization (ACO) · Bee colony optimization (BCO) · Average efficiency (AE) · Average percentage of test suite size reduction (ASR) · Percent average execution time reduction (AETR)

S. Singhal (✉) · B. Suri
University School of Information and Communication Technology,
G.G.S. Indraprastha University, Dwarka, New Delhi, India
e-mail: miss.shweta.singhal@gmail.com

B. Suri
e-mail: bhartisuri@gmail.com

S. Gupta · S. Panda
The Northcap University, Gurgaon, Haryana, India
e-mail: goyal_shivangi@yahoo.co.in

S. Panda
e-mail: supriyappanda@ncuindia.edu

© Springer Science+Business Media Singapore 2016
R.K. Choudhary et al. (eds.), *Advanced Computing and Communication Technologies*,
Advances in Intelligent Systems and Computing 452,
DOI 10.1007/978-981-10-1023-1_19

1 Introduction

Any project, be it commercial or professional, hardware or software, civil or mechanical has certain budget allocated to it in terms of finance, time, man power etc. Out of these sanctioned budget factors, the majority of the portion (80–90 %) is spent in its development that is from planning to delivery of the product and a small portion (10–20 %) is left for the maintenance phase. On the contrary, the Project Development phase lasts negligibly for a shorter duration of time as compared to its maintenance phase. Thus, thorough and faster mechanisms are required to check the modified project with minimal possible rework to be done in the maintenance phase. In software engineering, prioritization of existing test suite is done to expedite the maintenance phase processes. Many researchers have done enormous work in this context so as to make the software maintenance phase easy and efficient. Ant Colony and Bee Colony, both these evolutionary computation techniques have been explored by the researchers for solving problems in many diverse domains. Time-constrained test case prioritization is a deterministic problem. This paper compares the already applied ACO and BCO for test case prioritization [1, 2]. These techniques have been applied with added constraint of maximum fault coverage and minimum execution time taken by the prioritized regression test suite. Hence, we call this as multi-deterministic prioritization. The current study includes experimental comparison of ACO and BCO applied to multi deterministic prioritization for eight sample programs averaged over ten sample runs on each.

The paper is organized in the following sections: Section 2 gives the background and related work ensuing in ACO, BCO and Regression Testing. Sections 3 and 4 provide brief explanation of the ACO and BCO techniques. This is followed by Sect. 5 elaborating on the comparison and finally the last section concludes the findings.

2 Related Work

Prioritization for software development has been tackled in various ways by Rothermel et al. [3]. Li et al. [4] performed an empirical study by using various greedy algorithms for test case prioritization. Time-aware test suite prioritization was proposed by Walcott et. al. [5] where testing is bounded within a predefined period of time.

ACO is a nature inspired meta heuristic approach introduced by Dorrigo in [6]. It has been applied successfully to solve various NP hard optimization problems. Artificial ants find their applicability to a considerable number of optimization problems. Quadratic assignment [5], prioritization [7], scheduling [8], sequential ordering [9], routing in networks etc. have benefited from the ACO technique.

Singh et al [10] proposed an algorithm for Test Case Prioritization using ACO in a time restricted environment. This algorithm was then implemented as ACO_TCSP tool in [1]. This tool is one of the tools used to gather the comparison details in the current study. Suri and Singhal [11] also accomplished a survey of application of the ACO technique in various fields of software testing such as test case generation, selection, prioritization etc. ACO on test case prioritization problem has also been individually tested and analyzed on various data sets and in varying parameters in [12, 13]. The current paper has taken the earlier implementation [1] of the ACO technique for comparison.

An automated software testing application using the BCO by producing the CFG of the original code and the modified code for comparison of test suite has been presented in [14]. An Artificial Bee Colony Algorithm (ABC) was implemented in [15] in which cooperation and cooperative behavior of bees have been used to optimize the large scale projects. It optimizes the high dimensional variables considering them as graphs. Furthermore, A Test Case Generation algorithm for BPEL processes has been given in [16] where a transition coverage criterion has been used as the fitness function. Regarding regression test suite prioritization, for total fault coverage in minimum execution time of a test case and total code coverage in give time, algorithm has been designed and implemented in [2, 17] respectively. A secure Bee Colony optimization algorithm [18] for pairwise test set generation has been presented, implemented and analyzed w.r.t. the parallel techniques available. An efficient algorithm to generate test cases using Unified Modelling Language has been proposed in [19], where minimum possible timeframe of execution is ensured using (BCO-mGA), software testing technique ensures maximum coverage by executing the final minimized test suite. The technique proposed in [20] ensures full path coverage, by generating paths from cyclomatic complexity and hence it ensures of a reliable test of software under test. Furthermore, A Regression testing technique has been proposed in [21] which uses traceability parameter to check agent error of the modified code. Next, in [22] a novel approach using Bee Colony Optimization (BCO) based intelligent search algorithm for Criticality Analysis based on Sensitivity and Severity metrics and Testing has been proposed for the generating test cases efficiently.

3 ACO

ACO is a set of steps based on search algorithms of artificial intelligence for finding optimal solutions; this includes ANT System, proposed by Colorni, Dorigo and Maniezzo [23–25]. Natural ants are blind and tiny in size, yet still they are able to explore the shortest route between their nest and the food source. Ants can accomplish this by using their antennas and the pheromone liquid so as to keep informing their community ants about their followed paths. ACO technique is inspired from this very behavior of ants, i.e., capability of being in synchronization while exploring solutions for a local problem based on some previous information gathered and distributed by each ant to the community.

In addition, [14] ACO involves two important processes of Pheromone deposition and pheromone trail evaporation. Pheromone deposition, as the name suggests, is the process of ants laying down the pheromone on every path being followed. Pheromone trail evaporation, on the other hand, is the natural process of depleting the pheromone deposited by evaporation (in case of natural ants), on every path corresponding to the passing time since the pheromone had been deposited. This trail update is achieved when the ants either finish their search or have got the probable shortest path between their nest and the food source. Each multi-deterministic problem defines its own pheromone updating/deletion criteria based on their local search merged with the global search eventually.

Artificial ants are programmed to leave a virtual pheromone trail on the digital path segment they follow. The path to be followed by each ant is chosen at random including the support from the amount of "trail" deposited on the probable paths initiating from node of the current ant. Increased pheromone trail on any path would increase the probability of that particular path being followed. Ant would then reach the next node in the path and would again select the further path using the same process. This process would continue until the ant reaches back its starting node. This finished tour followed by each ant gives us a local solution for the possible shortest path which is then analyzed for optimality or explored further by other ants.

The complete algorithm and tool for ACO applied to test case selection and prioritization problem has been given by Suri and Singhal [1]. The same has been explained using various examples in [26, 27].

4 BCO

BCO is a population based search algorithm in which the bees are the main working agents. Any given problem is solved using the food foraging behavior of honey bees. There are three types of bees in a honey bee comb: **Queen Bee**—It is a unique bee holding the responsibility of laying eggs, which in turn helps in building the hive population; **Drone Bees**—These are few in number, which are responsible for mating with the queen bee and help in growing the hive; **Working Agents**—Thousands of Worker Bees are the main working agents in any bee colony. They do all the maintenance work, bring food for the hive members, guide the fellow bees to the rich food source etc. There are two types of worker bees namely Scouts and Foragers which follow the Exploration and the Exploitation process respectively [2].

The BCO algorithm for maximum fault coverage has already been developed [2] and has been implemented and analyzed in [10]. The same has been used as the basis of our BCO technique compared in the current paper.

5 Analysis and Comparison

The ACO and BCO algorithms have been analyzed and compared for the prioritization of fault based regression test suite on eight sample examples. Subsequently, the details are presented in Table 1 and the results for the both are plotted in Figs. 1, 2 and 3.

Table 1 Input programs details for ACO and BCO

Program no.	Program name	Language	Size (LOC)	No. of faults	Test suite size	Total test suite execution time
P1	CollAdmission	C++	281	5	9	105.32
P2	HotelMgmnt	C++	666	5	5	49.84
P3	triangle	C++	37	6	19	382
P4	quadratic	C++	38	8	19	441
P5	cost_of_pub	C++	31	8	19	382
P6	calculator	C++	101	9	25	82.5
P7	prev_day	C++	87	7	19	468
P8	railway_book	Java	129	10	26	177

Fig. 1 ACO versus BCO plot for ASR (average % of size reduction in test suite)

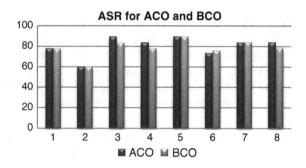

Fig. 2 ACO versus BCO plot for AETR (average % of execution time reduction for the test suite achieved using ACO and BCO)

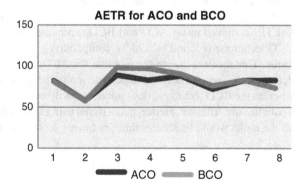

The proposed techniques have been analyzed by running them ten times for each example. We compute the Average Percentage of Test Suite Size Reduction (ASR), Percent Average Execution Time Reduction (AETR) and the Average Efficiency (AE) of the proposed algorithm (ACO/BCO) using following formulas (1), (2), and (3) respectively.

Fig. 3 Radar plot for ACO's versus BCO's AE (average efficiency % of efficient runs yielded perfect results on each ACO and BCO)

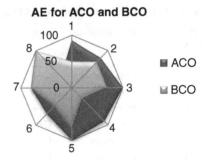

$$ASR = \left(\frac{|TS| - |PTS|}{|TS|} \right) * 100 \tag{1}$$

where, TS is the Total Test Suite Size, PTS is Prioritized Test Suite Size.

Figure 1 demonstrates the comparison for ASR computed on the sample eight programs using ACO and BCO respectively. As can be inferred from the graph, ACO and BCO, both techniques provide very high and almost similar average percentage of the reduction in the size of the test suite. This provides motivation for using these techniques in test case prioritization.

$$AETR = \left(\frac{TET - AET}{TET} \right) * 100 \tag{2}$$

where, TET is Total Execution Time, AET is Average Execution Time

A line graph comparison for the average percentage reduction in the execution time (AETR) achieved using ACO and BCO techniques have been plotted in the Fig. 2. BCO technique is found to achieve comparatively better AETR than the ACO technique. This result is due to the fact that the ACO technique includes complete fault detection as a second deterministic criterion in the path finding using the algorithm; whereas the BCO technique does not necessarily consider complete fault detection as a path finding criterion. Hence, paths found using BCO which are incapable of killing all the faults would be shorter thus are having lesser execution time.

$$AE = \left(\frac{BR}{TR} \right) * 100 \tag{3}$$

where, BR is Best Run of the respective algorithm, TR is the Total no. of Runs

Radar plot was chosen to represent the Average percentage efficiency of both the ACO and BCO techniques, multi-variation of the data being the sample eight programs, represented by 1–8 on the radar boundary in Fig. 3. The plot clearly represents better AE of the ACO technique for programs P1–P6, better AE of BCO technique for P8 and same AE for both the techniques for P7. Thus for the current sample data, ACO technique was in general found to represent better average percentage efficiency than the BCO technique for the problem of test case prioritization.

6 Conclusion and Future Work

Fault coverage based test case prioritization for minimum execution time has been analyzed and compared using ACO and BCO. It is clear from the above tables and plots that ASR, AERT and AE are mixed for both the approaches. ASR was found to be almost similar for both ACO and BCO. Hence a considerable reduction in the test suite size can be achieved using either of the techniques. AETR was found better for the BCO technique while AE was found better for the BCO technique. These techniques have been tested experimentally on the eight sample programs, and the very nature and size of these programs does not allow the authors to generalize these results. More examples need to be tested and analyzed to give a clearly better technique for test case prioritization. However, the results obtained still provide excellent results for both these techniques and thus the software researchers can benefit from either of the technique for solving the problem of test case prioritization.

References

1. Suri, B., Singhal, S.: Implementing ant colony optimization for test case selection and prioritization. Int. J. Comput. Sci. Eng. **3**(5), 1924–1932 (2011)
2. Kaur, A., Goyal, S.: A bee colony optimization algorithm for fault coverage based regression test suite prioritization. Int. J. Adv. Sci. Technol. Korea **29**, 17–29 (2011)
3. Rothermel, G., Untch, R.J., Chu, C.: Prioritizing test cases for regression testing. IEEE Trans. Softw. Eng. 929–948 (2001)
4. Li, H., Peng Lam, C.: Software test data generation using ant colony optimization. Trans. Eng. Comput. Technol. (2005)
5. Walcott, K.R., Soffa, M.L., Kapfhammer, G.M., Roos, R.S.: Time aware test suite prioritization. In: Proceedings of ACM/SIGSOFT International Symposium on Software Testing and Analysis, pp. 1–11 (2006)
6. Dorigo, M., Maniezzo, V., Colorni, A.: The ant system: optimization by a colony of cooperating agents. IEEE Trans. Syst. Man, Cybern. Part B: Cybern. **26**(1), 29–41 (1996)
7. Gambardella, L.M., Taillard, È.D., Agazzi, G.: A multiple ant colony system for vehicle routing problems with time windows. In: New Ideas in Optimization, pp. 63–76 (1999)
8. Stützle, T., Dorigo, M.: ACO algorithms for the quadratic assignment problem. New Ideas in Optimization McGraw Hill, pp. 33–50 (1999)
9. Singh, Y., Kaur, A., Suri, B.: Test case prioritization using ant colony optimization, association in computing machinery. In: Newsletter ACM SIGSOFT Software Engineering Notes, New York, USA, pp. 1–7 (2010)
10. Kaur, A., Goyal, S.: Implementation and analysis of the bee colony optimization algorithm for fault based regression test suite prioritization. Int. J. Comput. Appl. **41**, 1–9 (2012)
11. Suri, B., Singhal, S.: Literature survey of ant colony optimization in software testing. In: The Proceedings of the CSI Sixth International Conference on Software Engineering, Indore (2012)

12. Suri, B., Singhal, S.: Analyzing test case selection and prioritization using ACO. ACM SIGSOFT Softw. Eng. Notes **36**(6), 1–5
13. Suri, B., Singhal, S.: Understanding the effect of time-constraint bounded novel technique for regression test selection and prioritization. Int. J. Syst. Assur. Eng. Management. (2014)
14. Jeya Mala, D., Mohan, V., Kamalapriya, M.: Automated software test optimization framework—an artificial bee colony optimization based approach. Inst. Eng. Technol. **4**, 334-348 (2010)
15. Liang, Y., Liu, Y.: An improved artificial bee colony (ABC) algorithm for large scale optimization. Int. Symp. Instrum. Measur. Sensor Network Autom. **2**, 644–648 (2013)
16. Daghaghzadeh, M., Babamir, M.: An ABC based approach to test case generation for BPEL processes. In: International Conference on Computer and Knowledge Engineering, vol. 3 (2013)
17. Kaur, A., Goyal, S.: A bee colony optimization algorithm for code coverage based regression test suite prioritization. Int. J. Eng. Sci. Technol. **29**, 2786–2795 (2011)
18. Dahiya, S.S., Chhabra, J.K., Kumar, S.: Application of artificial bee colony algorithm to software testing. Australian Softw. Eng. Conf. **21**, 149–154 (2010)
19. Dalal, S., Chhillar, R.S.: A novel technique for generation of test cases based on bee colony optimization and modified genetic algorithm (BCO-mGA). Int. J. Comput. Appl. **68**(19), 0975–8887 (2013)
20. Karnavel, K., Santhoshkumar, J.: Automated software testing for application maintenance by using bee colony optimization algorithms (BCO). In: 2013 International Conference on Information Communication and Embedded Systems (ICICES), Chennai, 21–22 Feb 2013, pp. 327–330 (2013)
21. Srikanth, A., Kulkarni, N.J., Venkat, K., Singh, N., Ranjan, P., Srivastava, P.: Test case optimization using artificial bee colony algorithm, advances in computing and communications. Commun.Comput. Inform. Sci. **192**, 570–579
22. Dharmalingam, J., Balamuruga, M., Nathan, S.: Criticality analyzer and tester—an effective approach for critical components identification and verification. In: ICT and Critical Infrastructure: Proceedings of the 48th Annual Convention of Computer Society of India, Advances in Intelligent Systems and Computing, vol. I, pp. 663–670
23. Colorni, A., Dorigo, M., Maniezzo, V.: Distributed optimization by ant colonies. In: Proceedings ECAL'91, European Conference Artificial Life. Elsevier Publishing, Amsterdam (1991)
24. Dorigo, M.: Optimization, learning and natural algorithms. Ph.D. thesis, Politecnico diMilano, Milano (1992)
25. Dorigo, M., Maniezzo, V., Colorni, A.: The ant system: an autocatalytic optimizing process. Technical Report TR91-016, Politecnico di Milano (1991)
26. Dorigo, M., Socha, K.: An introduction to ant colony optimization. IRIDIA Technical Report Series, 10 (2006)
27. Suri, B., Singhal, S.: Test case selection and prioritization using ant colony optimization. In: International Conference on Advanced Computing, Communication and Networks, Chandigarh (2011)

Design of a Two Fingered Friction Gripper for a Wheel Mobile Robot

Nisha Bhatt and Nathi Ram Chauhan

Abstract A wheel mobile robot is used under tedious, dangerous and dirty conditions. These robots have a mobile platform, robotic arm and gripper. These mobile robots have numerous applications. A gripper of mobile robot is used for manipulation and pick and place tasks. The design of gripper is an active research topic within robotics and production field. The present work addresses a design of impactive two fingered gripper for a wheel mobile robot. The objective of paper is to design a gripper for pick and place task that can be utilized in any industry. The model of gripper is carried out using solidworks and Think-3 software. The paper discusses the mechanical design parameters like Von-Mises stress, deformation, and Torque requirement in computational analysis of gripper.

Keywords Friction gripper · ADAMS · ANSYS · Wheel mobile robot · Two fingers

1 Introduction

Usually, a gripper of industrial robots is a specialized device that grasps one or few objects of similar shape, size, and weight in repetitive operations. Grippers bear resemblance in the technical analogy for human grasping. They are used for automated industrial processes along with handling device to move, handle and sometimes even manipulate goods in industrial tasks. They use different principles to transfer the force of the handling device. Hence in such situation a well efficient gripper is worth of use. In industries, generally impactive grippers are used. These grippers utilize principle of penetrating or non penetrating contact for object

N. Bhatt (✉) · N.R. Chauhan
Indira Gandhi Delhi Technical University for Women, Kashmere Gate,
New Delhi 110006, India
e-mail: bhattnisha09@gmail.com

N.R. Chauhan
e-mail: nramchauhan@gmail.com

© Springer Science+Business Media Singapore 2016   195
R.K. Choudhary et al. (eds.), *Advanced Computing and Communication Technologies*,
Advances in Intelligent Systems and Computing 452,
DOI 10.1007/978-981-10-1023-1_20

retention. The manipulative operations are usually performed by using two-finger grippers, which are powered and controlled by one actuator only. In addition, two-finger grippers are used both for pick and place and assembling purposes since most of these tasks can be performed with a two-finger grasp configuration. Now a day's focus has been shifted to multi-fingered gripper. These grippers are active topic of concern in research filed. However for practical or industrial applications majority of industrial grippers are two fingered that are powered either pneumatically or electrically. Generally, electrical actuation is provided either from DC motor or stepper motor. This actuation is then converted into finger movements through transmission system like ball screws, gears, pulley or linkages.

In recent years various mechanisms of robotics gripper has attracted many researchers. Chen [1] described several mechanisms for different gripper functions. He also classified mechanical grippers according to the pair elements used in their construction such as linkage, gear and rack, cam, screw, rope and pulley types and miscellaneous. Chan and Cheung [2] presented an electro-magnetic two finger gripper actuated by variation of flux. The flux modeling of variable reluctance gripper was carried out to devise magnetic equations of actuator structure. Nagata [3] presented a linear sliding mechanism driven by rack and pinion to mount gripper fingers. Luo and Henderson [4] presented a pneumatically actuated gripper with parallel fingertips. Langde and Jaju [5] utilized the concept of a lever mechanism for their two fingered gripper. Mccormick and Geary [6] presented an electric gripper that used simple control logic and thus activated for a very short duration. The unique feature of this gripper was it used two mechanisms for each position. A permanent magnet locked the solenoid plunger in open position whereas for closed position a spring was used to lock plunger. Ohol and Kajale [7] showed simulation of multifinger gripper. Each finger was actuated by one servomotor. Each finger had three degree of freedom and thus adapts the object shape. All these grippers have different mechanical structure but are used for pick and place task. However in all these papers, it was observed that the mechanical design aspects of a gripper have not been discussed thoroughly. The importance of stresses, deformation, torque values etc. are still untouched. The authors have made an attempt to discuss mechanical design which is considered as backbone of a gripper design. The next section presents a computer aided design of two finger gripper proposed in this article.

2   Design Analysis

The computer aided design of a two finger gripper is shown in Fig. 1. The main aim of this paper is to discuss mechanical design of a gripper. The gripper is first modeled in 3D solid works CAD package and then surface modeling is applied on think 3 software. The CAD model is simulated into a complete working mechanism

Fig. 1 Model of proposed
gripper

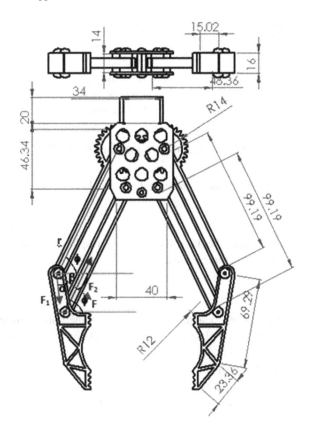

in ADAMS software. It has following components: (1) Gripper Jaws (2) Electric
actuator (DC motor) (3) Transmission element (two spur gears) (4) Base (5) Four
connecting links. Each gripper jaw is connected to the base via two connecting links
where these four connecting links form a parallelogram linkage. The reason of
choosing parallelogram linkage is that the gripping force will always be normal to
both the object and the finger at the point of contact. This insures that maximum
gripping force is applied at all the times on object. Also in this configuration, the
gripping force is independent of location of contact points. The fingers grasp an
object by following a curvilinear trajectory. These jaws exert frictional force on the
object for grasping. The jaws are actuated by a dc motor that drives spur gears. The
connecting link of linkage follows the rotation of these gears. The material selection
of components is as follows: Gripper jaw: alloy steel; Housing material (base): plain
carbon steel; Links: alloy steel; Gears: cast alloy steel. The weight of this gripper is
2.2 kg. Since it can carries up to 8 kg of object so payload to weight ratio is 3.3.
Table 1 presents the specifications of proposed design of gripper.

Table 1 Mechanical design specifications of gripper

Parameters	Numeric value
Length of jaws	98 mm
Dimension of base (length width)	50 * 40 mm
Thickness of jaws	16 mm
Degree of freedom	1
Outer diameter of gear	40 mm

2.1 Force Calculation for Gripper [8]

If m_o be the mass of object and g is acceleration due to gravity then weight of object

$$W = m_o g \tag{1}$$

As Friction between the fingers is responsible to hold the object.
Frictional force for each finger

$$F_f = \mu N \tag{2}$$

where μ is coefficient of friction & N is the normal component of frictional force.
For pick and place task of 8 kg object, the gripping force exerted by two fingered gripper will be
Equating Eqs. (1) and (2)

$$F_f = \frac{W}{2} \tag{3}$$

$F_f = 40\,\text{N}; \ \mu = 0.3; \ N = 133\ \text{N}$
Resolving forces horizontally

$$-F_1 \cos(\alpha + \beta) - F_2 \cos \varphi + N = 0 \tag{4}$$

where α, β and ϕ depends on position of connecting links depending upon size of object.

$$T_G = F * r \tag{5}$$

$$F_1 = F \cos \phi \tag{6}$$

Resolving forces vertically

$$F_1 \sin(\alpha + \beta) - F_2 \sin \phi = 0 \tag{7}$$

where r, F_1, F_2 and F are the gear radius, component of force in the direction of gripper acting on gear link, component of force in the direction of gripper acting on second link and force acting on gear link respectively.

Table 2 Variation of torque with object dimension

Object dimension in mm	Torque (T_G) in N m
03	2.20
20	2.88
30	2.95
40	3.02
50	3.11
60	3.18
70	3.12
80	3.05
90	2.67
100	2.39

After performing the required calculations, desired value of torque is calculated for rectangular object of different dimension. Table 2 shows the variation of gear torque with object dimension. The minimum opening of jaw is limited to 3 mm and maximum is 100 mm.

The object dimension shown in Table 2 is the diagonal information of rectangular object. The maximum value of torque observed is 3.18 N m for 60 mm of object dimension. The minimum value of torque observed is 2.20 N m for 3 mm of object.

Since the gripper discussed in this paper is inspired from the gripper used in defense research lab of India [7]. The difference between these two grippers is that our gripper is driven through gear train whereas the one used in defense lab is driven through screw gear train combination. The maximum torque value for our gripper is 3.18 N m for 60 mm of opening whereas the gripper of defense lab has maximum torque of 2.51 N m for 40 mm of opening. We have verified the above torque values from the published data of Pathak et al. [7]. The variation of torque with jaw angle in Fig. 2 follows same pattern as mentioned by Pathak et al. The design of proposed gripper is further verified through computational analysis.

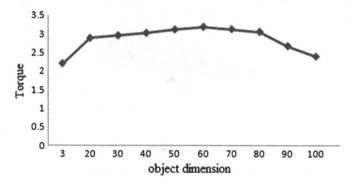

Fig. 2 Variation of torque with object dimension

3 Theoretical and Computational Analysis

The most crucial part of a gripper is its jaw because it comes in contact with an object. The normal component of frictional force is the gripping force applied on the object. The jaw is subjected to both normal and frictional forces of 133 and 40 N respectively applied near the tip of jaw.

The gripper jaw is considered as a cantilever beam subjected to 133 N of load in vertical plane. This force is considered to be a distributed force on the cantilever beam.

Deflection due to load

$$\delta = Fl^3/8EI \qquad (8)$$

$E = 204781.5$ N/mm^2; $F = 133$ N; $L = 98$ mm; $I = 965.44$ mm^4; $\delta = 7.628*$ 10^{-3} mm.

Stress induced due to external load

$$\sigma = Fl/Z \qquad (9)$$

where z is section modulus, $Z = 120.68$ mm^3 and stress is 101 MPa.

After the theoretical analysis, computational analysis is performed on ANSYS and ADAMS software.

Figures 3 and 4 shows static analysis results for jaw stress and deformation of gripper jaw. The material selected for jaw is alloy steel 4140. The yield strength for this material is 250 MPa. From static analysis in ANSYS software it was found the maximum value of Von-Mises stress is 101 MPa. Since the yield strength is greater than the maximum Von-Mises stress observed in gripper jaw. The design of jaw is

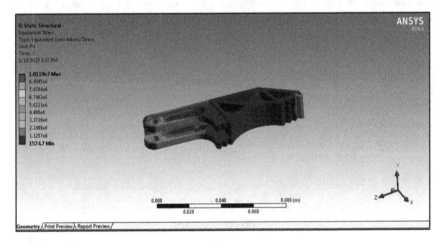

Fig. 3 Von-Mises stress observed in jaw

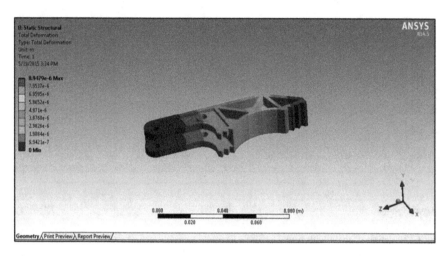

Fig. 4 Deflection of jaw

considered to be safe under static loading conditions. The maximum value of displacement is found out to be $8.9 * 10^{-3}$ mm.

After formulating the gripper model in ADAMS, simulation is performed and results of animation are plotted. The simulation is run for 5 s.

It can be seen in Fig. 5 that the maximum value of torque is 7.5 N m for coupler so for one gear it will be around 3.75 N m. The torque curve shows that torque increases while picking the object.

Similarly the variation of gripper jaw angle versus time is obtained for a simulation run of 5 s. The graph of jaw angle versus time shows numerical data during an interval of 0–2 s as shown in Fig. 5. This is because the gripper first opens its

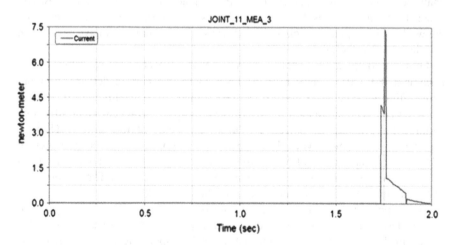

Fig. 5 Variation of torque with time

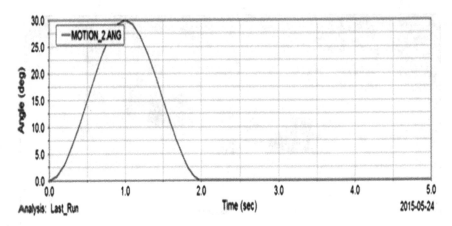

Fig. 6 Variation of theta with time

jaws (shown by gradual increase in jaw angle) and then closes to pick the object (shown by gradual decrease in jaw angle). During this motion the angular displacement of the gripper jaw varies from 0°–30°.

Figure 6 shows that the angular displacement decreases during closing of gripper jaws. The variation shows that jaws angle varies only while gripper is trying to pick the object and the jaws angle does not vary once it gripped the object.

4 Conclusion

The design and mechanics of a two finger gripper for rectangular object has been analyzed in the present paper. A gripper manipulates an object through its jaw and hence the design of jaw is the most critical phase in gripper's design. To understand the working of gripper in pick and place tasks, simulation results have been given and discussed in the present work.

From the computational analysis in ANSYS, the deflection of jaw is found to be maximum ($8.9 * 10^{-3}$ mm) near the tip because that portion deflects the more while lifting an object. The Von-Mises stress is found to be maximum (101 MPa) at the fixed end. The simulation in ADAMS verifies the theoretical value of torque required for gripping an object. The maximum value of torque (3.75 N m) is observed at the moment when the gripper picks an object. The error in theoretical and computational values of Von-Mises stress, deformation and Torque has been calculated as 7 N/mm^2, 0.0013 mm and 0.52 N m respectively. The performance of the design is verified by kinematic simulation in ADAMS.

The gripping torque of proposed gripper is found to be more as compared to gripper design implemented by Pathak et al. [9]. This is because the gripper discussed in this paper has gear train that is solely responsible for picking objects whereas in gripper [9] screw played the major role while lifting the object. It can be concluded that a two fingered friction gripper is best suited for a scenario where main task is pick and place of an object.

References

1. Chen, F.Y.: Gripping mechanisms for industrial robots. Mech. Mach. Theor. **17**, 299–311 (1982)
2. Chan, K.C., Cheung, N.C.: Magnetic modelling of a mutually-coupled two-finger variable reluctance gripper. In: Proceedings of the 28th Annual Conference of the IEEE Industrial Electronics Society IECON'02 (5–8 Nov 2002)
3. Nagata, K.: Manipulation by a parallel-jaw gripper having a turntable at each fingertip. In: Proceedings of IEEE International Conference on Robotics and Automation, vol. 2, pp. 1663–1670 (8–13 May 1994)
4. Luo, R.C., Henderson, T.C.: A servo controlled robotic gripper with multiple sensors and its logical specification. J. Rob. Syst. **3**(4), 409–420 (1986)
5. Langde, R., Jaju, S.B.: Designing and development of two finger compound gripper for internal and external gripping. Int. J. Innovative Res. Dev. **2**(3) (2013)
6. Geary, J.W., McCormick, P.E.: US Patent No. 20100171332A1, Wilmington, DE (US) (2010)
7. Ohol, S.S, Kajale, S.R.: Simulation of multifinger robotic gripper for dynamic analysis of dexterous grasping. In: Proceeding of world congress on engineering and computer science, October 22–24, San Francisco, USA (2008)
8. Weiss, M., Groover, M.P.: Industrial robotics: technology, programming and application. McGraw-Hill, New York (1986)
9. Pathak, M.K., Singh, P., Bansal, A.: Manipulator arm for handling of objects in confined spaces. In: Proceedings of Advances In Robotics, pp. 1–8 (04–06 Jul 2013)

Classification of Robot Swarm Models in Discrete Domain

Maitry Sinha and Srabani Mukhopadhyaya

Abstract An emerging and challenging area of research in swarm robotics is to consider swarms deployed in discrete domains. In the continuous domain, it has already been established that different computational and behavioral models of the robot swarm play an important role in solvability of different fundamental problems. Due to some basic differences, not all existing models in the continuous domain are relevant or significant with respect to the discrete case. In this paper we draw an analogy between the models already existing in both the domains and propose a few relevant models for the discrete domain.

Keywords Robot swarm · Discrete domain · Gathering · Multiplicity detection · CORDA model

1 Introduction

A swarm of robots consists of small, identical, anonymous, inexpensive mobile robots. The concept of swarm robots emerges from the collective behavior of small social insects such as bees and ants that function in a group. Each robot in a swarm is very simple and individually has limited capability; although collectively they can perform much more complex jobs. The current research trend is to design algorithms for a specific job in such a manner that each individual robot executes the same function and together they complete the job. These jobs are usually very basic in nature, for instance, jobs like gathering [1, 2], flocking [3], covering [3], converging [4] and so forth. By utilizing a swarm of robots to execute these basic jobs efficiently, often a far more complex task can be completed in a smooth,

M. Sinha (✉)
Calcutta Institute of Engineering and Management, Kolkata, India
e-mail: sinhamaitry@gmail.com

S. Mukhopadhyaya
Birla Institute of Technology, Mesra, Kolkata Extension Centre, Kolkata, India
e-mail: smukhopadhyaya@bitmesra.ac.in

© Springer Science+Business Media Singapore 2016
R.K. Choudhary et al. (eds.), *Advanced Computing and Communication Technologies*,
Advances in Intelligent Systems and Computing 452,
DOI 10.1007/978-981-10-1023-1_21

efficient, and time-saving manner. One of the most promising uses of swarm robotics is in rescue missions in disaster-struck areas. A swarm of robots can be sent to places, where rescue workers cannot reach, to assess the degree of damage, to detect the presence of life, or to send immediate relief [4, 5].

For theoretical research, these robots are considered as points on a 2-D plane. In continuous environment, the robots are deployed on a 2-D plane and they are free to move across the plane. However, there may be situations where the robots may not be allowed to move freely on the plane because of the presence of some obstacles or the whole plane might not be accessible to all the robots. Moreover, the situation may demand that the robots move through a particular route—this is where the idea of discrete domain comes in. In case of discrete domain, the robots are placed on the nodes of a graph on a 2-D plane and the robots can then move along the edges of the graph and rest only on some node [1]. The domain can also be some form of a network of computers and the robots are mobile agents. Thus, the robots have also been referred to as agents.

In continuous domain, researchers have proposed different computational models and features of the constituting robots and are investigating their suitability in solving several fundamental problems. It is now a well established fact that solvability of a problem greatly depends on the assumed models and features. As swarm robotics in discrete domain is a new research area, very few problems have been addressed in this domain. It has been observed that variations of models completely change the nature of the solutions and sometimes make the problem unsolvable under certain scenario.

Regarding computation and synchrony, in both continuous and discrete environment, researchers have accepted the following two models as the basic models:

(I) Computational model: In *CORDA model*, robots are assumed to execute a sequence of computational cycles as long as they are active and yet to complete their job. Each of the cycles consists of three phases, *LOOK*: when a robot senses its surroundings; *COMPUTE*: when the robot computes its next destination and route based on the information gathered in the look phase; and then *MOVE*: in which the robot actually moves to its destination.

(II) Models based on synchrony: Three models [3] are usually defined based on the activation scheduling of the robots: (i) in the *synchronous* model, all robots of the system are activated at the same instant of time and all of them execute the same phase of the computational cycles at the same time; (ii) In *semi-synchronous* model, some but not all the robots are activated at the same instant of time and all activated robots work synchronously; and (iii) in the *asynchronous* model, any robot can be activated at any time and the phases of their computational cycles are not necessarily synchronous.

Our Contribution: In this paper we summarize different models assumed in continuous as well as discrete domain and draw an analogy between these models. Finally we propose some models, which are relevant to the discrete domain.

In view of basic models and assumptions, a brief review of the main works done in the discrete domain is presented in the next section. In Sect. 3, we discuss and compare the existing models in both the domains and discuss their relevance in the discrete domain. We also propose new models relevant to the discrete domain.

2 Different Assumptions Used in Discrete Domain

Swarm robotics, operating in discrete environment, is a relatively new research domain. In this domain, researchers have mainly addressed the gathering problem on some particular types of graphs. The gathering problem is a very basic coordination problem. The robots need to gather at some particular node in the graph on which the robots are initially deployed. The node where the robots gather is not known a priory, rather it is decided during execution. So far, only gathering on ring [1, 6–8], grid [1, 9], tree [1], and regular bipartite graphs [10] have been addressed.

In the existing solutions of gathering problem in ring, it is observed that multiplicity detection capacity of robots plays a crucial role in deciding the solvability of the problem. Multiplicity refers to the situation when more than one robot occupy the same node [7, 8]. Depending on the multiplicity detection capabilities of the robots, four different variations are considered [1]:

(i)   **Local weak model**—A robot can detect whether multiple robots are present in the node where it is currently residing (local node). However the robot cannot perceive the exact number of robots present in the node.

(ii) **Local strong model**—In this case robots not only are able to detect multiplicity of its current node but they can also count the exact number of robots present.

(iii) **Global weak model**—In this case, a robot can identify whether there are multiple robots present in any node (in addition to the node where it is currently residing) of the graph or not. However it cannot count the exact number of robots present in a node in case there are more than one.

(iv) **Global strong model**—In this model a robot is able to perceive the exact number of robots present in any node of the graph.

In an undirected and un-oriented ring, the gathering problem has been solved for anonymous, oblivious robots with local weak multiplicity detection [7] capacity. The robots are assumed to be synchronous and their movements are instantaneous.

Insolvability of gathering problem is discussed in [7, 11]. The gathering problem in ring is unsolvable if (a) the number of robots is two, (b) robots have no multiplicity detection capability, (c) the initial configuration of robots over the ring is a periodic configuration, or (d) the initial configuration is symmetric in which axis of symmetry goes through two antipodal links. In these works, the initial configuration is also shown to be very important for solvability of gathering problem in ring.

Kamei et al. [8], addressed the gathering problem in a ring for asynchronous mobile robots having symmetric initial configuration, without global multiplicity detection capability. The authors give importance to two parameters while modeling the robots, synchrony and multiplicity detection capability.

In the literature, there are a few results regarding gathering on grids [1, 9]. An anonymous and undirected grid of $n \times m$ nodes is usually considered. The main difference in the assumed model that distinguishes the gathering problem on grid from the one on ring topology is non-requirement of the robots' multiplicity detection capability. It does not even need the local weak multiplicity detection capability. The only assumption in this regard is that initially each node in the grid can be occupied by at most one robot. It is shown that initially symmetric and periodic configurations are non-gatherable [9]. The authors have also shown that all the other cases are gatherable without any multiplicity detection capability of robots. Only in the case of 2×2 grid, the local weak multiplicity detection capability with asymmetric initial configuration is required for solvability of the problem.

Gathering on tree is also addressed in [1]. For the trees, with single center, the robots can gather at the center of the tree without any requirement of multiplicity detection capacity. However, the robots are assumed to have unlimited visibility. On the other hand, if the tree is bi-centric then it is not guaranteed that the robots can gather at a node. If the two sub-trees rooted at the centers along with the distribution of the robots are isomorphic, then gathering is not possible.

Guilbault et al. [10] addressed the gathering problem on regular bipartite graphs. They consider asynchronous oblivious robots with weak perception capability. Weak perception signifies that the robots can sense only their immediate neighborhood. The problem is solved assuming both versions of multiplicity detection capability, local as well as global. The researchers here have proved that an initial configuration of the robots on a regular bipartite graph is not gatherable if the configuration is not a star of size at least 3.

The rendezvous problem is the special case of gathering, where the number of robots is two. Kowalski et al. [12] introduce labels on agents while solving rendezvous problem on arbitrary graphs. To solve this problem with deterministic algorithms the researchers assume that the agents are synchronous with arbitrarily decided start up times and each agent has distinct labels. Every agent knows its own label but does not know the label of the other. This is a new direction with respect to anonymity of a robot in the swarm.

3 Proposed Models

Researchers have proposed different models and features of the constituting robots and have investigated their suitability in solving several fundamental problems in continuous domain. Some of the important parameters are (i) direction and orientation of local co-ordinate axes (ii) memory and (iii) visibility range of the robots.

3.1 Direction and Orientation of the Local Co-ordinate Axes

In the continuous domain, the robots are considered as points on a 2-D plane. Each robot has its own local co-ordinate system with itself at the origin, a specific unit of length, and a Cartesian co-ordinate system defined by the direction of the two co-ordinate axes. However, the unit length, the directions and orientations of the individual co-ordinate axis may differ. If the constituting robots do not agree on the orientation and direction of the axes, the problem intensifies, especially when robots do not have direct communication among themselves. In the existing literature, three models are defined depending on this feature [3].

1. **Total agreement** on local co-ordinate system, where all the robots of the system agree on the direction and orientation of their co-ordinate system.
2. **Partial agreement** on local co-ordinate system where all the robots agree on the direction and orientation of only one axis.
3. **No agreement** on the local co-ordinate system. This situation is much more complex compared to the previous two, though much closer to a practical scenario.

In discrete domain, the two dimensional plane is modeled by a graph and a robot can move only along the edges of the graph on which they are deployed. The robot's motion is always guided by an edge of the graph and so the direction and orientation of the local co-ordinate system is not much of an importance here. However, nodes of a graph may have degrees greater than one. A robot residing on such a node often needs to choose a particular route among the multiple links incident on that node. If the links are assumed to be anonymous, it would be difficult for a robot to take decision for the next move. In 2008, Fragniaud and Pelc addressed the gathering of two robots in trees [13], where they assumed that the edges incident to a vertex v are labeled as 1, 2, ..., d where d is the degree of v. Each edge $\{u, v\}$ thus has two labels from two ends, which are termed as port number at u and port number at v. Port numbering is local; i.e., there is no relation between port numbers at u and v. This kind of labeling of edges can be compared with the notion of direction in continuous domain.

Proposed Model

1. **Total agreement on link ordering**: All the robots agree on the ordering of the links incident on a node.
2. **Partial agreement on link ordering**: All the robots agree on the ordering (labelling) of the links incident on a node up to a reflection.
3. **No agreement on link ordering**: The robots do not agree on any kind of ordering among the links incident on a node.

3.2 Models Based on Memory

In continuous domain researchers have considered two types of robots based on their memory: (i) oblivious and (ii) non-oblivious. Oblivious robots have very limited memory by which they can only retain the information gathered in the current phase but cannot remember anything beyond the current phase. In discrete environment also researchers have considered mainly oblivious robots. In [5, 14], it is assumed that a bounded amount of memory is attached to the nodes, termed as *whiteboard*. Robots visiting a node can communicate among themselves through reading and writing on this board. This board can be accessed in a mutually exclusive way.

Proposed Model

Memory can be assigned to the links also. For an edge $\{u, v\}$, some information can be stored at both the ends u and v. However, restriction can be imposed on the amount of memory as required by the application. We propose the following whiteboard model at links:

Whiteboards at links: A bounded amount of memory is attached on both ends of an edge $\{u, v\}$. Robots present at node u (local w.r.t. u) can read or write on the white board of $\{u, v\}$ at u. However, robots residing at v (outsiders w.r.t. u) do not have any access on the white board of $\{u, v\}$ at u. This board can be accessed by the robots in a mutually exclusive way.

3.3 Visibility Range of the Robots

Visibility of a robot is a measure of how far in the plane a robot can view. Each robot of a swarm is equipped with a sensor, by which it can view the surroundings. For continuous case, the robots may be able to see the whole plane at a time or they may see just a portion of it. The former is referred to as *unlimited visibility* and the latter as *limited visibility* [3]. Limited visibility is measured by the visibility radius R, which means that the robot can view only a circular area of radius R. The notion of visibility range of robots is important for problem solving in discrete domain also. However, the terms *limited* and *unlimited visibility* are defined in a slightly different way. If a robot can sense the existence of all the other robots at different nodes of the graph, it is said to possess unlimited visibility. A robot with limited visibility senses the existence of other robots in the graph up to a certain distance from its current residing node. Shortest path length between two nodes is taken as the distance between two nodes. The word *weak perception* is used in [10] to indicate limited visibility.

Proposed Model

*Strong perception (**public**)*: Robots can view all other robots residing in different nodes of the graph. Moreover, robots can see the contents of the whiteboards (if any) placed at different nodes. In other words, whiteboards placed at nodes are assumed to be public with respect to reading, though writing right is preserved only for local nodes.

*Strong perception (**private**)*: Robots can view all other robots residing in different nodes of the graph. However, whiteboards (if any) placed at nodes are not visible to all the robots; reading and writing right is preserved only for those robots which are residing in that particular node.

*Weak perception (**public**)*: Robots can view only those robots which are residing within a distance, say R. Here, by distance we mean graph theoretic distance, i.e. the distance between two nodes is the shortest path length between those two nodes. Here, robots can also read the whiteboards (if any) placed at those nodes which are visible to it, i.e., which are within a distance R.

*Weak perception (**private**)*: Robots can view only those robots which are residing within a distance, say R, though whiteboards placed at different nodes are not visible to any outsider.

4 Conclusion

In this paper we have classified the existing assumptions for swarm robots in discrete domain into formal models and also proposed new models. These models would help us in finding minimal sets of capabilities for the robots to solve a problem. If a problem is unsolvable under a set of assumptions, the models would help us in choosing the next bigger set to attempt it.

References

1. D'Angelo, G., Di Stefano, G., Navarra, A.: Gathering asynchronous and oblivious robots on basic graph topologies under the look-compute-move model. HAL (2012). In: Alpern, S., et al. (eds.) Search Theory: A Game Theoretic Perspective. Springer, New York (2013)
2. Floccihni, P., Prencipe, G., Santoro, N., Widmayer, P.: Gathering of asynchronous robots with limited visibility. Theor. Comput. Sci. **337**(1–3), 147–168 (2005)
3. Floccihni, P., Prencipe, G., Santoro, N.: Distributed computing by oblivious mobile robots. Morgan and Claypool publisher (2012)
4. Ando, H., Oasa, Y., Suzuki, I., Yamashita, M.: A distributed memory less point convergence algorithm for mobile robots with limited visibility. IEEE Trans. Robot. Autom. **15**(5), 818–828 (1999)
5. Balamohan, B., Dobrev, S., Flocchini, P., Santoro, N.: Asynchronous exploration of an unknown anonymous dangerous graph with o(1) pebbles. In: Structural Information and Communication Complexity, pp. 279–290. Springer (2012)

6. D'Angelo, G., Di Stefano, G., Navarra, A.: Gathering of six robots on an anonymous asymmetric rings. In: Structural Information and Communicaion Complexity, pp. 174–185. Springer (2011)
7. Izumi, T., Kamei, S., Ooshita, F.: Mobile robots gathering algorithm with local weak multiplicity in rings. SIROCCO LNCS **6058**, 101–113 (2010)
8. Kamei, S., Lamani, A., Ooshita, F., Tixeuli, S.: Asynchronous Mobile robot gathering from symmetric configurations without global multiplicity detection. In: A. Kosowski and M. Yamashita (eds.) SIROCCO, 2011. LNCS, vol. 6796, pp. 150–161, Springer, Berlin Heidelberg (2011)
9. D'Angelo, G., Di Stefano, G., Navarra, A.: Gathering of robots on anonymous grids without multiplicity detection. In: 19th International Colloquium on Structural Information and Communication Complexity (SIROCCO), LNCS, vol. 7355, pp. 327–338 (2012)
10. Guilbault, S. and Pelc, A.: Gathering asynchronous oblivious agents with local vision in regular bipartite graphs. In: Structural Information and Communication Complexity, pp. 162–173. Springer (2011)
11. Klasing, R., Markou, E., Pelc, A.: Gathering asynchronous oblivious mobile robots in a ring. Theor. Comput. Sci. **390**(1), 27–39 (2008)
12. Kowalski, D.R., Pelc, A.: Polynomial deterministic rendezvous in arbitrary graphs. In: Proceedings of 15th Annual Symposium on Algorithms and Computation, ISSAC', LNCS, vol. 3341, pp. 644–656 (2004)
13. Fraigniaud, P., Pelc, A.: Deterministic Rendezvous in Trees with Little Memory. In: Taubenfeld (ed.) DISC 2008, LNCS, vol. 5218, pp. 242–256, Springer, Berlin Heidelberg (2008)
14. Dobrev, S., Flocchini, P., Prencipe, G., Santoro, N.: Mobile search for a black hole in an anonymous ring. Algorithmica **48**(1), 67–90 (2007)

3D Environment Reconstruction Using Mobile Robot Platform and Monocular Vision

Keshaw Dewangan, Arindam Saha, Karthikeyan Vaiapury
and Ranjan Dasgupta

Abstract Constructing a 3D map/perception model of an unknown indoor or outdoor environment using robotics is of compelling research nowadays because of the importance of the automatic monitoring system. Available IMU sensors and mobile robot kinematics allow 3D reconstruction to be finished in near real-time using a very low cost robotic platform. In this paper, we describe a framework for dense 3D reconstruction on an inexpensive robotic platform using a webcam and robot wheel odometry. Our experimental results show that our technique is efficient and robust to a variety of indoor and outdoor environment scenarios with different scale and size.

Keywords 3D environment reconstruction · Mobile robot · Robot operating system · Odometry · Camera calibration · Optical flow · Epipolar geometry · Robot locomotion

1 Introduction

The 3D technology is well established and accepted in manufacturing, chemical, automobile, construction industries and showing keen interest in investigating how this can be applied in practice. Robot based solution is in high demand in the market, especially manufacturing and chemical industries to inspect hazardous area where human cannot easily go. A variety of affordable mobile robots are available

K. Dewangan (✉) · A. Saha · K. Vaiapury · R. Dasgupta
TCS Innovation Labs Kolkata, Kolkata, India
e-mail: keshaw.dewangan@tcs.com

A. Saha
e-mail: ari.saha@tcs.com

K. Vaiapury
e-mail: karthikeyan.vaiapury@tcs.com

R. Dasgupta
e-mail: ranjan.dasgupta@tcs.com

© Springer Science+Business Media Singapore 2016   213
R.K. Choudhary et al. (eds.), *Advanced Computing and Communication Technologies*,
Advances in Intelligent Systems and Computing 452,
DOI 10.1007/978-981-10-1023-1_22

in the market due to the emerging advancement in robotics field. These mobile robots are equipped with different low cost sensors (like camera, IMU sensors etc.) including a light weight computing unit. So the possibility of environment monitoring and verification in 3D space using such low cost mobile robot is manifold. In fact, there is a quite powerful and stable structure-from-motion pipeline readily available for reconstructing 3D model from multiple 2D images as shown in [1–3].

In a recent work [4], Pradeep et al. has described a methodology for markerless tracking and 3D reconstruction in scenes of smaller size using RGB camera sensor. It tracks and re-localizes the camera pose and allows for high quality 3D model reconstruction using a webcam. Pizzoli et al. [5] proposed a solution by adapting a probabilistic approach in which depth map is computed by combining bayesian estimation and convex optimization techniques. All these implementations are limited to a small scene reconstruction and not suitable for an entire 3D environment creation.

The 3D reconstruction of an environment from multiple images or video captured by a single moving camera has been studied for several years and is well known as Structure-from-Motion (SfM). Recently, smart phones are used for image acquisition due to its low cost and easy availability. So researchers used smart phones sensors like accelerometer, magnetometer for data collection and 3D reconstruction, it reduces computation [6, 7] and few works such as [8, 9] have accomplished this, but the output is noisy due to a fast and course reconstruction.

A system capable of dense 3D reconstruction of an unknown environment in real-time through a mobile robot requires simultaneous localization and mapping (SLAM) [10]. In our system, localization of the robot is done from odometer and the robot movement is controlled by user. The estimation of accumulated error is done from expected next position and actual odometer value of left and right wheel, so the complexity is lesser in this case.

In this context to fulfill these requirements, we present an end to end framework capable of generating 3D reconstruction of an environment based on the image/video captured through a remote platform mounted on a two wheel based robot. This work is a core part of our system presented in [11]. Firebird VI robot [12] is used in our experiments which allows navigation across a given environment. Robot captures images, odometry and IMU data and sends to backend server where, 3D view of environment is constructed using some selected key frames from the captured images and poses information. The novelty of the proposed system is reconstructing an entire environment in near real-time using a very low cost user guided mobile robot platform.

We demonstrate the framework along with the performance of the approach with computation time details. We present different type of results to show the capability and robustness of the system to work in a wide range of scenes and environments. We also evaluate the accuracy of the reconstruction and compared with ground truth.

2 Robotic Platform Description

In our work, Firebird VI robot (refer Fig. 1) is used which controls all operations through ROS [13]. The framework is capable to work on any robotic platform that supports ROS and Firebird VI is chosen due to its low cost and readily availability. The block diagram of the entire system is provided in Fig. 2.

Fig. 1 *Left to right* figure shows the Firebird VI robot used in our experiments, sample data sets and corresponding reconstructions

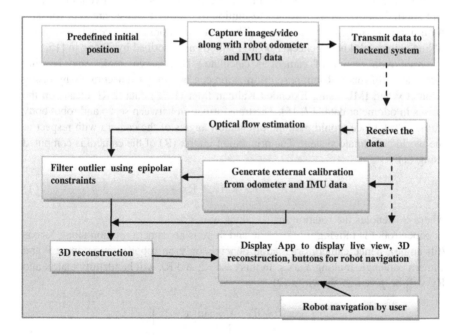

Fig. 2 Proposed system architecture: robotic vision

2.1 Data Acquisition

The robot starts navigation from its initial position and this is taken as the origin of world coordinate system by default in order to reduce the complexity. So the starting position is considered as predefined position. One very light weight and less computing power system is mounted on the top of robot, where master ROS handles tasks of robot navigation, pose estimation (using odometry and IMU sensor data) and Image capturing and transmitting to the backend server.

Image capturing is a major task in data acquisition, which is handled by the ROS package [14]. Another ROS program is running at the back-end system which is connected with the master ROS via, a wireless network and subscribes topics published by master ROS like images, odometry and IMU data. Back-end system uses odometry and IMU data to get pose parameters for some selected key frames and initiate the 3D reconstruction of the entire environment. The detail of 3D reconstruction process is explained in the next sections.

2.2 Camera Calibration

The camera mounted on the servo of the robot is an off-the-shelf webcam and it is fixed throughout our experiments. Pin-hole camera model [1] convention is used and zoom factor of the camera and resolution of captured images are kept constant for a fixed internal calibration matrix. The internal calibration process is performed offline using well known checker board methods as described by Zhang in [15, 16].

Orientation matrix of camera is calculated from servo angles and heading information of robot. Robot pose information can be estimated accurately by fusing odometry and IMU using Extended Kalman filter (EKF) data. EKF cleans up the noises in odometer data [17, 18]. The transformation between servo and robot body frame of reference would give the orientation angles of the camera with respect to the world coordinate system. The orientation matrix (R) of the camera is computed using (1).

$$R = Rz * Ry * Rx. \tag{1}$$

where R_i denotes the rotation matrix along axis i.

Since robot is moving in x–z plane and there is no camera rotation along X-axis (tilt angle), hence camera orientation depends only on robot rotation angle θ and camera pan angle α along Y axis. In this case Rz and Rx will be identity matrix and Ry will be given as shown in (2).

$$\begin{bmatrix} \cos(\alpha + \theta) & 0 & \sin(\alpha + \theta) \\ 0 & 1 & 0 \\ -\sin(\alpha + \theta) & 0 & \cos(\alpha + \theta) \end{bmatrix} \tag{2}$$

2.3 Dense Stereo Matching

The dense stereo matching is vast and we refer to [19] for a comparison of all existing methods. In fact, there are few relevant works available on real-time, dense reconstruction using a monocular moving camera.

Motion estimation by means of optical flow is standard technique for providing dense sampling in time. The predominant way of estimating dense optical flow in today's computer vision literature is by an approach of integrating rich descriptors into the variational optical flow setting as described in [20]. The main advantage of the selected approach is the ability to produce better results in a wide range of cases and also for large displacement.

Large displacement optical flow is a variational optimization technique which integrates discrete point matches with continuous energy formulation. The final goal is to find the global minima of the energy and for that the initial guess of the solution has to very close to the global minima. The entire energy is globally minimized and the details of minimization procedure are studied in [20].

The given approach is not directly applicable for any near real-time system because of its high computation. The running time between a pair of frames of 640 × 480 resolutions on a 2.13 GHz is about 55 s. The performance in terms of time is drastically improved by the use of general purpose graphics processing unit (GPU) as described in [21]. The parallel implementation on a GPU yields about 78 times faster performance compare to a serial C++ version. This implementation is further used in a dense 3D reconstruction with a hand held camera [22] where the system is not in real-time due to the tracking of every frame in captured video sequence.

The n-view point correspondence generation is carried out using the GPU implementation as described in [21]. The point trajectories are generated between some selected key frames from the captured video. Optical flow has an effect of accumulating errors in the flow vector. So, a long trajectory suffers from this error and leading to a significant drift. Short trajectories are almost free from the drift error, but the triangulation process suffers due to small base line measurements. Hence, we have not chosen consecutive frames rather selected frames that are having base line about 10 cm and the trajectory length chosen as less than 15 in our experimental setup.

2.4 Outlier Detection

Detecting outlier is a very primitive task before doing any further processing with the available information. Outlier detection process is very straight forward and it follows the epipolar constraints [1] as shown in (3). Accurate camera calibration estimation produces a better estimation of pair wise fundamental matrix (F) [1] which is used for noise cleaning. The corresponding points (x, x') ideally should follow the epipolar constraints as given in (3) [1]. In reality, the value never becomes zero rather it goes very close to zero. We used a dynamic threshold based on percentage of rejection because static threshold does not hold good for different type of scene. The threshold is always consider as below of 3 pixels.

$$x'^T Fx = 0 \tag{3}$$

2.5 3D Model Generation

The 3D point cloud is created using well known triangulation process as described in [23]. Each point is back projected onto the image plane to calculate the back projection error. Any 3D point with back projection error more than 3 pixels is considered as outlier.

The whole scene reconstruction is done in an incremental way. Images are divided into small sets such that the trajectory length is not more than 15 images. Each subset is merged after triangulation to get the final reconstruction. The scale rectification is done using IMU sensors [24].

3 Results

The implementation environment consists of Firebird VI robot as shown in Fig. 1 and a back-end system having Intel(R) Xeon(R) E5606 processor running at 2.13 GHz along with a NVIDIA Tesla C2050 Graphic Card. One ZOTAC ZBOXHD-ID11 is mounted on top of the robot. ROS hydro is installed inside Ubuntu 12.04 LTS in all the systems.

The entire capture task is running on the ZOTAC box mounted on the Firebird VI. Image is captured with 640 × 480 resolution using a Logitech C920 webcam. The 3D model reconstruction is carried out on the backend system due to less processing power of ZOTAC box.

3.1 Outputs

We presented two sample reconstruction sequences in different environment to demonstrate the robustness and usability of our solution. The presented samples contain several images which are affected by occlusion, motion blur.

Figure 3 shows a sample reconstruction of a wall in an indoor office environment of size 13 × 9.5 feet. The reconstruction is carried out with only 7 images in a single iteration. The reconstruction shows noisy output at the right bottom part and

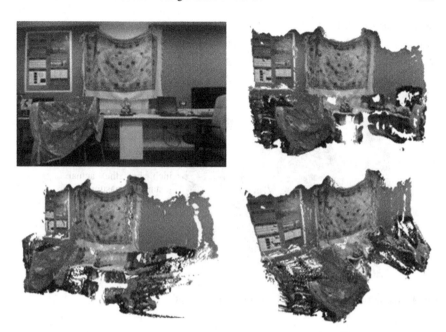

Fig. 3 *Topleft* sample image for reconstruction, *Rest* 3D point cloud after reconstruction

it is due to the fact that images that are used for reconstruction are taken only from the frontier side of the wall.

In Fig. 1 we presented another result where the data is captured in a living room. The three sides of the room are captured where different objects are placed. The dimension of the room is about 13 × 11 feet. Another data presented in Fig. 1 is captured of outdoor environment from more than hundred feet. The user can guide the robot to go closer and capture the frames to produce a more accurate and dense points in any required portion of the environment.

The presented samples justifies that our proposed solution is capable of reconstructing both indoor and outdoor environments without any size limit which is the basic requirement for any 3D reconstruction system. The robot locomotion is guided through user and this is advantageous to focus onto a specific object in the environment for better observation through high quality reconstruction.

3.2 Execution Time

The timing details of the samples presented above is given in Table 1.

Table 1 Timing details

Data sets	Frames	Tracking	Inlier detection and triangulation	Total
Wall	7	13	3 s	16 s
Living room	62	127	25 s	152 s

The distance on the reconstruction is measured as 21.2651 inch but the actual distance measured on the real structure as 21.25 inch.

Fig. 4 Reconstructed structure verification with ground truth

3.3 Ground Truth Verification

The accuracy of our results verified against the ground truth by measuring the distance between two points. One such measurement comparison is shown in Fig. 4 where outer distance between two legs of the reconstructed chair is 21.2651 in. and the actual measured value is 21.25 in. It shows our work accuracy is 98–99 %.

4 Conclusion

We have presented an approach for dense 3D reconstruction of any uncontrolled indoor and outdoor environment through monocular robotic vision, IMU and odometry. Our results shows that proposed work is useful for constructing 3D of indoor and outdoor environment in near real time. Further work is planned to integrate an IR thermal sensor and fuse thermal information onto the 3D structure to create an opto-thermal 3D and corrosion, erosion measurement.

References

1. Hartley, R., Zisserman, A.: Multiple view geometry in computer vision. Cambridge University Press. ISBN: 0-521-54051-8 (2003)
2. Newcombe, R.A., Lovegrove, S.J., Davison, A.J.: DTAM: Dense tracking and mapping in real-time, ICCV, pp. 2320–2327 (2011)

3. Pollefeys, M., Nister, D., Frahm, D.J.M., Akbarzadeh, A., Mordohai, P., Clipp, B., Engels, C., Gallup, D., Kim, S.J., Merrell, P., Salmi, C., Sinha, S., Talton, B., Wang, L., Yang, Q., Stewenius, H., Yang, R., Welch, G., Towles, H.: Detailed real-time urban 3D reconstruction from video. IJCV **78**(2–3), 143–167 (2008)

4. Pradeep, V., Rhemann, C., Izadi, S., Zach, C., Bleyer, M., Bathiche, S.: MonoFusion: Real-time 3D reconstruction of small scenes with a single web camera. In: The 13th IEEE International Symposium on Mixed and Augmented Reality, pp. 83–88 (2013)

5. Pizzoli, M., Forster, C., Scaramuzza, D.: REMODE: probabilistic, monocular dense reconstruction in real time. In: IEEE International Conference on Robotics and Automation (ICRA), Hong Kong, pp. 2609–2616 (2014)

6. Saha, A., Bhowmick, B., Sinha, A.: A system for near real-time 3D reconstruction from multi-view using 4G enabled mobile. In: IEEE International Conference on Mobile Services (MS), pp. 1–7, (2014)

7. Tanskanen, P., Kolev, K., Meier, L., Paulsen, F.C., Saurer, O., Pollefeys, M.: Live metric 3D reconstruction on mobile phones. In: ICCV, pp. 65–72 (2013)

8. Bhowmick, B., Mallik, A., Saha, A.: Mobiscan3D: A low cost framework for real time dense 3D reconstruction on mobile devices. In: IEEE 11th International Conference on Ubiquitous Intelligence and Computing, IEEE 11th International Conference on and Autonomic and Trusted Computing, and IEEE 14th International Conference on Scalable Computing and Communications and Its Associated Workshops (UTC-ATC-ScalCom), pp. 783–788 (2014)

9. Mallik, A., Bhowmick, B., Alam, S.: A multi-sensor information fusion approach for efficient 3D reconstruction in smart phone. In: International Conference on Image Processing, Computer Vision, and Pattern Recognition (IPCV), pp 291–298 (2015)

10. Davison, A.: Real-time simultaneous localisation and mapping with a single camera. In: IEEE International Conference on Computer Vision, pp. 1403–1410 (2003)

11. Deshpande, P., Reddy, V.R., Saha, A., Vaiapury, K., Dewangan, K., Dasgupta, R.: A next generation mobile robot with multi-mode sense of 3D perception. In: International Conference on Advanced Robotics (ICAR) Istanbul, pp 382–387, (2015)

12. Firebird VI. http://www.nex-robotics.com/fire-bird-vi-robot-platform.html (2015). Access 20 Oct 2015

13. Martinez, A., Fernández, E.: Learning ROS for robotics programming. PACKT Publishing Ltd. (2013). ISBN: 978-1-78216-144-8

14. ROS usb Camera Package. http://wiki.ros.org/usb_cam (2015). Accessed 20 Oct 2015

15. Zhang, Z.: Flexible camera calibration by viewing a plane from unknown orientations. In: International Conference on Computer Vision (ICCV'99), pp 666–673, (1999)

16. Zhang, Z.: A flexible new technique for camera calibration. IEEE Trans. Pattern Anal. Mach. Intell. **22**(11), 1330–1334 (2000)

17. ROS Robot Pose EKF Package. http://wiki.ros.org/robot_pose_ekf (2015). Accessed 20 Oct 2015

18. ROS Robot Localization Package. http://wiki.ros.org/robot_localization. Accessed 20 Oct 2015

19. Hirschmuller, H., Scharstein, D.: Evaluation of stereo matching costs on images with radiometric differences. IEEE Trans. Pattern Anal. Machine Intell. **31**(9), 1582–1599 (2009)

20. Brox, T., Malik, J.: Large displacement optical flow: descriptor matching in variational motion estimation. IEEE Trans. Pattern Anal. Mach. Intell. **33**(3), 500–513 (2011)

21. Sundaram, N., Brox, T., Keutzer, K.: Dense point trajectories by GPU-accelerated large displacement optical flow. In: European Conference on Computer Vision (ECCV), pp. 438–451, Crete, Greece, Springer, LNCS (2010)

22. Ummenhofer, B., Brox, T.: Dense 3D reconstruction with a hand-held camera. Springer, Berlin Heidelberg (2012)

23. Hartley, R.I., Sturm, P.: Triangulation. Comput. Vis. Image Underst. **68**(2), 146–157 (1997)

24. Nützi, G., Weiss, S., Scaramuzza, D., Siegwart, R.: Fusion of IMU and vision for absolute scale estimation in monocular SLAM. J. Intell. Robot Syst **61**(1–4), 287–299 (2011)

Real Time Pedestrian Detection Using CENTRIST Feature with Distance Estimation

Kaushal Joshi, R. Kavitha and Madhu S. Nair

Abstract Pedestrian Detection (PD) is an active research area for improving road safety. Most of the existing PD system does not meet the demanded performance. This paper presents a working PD system which improves performance. The system uses CENTRIST feature extractor and the linear Support Vector Machine (SVM) for training and detection of pedestrian. CENTRIST is very easy to compute without any preprocessing and normalization that makes it suitable for on-board system. During the training procedure, we exhaustively searched for negative samples. Detection results on INRIA dataset are more accurate compared to benchmark method HOG. We used monocular camera to estimate pedestrian distance which is fairly accurate. We apply our detector on real-time video without region of interest (ROI) selection and could achieve 7 fps detection speed.

Keywords ADAS · Pedestrian detection · CENTRIST · SVM · Distance estimation · Monocular camera

1 Introduction

Due to tremendous growth of automobile industry over the last century, road accidents have become an important cause of fatalities. In 2012, 4743 pedestrians were killed and an estimated 76,000 were injured in traffic crashes in the United States. On an average, a pedestrian was killed every 2 h and injured every 7 min in

K. Joshi · M.S. Nair (✉)
Department of Computer Science, University of Kerala, Kariavattom,
Thiruvananthapuram 695581, Kerala, India
e-mail: madhu_s_nair2001@yahoo.com

K. Joshi
e-mail: kaushalknack7210@gmail.com

R. Kavitha
Tata Elxsi Limited, ITPB Road Whitefield, Bangalore 560048, India
e-mail: kavitha.r@tataelxsi.co.in

© Springer Science+Business Media Singapore 2016
R.K. Choudhary et al. (eds.), *Advanced Computing and Communication Technologies*,
Advances in Intelligent Systems and Computing 452,
DOI 10.1007/978-981-10-1023-1_23

traffic crashes [1]. So we must have a reliable safety system which help to avoid the road accidents. Society also expect more intelligent vehicles, which are capable of assisting the driver in driving process. Such intelligent vehicles are equipped with Advanced Driver Assistance System (ADAS). ADAS is a system which control vehicle to improve road safety. While driving it assist driver by visual or audio or vibrate alarm. ADAS is a collection of subsystems like Pedestrian detection, Collision avoidance, Traffic sign recognition, Lane departure assistance, Adaptive cruise control and Parking assistance.

Pedestrian detection for ADAS is still a very challenging task. Pedestrian appears very different in different conditions like change in pose, different clothing style, carrying some object, having different size, weather conditions and cluttered background. So, developing a working system which overcome all of these challenges is not trivial. In the last decade, there has been a significant progress within pedestrian detection. The quality of pedestrian detection system depends on the features extracted, classifiers and datasets used. In 2003, Viola et al. [2] presented a method which uses intensity and motion information as features and trained a detector using AdaBoost. Dalal and Triggs [3] presented Histograms of Oriented Gradients (HOGs) feature and trained a liner SVM classifier. HOG became most popular feature for pedestrian detection. After HOG-SVM, other authors showed more detection accuracy using different features and classifier combination in [4–6]. Some recent works have got good pedestrian detection accuracy like in Zhang et al. [7], Benenson et al. [8] and Costea and Nedevschi [9], where [7] is based on informed haar feature and AdaBoost classifier with maximum detection accuracy on every pedestrian dataset. In [8] detector is based on HOG and color based feature using linear SVM classifier [7, 9] needs GPU for fast pedestrian detection. These detectors are still far from desired performance for on-board pedestrian detection system.

There is a lack of complete working PD system with balanced accuracy and speed of detection. After detection of pedestrian, we must give distance information to system to take appropriate action (i.e. give alarm to driver or decrease speed automatically). Combination of multiple features makes system complex and slow during detection. So we present here a working PD system which uses a single feature and achieve fast detection speed. We showed that proper training leads to better detection accuracy, and using monocular camera we can get almost correct distance.

2 Proposed Work

The paper describes a real-time PD scheme that uses CENTRIST feature proposed by Wu and Rehg [10]. To classify the extracted CENTRIST features, the method uses linear SVM [11]. The system uses Efficient Sub-window Search (ESS) by Lampert [12] which help to improve performance. It is also uses Non Maximum Suppression (NMS) algorithm to eliminate multiple detection. This system uses

(a) **(b)**

Fig. 1 **a** Sample image dataset from INRIA and **b** sample image dataset from MIT

camera geometry to estimate distance between camera and pedestrian using monocular camera. The system detect pedestrians and also estimate the distance in each video frame or image.

2.1 Training Phase

Training phase has two modules: feature extraction and classifier learning. The training samples are collected from INRIA [13] and MIT datasets [14], and some of the sample images are shown in Fig. 1.

This system used CENTRIST (CENsus TRansform hISTogram) [8] feature vector for training the linear SVM classifier. Earlier, CENTRIST was developed as a visual descriptor for recognizing topological places or scene categories. CENTRIST mainly encodes the structural properties within an image and suppresses detailed textural information. To get the Census Transform (CT) value of a pixel, compare the intensity with its eight neighboring pixels. If the center pixel is greater than or equal to one of its neighbors, a bit 1 is set in the corresponding location. Otherwise a bit 0 is set. The generated eight bit can be put together, which is consequently converted to a base-10 number (CT value). Figure 2 shows the conversion of a pixel value to CT value. Convert all pixel intensity values to CT values in the image and the resulting image is called a CT image.

For training, the method used small gray-scale image patches with 108×36 resolution and converted these image patches to CT images. The CT image is then divided into 9×4 blocks. Assuming 2×2 neighbor block as a super-block, extract 256 bin histogram of CT value from every super-blocks. Super-block is 50 % overlapped to its neighboring super-blocks. Concatenate all histograms and form a 6144 (24×256) dimension feature vector, which h represent an image patch. Generate 6144 dimension feature vectors for all training samples. Figure 3 shows an arrangement of blocks and super-blocks for an image patch.

Fig. 2 Convert a pixel value
to CT value

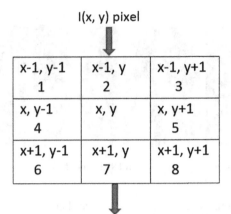

I(x, y) pixel

x-1, y-1	x-1, y	x-1, y+1
1	2	3
x, y-1	x, y	x, y+1
4		5
x+1, y-1	x+1, y	x+1, y+1
6	7	8

Compare the intensity of a pixel I(x, y) with
its eight neighboring pixels
If I(x, y) >= neighbor then, a bit 1 is set in
the corresponding location, else a bit 0.

$$(b_1\, b_2\, b_3\, b_4\, b_5\, b_6\, b_7\, b_8)_2 = (\text{Decimal})_{10} = CT(x, y)$$

Fig. 3 Divide 108 × 36
pixel CT image into 9 × 4
blocks = 24(8 × 3)
super-blocks

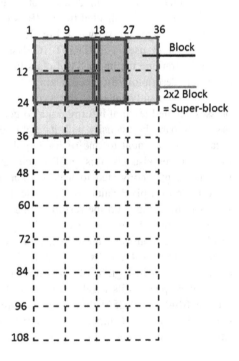

2.2 Detection Phase

The detection phase is comprised of five modules (i) image scaling, (ii) feature extraction, (iii) classification, (iv) NMS and (v) distance estimation as explained below.

Image Scaling To perform a full image detection, we slide a window over the whole image and resize the input image to detect pedestrians of different scales. Pedestrian close to the vehicle appear with more pixel height compared to the pedestrian far from the vehicle in the image. To detect pedestrians who are close to the vehicle, down-scale the input image and then search for pedestrians.

$$Scale(I, S) = I_1; (I_1, S) = I_2; \ldots (I_{n-1}, S) = I_n, \text{ where } I_n > \text{window size} \quad (1)$$

Scale (I, S) is a scaling function which resize the image, I. $S = (S_x, S_y)$ is a scaling factor along the horizontal axis (S_x) and vertical axis (S_y). One criterion to be satisfied here is that the size of the scaled image (In) must be greater than the detection window size.

Feature Extraction In the proposed method, we used CENTRIST feature for pedestrian detection, as it is easy to compute, fast and there is no need of any pre-processing. If we replace all bit 0 to 1 and 1 to 0 in Fig. 2 (8 bit binary sequence) then the intermediate result is Local Binary Pattern (LBP) [15]. The superiority of CENTRIST over well established conventional feature such as LBP is that, CENTRIST encode global structure of pedestrian better than LBP.

After rescaling the image, convert whole grayscale image into a CT image by modifying all pixel values to CT values as explained in Sect. 2.1 and Fig. 2. Now take the CT image patch of size 108 × 36 as search window, and extract CENTRIST feature from the search window. Here search window size is same as the training image patches shown in Fig. 3.

Classification Let C represent the scaled CT image corresponding to the input image I. Using a search window, extract 6144 dimension feature vector (i.e. f $\in$ R^{6144}). If we have already trained a linear classifier w $\in$ R^{6144}, then a search window is classified as an object of interest if and only if $w^T f \geq 0$. Inspired by Efficient Sub-window Search (ESS), Wu et al. [16] proposed an algorithm to compute $w^T f$ using a fixed number of machine instruction, without generating feature vector f. Generate an auxiliary image A by replacing the values of the CT image by its corresponding weight value w_i (obtained from trained classifier). Then the *score* is calculated as the sum of all the values of A, which is equivalent to w^T f. If the score is greater than 0, then a pedestrian is detected in the search window. So based on the coordinate of search window, draw a bounding box on the image, which surrounds the detected pedestrian.

Non Maximum Suppression (NMS) The system uses NMS to avoid multiple detections of same pedestrian. Multiple detection may occur because of (i) multi-scale search and (ii) sliding window approach, which leads to more than 90 % overlap to its neighbor. The method uses intersection based multiple detection removal and keep only a bounding box having highest detection score among all intersecting bounding boxes. Suppose we have 2 detection box A and B, where coordinate of box is (top, left, bottom, right = x_1, y_1, x_2, y_2) then:

$$
\begin{aligned}
\text{Intersection area}, I = {} & \max(0, \min(A \cdot x_2, B \cdot x_2) - \max(A \cdot x_1, B \cdot x_1)) \\
& \times \max(0, \min(A \cdot y_2, B \cdot y_2) - \max(A \cdot y_1, B \cdot y_1))
\end{aligned} \tag{2}
$$

Distance Estimation Distance estimation is a crucial step in PD systems as it is needed to take appropriate action in time (i.e. giving alarm to driver, automatically decrease the speed of vehicle etc.). The method uses monocular camera for pedestrian detection and distance estimation. Using camera similar triangle property in Eq. (3), we can find the distance from camera to pedestrian based on the following mathematical relationship:

$$
\frac{\text{Pixel height of pedestrian}}{\text{Focal length of camera}} = \frac{\text{Pedestrian height in real world}}{\text{Distance from camera to pedestrian}} \tag{3}
$$

Focal length is a camera specific parameter. To find the focal length of the experimental camera, we captured an object of known size at a known distance. In Fig. 4, the object (paper) of length 30 cm is placed at a distance of 100 cm from the camera. The pixel length of the object can be found out from the captured image (say x), then the focal length = $(x \times 100)/30$.

Fig. 4 Captured object at known distance to find the focal length of the camera

3 Experimental Analysis

To measure the performance of the proposed method, we used the ground truth of INRIA dataset and the matching criteria specified in [17]. A detection window R_d and a ground truth window R_g is considered as similar if:

$$\frac{Area\left(R_g \cap R_d\right)}{Area\left(R_g \cup R_d\right)} \geq 0.5 \tag{4}$$

For performance evaluation we used the standard metrics such as (i) False Positives Per Image (FPPI) and (ii) Miss Rate (MR). FPPI indicates the average number of false windows present in one image. Miss rate is the ratio of missed pedestrians and total pedestrians in the test dataset. MR and FPPI can be defined as:

$$MR = \frac{Total\ Missed\ Pedestrians}{Total\ Pedestrians} \tag{5}$$

$$FPPI = \frac{Total\ False\ Positives}{Total\ Number\ of\ Images} \tag{6}$$

The proposed method has been compared with the standard HOG based detector based on the quantitative metrics mentioned above. To compare the detectors we plot miss rate against false positives per image (using log-log plots) by varying the threshold on detection score, as shown in Fig. 5. It is evident from the figure that the proposed method have lesser miss rate (0.18) compared to HOG (0.23).

To check the detection speed we applied our method on a real video with 640×480 resolution. The quantitative analysis of standard HOG and the proposed method using different performance metrics are shown in Table 1. The proposed method achieved a detection speed of 7 fps and better detection accuracy rate of

Fig. 5 Performance on the INRIA dataset

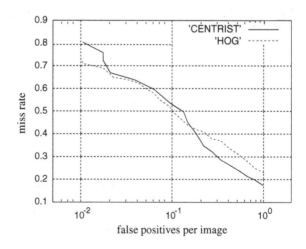

Table 1 Detection on 640 × 480 video frames

Methods	Miss rate	Detection accuracy	Average detection time (fps)
HOG	0.23	77	0.239
Proposed method	0.18	82	6.8

Fig. 6 Time to process one frame

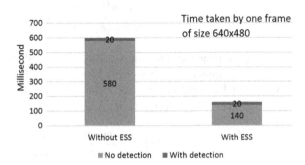

82 % compared to HOG based system with a speed of 0.239 fps and a detection accuracy rate of 77 %, for 640 × 480 resolution frames.

We trained a linear SVM classifier with 3341 positive samples from INRIA (2416) and MIT (925) datasets and an initial negative set of 12,180 patches. We used training methodology explained in [3]. During detection we used ESS technique which speed-up the detection process by 3–4 times as shown in Fig. 6.

To check the accuracy of the distance estimation, we found the average distance error, by comparing the estimated distance and the actual distance, which is approximately 0.4994 m. Figure 7 shows the difference between actual and estimated distances of some experimental samples. Detection result obtained based on

Fig. 7 Plot of actual and estimated distance

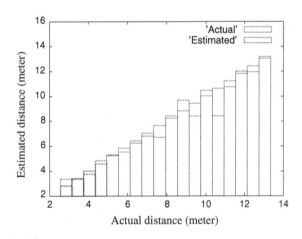

Fig. 8 Detection on real video frame and INRIA dataset image with low illumination

real video and INRIA dataset, shown in Fig. 8, proves that the system works fine in different illumination and blurring conditions.

4 Conclusion

In this paper a working PD system using CENTRIST feature is proposed, which is fast and doesn't need any pre-processing. Through experimental analysis it has been found that the proposed system is robust to illumination changes and blurring conditions. Combination of CENTRIST and linear SVM is sufficient enough to discriminate between pedestrian and non-pedestrian, if training is proper. Through experimental analysis it is proved that the proposed method works well in situation where the camera may lost its focus because of shake resulting in blurred video being recorded. As a future work, we are planning to restrict the search window to ROI so that the detection speed can be enhanced to 3–4 times faster, compared to the current detection speed.

References

1. Traffic Safety Facts. http://www-nrd.nhtsa.dot.gov/Pubs/811888.pdf (2015). Accessed on 05 Aug 2015
2. Viola, P., Jones, M.J., Snow, D.: Detecting pedestrians using patterns of motion and appearance. CVPR **2**, 734–741 (2003)
3. Dalal, N., Triggs, B.: Histograms of oriented gradients for human detection. CVPR **1**, 886–893 (2005)
4. Dollár, P., Tu, Z., Perona, P., Belongie, S.: Integral channel features. In: BMVC, pp. 91.1–91.11 (2009)
5. Dollár, P., Belongie, S., Perona, P.: The fastest pedestrian detector in the west. In: BMVC, pp. 68.1–68.11 (2010)

6. Benenson, R., Mathias, M., Timofte, R., Van Gool, L.: Pedestrian detection at 100 frames per second. In: CVPR, pp. 2903–2910 (2012)
7. Zhang, S., Bauckhage, C., Cremers, A.B.: Informed haar-like features improve pedestrian detection. In: CVPR, pp. 947–954 (2014)
8. Benenson, R., Mathias, M., Tuytelaars, T., Van Gool, L.: Seeking the strongest rigid detector. In: CVPR, pp. 3666–3673 (2013)
9. Costea, A.D., Nedevschi, S.: Word channel based multiscale pedestrian detection without image resizing and using only one classifier. In: CVPR, pp. 2393–2400. (2014)
10. Wu, J., Rehg, J.M.: CENTRIST: A visual descriptor for scene categorization. PAMI 33(8), 1489–1501 (2011)
11. Cortes, C., Vapnik, V.: Support-vector networks. Machine learning, vol. 20, no. 3, pp. 273–297. Springer (1995)
12. Lampert, C.H., et al.: Efficient subwindow search: a branch and bound framework for object localization. PAMI 31(12), 2129–2142 (2009)
13. INRIA Person Dataset. http://pascal.inrialpes.fr/data/human/ (2015). 05 Aug 2015
14. MIT Pedestrian Data. http://cbcl.mit.edu/software-datasets/PedestrianData.html (2015). Accessed 05 Aug 2015
15. Ojala, T., Pietikäinen, M., Mäenpää, T.: Multiresolution gray-scale and rotation invariant texture classification with local binary patterns. PAMI 24(7), 971–987 (2002)
16. Wu, J., Liu, N., Geyer, C., Rehg, J.M.: A real-time object detection frame work. IP. IEEE Trans. 22(10), 4096–4107 (2013)
17. Dollár, P., Wojek, C., Schiele, B., Perona, P.: Pedestrian detection: an evaluation of the state of the art. PAMI 34(4), 743–761 (2012)

Feature Based Reading Skill Analysis Using Electrooculogram Signals

D'Souza Sandra and N. Sriraam

Abstract Developing reading skills is an individualistic characteristic and it differs from person to person. Since the eye movements can be intimately correlated to the reading process, by critical observation of the movement of the eyes reading process can be analysed and studied. This research work conducts a pilot study to propose and investigate the importance of eye movements in reading skill analysis. We have considered Electrooculogram (EOG) recorded signals from a group of ten healthy volunteers, of which are five normal readers and five poor readers. Simulation results show a classification accuracy of 67.7 and 88 % using Yule-Walker's and Burg's estimation methods respectively for horizontal EOG and 74 and 81 % for vertical EOG. The Burg's estimation method stands out better for classification of reading skills. The results indicate the suitability of proposed scheme for identifying the poor readers and hence provide required assistance to people with reading disabilities.

Keywords EOG · Reading · AR model parameters · Classification · Artificial neural network

1 Introduction

Reading is a very important activity in the life of any individual. For many people, reading is a recreational activity. Though reading is an inevitable activity in most of the persons, in several others the reading skills are less developed. Eye movements

D. Sandra (✉)
Department of Instrumentation & Control Engineering,
Manipal Institute of Technology, Manipal, Karnataka, India
e-mail: sandra.dsouza@manipal.edu

N. Sriraam
Centre for Medical Electronics & Computing, M.S. Ramaiah Institute
of Technology, Bangalore, Karnataka, India
e-mail: sriraam@msrit.edu

© Springer Science+Business Media Singapore 2016
R.K. Choudhary et al. (eds.), *Advanced Computing and Communication Technologies*,
Advances in Intelligent Systems and Computing 452,
DOI 10.1007/978-981-10-1023-1_24

have a very important role in analysis of reading process. Analysis of reading traits of an individual by critical observation of his eye movements gives many insights about the reading process [1]. This is possible only when the eye movement analysis is automated. Reading skills have to be developed in childhood. Many a times, children have reading difficulty. Eye movements can be used to identify the specific problems of the reader. It is possible to identify which words or which part of the text is difficult for the reader by mapping eye fixations on the text [1]. This information is very helpful especially in child readers in two ways. Firstly, it can help in finding the difficulties of the child in reading and secondly, new teaching strategies and methods may be developed based on the child's personal strength or weaknesses. The teacher can monitor and further adjust teaching strategies based on the child's progress through the observation of eye movements [2]. Hence reading difficulties can be overcome and a great improvement will be guaranteed in a child's reading abilities.

Typically, during reading English text, the fixation duration is in the range 200–300 ms [3]. Normally saccades move through the text in the forward direction, backward saccades or regressions are exhibited by a proficient reader [3]. Basically, four classes of eye movements are identified in the literature. The saccades, are the movements made by the eyes when rapidly moving to a specific target, are prominently the movements which occur during reading or scanning any image. Saccadic eye movements or saccades are the most important eye movements that occur during reading [1–4]. Preceding and following the saccades are the pauses, also called as fixations. The brain obtains its information through the eyes only during fixations [3]. The second type of movements is pursuit eye movements which are smooth movements made while following a moving target. The third type of eye movement is conjugate eye movement. They are characterized by preserving an angular relationship between right and left eyes. When the subject focuses from a distant object to a near target and back to distant object, the electrical activity of the muscles causes the eye movements known as vergence movements.

Certain neuronal disorders like Schizophrenia, attentional disorders and hyper-activity, have been assessed based on the information on the types and character-istics of eye movements. Eye movements also contribute as indicators for assessing reading abilities [3, 4]. The eye movement tracking systems using infrared corneal reflections, video based methods, magnetic search coils etc., have been used. These methods are highly expensive, time consuming and complicated. They cause dis-comfort to the subjects. In this research work, a non-invasive technique called Electrooculography is used. Electrooculography is a technique that records the horizontal and vertical movements of the eyes by using separate electrodes for recording horizontal and vertical eye movements. Electrooculography system using virtual instrumentation is found to be economical and causes least discomfort to the subjects [5]. Several research articles highlight the fact that there is close correlation between the reading process and the eye movement pattern thus obtained. However, to the best knowledge of the authors, the artificial neural network based automated analysis of EOG for reading, to identify poor and a normal reader is a novel work.

Since reading involves both horizontal and vertical eye movements, the dual channel system of recording is used. The primary objective of this work is to evaluate the significance of autoregressive features in discriminating good and poor readers based on their eye movements. The wider aim extends to identifying the reading patterns of retarded or poor readers and hence find out the causes of their difficulty and provide strategies to assist them.

The major contribution of the work are (1) Electrooculogram data collection of 10 healthy volunteers during reading a passage for a duration of 10 min (2) Extraction of Autoregressive based features from the data set (3) Classification of poor readers from the normal readers using backpropagation artificial neural network and hence developing a software based tool for the application.

The paper is organized as follows: Sect. 2 includes the review of Electrooculography method, eye movements in reading, data acquisition and feature extraction and classification, Sect. 3 discusses the experimental results and performance evaluation of the proposed work and in Sect. 4 the conclusions are drawn.

2 Methods and Materials

This section explains the basic principles of Electrooculography, relation between eye movements and reading, the data acquisition, feature extraction and classification.

2.1 Electrooculography

The human eye is a source of bio-potentials. A potential exists between cornea and retina of the eye, where the cornea is positive and retina has negative polarity. This potential difference can be measured using a technique known as Electrooculography. Electrooculography makes use of non-invasive technique to measure the electric potential changes due to the movement of the eyes. The electrical signal corresponding to eye movements is known as Electrooculogram (EOG). This electrical signal varies in proportion to the movements of the eye balls [6]. There are several methods for acquiring eye movements; however, Electrooculography is a method that is preferred well due to its non-invasiveness, simplicity in usage, minimal discomfort to the subject and suitability for long time monitoring. Besides its use in clinical research, ophthalmological diagnosis and laboratories, EOG is currently used actively in developing assistive technologies and human computer interfaces based on the eye movements [7–11]. The ease and minimal discomfort to the subject, makes EOG a suitable choice for recording the eye movements.

Eye movements may be involuntary or sometimes voluntary and they occur continuously during any task. The movements can be horizontal eye movements and vertical eye movements. The horizontal movement of the eyes can be recorded by two electrodes placed on the canthi, and vertical movements can be recorded by placing electrodes just above and below eyes. Thus a dual channel data is obtained. A common reference electrode is placed on the forehead of the subject. With the eye at rest (no eye ball movement), the electrodes are effectively at the same potential and there is no change in potential. When the eye rotates, it causes a difference in potential. When the eye rotates to the right or left, it results in potential difference, with the electrode in the direction of movement, becoming positive with respect to the other electrode. Similarly when there is up and down movement potential difference occurs between top and bottom electrodes. The polarity of the signal is positive at the electrode towards which the eye is moving [12]. The observation of EOG recordings during reading gives a clear distinction between a normal reader and a retarded reader.

2.2 Eye Movement and Reading Development

As eye movements are an integral part of reading process, it is closely related to the process of reading. The development of reading skills is not immediate but a gradual process, and it is known that over the years the speed of reading and accuracy improves. An observation of the pattern of eye movements while reading shows a staircase pattern (Fig. 1). In the figure each staircase represents a line of text. The length of the staircase depends on the time taken by the reader to complete each line [1]. Fast reading shows shorter length staircase while a slow reading shows a longer staircase. The typical waveforms while reading are shown in the Fig. 1.

Saccades are the continual eye movements and between the saccades when the eye remains still, it is known as fixations, the typical fixation duration is 200–300 ms. During saccades eye movements has very high velocity up to 500° per second.

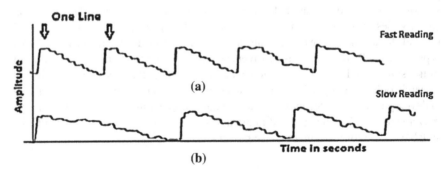

Fig. 1 Illustration of eye movements **a** normal reader **b** poor reader (*Source* Pavlidis G. Th.)

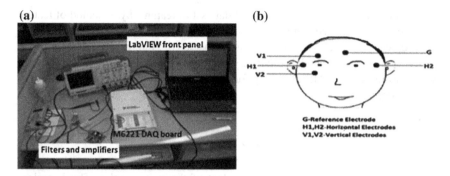

Fig. 2 EOG Acquisition **a** data acquisition system **b** electrode placement

2.3 Data Acquisition and Filtering

There has been continuous effort to design and develop cost effective EOG acquisition systems which are user friendly. In the current research, EOG signal acquisition is done using the virtual instrumentation based data acquisition system [5]. The typical amplitude range of EOG is 10–100 µV. Since the major components of EOG lie in the frequency range of 0.05–30 Hz, band pass filter with a cut-off frequency of 0.05 and 30 Hz is used for filtering. The 50 Hz notch filter used removes the power line interference. Silver (Ag)-Silver Chloride (AgCl) electrodes are attached after cleaning the skin surface using electrolytic gel for better contact and conductivity. Smoothing the waveform is done using finite impulse response filter designed using Bartlett window. National Instruments M Series USB-6221 is used as the data acquisition card to interface with PC. The signal acquisition and electrode placement is shown in Fig. 2.

The data acquisition is done at the rate 200 samples per second. We have considered a recording of 10 min for this study. Hence the data set contains 1, 20,000 samples per individual. Hence data reduction is done by selecting feature extraction of the data set. The recorded EOG pattern depends on the number of lines and the number of words in each line. Also important are the length of each line, and the formatting used for the reading text. The subject is instructed not to move the body and head while reading. Same text is given for all the volunteers. No physical movements like movement of the head and body are permitted during recording. Only eye movements are permitted during reading.

2.4 Extraction of Features

The EOG data set includes 1,20,000 samples per individual. Hence the data set is huge when the number of volunteers increases. The processing time is large, hence it is required to transform into a reduced data set of features by a method of feature extraction. The extracted features must contain the relevant information of the data. Auto regressive model based features are considered. Auto regressive model (AR model) is a stochastic model, especially useful in representing the biological signals. Auto-regression (AR) of order p for signal $s[n]$ is estimated using (1), where a_k are AR coefficients.

$$s[n] = -\sum_{k=1}^{p} a_k\, s[n-k] + e[n] \tag{1}$$

The noise error term $e[n]$ is independent of the past samples. The interclass variance of the features is high and the intra class variance is low [13–15]. Selection of the model order is done such that it should not over fit the data or lose any data [15]. Features are extracted from the acquired data using the autoregressive models based on Burg and Yule-Walker Model order is selected using Akaike Information Criteria (AIC) which is evaluated as in (2).

$$\mathrm{AIC}(p) = \ln\left(\sigma_p^2\right) + \frac{2p}{N} \tag{2}$$

Order of the model is p, the signal length is N and σ_p^2 is the variance of the error signal. It is required to compute AIC for different orders and the order for which AIC is minimum is the order of the optimized model.

After deciding the model order using Akaike Information Criteria, power spectral density is obtained using the AR coefficients and the error variance, as in (3)

$$s(f) = \sigma_p^2 T \left/ \left|\sum_{k=0}^{p} a_k e^{-j2\pi kT}\right|^2 \right. \tag{3}$$

The Yule-Walker method for AR parameters computation computes the biased estimate of the signal's autocorrelation function. Least squares minimization method is used for forward prediction error [13–15]. Using Burg's method, both the forward and backward prediction errors are minimized. Using Yule-Walker Eq. (4), the recursive estimate of parameters β1, β2, …, βp is done using autocorrelation lag $r_{yy}()$ and noise power variance $\sigma_{\varepsilon p}^2$.

$$\begin{bmatrix} r_{yy}(0) & r_{yy}(1) & \cdots & r_{yy}(p-1) \\ r_{yy}(1) & r_{yy}(0) & \cdots & r_{yy}(p-2) \\ \vdots & \vdots & \ddots & \vdots \\ r_{yy}(p-1) & r_{yy}(p-2) & \cdots & r_{yy}(0) \end{bmatrix} \begin{bmatrix} 1 \\ -\beta 1 \\ \vdots \\ -\beta p \end{bmatrix} = \begin{bmatrix} \sigma_{\varepsilon p}^2 \\ 0 \\ \vdots \\ 0 \end{bmatrix} \tag{4}$$

2.5 Back-Propagation Neural Network Classifier (BPNN)

The computing power of the neural network is credited to parallel distributed structure and its ability to learn. A multilayer artificial neural network (ANN) is a network with feedforward architecture. It consists of an input layer, an output layer and one or more hidden layers [16]. Each neuron has activation function and calculates its outputs and passes to the neuron in the next layer. The neurons in the hidden layers compute their output y_j as in (5),

$$y_j = f\left(\sum w_{ij}x_i\right) \tag{5}$$

where w_{ij} is the weights of the inter-connections and x_i is the input and f is the activation function used. The network needs to be trained. Back-propagation ANNs have been used in various applications. In the back-propagation algorithm errors are propagated backwards. The function of the learning algorithm is to update the weights of the inter connections. The solution to minimizing the nonlinear function is done using the Levenberg–Marquardt algorithm developed by Kenneth Levenberg and Donald Marquardt. The algorithm combines the high speed of the Gauss–Newton algorithm and the stability of the steepest descent method [17]. The weight updation is done according to (6).

$$\Delta w_{ij} = [(J^T(w)J(w) + \mu I]^{-1}J^T(w)E(w) \tag{6}$$

where J is the Jacobian matrix, μ is a constant, I is an identity matrix, and E(w) is an error function [18]. The back-propagation algorithm is chosen for its computational efficiency and accuracy due to back propagation of errors. The training algorithm minimizes the global error E as in (7), where P is the total number of patterns for training and Ep is the error for the training pattern p.

$$E = \frac{1}{P}\sum_{p=1}^{P} E_p \tag{7}$$

With N output nodes, and Oj and tj, are output and target at node j, Ep is calculated using Eq. (8).

$$E_p = \frac{1}{2}\sum_{j=1}^{N}(O_j - t_j)^2 \tag{8}$$

3 Results and Performance Evaluation

Eye movements have been studied for various purposes. Reading has been identified as an activity using Electrooculogram based analysis. Reading has been detected among several activities [19, 20]. Elman network has been used to classify the vertical and horizontal modes. The authors have used Electrooculogram based analysis [21]. In this work, the Electrooculogram signals have been recorded in horizontal and vertical channels for all 10 volunteers. We have considered 5 normal readers and 5 poor readers. The volunteers included both male and female adults of different ages. The experiment was conducted in the research laboratory of the department of Instrumentation and control Engineering. While conducting the experiments, it was observed that the poor readers had longer fixations and shorter saccades. Also the number of fixations were more compared to normal readers. While recording the EOG for normal readers, it was clear that they had shorter fixations, and longer saccades. The reading was faster compared to the poor readers. In this proposed work, the evaluation of the classifier results is performed using Classification matrix and Receiver operating Curves. We have used the classification matrix in terms of percentage and calculated the correct classification and

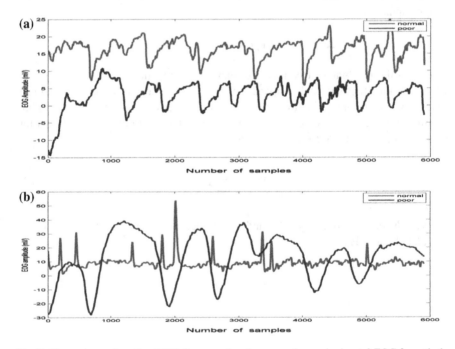

Fig. 3 Comparison of reading EOG for normal and poor readers: **a** horizontal EOG **b** vertical EOG

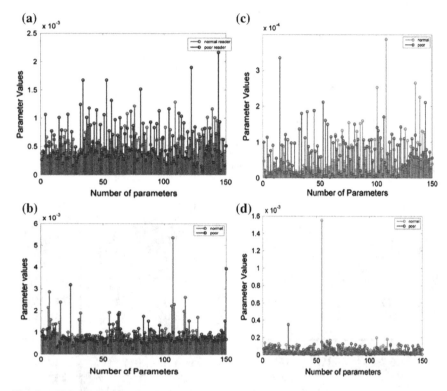

Fig. 4 **a** Yule-Walker's estimation for horizontal EOG, **b** Yule-Walker's estimation for vertical EOG, **c** Burg's estimation for horizontal EOG, **d** Burg's estimation for vertical EOG

incorrect classification percentage. Plotting false positive rate versus true positive rate, an extremely useful comparison of classification results is obtained. This plot is known as Receiver Operating Characteristics (ROC) plot. The two class classification is done considering the normal reader as class 1 and the poor reader as class 2. The raw data recorded while reading is shown is Fig. 3a, b. It can be observed that the waveform for normal reading has a consistent pattern while that of poor readers is irregular.

The feature extracted set contains Yule-Walker's and Burg's parameters. Figure 4 shows the variation of Yule-Walker's and Burg's features for poor and normal readers.

The ROC curves are a relation between sensitivity and specificity of the results. It is plotted using false positive rate versus true positive rate. For high accuracy, ROC curves appear close to the left top of the graph. Hence in this work the accuracy is moderate. Figure 5 shows a ROC curve for Horizontal EOG with the Yule-Walker's features considered.

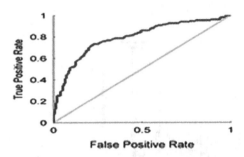

Fig. 5 Receiver operating curves for horizontal EOG

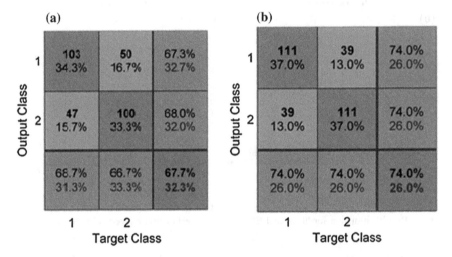

Fig. 6 Confusion matrix using Yule-Walker's estimation **a** horizontal EOG **b** vertical EOG

Confusion matrix is another way to express the classification accuracy. Figures 6 and 7 represent the classification accuracy using horizontal and vertical EOG for Yule-Walker's coefficient and Burg's coefficients respectively. From the ROC curve and confusion matrix it is observed that the classification accuracy is moderate and hence there is scope for further improvement in the system. The percentage indicated by green color in the last blue square of the confusion matrix represents the percentage of classification accuracy. It is observed from the four matrices that accuracy is better using Burg's parameters compared to Yule-Walker's parameters.

Table 1 summarizes the above results of the experimental procedures conducted. It is seen that the Burg's method proves more suitable for the current application.

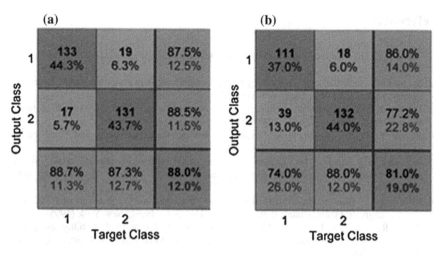

Fig. 7 Confusion matrix using Burg's estimation **a** horizontal EOG **b** vertical EOG

Table 1 Classification results

Estimation method	Classification accuracy (%)	
	Horizontal	Vertical
Yule-Walker's method	67.7	74
Burg's method	88	81

4 Conclusion

Reading skills and Eye movements are closely related. Automating the analysis of Electrooculogram signals for investigation of reading skills is a novel idea. The present work proposes a system for identification of poor readers among the normal readers. A two class classification problem of poor and normal readers has been considered. Back propagation neural network has been used for classification because of its robustness and speed. Feature extraction using auto regressive parameter estimation using Yule-Walker's and Burg's methods has been done. And it is found that the Burg's method is more effective for the current application. The work will be extended to identify better features and also classifiers for the current application. However, the pilot study has opened promising path for the future research in the area of reading skill analysis.

Acknowledgments The authors wish to acknowledge the volunteers for their cooperation in recoding the EOG signal.

References

1. Pavlidis, G.Th.: Eye movements in dyslexia: their diagnostic significance. J. Learn. Disabil. **18**(1) 1985
2. Solan, H.A.: Deficient eye-movement patterns in achieving high school students' three case studies. J. Learn. Disabil. **18**(2), 66–70 (1985)
3. Rayner, K.: Eye movements in reading and information processing: 20 years of research. Psychol. Bull. **124**(3), 372–422 (1998)
4. Rayner, Keith: Eye movements and attention in reading, scene perception, and visual search. Q. J. Exp. Psychol. **62**(8), 1457–1506 (2009)
5. D'Souza, S., Sriraam, N.: Design of EOG signal acquisition system using virtual instrumentation: a cost effective approach. Int. J. Measur. Technol. Instrum. Eng. **4**(1), 1–16 (2014)
6. Malmivuo, J., Plonsey, R.: Bioelectromagnetism: Principles and Applications of Bioelectric and Biomagnetic Fields, Chaps. 9 & 28. Oxford University Press, New York (1995)
7. Barea, R., Boquete, L., Mazo, M., Lopez, E.: System for assisted mobility using eye movements based on Electrooculography. IEEE Trans. Neural Syst. Rehabil. Eng. **10**(4), 209–218 (2002)
8. Postelnicu, Cristian-Cezar, Girbacia, Florin, Talaba, Doru: EOG-based visual navigation interface development. Expert Syst. Appl. **39**, 10857–10866 (2012)
9. Usakli, B., Gurkan, S., Aloise, F., Vecchiato, G., Babiloni, F.: On the use of electrooculogram for efficient human computer interfaces. Comput. Intell. Neurosci., Article ID 135629, 5 pages (2010)
10. Postelnicu, Cristian-Cezar, Girbacia, Florin, Talaba, Doru: EOG-based visual navigation interface development. Expert Syst. Appl. **39**, 10857–10866 (2012)
11. Usakli, B., Gurkan, S., Aloise, F., Vecchiato, G., Babiloni, F.: On the use of electrooculogram for efficient human computer interfaces. Comput. Intell. Neurosci., Article ID 135629, 5 pages (2010)
12. Trikha, M., Bhandari, A., Gandhi, T.: Automatic electrooculogram classification for microcontroller based interface design. IEEE Syst. Inf. Eng. Des. Symp. (2007)
13. Reddy, D.C.: Biomedical Signal Processing, Principles and Techniques, pp. 139–153. Tata McGraw-Hill Publishing Company Limited (2005). ISBN:0-07-058388-9
14. Palaniappan, R.: Introduction to Biological Signal Analysis, Chap. 5. Ventus Publishing (2010). ISBN:8776815943 9788776815943
15. Semmlow John L.: Biosignal and Biomedical Image Processing MATLAB Based Applications, Chap. 3. Marcel Dekker, Inc. (2004). ISBN:0-8247-4803-4
16. Haykin, S.: Neural Networks, a Comprehensive Foundation, Chap. 1. Pearson, Prentice Hall (1998). ISBN: 0132733501
17. Yu, H., Wilamowski, B.M.: Levenberg–Marquardt Training, The Industrial Electronics Handbook, Alburn University, vol. 5, Intelligent Systems, 2nd edn., Chap. 12, pp. 12–1 to 12–15, CRC Press (2011)
18. Guven, A., Kara, S.: Classification of Electrooculoram signals using artificial neural networks. Expert Syst. Appl. 199–205(2006)
19. Bulling, A., Ward, J.A., Gellersen, H., Troster, G.: Eye movement analysis for activity recognition using electrooculography. IEEE Trans. Pattern Anal. Mach. Intell. **33**(4) (2011)
20. Bulling, A., Ward, J.A., Gellersen, H.: Multimodal recognition of reading activity in transit using body-worn sensors. ACM Trans. Applications. Percept. 9, 1, Article 2, 21 pages, (2012). doi:10.1145/2134203.2134205
21. D'Souza, S., Sriraam, N.: Recognition of EOG based reading task using AR features. In: Proceedings of International Conference on Circuits, Communication, Control and Computing (I4C), pp. 113–117 (2014). ISBN:978-1-4799. doi:10.1109/CIMCA.2014.7057770

A Survey of Big Data in Healthcare Industry

Indu Khatri and Virendra Kumar Shrivastava

Abstract "Big Data" are data that are big not only in terms of "Volume" but also in terms of "Value". The exploration of big data in healthcare is increasing at an unprecedented rate. The credit goes to the advanced technologies that help to collect medical data from various sources and to the initiatives that bring deeper insights from these data. This paper presents the exceptional work done by corporations, educational institutions and governments leveraging big data to solve the problems and challenges pertaining to healthcare industry. This paper addresses the ongoing researches; researches that are in initial stages or that are mentioned in the Press Releases to show the advancement of big data in healthcare industry. The paper also proposes a common platform for healthcare analytics, aimed to reduce the redundancy in the techniques that are required in any kind of medical research.

Keywords Big · Data · Cognitive computing · Image processing · Data mining · Brain imaging · Medical imaging · Drug discovery · Diseases detection · Machine learning and deep learning algorithm

1 Introduction

"More Data = More Power + More Benefits" is the theme of this era. The power to retrieve and study heterogeneous healthcare data helps healthcare providers to deliver right intervention to the right patient at the right time and right cost. The healthcare industry generates huge amount of data i.e. in petabyte from Electronic

I. Khatri (✉)
Department of IP Operations & Program Management,
Cerner Healthcare, Bangalore 560045, Karnataka, India
e-mail: indu.khatri@hotmail.com

V.K. Shrivastava
Department of Computer Science & Engineering, APIIT SD India,
Panipat 132103, Haryana, India
e-mail: virendra@apiit.edu.in

© Springer Science+Business Media Singapore 2016   245
R.K. Choudhary et al. (eds.), *Advanced Computing and Communication Technologies*,
Advances in Intelligent Systems and Computing 452,
DOI 10.1007/978-981-10-1023-1_25

Health Records, Clinical Notes, Medical Images, Wearable sensors, Mobile Devices, Genomic Sequences and Social Media etc.

A study conducted in 2014 by EMC Digital Universe (with research and analysis by IDC) shows that healthcare data is increasing at the rate of 48 % per year, establishing Healthcare as one of the fastest growing segments of the market [1]. The data are estimated to increase up to 2,314 Exabytes (10^{18}) as per the study within next 5 years. The expanding volume of healthcare data unlocks novel opportunities to bring novel life changing insights while improving patient care [2].

The clinical data can be leveraged to fight against diseases by collecting small measurements and trying to find the diseases through machine centric scalable models than human-centric limited experience. Big data can be revolutionary for those who are struggling with deadly diseases. It takes a long time to detect the diseases. Using image processing and machine learning based classification models, diseases can be detected at very early stages. There are 6V's (Volume, Velocity, Variety, Veracity, Validity, and Volatility) of big data, evolving into value of data [3]. Many organizations have already started leveraging big data to solve their day-to-day problems. The other healthcare stakeholders can be encouraged with the means of this paper to accelerate the usage of big data for better insights in healthcare analytics. The quality of value can be increased while the costs can be reduced. The other set of industries such as Information and Technology, Electronics and Communication, and Analytics are bringing innovative technologies at a faster pace to help analyze these data.

The paper is organized as follows. Section 2 provides an overview of healthcare system. In Sect. 3, the paper discusses about big data initiatives in healthcare and recent innovations i.e. IBM Watson—an artificial intelligent computer system, Stanford and Google's collaboration on drug discovery, Calico, Frost and Sullivan's market research, Pittsburgh's Health Data Alliance and Human Brain Project. Section 4 talks about challenges in uplifting healthcare using big data. Section 5 explores future outlook: initiative—a common platform for healthcare big data and analytics. Last but not the least the study is concluded in Sect. 6.

2 Overview Healthcare System

Ferlie and Shortell [4, 5] have given healthcare system structure that is explained via four-level Model. According to this model, the healthcare system revolves around following four nested stages.

2.1 Patient

The first stage, patient is an element whose care defines the healthcare system. Lately, patients have started playing the role of active rather than passive consumers. An active patient is one who wants to be involved in the analysis, design, implementation and maintenance (coordination) of his/her care. The latest technologies enable patients to share their physiological real-time data with physicians (even from remote location), accelerates the speed of diagnosis and treatment. This is the stage which is encircled by all other stages.

2.2   Care Providers

The next stage, patient is an element whose care defines the healthcare system categorizes everyone who provides care to the patient. Healthcare professionals such as physicians, doctors, nurses, pharmacies and even the patients' family members who contribute to the delivery of care are designated as "Care Team". The care providers analyze the patients' data to standardize care, stratify patients and synthesize best decisions/care for their health.

2.3   Organization

The third stage is the organization that offers the infrastructure and other necessary resources to physically perform the care related work. The hospitals, clinic, community hospitals, nursing homes etc. are part of this stage. According to Ferlie and Shortell, "organization is an acute level that sets the culture for change via decision-making systems, operating systems and human resources."

2.4   Environment

The final stage includes political and economic environment that sets the regulatory, market, and policy framework. Public and private regulators, insurers, healthcare purchasers, research funders are few elements who act as pillars of healthcare organizations and other stages (Figs. 1, 2 and 3).

Fig. 1 Four-level healthcare model by Ferlie and Shortell shown in this sample caption [5]

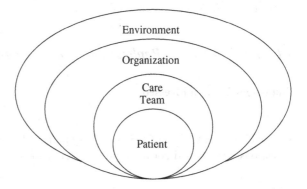

Fig. 2 Common cognitive framework used by humans and implemented for IBM Watson [8]

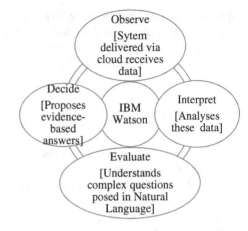

Fig. 3 Potential graph to show that prediction power improves as data; processes/tasks increase [12]

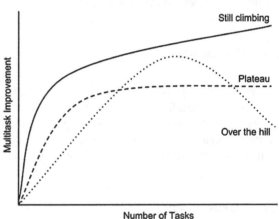

3 Big Data Initiatives in Healthcare

This paper provides a glimpse of few recent and innovative big data initiatives in healthcare. These initiatives are conducted either by corporations or by educational institutions. This section begins with mentioning what and how part of the challenges being addressed via each initiative.

3.1 IBM Watson—Replicating Human Cognition

Challenges being addressed

- Use either SI (MKS) or CGS as primary units. (SI units are Data collection from various sources and converts into structured form.

- Data analytics on top of structured data for medical insights.
- Applied advanced statistical, machine learning and natural language processing techniques for disease detection.

Description

The major challenge in medical field is to make smarter decisions at right time from very large volume of structured and unstructured data coming from heterogeneous sources. The doctors want to have an easy access to these data so that they can bring better insights for their patients (even for those who are in remote region). The clinicians want to shortlist eligible clinical trials for the patient [6]. The researchers want to study the data to develop effective drugs for the diseases [7].

IBM helps healthcare industry to overcome above cited challenges by providing centralized location to store heterogeneous data coming from different sources (electronic health records, mobile applications, clinical trials, social media etc.). These data can be de-identified, shared and combined with the frequently increasing observation of clinical, research and social health data (in a secure and private environment) via cloud computing. IBM Watson offers cognitive computing powers to be applied to the stored data. Watson's cognitive powers are similar to the powers that humans possess to inform their decisions: Observe; Interpret; Evaluate; and Decide [8].

IBM has recently joined hands with CVS Health to develop care management solutions for chronic diseases using predictive analytics and IBM Watson's cognitive computing. In the United States, chronic diseases such as heart disease, diabetes, hypertension and obesity are the leading cause of death and disability. The annual spending on these diseases represents 86 % of the United States' $2.9 trillion in total annual health spending [9]. With this collaboration, CVS Health's healthcare practitioners can use Watson's cognitive computing capabilities to achieve medical insights from blend of health information. Watson reads and understands information, interact in natural language and continuously learn by leveraging cognitive capabilities to bring patient centric primary care and meet health goals. The combined solution will focus on:

(a) Predict: Identifying patients at risk for declining health and conducting proactive and customized engagement programs for them.
(b) Adopt: Monitoring patients to adopt safe and healthy behaviors, adhering to the prescribed medicines and the healthy lifestyle.
(c) Suggest: Recommending proper use of cost-effective care.

3.2 Google–Stanford: Machine Learning for Drug Discovery

Challenges being addressed

- Create architecture for structured data.
- Obtain insights using Deep Learning, complex ML algorithms for drug discovery.

Description

Generally, the drug discovery takes about 12 years of research and costs almost over a billion dollars [10]. Even after such a time consuming and costly process, only a tiny fraction of drugs get approval for human use. Google Research in collaboration with Stanford's Pande Lab has conducted experiments by gathering a huge collection of publicly available data to achieve significant improvement over simple machine learning algorithms used for drug discovery and for decreasing costs involved in the process. The examination included how data from diverse sources can be explored to progress towards determining the effective chemical compounds that could be used as drug treatments for numerous diseases [11]. The research groups leveraged the experimental data from multiple diseases with multitask neural networks to improve the virtual drug screening effectiveness [12]. The teams gathered approximately 40 million measurements (from disparate sources) for over 200 distinct biological processes to achieve greater predictive power. Deep Learning algorithm provided a flexible approach for incorporating these data into predictive models. The groups trained multitasking neural architectures by providing a learning framework that gathers information from and allows information sharing across distinct biological sources for drug discovery. These techniques helped to discover the most effective drug treatments by leveraging more tasks and more data sets that can help in yielding better performance. With this whole process, the team concluded following major results of multitask framework:

- Better predictive accuracies than single-task methods.
- Predictive Power keeps improving as additional data and processes/tasks.
- Multitask improvement < −Total amount of data + Total number of tasks.
- Multi task networks permit restricted transferability to tasks that are not.

3.3 Calico—Control Your Age

Challenges being addressed

- Collects, visualizes and analyzes biological and genetic data.
- Gene structure analysis for delayed aging or increasing human life expectancy.

Description

Calico's mission is to harness advanced technologies to increase our understanding of the biology that controls lifespan [13]. With this stock pile of knowledge (big data), Calico unleashes ways with which people can live healthier and longer lives.

Calico has brought experts from different field of medicines and engineering together to perform researches by collecting, manipulating, analyzing and visualizing the biological big data. The recent research will be conducted via collaboration between AncestryDNA and Calico to understand the genetics of human lifespan. AncestryDNA plans to bring genetics data from its over one million genotyped customers. The team is planning to evaluate data with respect to public family trees. Just to elaborate, both the companies will together analyze and examine the genetics' role and influence in families enjoying unusual longevity. It is an endeavor to discover the genes that are serving few people to live longer. AncestryDNA will initially offer its huge volume of genomic data, tools and algorithms to undergo this analysis; Calico will later concentrate on formulating and marketing the potential therapeutics in case any emerge from the analysis [14].

Once genes for longevity are predicted, drug companies can develop genetically informed drugs. Big data is the magical phenomena that is helping and can help to an extent never imagined before. Imagine a social life where everybody is healthy and long living.

3.4 Frost and Sullivan—Big Data in Medical Imaging

Challenges being addressed

- Medical image data collection and processing for already existing images (CT Scans, X-ray reports, Magnetic Resonance Imaging).
- Real-time service oriented architecture development.

Description

Frost and Sullivan is especially interested in a special kind of biometric data, medical images for new studies [15]. Unlike big data in transaction data and human generated data, big data in medical imaging refers to the datasets (medical images and related datasets) that are in order of petabytes (10^{15}). US market for Big Data in medical imaging has been divided into Big Data Management and Big Data

Analytics. One of the major reasons for the increase in inflow of the medical imaging is practical implementation of multiple healthcare information technologies that have improved collection, inspection and even dissemination of medical images. Some of the imaging techniques that help in providing datasets are magnetic resonance imaging (MRI), X-ray, molecular imaging, ultrasound, computed tomography (CT) etc. To achieve plethora of information from medical images, a real-time approach via semantic technologies and service-oriented architecture model is being followed. Subsequently using advanced imaging techniques, this data will be debriefed and analyzed for clinical, operational and financial purposes. This increased data interoperability in a real-time environment will bring numerous medical advantages to healthcare. Through this combination of big data management and big data analytics solutions, medical images are expected to be interpreted with more precision. More correctness will lead to improved workflow efficiency, accurate diagnosis, appropriate treatment decisions and proper health management.

3.5 Pittsburgh Health Data Alliance—Carnegie Mellon University (CMU), The University of Pittsburgh and University of Pittsburgh Medical Center

Challenges being addressed

- Collection of raw clinical data, sensors and equipment's data and insurance records.
- Data analysis for optimization and higher efficiency in operations (service end).

Description

One of the major health data alliances of 2015 has been the Pittsburgh Health Data Alliance. The three Pittsburgh experts adore the impact of big data and are confident that together they can utilize the data to improve human health, revolutionizing the practice of medicine [16].

CMU contributes its distinctive capabilities with technology. University of Pittsburgh adds the health sciences proficiency and University of Pittsburgh Medical Center provides raw clinical data (electronic records, images, clinical notes, genomes, wearable sensors, and insurance records) to perform multiple researches. The move is towards fetching data, performing the analysis and implementing new healthcare technologies, products and services in real life scenarios immediately [17]. Center for Machine Learning and Health aims to solve healthcare related challenges using artificial intelligence and machine learning. Few examples include the techniques to identify the emergency room wait times of patients; mobile applications to keep track of a person's calories intake, exercise and sleep timings to prevent the onset of any disease. Other example may include monitoring patients who are at higher readmission risks, saving costs etc. The other center-Center for Commercial Applications of Healthcare Data—focuses on inventing new technologies based on engineered big data solutions. These technologies are strategized to be used in commercial theranostics and imaging systems for both doctors and patients. Both the centers as part of Pittsburgh Health Data Alliance aim to bring the knowledge from the data to the patient-centric solutions.

3.6 Human Brain Project (HBP)

Challenges being addressed

- Imitating brain, its structure and functioning for building efficient systems.
- Detecting brain diseases using simulated brain models combined with data mining brain related images.

Description

One of the most fascinating researches of the 21st century is the Human Brain Project, HBP. This international project began its journey in 2013 with an intention to understand the human brain. The marvelous power of human brain compels medical researchers and scientists around the world to know more about this tiny body part that contains enormous data. These brain insights are promised to diagnose and treat lethal brain diseases using innovative computing technologies. The HBP deals with these three major science areas for research: neuroscience, neuromedicine, and computing [18]. Through neuroscience, the project wants to collaborate, understand and stimulate information about human brain. Through neuromedicine, project aims to collect information about brain diseases by aggregating medical records from multiple sources. Through computing, the project concentrates on developing brain-inspired computing systems such as high-performing hardware and software. Keeping these target goals, project execution starts with building following six information and communications technology platforms: (a) Neuroinformatics Platform [19], (b) Brain Simulation Platform [20], (c) High Performance Computing Platform [21], (d) Medical Informatics Platform [22], (e) Neuro-morphic Computing Platform [23], (f) Neuro-robotics Platform [24].

4 Healthcare Big Data Challenges

Although Big Data is leading to significant healthcare performance improvement however there are still several challenges which persist [25, 26]:

(1) Understanding unstructured clinical notes in the right context: Medical information is collected by various examiners, however in different context. Based on the symptoms the test varies and so are corresponding reports and each medical examiner examines according to their own way. Lack of standardized approach makes the data unstructured and contextually right while aggregating the information.

(2) Handling large volumes of medical imaging data efficiently and extracting potentially useful information and biomarkers: Too much information is also not good as it could lead to lot of noise. Furthermore, data stored in unstructured form even worsens the situation.

(3) Complexity of the data: Analyzing genomic data is a computationally intensive task and combining with standard clinical data adds additional layers of complexity.

(4) Capturing the patient's behavioral data through several sensors; their various social interactions and communications: Build user specific database at one location when the information is distributed at many sources.

(5) Existing Electronic Health Records are limited to data acquisition than analytics: EHRs have greatly simplified data acquisition, but don't have the ability to aggregate, transform, or create actionable analytics from it. Intelligence is limited to retrospective reporting, which is insufficient for forward-looking healthcare data analytics that hold the key to performance improvement.

(6) Adapting changes: Institutions are notoriously resistant to change. This is especially true as they grow larger and existing processes and procedures become "the way it's always been done". The shift to an analytics-based culture requires everyone in the organization to use a single source of truth to guide choices, stop making gut decisions, and avoid "data shopping" to find data that supports a conclusion that has already been made.

5 Future Outlook: Common Platform for Healthcare Analytics

Almost all the researches mentioned above go through a phase of data collection, data mining and data analysis. The areas of application as well as approaches could be different, but the initial steps are similar. Much of the effort and time goes in data collection, cleaning and data preparation for models [27]. Since, most of these parts are standard and used by all the analytics firms or organizations, a common platform to pursue these common tasks can save a lot of time and effort. Analysts can concentrate on data-driven performance improvements by intending to make the Data Analysis process iterative, progressing level by level and focusing on early delivery. Maximum automation by leveraging Object Oriented Programming and robust design methodologies to handle any kind of data (textual, numeric etc.) can be utilized. Researchers can then spend more time on individual objective of building evidence for improved care delivery (based on brainstorming data from research, clinical care settings, and operational settings) [28]. Recently, a set of machine learning algorithms known as Deep Learning has solved this challenge.

Automatically extracting complex data representations (abstractions) is the basis for Deep Learning algorithms. The outcome of these algorithms is layered/ hierarchical architecture of learning and representing higher-level data in terms of lower-level data. Hierarchical learning approach of human brain is the inspiration of these algorithms. Deep Learning provides the power to extract representations from unsupervised data to computers, making machines independent of human knowledge [29, 30].

These algorithms take the data in any format (text, sound, video, numerical data etc.) in parallel or individually and build models corresponding to the defined objective. In deep learning, humans or researchers do not perform feature extraction directly, but these models themselves extract relevant features and generate insights. These features act together to present an intelligent behavior [31].

This kind of approach can also be used in Healthcare analytics through which supervised and unsupervised models can be built for drug discovery, disease detection and more generic insights or better services and optimization at application end(irrespective of the data formats-clinical data and logs, images, sound etc.). Healthcare companies can extract crucial consumer/patient behavioral patterns (current lifestyle habits and the consequent correlated future health risks) based on the information gathered from social networking websites. Since, healthcare is one of the biggest sources of data in the realm of analytics without organized and structured data. Therefore a common platform approach will be highly beneficial for researchers and analysts.

6 Conclusion

The role of big data is beyond the description. Healthcare stakeholders have begun experiencing the immense power that data possess. The researches, inventions, innovations and discoveries in the field are incomplete without the medical practitioners realizing their necessities. The promising advantages of big data such as evidence-based diagnosis and drugs; personalized care and treatment; decreased costs; faster and effective decisions can bring value into the lives of not only patients but also caregivers. The healthcare's future is clearly in real-time intelligent decision making from the data. Finally, after looking at the challenges in processing the data by healthcare analysts and researchers, it is proposed that there is a need of a common platform which can be leveraged by all the researchers to pursue common tasks of feature engineering and data preparation. This way more time will be spent on invention rather than on time-consuming task that can be automated.

References

1. EMC Digital Universe & IDC: The digital universe: driving data growth in healthcare, challenges and opportunities (2004)
2. Glaser, J.: Solving big problems with big data. In: Hospitals and Health Networks Daily, Dec (2014)
3. Khan, M.A., Uddin, M.F., Gupta, N.: Seven V's of big data. In: ASEE Zone 1, Proceedings of 2014 Zone 1 Conference of the American Society for Engineering Education (ASEE Zone 1) (2014)
4. Ferlie, E.B., Shortell, S.M.: Improving the quality of healthcare in the United Kingdom and the United States: a framework for change. **79**, 281–315 (2001)
5. National Academy of Sciences. http://www.ncbi.nlm.nih.gov/books/NBK22878/
6. Transforming clinical trial matching with cognitive computing. http://www.ibm.com/smarterplanet/us/en/ibmwatson/clinical-trial-matching.html. Accessed 12 Aug 2015
7. IBM Watson ushers in a new era of data-driven discoveries. https://www-03.ibm.com/press/us/en/pressrelease/44697.wss. Accessed 25 Aug 2015
8. What is watson. http://www.ibm.com/smarterplanet/us/en/ibmwatson/what-is-watson.html. Accessed 15 Aug 2015
9. CVS Health and IBM tap watson to develop care management solutions for chronic disease. http://www-03.ibm.com/press/us/en/pressrelease/47400.wss. Accessed 15 Aug 2015
10. Applications of physical simulation and Bayesian statistics/machine learning to biologically and biomedically important questions. http://pande.stanford.edu/projects/#ntoc1. Accessed 25 Aug 2015
11. Large-scale machine learning for drug discovery. http://googleresearch.blogspot.in/2015/03/large-scale-machine-learning-for-drug.html. Accessed March 2015
12. Massively multitask networks for drug discovery: a preliminary work on machine learning, unpublished. http://arxiv.org/pdf/1502.02072v1.pdf
13. Calico Labs: http://www.calicolabs.com/
14. AncestryDNA and calico to research the genetics of human lifespan. http://www.calicolabs.com/news/2015/07/21/. Accessed 25 Aug 2015
15. Frost, Sullivan: Big data opportunities in the US medical imaging market. Mounting data volumes in medical imaging driving the need for big data tools market engineering, April 2015
16. Pittsburgh Health Data Alliance: www.healthdataalliance.com
17. Pittsburgh Health Data Alliance: Pitt, CMU, UPMC form alliance to transform healthcare through big data, March 2015
18. Kasabov Nikola, K.: Springer handbook of bio-/neuro informatic, pp ix, ISBN:978-3-642-30574-0 (2014)
19. https://www.humanbrainproject.eu/neuroinformatics-platform1
20. https://www.humanbrainproject.eu/brain-simulation-platform1
21. https://www.humanbrainproject.eu/high-performance-computing-platform1
22. https://www.humanbrainproject.eu/medical-informatics-platform1
23. https://www.humanbrainproject.eu/neuromorphic-computing-platform1
24. https://www.humanbrainproject.eu/neurorobotics-platform1
25. Sun, J., Reddy, C.K.: Big data analytics for healthcare. In: SIAM International Conference on Data Mining (2013)

26. getting started with healthcare data analytics. http://www.mckesson.com/healthcare-analytics/healthcare-big-data-challenges
27. http://ventanaresearch.com/blog/commentblog.aspx?id=4716
28. Nambiar, R., Bhardwaj, R., Sethi, A., Varghese, R.: A look at challenges and opportunities of big data in healthcare. In: IEEE, International Conference on Big Data (2013)
29. http://www.journalofbigdata.com/content/2/1/1
30. http://www.iro.umontreal.ca/ ~ lisa/pointeurs/bengio+lecun_chapter2007.pdf
31. Bengio, Y., Goodfellow, I.J., Courville, A.: Deep learning. Book in Preparation for MIT Press, Chap. 12, Unpublished (2015)

Modeling Data Heterogeneity Using Big DataSpace Architecture

Vishal Sheokand and Vikram Singh

Abstract With the wide use of information expertise in advanced analytics, basically three characteristics of big data have been identified. These are volume, velocity and variety. The first of these two have enjoyed quite a lot of focus, volume of data and velocity of data, less thought has been focused on variety of available data worldwide. Data variety refers to the nature of data in store and under processing, which has three orthogonal natures: structured, semi-structured and unstructured. To handle the variety of data, current universally acceptable solutions are either costlier than customized solutions or less efficient to cater data heterogeneity. Thus, a basic idea is to, first design data processing systems that create abstraction that covers a wide range of data types and support fundamental processing on underlying heterogeneous data. In this paper, we conceptualized data management architecture 'Big DataSpace', for big data processing with the capability to combine heterogeneous data from various data sources. Further, we explain how Big DataSpace architecture can help in processing the heterogeneous and distributed data, a fundamental task in data management.

Keywords Big Data Mining · Dataspace System · Data Intensive Computing · Data Variety · Data Volume · Data Integration System

1 Introduction

The Evolution of 'Big' aspect of data not only imposed new processing and analytics challenges but also generates new breakthrough, as interconnected data with complex and heterogeneous content provides new insights. In modern day of technology,

V. Sheokand (✉) · V. Singh
Computer Engg. Dept., National Institute of Technology,
Kurukshetra 136119, Haryana, India
e-mail: vishalsheokand007@gmail.com

V. Singh
e-mail: viks@nitkkr.ac.in

© Springer Science+Business Media Singapore 2016   259
R.K. Choudhary et al. (eds.), *Advanced Computing and Communication Technologies*,
Advances in Intelligent Systems and Computing 452,
DOI 10.1007/978-981-10-1023-1_26

big data is considered a greatly expanding asset to humans. All what we need then is to develop the right tools for efficient storage, access and analysis of big data, and current tools have failed to do so. Mining on big data requires scalable approaches and techniques, effective preprocessing, superior parallel computing environments, and intelligent and effective user interaction. It is widely acknowledged fact that in big data processing, multiple challenges involve not just the Volume of data. The Variety of data and Velocity of data are other aspects of big data processing imposing huge and complex tasks. Variety of data is typically due to the different data forms, different data representation, and data semantic interpretation. Data velocity is the rate of data or information generated by a sources or simply the time in which it must be acted upon. As lot of researcher and industry analyst are involve in the development of application or advanced analytics, some additional features are evolving e.g. Data Veracity, Data Values, Data Visualization, Data Variability etc.

From the mining viewpoint, mining big data has opened many new pre-processing and storage challenges and list of opportunities to data analytics for wide spectrum of scientific and engineering applications. Big data mining provides greater assessment, as it generates hidden information and more valuable insights. This imposes an incredible challenge to pull out these insights and hidden knowledge, since currently used knowledge discovery tools are unable to handle big data. Similar is the case with data mining tools [1]. These traditional approaches have failed to cope with inadequate scalability and parallelism, other challenges are like unprecedented heterogeneity, volume, velocity, values, veracity, and trust coming along with big data and big data mining [2, 3].

1.1 Big Data, Variety Bigger Hurdle than Volume

In a recent report on the big data related advancement, it was suggested that the real potential of big data is hidden in the variety (heterogeneity or diversity) of the data [4]. To get the full benefit of the actual thrust of big data analytics one has to uncover the power of diverse data from different data sources. Big data sure involves a great variety of data forms: text, images, videos, sounds, etc. Big data frequently comes in the form of streams of a variety of types. In modern world of information, data contributed or generated by user is not fully structured. Examples of loosely structured data are tweets, blogs, etc. Multimedia content is also not structured well enough for search operations. Thus a transformation of loosely structured content into structured one is required.

In past few years development in advance analytics continue to focus on the huge volume of big data in various domains. But it's the variety of data which impose bigger challenge to data scientists/analyst. According to a survey in the Waltham, by a group of data scientists, it is stated that [3, 4]:

- Big data has made related analytics more complicated but it is data variety, not volume of data, which is to be blamed.

Fig. 1 Big Data Analytics challenges in various domain applications used in major cities

- Analysts are using complex analytics on their big data now, or are planning to do so within next 2 years. They find it more difficult to fit their data into relational database tables. The various other challenges for big data analytics are depicted and can be summed up as in below graph [4].

Figure 1 clearly suggests that, data variety is the biggest hurdle for data analysts. Data being of heterogeneous types is one of the major components of Big Data complexity. Variety in big data is resultant of the fact that data is produced by vast number of sources [5]. Hence the data types of produced data can also be different for different sources, resulting in great deal of variety. The daunting task then is to mine such vast and heterogeneous data. Hence data heterogeneity in big data becomes an obligation to agree to. Also the data present can be in any of the structured, semi-structured or completely unstructured form. While first one can be fit well into database systems, rest two pose quite a challenge, especially the last one. In modern systems, unstructured data is stored in files Semi-structured data too, if couldn't be mapped easily onto structured one, is stored same as unstructured one, is stored same as unstructured one, especially in data intensive computing and scientific computational areas [6].

1.2 Big Data Mining

Big Data has high veracity, is dynamic (different values), heterogeneous (variety) and inter-related. These properties of big data can be used to produce better mining results than those obtained from individual smaller databases. This is because universal data obtained from different sources can counter fluctuations. Hence pattern and knowledge revealed is much more reliable. Also the undesirable features of databases like redundancy become useful in case of big data. Redundancy can be exploited to calculate missing data and validate true value in case of inconsistency.

Big data mining demands scalable algorithms along with huge computing powers. The process of mining gets equally benefit from these [7]. Big data analysis can be used to get real time answers required by the analysts. Big data analysis can also be used to automatically generate content from user blogs and websites [8].

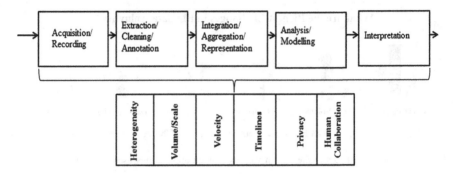

Fig. 2 Big Data Mining process and its challenges

A typical big data mining process is illustrated in Fig. 2, various inherent challenges also listed.

One of the major problems faced by big data analysis is poor coordination among the local DBMSs. They provide their own querying mechanism for data retrieval and moreover, data mining and analysis is very different from data retrieval [7, 9]. The data transformation and extraction process is complex and an obstructive when it carried out over the interactive sophistication [10]. Traditional database models are incapable of handling complex data in the context of Big Data. Currently, there is no acknowledged, effective and efficient data model to handle big data. In the next section, a proposed system is discussed for the purpose of big data mining. The architecture is purely based on the DataSpace management systems.

1.3 Related Work and Outline

In [11, 12], a conceptual data management architecture to unify heterogeneous, disperse local databases (Relational DB, text DB etc.) into a single global schema. The success of system over heterogeneous DB depend on the efficiency of its database processing utilities [5], such as query processing [13], indexing [14] and semantic integration of local DB's. Big Data is a set of loosely connected heterogeneous and distributed data sources [6], huge data generated on high processing speed of high variety [15]. In [2, 3], it is highlighted big data variety is bigger challenges than volume in big data processing. Current universally available data management systems fail to harness data heterogeneity inherent in big data [15], thus a new architecture can be adapted. In this paper, we personalized data management architecture to model data heterogeneity based on the schema-less design approach [16].

In Sect. 2, the fundamentals of DataSpace systems and its data processing components are illustrated. The DataSpace support services (DSSP) is set of database utilities in DataSpace, for proposed Big DataSpace architecture and it is illustrated in Sect. 2.1. Further, Sect. 2.2 explains the query processing and its inherent challenges.

2 DataSpace System

Most modern data management scenarios often does not have an architecture in which all type of data (structured, semi-structured, unstructured) can fit nicely into a single management system [11]. Instead, data administrator or manager needs to handle loosely connected data sources. The challenges then faced are as following [11, 12, 17]:

- User searches over all data sources without knowing about them individually.
- Providing a data management organization which enforces integrity constraints and track data flow and lineage between participant systems.
- Recovery, availability and redundancy are required to be controlled.

In year 2005, a conceptual model of architecture was proposed for Dataspace for solving these problems. Data integration and some of the data exchange systems along with data integration systems have tried to provide DataSpace management systems are considered as the next development of data integration architectures [10, 16, 18]. Dataspace systems differ from traditional data integration systems, as in dataspace semantic integration among mediated schema and data sources are not required before any service can be provided. Thus initialization of data integration system is not required upfront [6].

Dataspaces is not a data integration based approach (from local data sources); rather, it is based on data co-existence. There are set of support functionality over the designed dataspace called DSSP. DSSP is to support basic functionalities over data sources, irrespective of their integration. Participants are integrated in an incremental way, commonly known as "pay-as-you-go" fashion [19]. The integration of data or schema lead to an integration cost. In the proposed architecture, first we define semantic mapping between unified mediated schema and individual data schema and then data is uploaded for the analytics purpose in pay-as-you-go-approach fashion [5, 20]. To the best of our knowledge, no attempt has been made yet in literature to solve heterogeneity in big data using the concept of dataspaces. In our view, heterogeneity of big data can be better resolved using modified dataspace architecture. In the next section, we propose architecture for the same.

2.1 Proposed Big DataSpace Architecture

Big DataSpace System is as new data management architecture for huge and heterogeneous data [12]. In this system, data reside in the individual data sources and the semantic mappings are created among data sources and a mediated schema is created. The data processing proceeds by reformulating queries over a mediated schema onto appropriate data instance on the individual data sources. Creating these semantic mappings can be a major bottleneck in building these systems because they require significant upfront effort. DSSP provides a developer with data

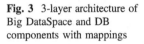

Fig. 3 3-layer architecture of Big DataSpace and DB components with mappings

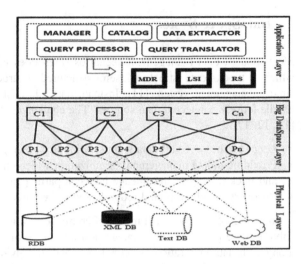

processing services such that she can focus on the challenges involved with development of the application rather than managing the data.

Big DataSpace is a 3-layered conceptual architecture, also shown in Fig. 3. In architecture, physical layer is a set of individual data sources or participant data sources; e.g. HTML DB, text DB, relational DB, web DB etc. The mapping between dataspace schema and each of the participant databases' schemas is defined in best-effort approach and according to the semantics of both the schemas. Each data source is made available via participant objects (P_1, P_2, P_3, ..., P_n) in Big DataSpace layer. A data source supplies its operational data to only mapped participant objects using mapping.

Big DataSpace layer consists of the DB objects of each underlying Database's. The participants are the primary entities to have direct access on logical data of specific database. The database in the participant nodes is developed on the pay-as-go-approach; semantic mapping between the different data sources and participant objects is purely based on the query access pattern. The Big DataSpace consists of participant and computing nodes (C_1, C_2, C_3, ..., C_n), also shown in Fig. 3. Each computing node represents the clustered data sources or data coming from various database. Each DB objects is defined as database mapping.

The mapping from physical layer to participants handles heterogeneity. All the data (heterogeneous) from physical layer can be mapped to participant nodes in homogeneous form. A data model is required for such homogeneous representation, one of the example data model representation being iDM (integrated Data Mode). We also need semantic integration rules for such mappings. Much work has been done in the literature to define such rules. Different databases require different rules of integration. Once heterogeneity is handled, conventional big data algorithms and approaches can be used.

In the application layer, various data management related components are placed. A casual user performs some fundamental database operation e.g. query processing. Participant catalog maintains the list of participants of each of the DB and also identifies the new participants. For query processing and search related operations a module is added into the application layer. Relationship manager manages the mapping of semantics between various data sources and participants.

The application layer consists of DBMS repository components, which are important on data modeling and query processing, e.g. MDR (Meta Data Repository), LSI (Local Store and Index), RS (Replication Storage), query processor, query translator, data extractor etc. Each component will play an important role in heterogeneous data processing the analytics. The MDR component is the fundamental relationship among the various query processing related functions. The LSI will maintain the indexing among participant local data sources. The indexing among the data sources is based on their semantic relationship.

Local Store and Index (LSI), manages the efficient retrieval of searched results from big data space layer without accessing the actual data. Data items are associated by queryable association. This further improves access to the data sources. The data items in each of participant data sources are index to the mediated schema created in dataspace. Thus index must be an adaptive one. Final component is RS (Replication Storage) handles data replication to increase the availability and reliability of access performance.

2.2 Query Answering on Big DataSpace

Querying on Big DataSpace opens new possibility on three orthogonal dimensions of big data analytics, first achieving a more complete view, second integrating fragment information and finally aggregating data from different sources. In any data integration system, queries in variety of languages are anticipated. The keyword or structured queries over the global schema belongs to the major set of database activities, but in many modern applications the user queries are being generated through filling forms(queries with multiple predicates selection) [14]. In Fig. 4, typical relationship among computing nodes and different participant nodes are shown. When user interacts more intensely a specific participant data sources a complex queries is required [21]. Regardless of the data model or the participant schema used for storing or organizing the dataspace respectively, when a query is posed by user, the results are expected to be produced after considering all the relevant data in the dataspace unless specified explicitly. The user expects to obtain the results from other sources as well even if the terms of its schema processing in Big DataSpace architecture. There are numerous approaches for answering user query in traditional data integration and dataspace systems are used, thus the similar

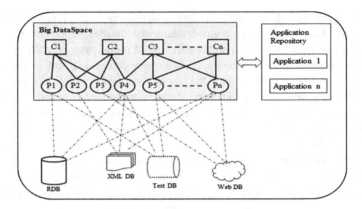

Fig. 4 Big DataSpace layer and application repository

methods are used for employed in proposed Big Dataspace architecture also. Following are some approaches for query answering:

Ranked Answers For a user query, the query relevant data is retrieved from various participant data sources. A ranking mechanism is required to rank the retrieved results for user query based on the relevance to the user query. Ranking of answers will minimize the data heterogeneity also.

Sources as Results Besides ranking query answers as documents, tuples, etc., a data source can also be used as the answer in dataspace. Answer can be pointers to other sources where answers may be found.

Iterative Querying Dataspace interactions are different than traditional "pose a query and get the answer" mechanism. It involves posing iterative queries where refined results are used to pose further queries finally leading to results.

Reflection on Coverage and Accuracy Similar to the dataspace systems, a DSSP is anticipated to provide answers which are complete in its coverage, also accurate in its data.

In this paper, we have tried to highlight the challenges imposed by data heterogeneity on big data analytics or processing and address the challenges via proposing data management architecture. Our focus has been on identifying the various DB components related mappings and their placement in layered architecture along with their responsibilities. We also highlighted possible querying mechanism in a schema-less environment. However, the proposed architecture opens up new challenges of mapping different data sources to participant objects, which requires to be explored further.

3 Conclusion

We today live in the generation of big data where enormous amount of unstructured, semi-structured and heterogeneous data are being continually generated. Big data discloses the limitations of existing data mining techniques, and opens up new challenges related to big data mining. In spite of the limited work done, it is believe that vast work is required to overcome the challenges related to heterogeneity, scalability, accuracy, privacy, speed, trust etc. of big data. Most of modern applications or techniques today rarely have a unified global data management system to that can nicely adapt various heterogeneous databases. In this paper, a conceptual architecture for modelling heterogeneous big data is discussed. The DSSP is suit of components to support the basic mining tasks over differently structured databases such as web DB, text DB, Image DB, relation DB etc. The Big DataSpace architecture primarily defines the semantics integration between data sources and mediated schema and performs all data mining related tasks on the mediated schema. Our future direction involves the development of an efficient semantic mapping and an effective query processing mechanism for the proposed architecture.

References

1. Anis, D.S., Dong, X., Halevy, A.Y.: Bootstrapping pay-as-you-go data integration systems. In: Proceedings of the 2008 ACM SIGMOD International Conference on Management of Data, pp. 861–874. ACM, USA (2008)
2. Divyakant, A., Bernstein, P., et. al.: Challenges and Opportunities with Big Data. A community white paper. Feb, USA (2012)
3. David, L., Alex, P., et. al.: Computational Social Science. A technical report on Science, vol. 323(5915), pp. 721–723. USA (2009)
4. Daizy, Z., Dong, X., Sarma, A.D., Franklin, M.J., Halevy, A.Y.: Functional dependency generation and applications in pay-as-you-go data integration systems. In: WebDB (2009)
5. Steve, L.: The age of Big Data. A technical report. New York Times, Feb (2012)
6. Singh, G., Bharathi, S., Chervenak, A., Deelman, E., Kesselman, C., Manohar, M., Patil, S., Pearlman L.: A metadata catalog service for data intensive applications. In: Proceedings of International Conference on Supercomputing, pp. 20–37. IEEE/ACM, USA (2003)
7. Vagelis, H., Gravano, L., Papakonstantinou, Y.: Efficient IR-Style keyword search over relational databases. In: Proceedings of the International Conference on VLDB, pp. 850–861. Berlin, Germany (2003)
8. Vagelis, H., Papakonstantinou, Y.: DISCOVER: Keyword search in relational databases. In: Proceedings of the International Conference on VLDB, pp. 670–681. Berlin, Germany (2002)
9. Dittrich, J.P.: iDM: A unified and versatile data model for personal dataspace management. In: Proceedings of the International Conference on VLDB, pp. 367–378. Seoul, Korea (2006)
10. Salles, M.A., Dittrich, V.J., Blunschi, L.: Intentional associations in Dataspaces. In: Proceedings of International Conference of Data Engineering, pp. 30–35. IEEE, USA (2010)
11. Franklin, M., Halevy A., Maier, D.: From databases to dataspaces: A new abstraction for information management. In: Proceedings of the 2005 ACM SIGMOID Record, vol. 34(4), pp. 27–33, ACM USA (2005)

12. Ibrahim, E., Peter, B., Tjoa, A.M.: Towards realization of dataspaces. In: Proceedings of the 17th International Conference on Database and Expert Systems Applications, pp. 266–272. IEEE, USA (2006)
13. Bhalotia, G., Nakhey, C., Hulgeri, A., Chakrabarti, S., Sudarshanz S.: Keyword Searching and browsing in databases using BANKS. In: Proceedings of the International Conference of Data Engineering, pp. 431–441. IEEE, USA (2002)
14. Xin, D., Halevy, A.: Indexing dataspaces. In: Proceedings of 2007 ACM SIGMOD International Conference on Management of Data, pp. 32–45. ACM, USA (2007)
15. Manyika, J., Chui, M., et.al.: Big data: the next frontier for innovation, competition, and productivity. A Technical Report. McKinsey Global Institute (2011)
16. Marcos, A., Salles M.A., Dittrich J.: iTrails: pay-as-you-go information integration in dataspaces. In: Proceedings of International Conference of VLDB, pp 663–674. Vienna, Austria (2007)
17. Dittrich, J.P.: iMeMex: A platform for personal dataspace management. In: Proceedings of 2nd Invitational Workshop for Personal Information Management, pp. 292–308. USA (2006)
18. Salles, M.V.: Pay-as-you-go information integration in personal and social dataspaces. Ph.D. Dissertation, ETH Zurich (2008)
19. Sanjay, A., Chaudhuri, S., Das, G.: Dbxplorer: a system for keyword-based search over relational databases. In: Proceedings of the International Conference on Data Engineering, pp. 1–5. IEEE, USA (2002)
20. Shawn, R.J., Franklin, M.J., Halevy, A.Y.: Pay-as-you-go user feedback for dataspace systems. In: Proceedings of the SIGMOD Conference, pp. 847–860. ACM, USA (2008)
21. Yuhan, C., Xin, L.: Personal information management with SEMEX. In: Proceedings of 2005 ACM SIGMOD International Conference on Management of Data, pp. 921–923. ACM, USA (2005)

Smart Cities [Meixi (China) × Kochi (India)] Notions (Sustainable Management Action Resource Tools for Cities)

Somayya Madakam and R. Ramaswamy

Abstract Smart Cities is one of the emerging concepts in different continents including Europe, US, Africa, Australia and Asia too. Different people are using for different purposes, including India's Prime Minister Shree Narendra Modi for political strategy with 100 new Smart Cities, IBM for business with "Smarter Planet", rich people are looking for the Quality of Life, UAE people in need of sustainable city and a common man looking for employment. The tools using in these cities are Internet of Things Technologies, which brings the operational efficiency. The objective of this article is to explore importance of Smart Cities. This article also clearly depicting Smart Cities conceptual model [(U/I: IoT: S/D: S/H: A)]/L-M based on some primary data on observations and technical discussions along with secondary data. Moreover, this manuscript discussing current scenario of Smart Cities, sightsee both Meixi (China) and Kochi (India) cities.

Keywords Smart cities · SmartCity Kochi · Smart cities in India · Meixi lake smart city · Smart cities in China

1 Introduction

China and India are two of the world's oldest civilizations and have co-existed in peace for millennia. Both of the countries have seen dramatic economic growth in recent decades. However, each is still overcome by widespread poverty. In 1950, India (17 %) was a more urban nation than China (13 %). However, from 1950 to 2005, China urbanized far more rapidly than India ever before, to an urbanization

S. Madakam (✉) · R. Ramaswamy
IT Applications Group, National Institute of Industrial Engineering (NITIE),
Mumbai, India
e-mail: somu4smart@gmail.com

R. Ramaswamy
e-mail: ramaswamy2008@gmail.com

© Springer Science+Business Media Singapore 2016
R.K. Choudhary et al. (eds.), *Advanced Computing and Communication Technologies*,
Advances in Intelligent Systems and Computing 452,
DOI 10.1007/978-981-10-1023-1_27

growth rate of 41 %, compared with India that of 29 %. That means the statistics indicate that China has embraced and shaped the process very systematically, while India is still waking up to its urban realities, issues, opportunities and interventions. Urbanization is an index of transformation from traditional rural economies to modern industrial one. It is progressive concentration [1] of population in an urban unit. Quantification of urbanization is very difficult. It is a long-term process. Kingsley Davis has explained urbanization as process [2] of a switch from spread out pattern of human settlements to one of concentration in urban centers. Urban Regions are the places, where the man can live, work, learn and play. These are the places in which urban citizen can find all kinds of business activities, productions, and services including medical, hotel, home appliances, FMCG, chemical industrial material are among others. These are the central knots for higher educational hubs, Research & Development and national importance training and learning centers. So we can say that cities are vivacious nodes for the national economic development. They bring the happy life to citizens.

2 Need of the Study

On the other hand, cities are also suffering from many hitches including lack of adequate transportation amenities during peak office hours, illegal slums built up, which may cause for handicapped in providing municipality services, insufficient infrastructure, security problems in the crowdy places to name a few. We know that even though 2 % of the earth is geographically occupied by the cities but they are emitting 80 % of carbon by-products, which may cause for the global warming. In directly it will also cause the natural calamities. The best paradigm is the recent Nepal country earthquake, which left with many lives with handicapped in both socially, economically and even politically. And rural migration in exponential rate is one more pathogenic problem-the cities fronting around the world. So it is inevitable that we need to modernize the existing Brown Field Cities with the help of Internet of Things (IoT) technologies or build new Smart Cities in the enriched natural resource regions like country sides.

3 Smart Cities

The word "Smart City" is in existence since 1992 in the form of "Digital Cities". However, it was not given much attention in the urbanization until 2008, when IBM exposed it in the form of "Smarter Planet Vision". There is no standard operational explanation for the word "Smart Cities" and there is a huge dilemma about Smart Cities definitions. The pretty understanding of Smart City would be "The cities which use Internet of Things technologies for their city operational efficiency, to provide a better life to citizens and sustainable for the future generations". Similarly, there are some working definitions which were given by policy makers, mayors, business tycoons, researchers, academicians, practitioners, developers based on the city requirement, technologies deployed in different applications. The author [3] said that a Smart City is a city that monitors and integrates conditions of all of its critical infrastructures, including roads, bridges, tunnels, rails, subways, airports, seaports, communications, water, power, even major buildings, can better optimize its resources, plan its preventive maintenance activities, and monitor security aspects while maximizing services to its citizens. But when it comes to the author [4], A Smart City is well performing in a forward—looking way in 6 dimensions economy, people, governance, mobility, environment, and living, built on the smart combination of endowments and activities of self—decisive, independent and aware citizens. Some people prefer a broader definition: Smart Cities use Internet of Things (IoT) to be more intelligent and efficient in the use of resources, resulting in cost and energy savings, improved service delivery and Quality of Life, and reduced environmental footprint—all supporting innovation and the low—carbon economy. Charles Dickens portrayed the 18th century as a tale of two cities; 21st century though will be a tale of Smart Cities. Smart Cities are planned cities which focus on the development of infrastructure and timely delivery of services. In the course of the of the last decade, the concept of the 'Smart Cities', considered by many as new century's stage of urban development, has become fairly trendy in the policy and business arenas [5]. Smart Cities are the best urban solutions for the future generation with new urbanism principles using Internet of Things (IoT) technologies, can date with city components for providing a good life to city dweller in the day to day life.

Walters and Brown [6] provide a list of features of smart and sustainable community planning and design, ranging from municipal polices to planning strategies to detailed urban design concepts. These can be briefly summarized as promoting diverse, compact, and mixed use neighborhoods that are walkable and transit supportive, with physically defined and accessible public spaces. Both urban and natural are comprised of energy efficient buildings that follow the premise of 'Long Life, Loose Fit' that is adaptable to changing patterns of use without major disruption to themselves. A Smart City, at a simple refers to a meticulously planned city that relies on the IoT technologies as an enablers to solve many of urban issues —from use of sensors, actuators, RFID, smart grids, cloud and data analytics that allow city infrastructure and services to meet the city problems and citizen demands efficiently and reliably. A Smart City Projects from scratch, some brand new cities are currently being developed worldwide as completely new projects such as Songdo (South Korea), Meixi Lake (China), Masdar (UAE) and Lavasa (India) [7]. These are the most desirable and entertain places to urban citizens with galore of modernized amenities to provide Quality of Life.

4 Smart City 360° Conceptual Model

Still now there are no models that can represent the structural components and its allied functions. Here the Smart Cities 360° Conceptual Model: [(U/I: IoT: S/D: S/H: A)]/L-M is shown next page in Fig. 1. In this model, there are mainly 6 components (1) U/I: Urban Issues (2) IoT: Internet of Things (3) S/D: Six Dimensions (4) S/H: Stake Holders (5) A: Advantages (6) L-M: Land and Money. This model works at the bottom-up approach level depicting that Smart Cities need to be constitute to resolves the Urban Issues like rural migration, inadequate infrastructure, suitable housing, pathogenic road ways, sewage water problems, electricity theft, leakages of gas, poor governance, pollution etc. So different identification, wireless, security, ambient intelligence, and other embedding IoT technologies will be used to connect to Smart Devices for auto identification, automation, monitoring and controlling of city Six Dimensions like transportation, governance, environment, people, economy and living services. For this purpose, a huge investment and cooperation is required by different stake holders like, corporates, government, real estate, citizens who are going to reside and Non-Government Organizations. These things to be properly put in place to constitute the Smart Cities in vertical or horizontal manner. Off course to the base Land and money. Investment from collaborative companies essential.

5 Smart Cities in China

Economic opportunities in China's cities are fueling rural to urban migration on a massive scale. The resettlement is thought to be one of the largest sustained migrations the world has ever seen, involving the relocation of more than 15 million people from rural to urban areas each year on an average of 40,000 people every day [8]. The garbage piles around the cities, motor vehicle emissions, sound, dust and smoke pollution are a cause of much aggravation in most of the cities. Besides with environmental effects and economic pressures are mounting Chinese cities need to use all smart solutions for urban issues resolving. Needless to say that the total quantity of contaminants discharged in some parts of China's cities are in excess of ecological capacities. China's massive fast economic growth is too smearing gigantic pressure on the china country's societal and natural environment. In this line, China started construction of 'Smart Cities'. In its 12th Five Year Plan, China's major focus is for boosting of its national economy by means of Smart City projects. The number of Smart City projects in China as of Sept., 2013, are around 311 cities. These cities include cities of all size through the entire country, touching 52.6 % of the population of 1.4 billion. The capital city of central China's Hubei Province (Wuhan) is listed among the one of the first nine Smart Cities, in addition to the other eight cities as Taiyuan (Shanxi), Guangzhou (Guangdong), Xuzhou and Wuxi (Jiangsu), Linyi and Zibo (Shandong), Zhengzhou (Henan) and Chongqing.

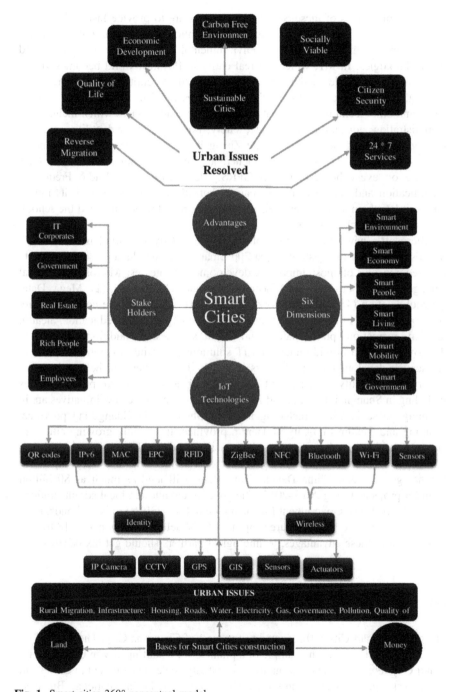

Fig. 1 Smart cities 360° conceptual model

The main objectives of these Smart City projects are to provide fast and effective services (24 × 7) in areas such as security, education, governance, public utilities, traffic management, healthcare and environmental protection by using advanced IoT technologies. In IoT, which the real entity in Physical world becomes virtual entity in Cyber world, and both physical and digital entities are enhance with sensing, processing, and self-adopting capabilities [9]. These cities uses internet, cloud and other technologies to make infrastructure and services which include city administration, education, health care, safety, transportation and utilities more in intelligent on in efficient manner. The Zhengzhou Yuntong bus company, one of main suppliers of school buses launched a project in China. It installed GPS tracking on every bus along with GPRS, Identification like Radio Frequency Identification and launched an application that allows parents to track its movements: it lets them know when their child has boarded the bus, reached the school, and is on way back home.

Microsoft's new global urban solutions [Smart City initiative] is "CityNext". Microsoft is going to constitute circa 200 Smart Cities in China parse. "CityNext" has become one of most important development strategies. Microsoft's general strategies will focus on four facets (1) Cloud Computing, (2) Mega Data, (3) Mobilization (4) Social Networking. CityNext objects include connecting city components like infrastructure, transportation, energy, water, education, health, tourism, recreation, public safety and social services. In addition assisting to the government to provide integrated IT solutions, from information collection to application development, at all levels in China. These cities will help to improve governance, Quality of Life and facilitate citizen's access to public utilities including in Shanghai, Beijing, and Xiamen cities. Many more new initiatives are in planning phase in China including north of Shenzhen in Guangdong province, Ganging and Tianjin Eco-City in Tianjin province. It's rapidly growing cities are taking hub phase in the country's plans, design and build for low-carbon development and sustainable. Financing such mega projects are also very expensive for the government. China Development Bank will lend as much as 80 billion yuan for projects during 2013–2015. The government allowed local administrations to introduced Local Government Financing Vehicles (LGFVs) to issue bonds, raise funds for large-scale infrastructure projects. LGFV debt has ballooned to 18 trillion yuan. Besides these advantages, technological partners should get tax benefits too.

5.1 Meixi Lake Smart City

Meixi Lake Smart City [10], located to the west of Changsha City. The Smart City project at Meixi Lake is an example of a planned for a start-up city in China. Meixi Smart City represents Gale International's first city-scale development in China and its largest international commitment since the Songdo International Business District development in South Korea began in 2001. Meixi Smart City aims to establish a new model of living in connection with nature. The high-density urban

Picture 1 Meixi lake smart
city

plan is enriched by a wide variety of different types of buildings for different
functions and use that integrate with mountains, parks, lakes and canals. Meixi
Lake city will include resource-saving residential developments, schools, healthcare
facilities, retail shopping complexes, high-tech office space, a five-star hotel and a
world-class convention center, all set within an integrated, people-friendly lake
environment that will be surrounded by abundant green space. This level of
diversification defines an environmental context that promotes both healthier and
wealthier lifestyle. The city maximizing opportunities through transport infrastructure, promoting a development strategy of high density core areas, a new central
business district as well as seven character districts. Meixi Lake development seeks
to provide an unparalleled Quality of Life (QoL) for its inhabitants. Gale
International and Cisco [11], working with other global partners, such as UTC and
3M, intend to deliver sustainable. Meixi City of the future that will allow those who
live and work there to thrive for generations to come. Meixi Lake will make
extensive use of new technologies [12]. Advanced environmental engineering,
pedestrian planning, cluster zoning, and garden integration, are all made part of a
holistic strategy of design in this healthy city by Kohn Pedersen Fox Associates.
The Meixi Smart City project plans include (1) 1,675 Acre CBD in Changsha
(2) 500 Acre Lake (3) Eco-friendly Design (Targeting LEED and China 3 Star
Standards) (4) 1.5 M SF Convention Centre (5) R&D Centre (6) Cultural Program
Islands (7) Integrated Subway, Tram, Canal and Water Taxi Transportation Systems
(Picture 1) [http://www.phaidon.com/resource/ml3.jpg].

6 Smart Cities in India

As India's population continues to grow, more citizens will move to cities in the
fourth coming years. It is also estimated that by the end of 2050, the number of
people living in Indian cities will be approximately 843 million. To house this
gigantic urbanization, India needs to find smarter new urban solutions. There is
tremendous potential in India to build an effective ecosystem to enable our

burgeoning urban areas to become smart by using digital technology which is now a days in the form of IoT. In addition build the new Smart Cities in the ecologically balanced regions or outskirts of urban spaces. In line to this India plans to constitute 100 New Smart Cities in the future and will develop modern satellite towns around existing cities. Under the Smart City programme, Government of India had already allocated initial Rs. 7060 crores for Smart Cities in budget 2014–15. Mr. Modi set out GIFT Smart City project in 2007 near Ahmedabad. This is going to boast financial services in India (8–10 %) along with providing the better life to city dwellers. A few more Smart City projects also, which were already started in India, way back between 2000–2010. Lavasa Smart City (2007) is a sustainable city with complete home automation. Similarly SmartCity Kochi and Mahindra World Cities and Wave City. The Delhi-Mumbai Industrial Corridor (DMIC) will serve as a 1,483 km manufacturing hub on the alignment of dedicated freight corridor. Totally eight industrial zones will come up along the DMIC corridor including states of Uttara Pradesh, Haryana, Rajasthan, Madhya Pradesh, Gujarat and Maharashtra. The initial phase of the new cities will be completed by 2019. Ministry of Urban Development has plans to develop 2 Smart Cities in each of India's 29 states to create 10–15 % rise in employment. India and US going to sign on agreement in construction of three Smart Cities (Ahmedabad, Ajmer, and Visakhakaptam). With the collaboration of Japan, the city of Banaras is renovating as Smart City. Singapore is going to develop a new state Smart Capital for the newly formed state of AP. Several countries like Japan, Germany, Sweden, Singapore, Israel, United Kingdom, the USA, Hong Kong and the Netherlands besides multinational corporate have shown keen interest in partnering in building Smart Cities in India. Many private players like Microsoft, IBM, Essel Group and Prince of Qatar shown interest in developing these cities. "Smart Cities: India 2015" at New Delhi conference is a good sign of implementation in India.

6.1 SmartCity Kochi (SCK)

SmartCity Kochi (SCK) is a sustainable Smart City project for knowledge based companies located at Kakkanad in the city of Kochi. The SmartCity Kochi is an IT —SEZ. SmartCity Kochi Infrastructure Pvt. Ltd. is a joint venture company formed by Kerala Government (16 %) and TECOM investments (84 %) to house ICT, media, finance, research and innovation giants. The TECOM is a subsidiary of Dubai holding Company. The initial investment for the SCK, being executed by the (DIC) management will be over $400 million. The Government of Kerala and Dubai based Technology and Media Free Zone Authority (TECOM) went for agreement on 13-5-2007 for setting up the IT Infrastructure Park. The project predicts an investment to the tune of Rs. 1,500 crore. SCK will provide an infrastructure [13, 14] automated systems using IoT, environment and support systems to promote the growth of knowledge-based industry firms. Smart City project committed to create circa 90,000 jobs. The project will have substantial land for greenery, IoT, finance and research clusters, attracting companies both in India and across the world. SCK01 is aimed to be one of the largest LEED platinum rated IT Buildings in India as all of the forthcoming structures coming up here are being developed as an eco-friendly by protecting the local flora and fauna. During piling works, an intense replanting drive was undertaken and trees were replanted elsewhere. The SCK will consisting of (1) Intelligent Building Management System (2) Adequate vertical transportation system (3) Centralized HVAC system through energy efficient chillers plant (4) Adequate power supply intended 10 VA/sq. ft. (5) Its own sewage treatment plant for treating the effluent (6) Provision for Server Rooms on each floor plate (7) Fully secured buildings with Access Control systems. SCK will provide an infrastructure, automated systems using IoT, environment.

References

1. Davis, K.: The urbanization of the human population. Sci. Am. **213**(3), 41–53 (1965)
2. Davis, K.: Urbanization in India—past and future. In: Turner, R. (ed.) India's Urban Future. University of California Press, Berkley (1962)
3. Hall, R.E.: The vision of a smart city. In: Proceedings of the 2nd International Life Extension Technology Workshop, Paris, France, 8 Sept 2000
4. Giffinger, R., Fertner, C., Kramar, H., Kalasek, R., Pichler-Milanović, N., Meijers, E.: Smart Cities: Ranking of European Medium-Sized Cities, Vienna, Austria: Centre of Regional Science (SRF), Vienna University of Technology (2007)
5. Paskaleva, K.A.: The smart city: a nexus for open innovation. Intell. Build. Int. 3.3, 153–171 (2011)
6. Walters, D., Brown, L.L.: Design First: Design-Based Planning for Communities. Routledge (2004)
7. Pellicer, S., Santa, G., Bleda, A.L., Maestre, R., Jara, A.G., Gomez Skarmeta, A.: A global perspective of smart cities: a survey. In: Innovative Mobile and Internet Services in Ubiquitous Computing (IMIS) (2013)
8. Appleyard, B., Zheng, Y., Watson, R., Bruce, L., Sohmer, R., Li, X., Qian, J.: Smart Cities: Solutions for China's Rapid Urbanization (2007)
9. Ning, H.: RFID Major Projects and State Internet of Things, 2nd edn. China Machine Press, Beijing (2010)
10. The KPF design for the Meixi Lake. http://www.kpf.com/project.asp?ID=135. Accessed 21 Dec 2015
11. Very Smart Cities. http://www.forbes.com/2009/09/03/korea-gale-meixi-technology-21-century-cities-09-songdo.html. Accessed 1 Jan 2015
12. Alusi, A., Eccles, R.G., Edmondson, A.C., Zuzul, T.: Sustainable cities: oxymoron or the shape of the future? Harvard Business School Organizational Behavior Unit Working Paper 11–062 (2011)
13. Al-Hader, M., Rodzi, A.: The smart city infrastructure development and monitoring. Theoret. Empirical Res. Urban Manage. 4(2), 87–94 (2009)
14. Bartoli, A., Hernández-Serrano, J., Soriano, M., Dohler, M., Kountouris, A., Barthel, D.: Security and privacy in your smart city. In: Proceedings of Barcelona Smart Cities Congress (2011)

An In Silico Approach for Targeting Plasmodium Phosphatidylinositol 4-Kinase to Eradicate Malaria

Kamal Kumar Chaudhary, Sarvesh Kumar Gupta and Nidhi Mishra

Abstract Accomplishing the destination of malaria evacuation will depend upon directing Plasmodium pathways necessity throughout all life stages. Here, we selected a lipid kinase, phosphatidylinositol 4-kinase (PI4K), as the potential drug target, In order to achieve a novel antimalarial compound that inhibits the intracellular development of multiple Plasmodium species at each stage of infection in the vertebrate host. Virtual screening was performed against more than thousands of compounds from ZINC database to get some potent natural compounds which are able to inhibit PI4K. Binding affinity of screened compounds was compared with well-known inhibitor like Primaquine as a reference molecule, by analyzing their docking score and binding efficiency with the receptor. ADMET properties of the obtained screened compounds were analyzed to check drug like property. Based on the aforementioned analysis, it has been suggested that these screened potent compounds are capable to inhibit PI4K for the prevention, treatment and elimination of malaria.

Keywords Phosphatidylinositol 4-kinase (PI4K) · Virtual screening · ADME/T

1 Introduction

To eradicate malaria, broadly acting medicines must be formulated that therapeutic the diagnostic asexual blood stage, clear the coming before liver phase transmission that can induce relapses and block parasite infection to mosquitoes [1]. Relapse prevention is particularly significant for *P. falciparum* and *P. vivax*, which makes

K.K. Chaudhary · S.K. Gupta · N. Mishra (✉)
Division of Bioinformatics and Applied Science, Indian Institute
of Information Technology, Allahabad 211012, UP, India
e-mail: nidhimishra@iiita.ac.in

© Springer Science+Business Media Singapore 2016
R.K. Choudhary et al. (eds.), *Advanced Computing and Communication Technologies*,
Advances in Intelligent Systems and Computing 452,
DOI 10.1007/978-981-10-1023-1_28

intra-hepatic hypnozoites that, can remain for year earlier reinitiating growth and activating blood stage infection. Primaquine is the drug of choice for treating malaria and only accredited marketed antimalarial drug capable of eradicating the hypnozoite reservoir and perform a radicalcure. Nevertheless, side-effects and weak activity against blood stages prevent widespread usage of primaquine [2]. Subsequently primaquine's target and mechanism of action are not well known, the explore for new revolutionary therapeutic drugs has been limited to concerned analogues, such as Tafenoquine [3]. There is a clear demand for druggable and chemically related validated targets that are necessity in all lifecycle stages of the malaria parasite. Here we describe that a parasite phosphatidylinositol 4-kinase, anomni present eukaryotic enzyme that phosphorylates lipids to govern intracellular signaling and trafficking, is inhibited by screened molecules. In blood stages, inhibitors block a late step in parasite growth by interrupting plasma membrane ingression throughout growing daughter merozoites. This probably stems from varied phosphatidylinositol 4-phosphate (PI4P) pools and disrupted Rab11A-mediated membrane trafficking. Determinations corroborate PI4K as the first known drug target required across all Plasmodium lifecycle stages [4].

2 Methodology

All molecular source and computational analysis was performed by using Schrodinger Maestro version 2014-1, 9.7 build panel. Ligands were prepared using LigPrep application and condition by means of the Optimized Potentials for the Liquid Simulations of electrostatic force field. The protein was modeled before virtual screening and molecular docking the ligands into the active site of the protein molecule. Further the modeled structure of the protein molecule was used to predict the active site of the protein molecule which was then introduced for generation of grid. The energy minimized inhibitor were docked into the generated grid using Glide (Glide, version 6.2, Schrödinger, Inc.) on a Windows 7& Linux based (CentOS release 6.5 Linux-86x-64 platform in Lenovo Intel(R) core i3-3220 (TM) CPU @ 3.30 GHz processor 6 GB RAM workstation. The QikProp program (QikProp, v3.9, Schrödinger software) was used to receive the ADME/T characteristics of the screened compounds along with reference compound Primaquine.

2.1 Protein Preparation

The X-ray crystal structure of PI4K protein was not available in the Protein Data Bank (PDB) therefore we modeled the structure with the help of SWISS-MODEL (http://swissmodel.expasy.org) [5] followed by NCBI Blast (http://blast.ncbi.nlm. nih.gov/Blast.cgi) [6]. The structural features, catalytic residues and active site residues of PI4K were analyzed. For further studies, preparation of protein structure was processed through "Protein Preparation Wizard" [7] of Maestro v9.7 interface of Schrodinger [8]. Protein preparation process involved assigning bond orders, addition of hydrogen bonds, creation of disulphide bonds, conversion of selenomethionine to methionine, filling of missing residues using Prime, capping of termini, deletion of waters and optimization. Energy was minimized using the OPLS_2005 molecular mechanics force field with default value of cut off RMSD (Root Mean Square Deviation).

2.2 Ligand Preparation

Approximately two thousands compounds were extracted from the ZINC databases (http://zinc.docking.org/browse/catalogs/natural-products/) [9] and then processed with Phase v3.8 module of Schrodinger suite [10] for creating a Phase database format (.phdb) of the Schrödinger software for virtual screening followed by LigPrep v2.9 wizard of Maestro v9.7 interface of Schrodinger [11]. It involved generation of maximum possible isomeric and ionization variants. Applying Lipinski's filter, the ligands having poor pharmacological properties were discarded to prepare a virtual library having pharmacologically preferred ligands [12].

2.3 Grid Generation

There may not be any information about the potential ligand binding site for PI4K protein, while in few literatures a putative binding site has been identified by experimental means, but the druggability of the target is not known [4]. Hence it is predicted by SiteMap, which is used for identifying and analyzing binding sites and for predicting target druggability [13]. A receptor grid was generated in the region of these residues of PI4K using Glide v6.2 of Maestro v9.7 interface of Schrodinger with default parameters. Grid point scale for X-axis, Y-axis and Z-axis (-5.86, 364.88 and 90.24) at 10 Å respectively within the grid parameters and grid generation was performed using OPLS 2005 [14].

2.4 Virtual Screening and Molecular Docking Studies

A lead molecule with best docking score was retained through implementation of three subsequent docking operations such as HTVS, SP and XP were applied using

Glide v6.2 of Maestro v9.7 interface of Schrodinger. Based on XP GScore, favorably docked ligands were ranked [15].

2.5 ADME/T Prediction Analysis

To evaluate the ADMET properties QikProp v3.9 module of Maestro v9.7 interface of Schrodinger was used [16]. Various physio-chemical descriptors were calculated to further account for the potential of the lead molecule to act as efficient drug candidate. Violation of Lipinski's rule, if any, was assessed using obtained values for these physio-chemical descriptors. With reference to these values for proposed lead molecule PI4K inhibitors, a comparative study was performed.

3   Result and Discussion

Computational Simulation and Virtual screening was carried out for finding screening potent inhibitors of PI4K including library thousand chemical molecules from generated library. Top eight compounds are presented here along with their glide score energy in drug molecule of hydrogen bond, Van Der Waals and electrostatic energy.

3.1 Validation of Receptor (Target) Protein

The PI4K protein was validated by Ramachandran plot through RAMPAGE by Paul de Bakker and Simon Lovell (http://mordred.bioc.cam.ac.uk/). It showed that 95 % residues were in the most common favorable zone, 2.7 % residue in the generously allowed region, 2.3 % residues in the outlier regions. Hence the protein structure has been selected from the stable structure for the study [17].

3.2 Virtual Screening and Molecular Docking Studies

The prepared protein structure of PI4K was virtually screened against a library of approximately two thousands compounds extracted from ZINC database

Table 1 Molecular docking energy calculations of top eight screened compounds along with primaquine as a reference compound

S. No.	Compound ID	Popular name	Docking score	Glide g score	Glide E-model
1	ZINC00056474	Nordihydroguaiaretic acid	−9.986	−9.986	−66.327
2	ZINC04098166	Picrotoxinin	−9.255	−9.255	−38.587
3	ZINC03881558	Morin	−9.010	−9.015	−55.349
4	ZINC00389747	Naloxone hydrochloride	−8.588	−8.602	−46.325
5	ZINC05037532	Cryptomeridiol	−8.445	−8.445	−37.798
6	ZINC96221670	2-[(4S)-1-(1,3-benzodioxol-5-ylmethyl)-2,5-dioxo-imidazolidin-4-yl]-N-(2-thiazol-2-ylethyl)acetamide	−8.426	−8.426	−62.239
7	ZINC13481200	Xanthopurpurin	−8.221	−8.253	−51.606
8	ZINC04098133	Lactucin	−7.681	−7.705	−39.714
9	PRIMAQUINE	8-(4-Amino-1-methylbutylamino)-6-methoxyquinoline	−4.376	−4.376	−41.636

(http://zinc.docking.org). Three subsequent docking procedures such as HTVS, SP and XP were implemented using Glide v6.2 of Maestro v9.7 interface of Schrodinger. Based on XP G Score, favorably docked ligands were ranked. To find the top poses of the ligands, Glide E-model was used. The lead compound ID ZINC00056474 had a Glide Score of −9.986 and had better binding affinity against PI4K receptor. Similar docking parameters against PI4K are also reported in Table 1 for the comparative purpose. The docking studies indicated that the proposed lead compound ZINC00056474 showed strong hydrogen bond and hydrophobic interactions with the important binding residues of PI4K. The lead compound ZINC00056474 occupies the better binding efficiency against PI4K with lowest docking score and strong interaction in comparison of some other screened compounds against PI4K along with Primaquine as a reference compound (Fig. 1).

3.3 ADME/T Analysis

ADMET properties such as Molecular weight, Hydrogen bond donors, Hydrogen bond acceptors, log P (Octanol/water partition coefficient), percentage of human oral absorption, CNS activity and BBB (blood brain barrier) partition coefficient are

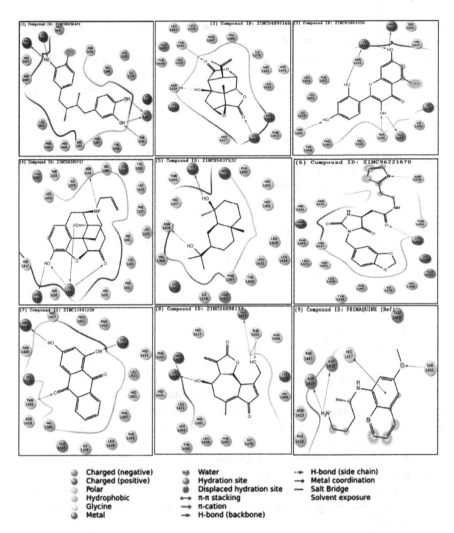

Fig. 1 This shows a figure consisting of different types Protein Ligand Interaction *1* ZINC00056474, *2* ZINC04098166, *3* ZINC03881558, *4* ZINC00389747, *5* ZINC05037532, *6* ZINC96221670, *7* ZINC13481200, *8* ZINC04098133 and *9* PRIMAQUINE

important for ADME estimation. All these values for Screened molecule are following the recommended ranges determined below in Table 2 for a drug with acceptable pharmacological properties. This depicts the excellent potential of ZINC00056474 as prospective lead PI4K inhibitor.

Table 2 ADMET Properties analysis of top eight screened compounds along with primaquine as a reference compound

ADMET	ZINC00056474	ZINC04098166	ZINC03881558	ZINC00389747	ZINC05037532	ZINC96221670	ZINC13481200	ZINC04098133	Primaquine (ref)
Mol. Wt.	302.3	292.2	302.2	327.3	240.3	402.4	240.2	276.2	259.347
Dipole	5.457	9.505	2.712	7.840	2.833	5.239	1.269	4.304	1.218
H-bond donor	4.000	0.000	4.000	2.000	2.000	1.000	1.000	2.000	3.000
H-bond acceptors	3.000	7.750	5.250	6.250	1.500	8.000	4.500	8.400	3.750
Potential energy	62.689	320.739	70.638	352.178	128.586	97.683	86.611	127.512	82.393
QPPcaco	118.645	712.910	23.713	159.174	2673.3	169.838	182.498	158.885	435.057
QPlogPw	10.487	8.816	14.329	10.848	5.036	13.705	9.198	12.976	9.415
QPlogPo/w	2.414	-0.313	0.414	1.441	3.398	1.949	1.040	0.039	2.161
QPlogS	-3.264	0.656	-2.749	-1.699	-3.741	-3.577	-2.442	-2.088	-2.415
QPlogHERG	-4.635	-1.318	-4.922	-4.135	-3.003	-4.264	-4.432	-3.663	-5.553
QPlogBB	-1.802	-0.331	-2.236	-0.267	-0.101	-1.409	-1.077	-1.276	-0.188
QPlogKP	-3.762	-3.481	-5.315	-5.078	-2.337	-3.100	-3.804	-4.382	-3.695
QPPMDK	49.399	343.161	8.668	75.084	1432.0	220.103	78.679	67.735	222.608
CNS	-2	-1	-2	1	0	-2	-2	-2	0
% HOA	78.204	76.175	53.978	74.793	100.00	78.272	73.505	66.568	86.825

Recommended ranges:
Mol. Wt.: <500; Hydrogen Bond Donor: <5; Hydrogen Bond Acceptor: <10; Log P (Octanol/water partition coefficient): <5; % Oral Absorption: >80 % High, <25 % Poor; BBB Partition Coefficient: −3; CNS Activity: −2 Inactive, 2 Active

4 Conclusion

Virtual screening method is widely used for reducing cost and time of drug discovery process. It has been clearly shown that the Structure based virtual screening approach utilized in this study successfully find eight potentially active compound on the basis of virtual screening against PI4K from the ZINC database which may be potential inhibitors of PI4K. The docked poses of these compounds resembles similar orientation as observed ligand. Therefore this kind of study shows the importance of large libraries of molecule and their use to intensify drug development process former synthesis. PI4K plays an important role in intracellular development of multiple Plasmodium species at each stage of infection in the vertebrate host. Thus, it can act as a therapeutic target for the treatment of malaria. Various safety, tolerability and immunological concerns related with already documented PIK4, making them unfit for clinical use motivated us to discover safe compound with acceptable pharmacological properties. We have proposed top eight inhibitors against PI4K, based on rational drug design. Docking studies revealed that the better binding interaction of ZINC00056474 against PI4K with reference to Primaquine. ZINC00056474 is having acceptable pharmacological properties thus it could be a futuristic perspective chemical compound for the treatment of malaria.

References

1. Greenwood, B.M., Fidock, D.A., Kyle, D.E., Kappe, S.H., Alonso, P.L., Collins, F.H., Duffy, P.E.: Malaria: progress, perils, and prospects for eradication. J. Clin. Invest. **118**, 1266–1276 (2008)
2. Vale, N., Moreira, R., Gomes, P.: Primaquine revisited six decades after its discovery. Eur. J. Med. Chem. **44**, 937–953 (2009)
3. Wells, T.N., Burrows, J.N., Baird, J.K.: Targeting the hypnozoite reservoir of *Plasmodium vivax*: the hidden obstacle to malaria elimination. Trends Parasitol. **26**, 145–151 (2010)
4. McNamara, et al.: Targeting plasmodium phosphatidylinositol 4-kinase to eliminate malaria. Nature **504**(7479), 248–253 (2013)
5. Biasini, M., Bienert, S., Waterhouse, A., Arnold, K., Studer, G., Schmidt, T., Kiefer, F., Cassarino, T.G., Bertoni, M., Bordoli, L., Schwede, T.: SWISS-model: modelling protein tertiary and quaternary structure using evolutionary information. Nucleic Acids Res. **42**(1), 252–258 (2014)
6. Altschul, S.F., Gish, W., Miller, W., Myers, E.W., Lipman, D.J.: Basic local alignment search tool. J. Mol. Biol. **215**, 403–410 (1990)
7. Schrödinger Release 2014-1: Schrödinger Suite 2014-1 Protein Preparation Wizard; Epik version 2.7, Schrödinger, LLC, New York, NY, 2013; Impact version 6.2, Schrödinger, LLC, New York, NY, 2014; Prime version 3.5, Schrödinger, LLC, New York (2014)
8. Schrödinger Release 2014-1: Maestro, version 9.7, Schrödinger, LLC, New York (2014)
9. Irwin, J.J., Sterling, T., Mysinger, M.M., Bolstad, E.S., Coleman, R.G.: ZINC: a free tool to discover chemistry for biology. J. ChemInf Model **52**(7), 1757–1768 (2012)
10. Dixon, S.L., Smondyrev, A.M., Knoll, E.H., Rao, S.N., Shaw, D.E., Friesner, R.A.: PHASE: a new engine for pharmacophore perception, 3D QSAR model development, and 3D database screening. 1. Methodology and preliminary results. J. Comput. Aided Mol. Des. **20**, 647–671 (2006)

11. Schrödinger Release 2014-1: LigPrep, version 2.9, Schrödinger, LLC, New York (2014)
12. Lipinski, C.A., Lombardo, F., Dominy, B.W., Fenney, P.J.: Experimental and computational approaches to estimate solubility and permeability in drug discovery and development settings. Adv. Drug Deliv. Rev. **46**, 3–26 (2001)
13. Halgren, T.: Identifying and characterizing binding sites and assessing druggability. J. Chem. Inf. Model. **49**, 377–389 (2009)
14. Halgren, T.A., Murphy, R.B., Friesner, R.A., Beard, H.S., Frye, L.L., Pollard, W.T., Banks, J. L.: Glide: A new approach for rapid, accurate docking and scoring. 2. Enrichment factors in database screening. J. Med. Chem. **47**, 1750–1759 (2004)
15. Friesner, R.A., Murphy, R.B., Repasky, M.P., Frye, L.L., Greenwood, J.R., Halgren, T.A., Sanschagrin, P.C., Mainz, D.T.: Extra precision glide: docking and scoring incorporating a model of hydrophobic enclosure for protein-ligand complexes. J. Med. Chem. **49**, 6177–6196 (2006)
16. Small-Molecule Drug Discovery Suite 2014-1: QikProp, version 3.9, Schrödinger, LLC, New York (2014)
17. Lovell, S.C., Davis, I.W., Arendall III, W.B., de Bakker, P.I.W., Word, J.M., Prisant, M.G., Richardson, J.S. and Richardson D.C.: Structure validation by Calpha geometry: phi, psi and Cbeta deviation. Proteins Struct. Funct. Genet. **50**, 437–450 (2002)

Effect of Binding Constant on Phase Diagram for Three-Lane Exclusion Process

Atul Kumar Verma and Arvind Kumar Gupta

Abstract The goal of this paper is to analyze the effect of Langmuir-kinetics on phase diagrams of three-lane totally asymmetric simple exclusion process under the fully asymmetric coupling environment. We introduce a new parameter, namely binding constant (K) which signifies the ratio of attachment and detachment rates. The steady-state dynamics of the system is analyzed for different value of K. Phase diagrams and density profiles are obtained using mean-field theory which are found to be in good agreement with Monte-Carlo simulation results.

Keywords Motor proteins · TASEP · Langmuir-kinetics · Phase diagram · Density profile

1 Introduction

Motor-proteins are a type of molecular motors that can move along filaments to deliver cargoes to specific locations in cells and in this intracellular trans-port filaments serve as macromolecular highways [1]. Intracellular transport of motor-proteins is an important aspect not only in understanding the physical and bio-logical properties of a cell but also to get insight about some diseases like neurodegenerative diseases, ciliary dyskinesias, situs inversus, retinitis pigmentosa, tumor suppression, left right body determination etc. [2]. Motor-protein transport is a far from equilibrium process which in past decades is studied by using totally asymmetric simple exclusion process (TASEP) model [3] that is a paradigm of non-equilibrium thermodynamics. TASEP is a discrete non-equilibrium model first used for describing the kinetics of biopolymerization in 1968 [4]. It can also explain

A.K. Verma (✉) · A.K. Gupta
Department of Mathematics, Indian Institute of Technology Ropar,
Rupnagar 140001, Punjab, India
e-mail: atulkv@iitrpr.ac.in

A.K. Gupta
e-mail: akgupta@iitrpr.ac.in

© Springer Science+Business Media Singapore 2016
R.K. Choudhary et al. (eds.), *Advanced Computing and Communication Technologies*,
Advances in Intelligent Systems and Computing 452,
DOI 10.1007/978-981-10-1023-1_29

many non-equilibrium phenomena occurring not only in physics but also in many other applied areas [5] like protein synthesis, translation of mRNA and vehicular flow in traffic etc. TASEP is defined on one dimensional lattice on which each site either can be empty or occupied by one particle. Particles can move to their neighboring site in one direction if target site is empty. Recently, many researchers have given lot of attention to TASEP with particle creation and annihilation [Langmuir Kinetics (LK)] in bulk. It is important to study such processes with LK not only for understanding non-equilibrium phenomena, but also the intracellular transport of molecular motors where attachment and detachment of motors occurs between the cytoplasm and the filament. Single-channel TASEP with LK has been studied comprehensively [6] and it is observed that LK affects system properties significantly.

In real world, there are wide occupancy of multi-lane transport processes such as a motor protein can move along multi parallel filaments [1] forming a multi-lane systems so it is important to study multi-lane exclusion processes. Looking at the importance of multi-lane transport processes some researchers have studied many two-lane TASEP models with [7, 8] and without [9] LK. In the recent years, few researchers have focused on three-lane TASEP [10–12]. In this trend, Wang et al. [10] proposed a weakly and asymmetrically coupled three-lane TASEP and observed bulk induced phase transitions. Cai et al. [11] studied multi-lane TASEPs with fully asymmetric coupling and noticed that the number of lanes affect structure of phase diagram significantly. Jiang et al. [12] proposed a strong asymmetrically coupled multi-lane partially asymmetric simple exclusion process (PASEP) to make the steady-state phase diagram.

It should be noted that all above studies [10–12] in three-lane system have focused on only TASEP without particle attachment-detachment process. Our main purpose is to generate phase diagrams for three-lane TASEP with LK for various values of binding constant.

2 Model Description

We consider particle transport along three parallel one dimensional coupled lattice channels each with N sites denoted by 1, 2 and 3 (Fig. 1). The state of a site i (i = 1, 2, 3, ..., N) in lane j (j = 1, 2, 3) is characterized by an occupation variable τ_j^i where $\tau_j^i = 1$ (or $\tau_j^i = 0$) according to the site is occupied or vacant, respectively. Particles are inserted in the system under hard-core exclusion principle which insures that one site cannot be occupied by more than on particles. For each time step, a lattice site (i, j) is chosen at random and random sequential updates rules are adopted. Following are the dynamical rules:

(a) At entrance (i = 1), a particle enters the lattice with rate α provided $\tau_j^1 = 0$. If $\tau_j^1 = 1$ and $\tau_j^2 = 0$, then the particle in the site of (1, j) moves to (2, j) with unit rate. Neither lane changing nor attachment-detachment takes place at i = 1.

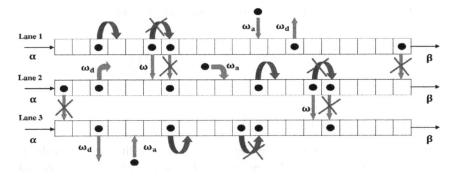

Fig. 1 Sketch of three lane TASEP with LK. *Arrows* (*crosses*) indicate the allowed (prohibited) hopping

(b) At the exit (i = N), a particle can leave the system with rate β. No lane changing and attachment-detachment occur here.

(c) In the bulk (1 < i < N), if $\tau_j^i = 1$, then the particle firstly tries to leave the system with detachment rate w_d. If it fails, then it moves forward to site (i + 1, j) with unit rate provided $\tau_j^{i+1} = 0$; otherwise it jumps to the next lane (j + 1) with rate w only if the target site is vacant. Note that due to consideration of fully asymmetric coupling environment particles of lane j cannot move on lane (j − 1). Further, if $\tau_j^i = 0$ then a particle can enter the system with attachment rate ω_a.

2.1 Master-Equations

The master-equations for the evolution of the particle densities of bulk sites in each lane (j = 1, 2, 3) are given by following equations:

$$\frac{d\langle \tau_1^i \rangle}{dt} = \langle \tau_1^{i-1}(1 - \tau_1^i) \rangle - \langle \tau_1^i(1 - \tau_1^{i+1}) \rangle + \omega_a \langle (1 - \tau_1^i) \rangle - \omega_d \langle \tau_1^i \rangle$$
$$- \omega \langle \tau_1^i \tau_1^{i+1}(1 - \tau_2^i) \rangle, \tag{1}$$

$$\frac{d\langle \tau_2^i \rangle}{dt} = \langle \tau_2^{i-1}(1 - \tau_2^i) \rangle - \langle \tau_2^i(1 - \tau_2^{i+1}) \rangle + \omega_a \langle (1 - \tau_2^i) \rangle - \omega_d \langle \tau_2^i \rangle + \omega \langle \tau_1^i \tau_1^{i+1}(1 - \tau_2^i) \rangle$$
$$- \omega < \tau_2^i \tau_2^{i+1}(1 - \tau_3^i), \tag{2}$$

$$\frac{d\langle \tau_3^i \rangle}{dt} = \langle \tau_3^{i-1}(1 - \tau_3^i) \rangle - \langle \tau_3^i(1 - \tau_3^{i+1}) \rangle + \omega_a \langle (1 - \tau_3^i) \rangle$$
$$- \omega_d \langle \tau_3^i \rangle + \omega \langle \tau_2^i \tau_2^{i+1}(1 - \tau_3^i) \rangle. \tag{3}$$

At the boundaries, equations of particle density are given by:

$$\frac{d\langle \tau_j^1 \rangle}{dt} = \alpha \left\langle \left(1 - \tau_j^1\right)\right\rangle - \left\langle \tau_j^1 \left(1 - \tau_j^2\right)\right\rangle, \tag{4}$$

$$\frac{d\langle \tau_j^1 \rangle}{dt} = \left\langle \tau_j^{L-1}\left(1 - \tau_j^L\right)\right\rangle - \beta \left\langle \tau_j^L \right\rangle. \tag{5}$$

where $\langle \ldots \rangle$ stands for statistical average. The continuum limit of the model is derived by coarse-gaining a discrete-lattice with lattice constant $\epsilon = 1/N$ and rescaling the time as $t' = t\epsilon$. We rescale various kinetic rates like attachment, detachment and lane changing rates as $\omega_a N = \Omega_a, \omega_b N = \Omega_b$ and $\omega N = \Omega$ to inspect the competing interplay between bulk and boundary dynamics. Our major purpose is to investigate the effect of ratio of attachment and detachment rates on system dynamics for maximum possible order of lane changing rate. In order to do so we define a new parameter $K = \frac{\omega_a}{\omega_d} = \frac{\Omega_a}{\Omega_d}$, which denotes ratio of attachment and detachment rates and can be called binding constant. Clearly K is always positive and an important factor that can play an important role in characterizing phase diagrams. In this paper, we focus on effect of K on system properties.

Replacing binary discrete number $\langle \tau_j^i \rangle$ with continuous variable ρ_j^i and dropping the superscript $_i$ because of independency of all three lanes from any kind of spatial inhomogeneity, we get

$$\frac{\partial}{\partial t'} \begin{bmatrix} \rho_1 \\ \rho_2 \\ \rho_3 \end{bmatrix} + \frac{\partial}{\partial x} \begin{bmatrix} -\frac{\varepsilon}{2}\frac{\partial \rho_1}{\partial x} + \rho_1(1 - \rho_1) \\ -\frac{\varepsilon}{2}\frac{\partial \rho_2}{\partial x} + \rho_2(1 - \rho_2) \\ -\frac{\varepsilon}{2}\frac{\partial \rho_3}{\partial x} + \rho_3(1 - \rho_3) \end{bmatrix} = S \tag{6}$$

where

$$S = \begin{bmatrix} K\Omega_d(1 - \rho_1) - \Omega_d\rho_1 - \Omega\rho_1^2(1 - \rho_2) \\ K\Omega_d(1 - \rho_2) - \Omega_d\rho_2 + \Omega\rho_1^2(1 - \rho_2) - \Omega\rho_2^2(1 - \rho_3) \\ K\Omega_d(1 - \rho_3) - \Omega_d\rho_3 + \Omega\rho_2^2(1 - \rho_3) \end{bmatrix}.$$

Here S represents non-conservative terms formed by combination of different transitions like lane changing and LK. The boundary conditions for above system of non-linear coupled Eq. (6) are $\rho_1(0) = \rho_{2(0)} = \rho_{3(0)} = \alpha$ and $\rho_1(1) = \rho_2(1) = \rho_3(1) = 1 - \beta = \gamma$ (say).

3 Steady-State Solution: Methodology

Above system of equations are difficult to solve analytically as equations are coupled and non-linear, so we solve these equations numerically. In order to get density profiles of all three-lanes we discretize system of Eq. (6) as

$$\frac{\rho_{d,i}^{j+1} - \rho_{d,i}^{j}}{\Delta t'} = \frac{1}{2N} \frac{\rho_{d,i-1}^{j} - 2\rho_{d,i}^{j} + \rho_{d,i+1}^{j}}{\Delta x^2} + (2\rho_{d,i}^{j} - 1) \frac{\rho_{d,i+1}^{j} - \rho_{d,i-1}^{j}}{2\Delta x} - \Omega_d \left((K+1)\rho_{d,i}^{j} - K \right)$$
$$+ \Omega \left(\rho_{d-1,i}^{j} \right)^2 \left(1 - \rho_{d,i}^{j} \right) - \Omega \left(\rho_{d,i}^{j} \right)^2 \left(1 - \rho_{d+1,i}^{j} \right).$$

$$(7)$$

where $i, j, \Delta x, \Delta t'$ and d are space index, time index, space-step, time-step and lane number respectively. Also, the second last term is only for lane 1 and 2 while second last term is for lane 2 and 3. The density profile at steady-state can be calculated from Eq. (7) when $j \to \infty$.

4 Results

We wish to examine the effect of binding constant $\left(K = \frac{\Omega_a}{\Omega_d} \right)$ on phase diagram of proposed model. In order to do so, we determine phase diagrams for various K lying between 0 and 1. We validate our mean-field results by Monte-Carlo simulation results for system size $N = 1000$. Simulations are carried out for 10^{10} time-steps and first 5 % are neglected to be confident that system reached the steady-state. A time interval of 10 N between each average step is taken to calculate densities in all three lanes. To begin with we take into account a pinnacle case: $K = 0$.

Obviously $K = 0$ infers $\Omega_a = 0$, which is a case when no particles can attach to bulk. The steady-state phase diagram of proposed model under the case $K = 0$ is given in Fig. 2a. As one can see in phase diagram apart from L, H and S phases one mixed phase C_2 is detected where C_2 is monotonically increasing density profile that crosses density 0.5 once. From Fig. 2a it is clear that there exist 8 distinct steady-state phases namely LLL, LLC$_2$, LC$_2$C$_2$, LLS, LLH, LC$_2$S, LC$_2$H and LSH. Maximum part of phase plane is covered by LC$_2$C$_2$ phase. It should be noted that phases like LLC$_2$, LC$_2$C$_2$ and LC$_2$S are not observed before this study as per our knowledge. Figure 3 shows some typical density profiles. One can easily notice that we always have $\rho_3 > \rho_2 > \rho_1$ because of fully asymmetric coupling conditions. As a outcome we never have H phase in first lane. It should be noted that some there is a little difference between theoretical and Monte-Carlo results this is because of mutual interactions between all three lanes (as lane changing rate is of higher order). Further it has been seen that for lower order of lane changing rate theoretical and simulation results correlate well. Now we discuss effect of K on phase diagram

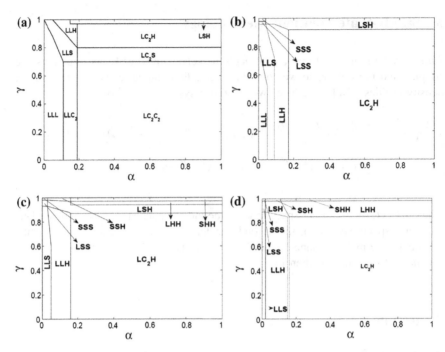

Fig. 2 Phase diagrams for **a** K = 0, **b** K = 0.35, **c** K = 0.70, **d** K = 1. Here Ω_d = 0.2 and Ω = 1000

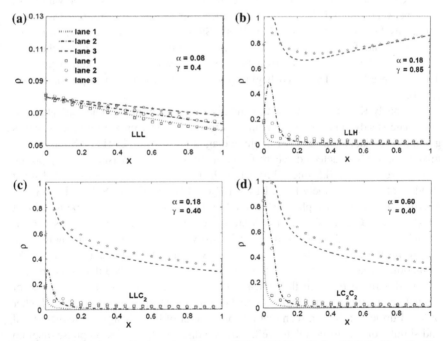

Fig. 3 Typical density profiles in various phases **a** LLL, **b** LLH, **c** LLC$_2$, **d** LC$_2$C$_2$. Here $\Omega_a = \Omega_d$ =0.2 (color online)

(Fig. 2a). From Fig. 2b–d it is clear that K affects phase diagram significantly. Firstly we discuss the case when K is increased from 0 to 0.35. Easily one can see that there is a great structural change in phase diagram for K = 0.35. There exists 8 distinct steady-state phases for K = 0.35 (see Fig. 2b). In these 8 phases LSS and SSS are two new phases while others were also present in phase diagram for K = 0 but increase in K leads to disappearance of some phases like LLC_2, LC_2C_2, LC_2S and LC_2H. Also with increase in K phase LLL shrinks while LSH, LLS, LLH and LC_2H expands. If we further increase value of K from 0.35 to 0.70, we get 9 distinct steady-state phases. It should be noted that LLL phase disappeared, this is justified because on increase K attachment rate Ωa also increase so more particles can attach to bulk so density in both lanes increase that leads to disappearance of LLL phase. Moreover, two new phases SSH and SHH appear in phase diagram for K = 0.70. Interestingly there is no change in region covered by LC2H phase. Further, phases LLS and LSH shrink while LLH expands with increase in K. If K is again increased from 0.70 to 1, we get 9 distinct steady-state phases in phase diagram for K = 1. It is also notable that number of steady-state phases decreases firstly then increases with increase in K.

5   Conclusions

We have proposed a three-lane TASEP model with LK under the fully asymmetric coupling conditions. A new parameter binding constant (K) which denotes ratio of attachment and detachment rates is introduced. Mean-filed theory is used to study the proposed model while computer Monte-Carlo simulations is performed to validate our mean-field results which agree with simulation results. We have obtained phase diagrams for various values of K. It is found that binding constant affects system properties significantly. Further, structure of phase diagrams depend on binding constant. Additionally, number of steady-state phases vary non-monotonically with increase in binding constant It is also seen that LSH phase firstly expands then shrinks with increase in binding constant. Moreover, we found that region of LC_2H phase is independent of binding constant greater than 0.35.

References

1. Howard, J., Clark, R.: Mechanics of motor proteins and the cytoskeleton. Appl. Mech. Rev. **55**, 39 (2002)
2. Kinesin molecular motors: transport pathways, receptors, and human disease. Proc. Natl. Acad. Sci. U.S.A. **98**(13), 69997003 (2001)
3. An exactly soluble non-equilibrium system: the asymmetric simple exclusion process. Phys. Rep. **301**(1), 6583 (1998)
4. MacDonald, C.T., Gibbs, J.H., Pipkin, A.C.: Kinetics of biopolymerization on nucleic acid templates. Biopolymers **6**(1), 125 (1968)

5. Chowdhury, D., Schadschneider, A., Nishinari, K.: Physics of transport and traffic phenomena in biology: from molecular motors and cells to organisms. Phys. Life Rev. 2(4), 318352 (2005)
6. Parmeggiani, A., Franosch, T., Frey, E.: Phase coexistence in driven onedimensional transport. Phys. Rev. Lett. 90(8), 086601 (2003)
7. Jiang, R., Wang, R., Wu, Q.S.: Two-lane totally asymmetric exclusion processes with particle creation and annihilation. Phys. A 375(1), 247256 (2007)
8. Gupta, A.K., Dhiman, I.: Asymmetric coupling in two-lane simple exclusion processes with langmuir kinetics: Phase diagrams and boundary layers. Phys. Rev. E 89(2), 022131 (2014)
9. Pronina, E., Kolomeisky, A.B.: Asymmetric coupling in two-channel simple exclusion processes. Phys. A 372(1), 1221 (2006)
10. Wang, Y.Q., Jiang, R., Kolomeisky, A.B., Hu, M.B.: Bulk induced phase transition in driven diffusive systems. Sci. Rep. 4 (2014)
11. Cai, Z.P., Yuan, Y.M., Jiang, R., Hu, M.B., Wu, Q.S., Wu, Y.H.: Asymmetric coupling in multi-channel simple exclusion processes. J. Stat. Mech. Theor. Exp. 2008(07), P07016 (2008)
12. Shi, Q.H., Jiang, R., Hu, M.B., Wu, Q.S.: Strong asymmetric coupling of multilane paseps. Phys. Lett. A 376(40), 26402644 (2012)

SIR Model of Swine Flu in Shimla

Vinod Kumar, Deepak Kumar and Pooja

Abstract In this work, an attempt has been made to study the SIR model on swine flu, which produced by viral infection agent in Himachal Pradesh. The SIR model is analyzed on the numerical data obtained from IGMC, Shimla, H.P. This Mathematical model is based on ordinary differential equations for communicable disease. The concept of basic reproduction number R_0 is used to analyze the stages of the disease. The result is helpful for health policy makers and reducing the number of deaths spread from the Swine flu.

Keywords Epidemic disease · Stability analysis · Swine flu · Basic reproduction number · Mathematical modelling

1 Introduction

Swine Flu is highly contagious respiratory disease which "refers to any strain of the influenza of infectious virus, called the swine influenza virus that is epidemic infection in pigs". The signs and symptoms of swine flu are fever with a body temperature over 100 °F, body weakness, cold and runny nose, cough or sore throat, body aches, chills; fatigue, diarrhea and vomiting sometimes, but more commonly seen with seasonal flu. The subtypes of influenza A disease are also known as H1N1, H1N2, H2N3, H3N1 and H3N2. This work is representing that,

V. Kumar (✉) · D. Kumar
Department of Mathematics, Manav Rachna International University,
Faridabad, India
e-mail: vinod.k4bais@gmail.com

D. Kumar
e-mail: deepakkumar.fet@mriu.edu.in

Pooja
Department of Microbiology, IGMC, Shimla, HP, India
e-mail: drpoojasharma@gmail.com

© Springer Science+Business Media Singapore 2016
R.K. Choudhary et al. (eds.), *Advanced Computing and Communication Technologies*,
Advances in Intelligent Systems and Computing 452,
DOI 10.1007/978-981-10-1023-1_30

how changes in human behaviour and social mixing influence the epidemic by the study of basic reproduction number, equilibrium points and stability condition [1, 2] (Figs. 1, 2 and 3).

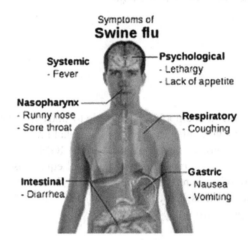

Fig. 1 Symptoms of swine flu

Fig. 2 Structure of SIR

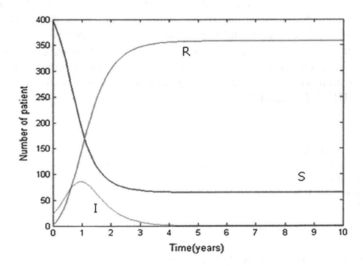

Fig. 3 In this graph, S represents the susceptible to the infectious stage with the *blue color*, I represent the total number of infected individuals with *green color* and R represents the recovery rate with respect to time with *red color*

In the past few months, the number of diagnosed swine flu and influenza A (H1N1) cases has raised, in area of Himachal Pradesh. In this region, the number of confirmed infections has increased, but this only covers a small proportion of the cases which have occurred in that region [3]. The world health organization defined the situation for pandemic from infectious flu and describing an optimal control policy to protect [2]. The application of mathematical modeling on chronic disease through SEIR, SIR and many other models providing the concept of secondary infection and analyzing various stages of epidemic disease with their mathematical ecology in the population [4, 5]. The mechanism of swine flu infection spreading and procedure of providing the treatment is given by a new algorithm [6]. The compartment model SIR discussed by Kermack and Mckendrick in 1927 and many mathematical models are used by many mathematicians to discussed various stages of endemic, epidemic and pandemic diseases with their serious effects on living things [7].

2 Mathematical Model

The SIR model for mathematical epidemiology was proposed by Kermack and Mckendrick. The governing equations on swine flu are given below:

$$\frac{dS}{dt} = -\beta SI \tag{1}$$

$$\frac{dI}{dt} = \beta SI - \alpha I \tag{2}$$

$$\frac{dR}{dt} = \alpha I \tag{3}$$

2.1 Diagram of the Model

2.1.1 Variables and Parameter Description

S = Susceptible class, I = infective class, R = Total recovered, β = rate of transmission, α = rate of recovery.

3 Analysis of the Disease Free Equilibrium

If the individual contains no symptoms of swine flu in the susceptible class. Then the numbers of infective will be zero. i.e. $I = 0$. This state of the condition in the absence of the infection is said to be as disease free equilibrium in the population. Therefore, we have

$$\beta SI - \alpha I = 0 \Rightarrow S_0 = \frac{\alpha}{\beta} \tag{4}$$

Thus, the disease free equilibrium point is given as $(S_0, 0, 0, 0, 0)$.

3.1 The Basic Reproduction Number

The basic reproduction number is defined for an epidemic SIR compartmental model for the Swine flu. If $R_0 < 1$, then the disease free equilibrium is stable; otherwise, it is unstable for $R_0 > 1$. Mathematically defined as,

$$R_0 = \frac{The\ transmission\ rate}{The\ rate\ of\ recovery} \tag{5}$$

This basic reproduction ratio has important role in determining the rate of infection. Hence, the epidemic outbreak of the disease is stable or unstable and the host will recover or not recover from the disease.

3.2 Equilibrium Analysis of the Disease

At equilibrium,

$$\frac{dS}{dt} = \frac{dI}{dt} = \frac{dR}{dt} = 0 \tag{6}$$

$$\text{Case 1: } \frac{dS}{dt} = \frac{dI}{dt}$$

$$\Rightarrow S \propto \frac{k_1}{R_0}; \tag{7}$$

$$\text{Where } k_1 = \frac{1}{2}$$

Therefore, the susceptibility increases as the basic reproduction ratio decreases and vice versa.

$$\text{Case 2:} \frac{dI}{dt} = \frac{dR}{dt}$$

$$\Rightarrow S \propto \frac{k_2}{R_0}; \tag{8}$$

$$\text{Where } k_2 = 2$$

Therefore, the susceptibility increases as the basic reproduction ratio decreases and vice versa.

$$\text{Case 3:} \frac{dS}{dt} = \frac{dR}{dt}$$

$$\Rightarrow S \propto \frac{k_3}{R_0}; \tag{9}$$

$$\text{Where } k_3 = -1$$

Therefore, the susceptibility increases as the basic reproduction ratio decreases and vice versa.

3.3 Stability Analysis of the SIR Model

Jacobian matrix of the governing equation of the system is given as,

$$J = \begin{bmatrix} -\beta I & -\beta S & 0 \\ \beta I & \beta S - \alpha & 0 \\ 0 & \alpha & 0 \end{bmatrix} \cdot \begin{bmatrix} S \\ I \\ R \end{bmatrix} \tag{10}$$

$$\text{Now, Det(J)} = \begin{vmatrix} -\beta I - \lambda & -\beta S & 0 \\ \beta I & \beta S - \alpha - \lambda & 0 \\ 0 & \alpha & 0 - \lambda \end{vmatrix} = 0 \tag{11}$$

$$\Rightarrow (-\lambda) \left[\lambda = \frac{-(\beta I - \beta S + \alpha) \pm \sqrt{(\beta I - \beta S + \alpha)^2 - 4\beta \alpha I}}{2} \right] = 0 \tag{12}$$

i.e. $\lambda_1 < 0$, $\lambda_2 < 0$ and $\lambda_3 < 0$; if $-(\beta I - \beta S + \alpha) > \sqrt{(\beta I - \beta S + \alpha)^2 - 4\beta \alpha I}$

Since, all the eigen values are negative then given model is stable, otherwise the model is unstable.

Table 1 Numerical values of the parameters

Parameters	B	α	S(0)	I(0)	R_0
Values	0.012531 per year	2.46269 per year	399 per year	23 per year	0.00125 per year

$$\lambda_1 < 0, \lambda_2 < 0 \text{ and } \lambda_3 > 0; \text{ if } -(\beta I - \beta S + \alpha) < \sqrt{(\beta I - \beta S + \alpha)^2 - 4\beta\alpha I}.$$

It is observed that the stability condition of disease free equilibrium exists i.e. the value of $R_0 < 1$ i.e. basic reproduction number is less than 1, which shows that the total number of susceptible decreases.

4 Numerical and Graphical Analysis of the Model

Data for the influenza cases from January 2013 to December 2013, is collected from IGMC Shimla, HP. Numerical analysis for the disease on the population is given below with the help of the basic reproduction number.

From the Eq. 7, we have $R_0 = \dfrac{k_1}{S(t)}$; for all $k_1 > 0$. Consider $k_1 = 1/2$, then R_0

$$= 0.00125 > 0$$

$$(13)$$

Therefore, the disease free equilibrium is unstable and the infected class will not recover completely from the infection (Table 1).

The graph shows that susceptible class is decreasing with respect to time and removal class is increasing with respect to time due to treatment on time. The infected class is increasing initially but it is approaching to zero after getting treatment.

5 Conclusion

In this study, we conclude that the recovery rate from the epidemic disease is high due to the strong immunity in the population. The SIR model helps to explain and understand the various stages of the disease, which are given graphically through MATLAB analysis. The SIR model has used the basic reproduction number to the study of stability of the disease. We found the basic reproduction number $R_0 < 1$, this value shows that epidemic will not spread. This model approaches that if the social awareness about the diseases among the susceptible, it may help to decrease the infection rate of epidemic disease.

References

1. Das, P., Gazi, N.H., Das, K., Mukherjee, D.: Stability analysis of swine flu transmission—a mathematical approach. Comput. Math. Biol. **3**(1), (2014)
2. Sebastian, M.R., Lodha, R., Kabra, S.K.: Swine origin influenza (Swine Flu). Indian J Pediatr **76** (2009)
3. Numerical Data Collected from Indira Gandhi Medical College, Shimla, January, 2013 to December, 2013
4. Herbert, W.: Hethcote: the mathematics of infectious diseases. SIAM Rev **42**(4), 599–653 (2000)
5. Henneman, K., Peursem, D.V., Huber, V.C.: Mathematical modeling of influenza and a secondary bacterial infection. WSEAS Trans. Biol. Biomed. 10(1) (2013)
6. Aldila, D., Nuraini, N., Soewono, E.: Optimal control problem in preventing of swine flu disease transmission. AMS **8**(71), 3501–3512 (2014)
7. Singh, M., Sharma, S.: An epidemiological study of recent outbreak of influenza A H1N1 (Swine Flu) in Western Rajasthan region of India. JMAS **3**(2), 48–52 (2013)
8. kumar, V., Kumar, D.: SITR dynamical model for influenza. IJETSR **2** (2015)
9. Lin, F., Muthu Raman, K., Lawley, M.: An optimal control theory approach to non-pharmaceutical interventions. BMC Infect Dis **10**, 32 (2010)

Higher Order Compact Numerical Simulation of Shear Flow Past Inclined Square Cylinder

Atendra Kumar and Rajendra K. Ray

Abstract The numerical simulation of 2D unsteady, incompressible shear flow past square cylinder with an angle of incidence ($\alpha = 45°$) is carried out in this paper. Simulations are performed using ψ-ω formulation of Navier-Stokes equations on compact uniform grid. Higher Order Compact (HOC) formulation is used to discretize the governing equations. Numerical results are presented for Reynolds number $Re = 100$ with three different shear parameters $K = 0.0$, 0.05 and 0.1. Computed results are compared with existing numerical results. Our computed results are found to produce better approximation of the exact flow dynamics. Shedding frequency is seen to decrease with increasing K.

Keywords Shear rate · Inclined square cylinder · Vortex shedding · Reynolds number · Higher order compact formulation

1 Introduction

The evolution of incompressible viscous flow induced by an impulsively started square cylinder is one of the most studied problems. The problem is attractive due to its academic importance as well as its associated technical issues, related with energy conservation etc. Although its practical importance in various fields like vortex flowmeters, buildings, long spanned bridges, towers etc., mostly by experimental study but now a days also by using numerical simulation. Yoon et al. [1] numerically investigate the characteristics of laminar flow past inclined square cylinder immersed in uniform freestream. Sohankar et al. [2] proposed numerical study of 2D unsteady laminar flow around a square cylinder with various incidence

A. Kumar (✉) · R.K. Ray
School of Basic Sciences, Indian Institute of Technology Mandi,
Mandi 175001, HP, India
e-mail: atendra.iitd@gmail.com

R.K. Ray
e-mail: rajendra@iitmandi.ac.in

© Springer Science+Business Media Singapore 2016   305
R.K. Choudhary et al. (eds.), *Advanced Computing and Communication Technologies*,
Advances in Intelligent Systems and Computing 452,
DOI 10.1007/978-981-10-1023-1_31

angles ($\alpha = 0° - 45°$) and Reynolds number ranges from $Re = 50 - 200$. The flow is computed by using an incompressible SIMPLEC code with non-staggered mesh. The Stuart-Landau equation used to investigate the vortex shedding phenomenon at different inclinations of cylinder. Later on Dutta et al. [3] gave experimental study about the flow past square cylinder with the use of particle image velocimetry, hot wire anemometry. In their study, the Reynolds number (Re) is 410, based on the cylinder side and the centerline velocity at inlet. Experimental results shown in terms of coefficients of drag, streamlines, mean-time velocity, strouhal number, power spectra and vorticity contours. The finite difference method based on large eddy simulation results, are shown in Li et al. [4], for incompressible heat and fluid flows around square cylinder at high Reynolds number.

In this paper, we study the above mentioned problem by extending a newly developed HOC scheme. The usefulness of HOC scheme to reproduce the almost accurate flow phenomena is already established for uniform flows [5, 6]. But the efficiency of the scheme is not properly tested for shear flows. Therefore, this work is an important addition to the existing works in terms of establishing the usefulness of the HOC scheme and in depth study the present flow problem.

2 Problem Formulation and Discretization

2.1 Governing Equations

Consider the problem of unsteady, incompressible shear flow past inclined square cylinder of width 'a' (Fig. 1). Flow is governed by the incompressible Navier-Stokes equations. The dimensionless stream function-vorticity ($\psi - \omega$) form of N-S equations in Cartesian coordinates (x, y) are given by

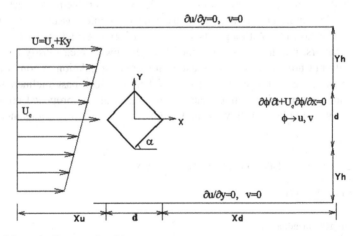

Fig. 1 Schematic diagram of problem

$$\frac{\partial^2 \omega}{\partial x^2} + \frac{\partial^2 \omega}{\partial y^2} = \text{Re}\left(u\frac{\partial \omega}{\partial x} + v\frac{\partial \omega}{\partial y} + \frac{\partial \omega}{\partial t} \right) \tag{1}$$

$$\frac{\partial^2 \psi}{\partial x^2} + \frac{\partial^2 \psi}{\partial y^2} = -\omega \tag{2}$$

Here u, v, ψ, ω represents velocity in x-direction, y-direction, stream function and vorticity, respectively. The velocity components u and v in terms of stream function can be given as,

$$u = \frac{\partial \psi}{\partial y}, \qquad v = -\frac{\partial \psi}{\partial x} \tag{3}$$

And vorticity ω is

$$\omega = \frac{\partial v}{\partial x} - \frac{\partial u}{\partial y} \tag{4}$$

At inlet we apply a linear shear velocity profile,

$$u = U_c + Ky, \qquad v = 0 \tag{5}$$

where U_c represents the centerline velocity at inlet and K is the shear parameter. And, at the outflow a convective boundary condition [7] is used as

$$\frac{\partial \phi}{\partial t} + U_c \frac{\partial \phi}{\partial x} = 0 \tag{6}$$

applied for all velocity components. This provides smooth discharge of vortices at the outlet.

In the problem, artificial boundaries of the computational domain are sufficiently far from the cylinder boundary. Therefore, their effect on the characteristics of the flow near the square cylinder is negligible. The boundary conditions on the top and bottom boundaries can be given as

$$\frac{\partial u}{\partial y} = K, \qquad v = 0 \tag{7}$$

Now, on the surface of the cylinder, no-slip conditions have been used.

2.2 Higher Order Compact Discretization

The computation is carried out by extending the recently developed transformation-free higher order compact (HOC) numerical scheme [5], for unsteady two dimensional (2D) convection-diffusion equations on Cartesian mesh. Thus the HOC discretization of governing equations at (i, j)th mesh point of the computational domain can be written as

$$
\begin{aligned}
&\Big[\mathrm{Re} + A11_{ij}\delta_x^2 + A12_{ij}\delta_y^2 + A13_{ij}\delta_x + A14_{ij}\delta_y + A15_{ij}\delta_x\delta_y \\
&\qquad + A16_{ij}\delta_x\delta_y^2 + A17_{ij}\delta_x^2\delta_y + A18_{ij}\delta_x^2\delta_y^2\Big]\omega_{ij}^{n+1} \\
&= \Big[\mathrm{Re} + A21_{ij}\delta_x^2 + A22_{ij}\delta_y^2 + A23_{ij}\delta_x + A24_{ij}\delta_y + A25_{ij}\delta_x\delta_y \\
&\qquad + A26_{ij}\delta_x\delta_y^2 + A27_{ij}\delta_x^2\delta_y + A28_{ij}\delta_x^2\delta_y^2\Big]\omega_{ij}^{n}
\end{aligned}
\tag{8}
$$

And

$$
[\delta_x^2 + \delta_y^2 - (H2 + K2)\delta_x^2\delta_y^2]\psi_{ij} = [-1 + H2\delta_x^2 + K2\delta_y^2]\omega_{ij}
\tag{9}
$$

where the coefficients are defined as

$$
\begin{aligned}
aligned A11_{ij} &= -H12\mathrm{Re} - 0.5\Delta t A1_{ij},\ A12_{ij} = -K12\mathrm{Re} - 0.5\Delta t A2_{ij}, A13_{ij} \\
&= -H11\mathrm{Re} - H12u_{ij}\mathrm{Re}^2 - 0.5\Delta t A3_{ij}, A14_{ij} \\
&= -K11\mathrm{Re} - K12v_{ij}\mathrm{Re}^2 - 0.54_{ij}, A15_{ij} = -0.5\Delta t A5_{ij}, A16_{ij} \\
&= -0.5\Delta t A6_{ij}, A17_{ij} = -0.5\Delta t A7_{ij}, A18_{ij} = -0.5\Delta t A8_{ij}, aligned
\end{aligned}
$$

where,

$$
\begin{aligned}
A1_{ij} &= 1 + H11\mathrm{Re}u_{ij} + H12\mathrm{Re}^2 u_{ij}^2 + 2H12\mathrm{Re}(u_x)_{ij}, \\
A2_{ij} &= 1 + K11\mathrm{Re}v_{ij} + K12\mathrm{Re}^2 v_{ij}^2 + 2K12\mathrm{Re}(v_y)_{ij}, \\
A3_{ij} &= -\mathrm{Re}u_{ij} + H11\mathrm{Re}(u_x)_{ij} + K11\mathrm{Re}(u_y)_{ij} + H12\mathrm{Re}^2 u_{ij}(u_x)_{ij} \\
&\quad + H12\mathrm{Re}(u_{xx})_{ij} + K12\mathrm{Re}(u_{yy})_{ij} + K12\mathrm{Re}^2 v_{ij}(u_y)_{ij}, \\
A4_{ij} &= -\mathrm{Re}v_{ij} + H11\mathrm{Re}(v_x)_{ij} + K11\mathrm{Re}(v_y)_{ij} + H12\mathrm{Re}^2 u_{ij}(v_x)_{ij} \\
&\quad + H12\mathrm{Re}(v_{xx})_{ij} + K12\mathrm{Re}(v_{yy})_{ij} + K12\mathrm{Re}^2 v_{ij}(v_y)_{ij}, \\
A5_{ij} &= H11\mathrm{Re}v_{ij} + K11\mathrm{Re}u_{ij} + H12\mathrm{Re}^2 u_{ij}v_{ij} + 2H12\mathrm{Re}(v_x)_{ij} \\
&\quad + 2K12\mathrm{Re}(u_y)_{ij} + K12\mathrm{Re}^2 u_{ij}v_{ij}, \\
A6_{ij} &= -H11 - H12\mathrm{Re}u_{ij} + K12\mathrm{Re}u_{ij}, \\
A7_{ij} &= -K11 + H12\mathrm{Re}v_{ij} - K12\mathrm{Re}v_{ij}, A8_{ij} = -H12 - K12,
\end{aligned}
$$

$$H2 = -\frac{h^2}{12}, \; K2 = -\frac{h^2}{12}, \; H11 = \text{Re}u_{ij}\frac{h^2}{6}, \; K11 = \text{Re}v_{ij}\frac{h^2}{6}, \; H12 = -\frac{h^2}{12}, \; K12 = -\frac{h^2}{12}$$

The detail of the discretization and the finite difference operators can be found in [5].

3 Results and Discussion

We compute the problem of the flow past square cylinder inclined at angle of $(\alpha = 45°)$ at $Re = 100$ with shear rate $K = 0.0, 0.05$ and 0.1. The Reynolds number defined as $\text{Re} = \frac{U_c d}{\nu}$, where $d/a = \cos\alpha + \sin\alpha$.

Figure 2 plots the streamline pattern for $K = 0.0$ at different time instants for one non-dimensional time period and shows the symmetric flow pattern, with respect to the centerline and generating vortices closely circulates behind the cylinder. The Kármán vortex street clearly seen behind the cylinder along with constant gap between alternate vortices shed. In Fig. 3, the similar streamline pattern can be observed for $K = 0.05$, with different strength and size of the vortices.

Oscillatory vortex shedding phenomenon still preserve Kármán vortex street when $K = 0.1$, as shown in Fig. 4. From this figure, it can be observed that size of vortices is more than from the previous K values. Also the shedding frequency decreased, due to the shear effect, but still alternate shedding of vortices present and settled in two different lines.

Further, effect of shear rate on the flow phenomenon can be observed from the vorticity contours as shown in Figs. 5, 6 and 7. For uniform flow past inclined square cylinder (K = 0.0), alternate vortices of the same strength are generated from the upper and lower side, as shown in Fig. 5. In this case, the location of the stagnation point is situated at the front corner of square cylinder. Figure 5a shows

Fig. 2 Streamline pattern for $K = 0.0$, $Re = 100$, **a** t = 295, **b** t = 298, **c** t = 301

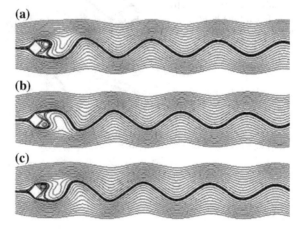

Fig. 3 Streamline pattern for
K = 0.05, *Re* = 100,
a t = 370, **b** t = 373,
c t = 376

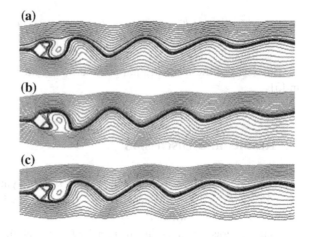

Fig. 4 Streamline pattern for
K = 0.1, *Re* = 100, **a** t = 297,
b t = 300, **c** t = 303

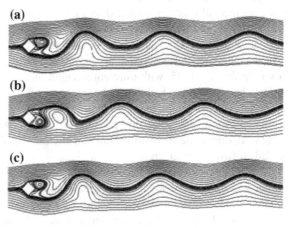

Fig. 5 Contours of vorticity
for one cycle of vortex
shedding, K = 0.0, *Re* = 100,
at different time levels,
a t = 295, **b** t = 298,
c t = 301

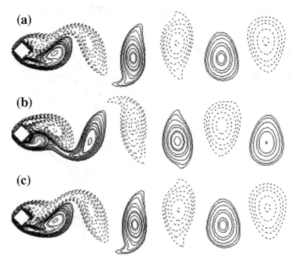

Fig. 6 Vorticity contours of vortex shedding at K = 0.05, *Re* = 100, different time instants, **a** t = 370, **b** t = 373, **c** t = 376

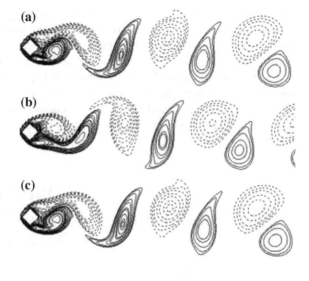

Fig. 7 Vorticity contours of vortex shedding at K = 0.1, *Re* = 100, different time instants, **a** t = 297, **b** t = 300, **c** t = 303

that an anticlockwise vortex appear from the lower corner of the cylinder and a clockwise vortex develops from the upper corner of the cylinder, at t = 295. Whereas, at t = 298, a clockwise vortex shed from the upper of the cylinder and the anticlockwise vortex rebuild from the other corner of the cylinder.

In Fig. 6, the vorticity contours for shear rate K = 0.05 at different time instants are shown. The similar structure of vortex shedding behind the square cylinder occurs as for K = 0.0. Figure 6a shows, at t = 370, a clockwise vortex dissociate from upper corner of the cylinder, while an anti-clockwise vortex developed from the lower corner of the cylinder. At t = 373, anti-clockwise vortex has been developed and can be detach from the cylinder, as illustrated in Fig. 6b. The flow is completely periodic and maintain alternate vortices phenomenon.

Moreover when K = 0.1, the flow phenomenon e.g. size and strength of the vortices, is different from previous K values. The anticlockwise vortices are of slender type, in shape whereas the clockwise vortices are rolling over them. The alternate vortex shedding phenomenon still remains but vortex shedding frequency decreases. From the numerical study for K = 0.1, it is confirmed that vortices are shed alternatively, with anti-clockwise vortices being flush in size and the clockwise vortices being rotated. Variation of velocity components behind the cylinder with time at point (2.58, 0.33), also depicts the flow phenomenon (Figs. 8, 9 and 10).

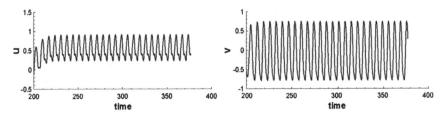

Fig. 8 Velocity fluctuation with time at point (2.58, 0.33) for *Re* = 100, K = 0.0

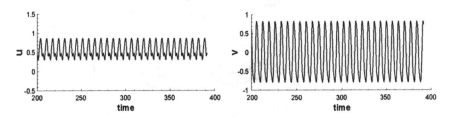

Fig. 9 Velocity fluctuation with time at point (2.58, 0.33) for *Re* = 100, K = 0.05

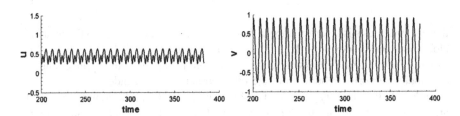

Fig. 10 Velocity fluctuation with time at point (2.58, 0.33) for *Re* = 100, K = 0.1

4 Conclusion

This paper presents the HOC finite difference simulation of incompressible linear shear flow past inclined square cylinder with inclination of ($\alpha = 45°$) at $Re = 100$ with three different shear parameter values $K = 0.0$, 0.05, 0.1. From numerical study, it is observed that the wake improvement and vortex shedding are considerably depend of the shear rate (K). The growth of K is more liable for the disparity in the shape and strength of the alternately shed vortices.

Acknowledgments The work has been done under the SERB funded project (No.: SERB/F/7046/2013-2014 dated 12.02.2014). The authors kindly acknowledge the financial support of SERB.

References

1. Yoon, D.H., Yang, K.S., Choi, C.B.: Flow past a square cylinder with an angle of incidence. Phys. Fluids **22**, 043603 (2010)
2. Sohankar, A., Norberg, C., Davidson, L.: Low-reynolds-number flow around a square cylinder at incidence: study of blockage, onset of vortex shedding and outlet boundary condition. Int. J. Numer. Meth. Fluids **26**, 39–56 (1998)
3. Dutta, S., Panigrahi, P.K., Muralidhar, K.: Experimental investigation of flow past a square cylinder at an angle of incidence. J. Eng. Mech. **134**, 788–803 (2008)
4. Li, Y., Chan, C.K., Mei, B., Zhu, Z.: LES of incompressible heat and fluid flows past a square cylinder at high reynolds numbers. Int. J. Comput. Fluid Dyn. **29**, 272–285 (2015)
5. Dalal, D., Kalita, J.C., Dass, A.K.: A transformation-free hoc scheme for steady-state convection diffusion on non-uniform grids. Int. J. Numer. Meth. Fluids **44**, 33–53 (2004)
6. Kalita, J.C., Ray, R.K.: A transformation-free HOC scheme for incompressible viscous flows past an impulsively started circular cylinder. J. Comput. Phys. **228**, 5207–5236 (2009)
7. Lankadasu, A., Vengadesan, S.: Shear effect on square cylinder wake transition characteristics. Int. J. Numer. Meth. Fluids **67**, 1112–1134 (2010)

A Cartesian Grid Method for Solving Stokes Equations with Discontinuous Viscosity and Singular Sources

H.V.R. Mittal and Rajendra K. Ray

Abstract In this paper, we present a new finite difference scheme for solving incompressible, steady stokes equations in discontinuous domains. While solving two phase Stokes equations, across some interfaces, there are discontinuities in the pressure, viscosity and velocity. Since, these jump conditions are coupled together, it is difficult to discretise and solve the system of equations accurately. We apply the augmented approach introduced by Li et al. (Int J Numer Meth. Fluids 44:33–53, 2004) to decouple these jump conditions. A new finite difference method is then developed and presented to solve the resulting augmented system of Stokes equations. Points of intersection of grid lines and the interface are used as a node in the finite difference stencil and jump conditions are then used to determine the values at these intersection points. Numerical solutions are compared with the corresponding analytical solutions and those of the augmented immersed interface method. The method is found to be second order accurate for almost all the variables in the infinity norm.

Keywords Finite difference · HOC scheme · Immersed interface · Level set method · Stokes equations

1 Introduction

The incompressible, two phase Stokes equations with discontinuities in the viscosity and with singular source terms are often encountered in many practical applications in CFD. Over the last few decades, there has been an ongoing quest for new numerical methods to solve fluid problems with interfaces. Most of the numerical works on such problems involved mainly the use of immersed boundary (IBM) [1] or immersed interface (IIM) methods [2] on Cartesian grids. In the

H.V.R. Mittal (✉) · R.K. Ray
School of Basic Sciences, Indian Institute of Technology Mandi,
Kamand 175005, Himachal Pradesh, India
e-mail: hvrmittal@gmail.com

© Springer Science+Business Media Singapore 2016   315
R.K. Choudhary et al. (eds.), *Advanced Computing and Communication Technologies*,
Advances in Intelligent Systems and Computing 452,
DOI 10.1007/978-981-10-1023-1_32

present work, a new immersed interface finite difference scheme, based on Cartesian coordinates, is presented for solving Stokes equations with arbitrary immersed interfaces in two dimensions. The originality of the scheme lies in the use of additional values at interfacial points (points at which grid lines intersect the interface) as nodes in finite difference stencil, which allows straightforward use of the standard finite difference approximations. Jump conditions of the variable and flux are then used to determine these interfacial values.

The paper is arranged as follows. Section 2 deals with mathematical formulation which covers governing equations, jump conditions and their discretisation procedure. Section 3 deals with the numerical results and Sect. 4 with the conclusion.

2 Mathematical Formulation

2.1 Governing Equations and the Jump Conditions

Assume that Ω is a domain in R^2 and Γ is an arbitrary piecewise smooth curve in Ω as shown in Fig. 1a. Level set formulation introduced by Osher and Sethian [3] is used to represent the interface Γ. Square domain consists of the union of two sub-domains Ω^+ and Ω^- separated by a closed interface Γ. Two dimensional steady Stokes equations can be represented as

$$\nabla p = \mu \Delta u + G + \int_{\Gamma} F(s)\delta(x - X(s))ds, \quad x \in \Omega, \tag{1}$$

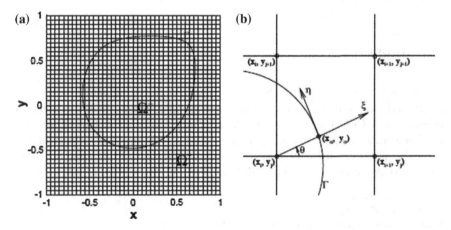

Fig. 1 a Uniform cartesian grid in two-dimensions with interface Γ separating the whole domain into two separate sub domains Ω^- and Ω^+. **b** Geometry at an irregular point (x_i, y_j) and its normal projection (x_α, y_α) on Γ; (ξ, η) is the local coordinate system aligned with Γ at (x_α, y_α)

$$\nabla u = 0, \quad x \in \Omega, \tag{2}$$

where $u = (u, v)$ denotes the velocity vector, p denotes the pressure, F(s) is the singular source term, G(x) denotes the bounded force which may be discontinuous across Γ as well when Γ is parameterized by s. The viscosity μ is supposed as a piecewise constant

$$\mu = \begin{cases} \mu^+, & \Gamma > 0 \\ \mu^-, & \Gamma < 0 \end{cases}$$

Assume $\Gamma \in C^2$, $\hat{F}_1(s) \in C^1$, and $\hat{F}_2(s) \in C^1$ where $\hat{F}_1(s) = F(s) \cdot n$ and $\hat{F}_2(s) = F(s)$; n is the normal and τ is tangential to the interface, respectively. Jump conditions across Γ can be written as.

$$[p] = 2\left[\mu \frac{\partial u}{\partial \xi} \cdot n\right] + \hat{F}_1, \tag{3}$$

$$\left[\frac{\partial p}{\partial \xi}\right] = [G \cdot n] + \frac{\partial \hat{F}_2}{\partial n} + 2\left[\mu \frac{\partial^2}{\partial \eta^2}(u \cdot n)\right], \tag{4}$$

$$\left[\mu \frac{\partial u}{\partial \xi} \cdot \tau\right] + \left[\mu \frac{\partial u}{\partial \eta} \cdot n\right] + \hat{F}_2 = 0, \tag{5}$$

$$[\mu \Delta \cdot u] = 0. \tag{6}$$

Balancing force along the n-axis at Γ gives us the jump condition (3) while the balancing forces along the τ-axis at Γ gives the jump condition (5). Applying the divergence operator to Eq. (1) gives jump condition (4) while the last jump condition (6) is derived from incompressibility condition $\nabla \cdot u = 0$.

Since, these jump conditions are coupled together, it is difficult to discretise and solve the system of equations accurately. We therefore apply the augmented approach introduced by Li et al. in 2007 [4] to decouple these jump conditions. Jumps $([\mu u](s), [\mu v](s)) = ([\mu u](s), [\mu v](s))$ are introduced as two augmented variables so that these jump conditions can be decoupled. Denote

$$q = (q_1, q_2)^T = ([\tilde{u}], [\tilde{v}])^T$$

Let p, u and v denote the solutions of the system of Eqs. (1)–(2). Then, the variables $\tilde{u}, \tilde{v}$, p and q satisfy the following set of equations:

$$\begin{cases} \Delta p = \nabla \cdot \boldsymbol{G} \\ [p] = \hat{F}_1 - 2\dfrac{\partial q}{\partial \tau} \cdot \boldsymbol{\tau} \\ \left[\dfrac{\partial p}{\partial n}\right] = \dfrac{\partial \hat{F}_2}{\partial \tau} + 2\dfrac{\partial^2}{\partial \eta^2}(\boldsymbol{q} \cdot \boldsymbol{n}) + [\boldsymbol{G} \cdot \boldsymbol{n}] \end{cases} \tag{7}$$

$$\begin{cases} \Delta \tilde{u} = p_x - \boldsymbol{G}_1, \\ [\tilde{u}] = q_1, \quad \left[\dfrac{\partial \tilde{u}}{\partial n}\right] = \left(\hat{F}_2 + \dfrac{\partial q}{\partial \tau} \cdot \boldsymbol{n}\right)\sin\theta - \left(\dfrac{\partial q}{\partial \tau} \cdot \boldsymbol{\tau}\right)\cos\theta, \end{cases} \tag{8}$$

$$\begin{cases} \Delta \tilde{v} = p_y - \boldsymbol{G}_2, \\ [\tilde{v}] = q_2, \quad \left[\dfrac{\partial \tilde{v}}{\partial n}\right] = -\left(\hat{F}_2 + \dfrac{\partial q}{\partial \tau} \cdot \boldsymbol{n}\right)\cos\theta - \left(\dfrac{\partial q}{\partial \tau} \cdot \boldsymbol{\tau}\right)\sin\theta, \\ \left[\dfrac{\tilde{u}}{\mu}\right] = 0, \quad \left[\dfrac{\tilde{v}}{\mu}\right] = 0 \end{cases} \tag{9}$$

Here, θ denotes the angle between x-axis and n-axis. These jump conditions are derived and explained in detail in [4]. It is worth mentioning that this system of partial differential equations is an elliptic system in which we have to solve all the equations simultaneously. If q is known to us, pressure jump conditions can be calculated. Hence pressure Eq. (7) can be solved independently. After solving for the pressure, u and v can be solved from the Eqs. (8) and (9) respectively.

2.2 Discretisation Procedure

Consider a uniform cartesian grid [a, b] × [c, d] in the whole domain as shown in Fig. 1a. The grid points are denoted by (x_i, y_j) for $0 \leq i \leq m$, $0 \leq j \leq n$. These grid points are categorised into two classes: regular and irregular grid points. A grid point is classified as irregular if the finite difference stencil resulting from the corresponding derivatives intersects Γ. Otherwise; it is termed as a regular grid point.

For regular grid points, finite difference approximations of arbitrary order, derived from a Taylor series expansion can be directly used. We use a Higher Order Compact (HOC) finite difference scheme proposed by kalita et al. [5] for discretisation. The idea is to use the original partial differential equation to substitute the leading truncation error terms in the finite difference equation. We begin by considering a general poisson equation $\nabla^2 u = f$ in variable u and forcing function f defined on the whole domain. For example, HOC finite difference formulation at a regular point (x_i, y_j) for $\frac{\partial^2 u}{\partial x^2}$ can be written as

$$\frac{\partial^2 u}{\partial x^2} = \delta x^2 u_{i,j} - \frac{dx^2}{12} \left(\frac{\partial^2 f}{\partial x^2} - \delta x^2 \delta y^2 u_{i,j} \right) \tag{10}$$

Here, δx^2 and δy^2 denote the standard central difference approximations respectively. The jump conditions across the interface can be written as

$$[u] = u^+ - u^- = \hat{C} \tag{11}$$

$$[u_\xi] = u_\xi^+ - u_\xi^- = \sigma, \qquad [u_\eta] = u_\eta^+ - u_\eta^- = \lambda. \tag{12}$$

with [.] denoting the jump, the superscripts "+" and "−" represent the variables in the Ω^+ and Ω^- regions respectively. Since the jump conditions are usually defined in normal (ξ-axis) or tangential (η-axis) direction of the interface, we introduce local coordinates (ξ, η) aligned with the interface at (X, Y) (see Fig. 1b), as

$$\xi = (x - x_\alpha) \cos \theta + (y - y_\alpha) \sin \theta,$$
$$\eta = -(x - x_\alpha) \sin \theta + (y - y_\alpha) \cos \theta,$$

where θ is the angle between +ve x-axis and the normal direction. Transforming the jump condition on u through local coordinate system and then differentiating it with respect to x and y yields

$$[u_x] = [u_\xi] \cos \theta - [u_\eta] \sin \theta \tag{13}$$

$$[u_y] = [u_\xi] \sin \theta + [u_\eta] \cos \theta \tag{14}$$

Since, Taylor series expansion, which provides the platform for finite difference approximations, is not valid for non-smooth functions, standard approximations fail to yield correct numerical solutions near the interface. Hence, a new modified finite difference is developed to discretise the governing equation at irregular points. We use standard central difference formulas including interfacial points (points at which grid lines intersect the interface) nodes in the stencil. The two jump conditions are then used to determine the values at these interfacial points at a desired order of accuracy. We consider a dimension by dimension splitting approach for discretization. We will only derive the one dimensional formula on the segment $[x_{i-1}, x_{i+2}] \times \{y_j\}$. Other cases are duplication. For example, consider the situation shown in Fig. 2.

For the irregular grid point "x_i" depicted in Fig. 2, we discretise the derivative using the nodes x_{i-1}, x_i and the interfacial point x as

$$\frac{\partial^2 u}{\partial x^2} = \frac{\frac{u^- - u_i}{x_\alpha - x_i} - \frac{u_i - u_{i-1}}{dx}}{\frac{x_\alpha - x_i}{2} + \frac{dx}{2}} \tag{15}$$

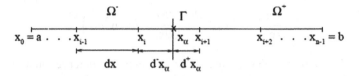

Fig. 2 Schematic diagram of general uniform grid stencil on the segment $[x_{i-1}, x_{i+2}] \times \{y_j\}$ with an interface Γ located at point x_α between irregular nodes x_i and x_{i+1}

Second order accurate one sided finite difference formula for u_x^- using three points on the left side of Γ can be written as

$$u_x^- = \frac{dx\left(dx + 2dx_\alpha^-\right)u^- - \left(dx + dx_\alpha^-\right)^2 u_i + \left(dx_\alpha^-\right)^2 u_{i-1}}{dx\,dx_\alpha^-\left(dx + dx_\alpha^-\right)} \qquad (16)$$

Similarly expression for u_x^+ can be written as

$$u_x^+ = \frac{-dx\left(dx + 2dx_\alpha^+\right)u^+ + \left(dx + dx_\alpha^+\right)^2 u_{i+1} - \left(dx_\alpha^+\right)^2 u_{i+2}}{dx\,dx_\alpha^+\left(dx + dx_\alpha^+\right)} \qquad (17)$$

Substituting the values of u_x^- and u_x^+ into Eq. (13), we get an equation in two unknowns u$^+$ and u$^-$. The first jump condition Eq. (11) is the second equation in variables u$^+$ and u$^-$. This system of two equations in two unknown variables can be easily solved analytically to find initial values of u$^+$ and u$^-$. These values can be substituted in the approximation given in Eq. (16). It is worth mentioning that we have considered the quantities u$_{i-1}$, u$_i$, u$_{i+1}$ and u$_{i+2}$ as constants in our system of equations. The reason is that, the initial guess is normally set equal to zero at all grid nodes. These values are updated at every iteration and the numerical solution is iterated till it converges towards the correct solution.

Special cases might arise when there is a scarcity of irregular points or when the grid point coincides with the interfacial point. In the former case, one can use a first order accurate formula for u_x^- or u_x^+ that will only involve u$_i$ and u$_{i+1}$ as nodes in the stencil. In the latter case, the grid point may be considered lying in either Ω^+-region or Ω^--region and the modified difference scheme can be employed.

3 Results and Discussion

Two numerical examples are presented to check the efficiency of our methodology. Level set function ϕ is given as $\phi = x^2 + y^2 - 1$ and the analytical solution is known under different situations. Denote by L_p^∞ and L_u^∞, infinity norm of error at all the grid nodes for pressure and velocity, respectively and $L_u^\infty = \left(\frac{L_u^\infty + L_v^\infty}{2}\right)$.

Example 1 We start with an example where the velocity is smooth and pressure, viscosity are discontinuous at the interface. The exact solution of pressure and velocity is given by

$$p = \begin{cases} 0, & \Gamma \leq 0 \\ 1, & \Gamma > 0 \end{cases}$$
$$u = y(x^2 + y^2 - 1), \quad (x,y) \in \Omega, \qquad v = -x(x^2 + y^2 - 1), \quad (x,y) \in \Omega.$$

and the viscosity μ is

$$\mu = \begin{cases} 0.5, & \Gamma \leq 0 \\ 1, & \Gamma > 0 \end{cases}$$

The normal and tangential force density components can be calculated from the jump conditions as $\hat{F}_1 = -1$ and $\hat{F}_2 = -1$. The external forcing term G is given by $G = (-4y, 4x)$ outside and $G = (-8y, 8x)$ inside Γ.

Table 1 shows the grid refinement analysis of the error infinity norm of our computed results and compares the respective orders of accuracies with the results of Li et al. [4]. Second order accuracy is reported by our numerical simulations for both pressure and velocities, which are comparable to those reported by Ref. [4].

Example 2 In the second example, the quantities u, v and p are kept unchanged inside the interface but they are all set equal to zero outside. The velocity is now discontinuous and force density components are $\hat{F}_1 = -1$ and $\hat{F}_2 = -2$. The external forcing term G is given value zero outside Γ.

We compare our computed results for both pressure and velocity with Le et al. [4] in Table 2. One can see from the table that our computed error infinity norm, for both pressure and velocity, is better than those of Le et al. [4]. Least square fit line of the absolute errors in log-log plots in Fig. 3 shows second order accuracy of our scheme, for both pressure and velocity.

Table 1 Grid refinement analysis and comparison of orders of accuracy of the present scheme with those of Li et al. [4]

N	Present scheme				[4]		
	L_p^∞ error	p-order	L_u^∞ error	u-order	p-order	u-order	
32	3.6271×10^{-3}		3.3283×10^{-3}				
64	9.0124×10^{-4}	2.0090	2.2846×10^{-4}	2.0065	1.4353	1.9242	
128	2.2372×10^{-4}	2.0140	2.1124×10^{-4}	1.9716	1.6883	2.1367	
256	5.0814×10^{-5}	2.1386	5.2241×10^{-5}	2.0518	1.8183	2.2635	
512	1.2421×10^{-5}	2.0326	1.3024×10^{-5}	2.0042	1.8541	1.7155	

Table 2 Grid refinement analysis and comparison of infinity error norms of the computed results of the present scheme with those of Li et al. [4]

N	Present scheme		[4]	
	L_p^∞ error	L_u^∞ error	L_p^∞ error	L_p^∞ error
32	3.8533×10^{-3}	4.0338×10^{-3}	8.4430×10^{-3}	3.4549×10^{-3}
64	6.9871×10^{-4}	9.5724×10^{-4}	2.8405×10^{-4}	8.8800×10^{-4}
128	2.5818×10^{-4}	2.3942×10^{-4}	8.0952×10^{-4}	2.2666×10^{-4}
256	6.2649×10^{-5}	5.8108×10^{-5}	2.5417×10^{-5}	4.7693×10^{-5}
512	1.4171×10^{-5}	1.3448×10^{-5}	5.8083×10^{-5}	1.4086×10^{-5}

Fig. 3 Linear regression analysis using log-log scale for both pressure and velocity. Negative of slopes of these regression lines show order of accuracy

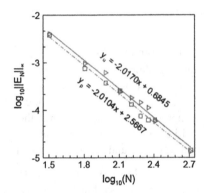

4   Conclusion

In this paper, we present a new finite difference scheme for Stokes flows in domains with discontinuities due to the presence of immersed circular interfaces. Augmented approach developed by Li et al. [4] is employed to decouple the jump conditions of pressure and velocity. Recently developed HOC scheme is then used for discretisation at the points away from the interface which produces least third order accuracy on a compact stencil. A new modified finite difference scheme is developed and implemented at irregular points. Interfacial points (points where interface intersects the grid lines) are used as nodes in the finite difference stencil and jump conditions across the interface are used to determine the values at these interfacial points. Studies confirm that our scheme has overall second order of accuracy and it produces better results than the augmented immersed interface method. However, order of accuracy can be increased further by involving more number of grid nodes in the finite difference stencil used at irregular points. As stokes flow represents a two-phase flow, through the two numerical experiments presented in the current study, our scheme can be easily extended to solve multiphase flows with both fixed and moving boundaries.

References

1. Peskin, C.S.: Numerical analysis of blood flow in the heart. J. Comput. Phys. **25**, 220–252 (1977)
2. Leveque, R.J., Li, Z.: The immersed interface method for elliptic equations with discontinuous coefficients and singular source. SIAM J. Numer. Anal. **31**, 1019–1044 (1994)
3. Osher, S., Sethian, J.A.: Fronts propagating with curvature dependent speed: Algorithms based on Hamilton-Jacobi formulations. J. Comput. Phys. **79**, 12–49 (1988)
4. Li, Z., Ito, K., Lai, M.C.: An augmented approach for stokes equations with a discontinuous viscosity and singular sources. Comput. Fluids **36**, 622–635 (2007)
5. Kalita, J.C., Dass, A.K., Dalal, D.C.: A transformation-free HOC scheme for steady state convection-diffusion on non-uniform grids. Int. J. Numer. Meth. Fluids **44**, 33–53 (2004)

Shor's Algorithm for Quantum Factoring

Manisha J. Nene and Gaurav Upadhyay

Abstract The strength of the famous RSA algorithm depends on the difficulty of factoring large integers in polynomial time. Shor's quantum algorithm has the potential to break RSA in reasonable time. Although, development and commercial availability of high power (16-qubit and more) quantum computers is still some years away, efforts in the direction of simulating quantum algorithms on a classical system are welcome. This paper presents a systematic approach in the direction of factoring integers using the Shor's quantum algorithm on a classical system, where the results of simulation are corroborated with theoretical results.

Keywords Quantum computing · Shor's algorithm · Factorization

1 Introduction

In near future a more fundamental science, that of quantum physics, will catapult the world of computing at such a pace which had been unprecedented. As the transistors on the chips get smaller, the weird and fascinating effects of quantum physics dominate the manner in which interactions between various components take place. Of particular interest are the phenomenon of *entanglement* and *superposition* which can be made use of in Quantum Information Processing (QIP). Construction of quantum information theory on the lines of Shannon's classical information theory is a landmark event in the history of QIP. Unique properties of quantum states as highlighted by the quantum information theory are: "Quantum states are impossible to copy" and "Quantum states cannot be measured with certainty". In order to fully appreciate the power of QIP, an understanding of the

M.J. Nene (✉) · G. Upadhyay
Defence Institute of Advanced Technology, Girinagar, Pune 411025, India
e-mail: mjnene@diat.ac.in

G. Upadhyay
e-mail: gaurav_signals@yahoo.co.in

© Springer Science+Business Media Singapore 2016
R.K. Choudhary et al. (eds.), *Advanced Computing and Communication Technologies*,
Advances in Intelligent Systems and Computing 452,
DOI 10.1007/978-981-10-1023-1_33

concepts of No-cloning theorem, Holevo bound, quantum error-coding, quantum data compression and communication over noisy quantum channel is required [1]. Fascinating advances have been made in the field since the First Conference on the Physics of computation held at MIT in which Richard Feynman drew a parallel between an exponentially powerful quantum computation methodology and a multi-particle interference experiment and its measurement [2]. Numerous algorithms and protocols have been developed which have strengthened the foundations of quantum computing. Deutsch algorithm [3], Simon algorithm [4], Shor's quantum algorithm [5] and BB84 protocol [6] are some of them. Related work in the field of factoring large integers and particularly Shor's algorithm can be found in [5, 7–14].

2   Quantum Computing and Shor's Algorithm

The fundamental unit in quantum computing is a qubit, a quantum bit. A qubit, unlike a classical bit can take on infinite values and be in superposition. This is represented by linear combination of vectors, which are called states in quantum mechanics. Dirac-ket notation, which is standard in quantum mechanics, is used to represent the states of a qubit. The classical bit '0' and '1' are replaced by the qubit $|0\rangle$ and $|1\rangle$. A 2-qubit quantum state will have its resultant vector in 4-dimensional space. These states form an ortho-normal computational basis for the computer. Measurement of the state of the register will collapse onto any one of the four states non-deterministically. One needs to have many copies of the same state in order to access the state via measurement [9]. It can be seen that an n-qubit quantum computer can be in a superposition of 2^n states or computational basis, for such a computer will have 2^n values. With the Shor's quantum algorithm the problem of factoring and finding discrete logarithm becomes computationally feasible. The problem of finding factors is reduced to the problem of finding order of an integer x less than N. The number x is selected randomly and if they have common factors, then the GCD will bring out the factors for N. If x is co-prime to N then the case is investigated and a least positive integer r is found which is the order of x modulo N.

3   Problem Statement and Proposed Solution

3.1   Problem Statement and Scope

To simulate Shor's algorithm for quantum factoring on classical computer. To analyze the periodicity of the values or amplitudes of the computational basis for up to 3-digit integer values of N.

3.2 Simulation Platform

The simulation is made on a classical computer running Microsoft Window® 7 Ultimate (64-bit) running on an a Intel® core™ i5-4260U CPU@1.40 GHz having 4 GB RAM. The software platform used is MATLAB 7.12.0 (R2011a) with Quantum Computing Function (QCF) Tool box.

3.3 Algorithm

Step 1. Choose N and x, $\ni$ gcd $(N, x) = 1$.
Step 2. Choose q, $\ni N^2 < 2^q < 2N^2$. Let $2^q = Q$.
Step 3. Generate computational basis vector (C) having values from 1 to Q.
Step 4. Generate a zero vector (Z) of same length Q, first element initialized to x.
Step 5. Generate elements 2 to Q for $Z \ni Z(i) = x \times Z(i-1) \bmod N$.
Step 6. Find the Quantum Discrete Fourier Transform (QDFT) of Z and normalize the values to 1.
Step 7. Select the values in C, where Z is having a peaks i.e local maximum.
Step 8. Calculate Continued Fraction Expansion of selected values of C and Q.
Step 9. Check if r is even. If No, then choose another x and goto Step 2.
Step 10. The factors of N are given by $\gcd(x^{(r/2)} - 1, N)$ and $\gcd(x^{(r/2)} + 1, N)$

4 Simulation Results

4.1 Qu-bit Simulation

Simulation of registers containing 2- and 9-qubits was carried out and measurements made 10 times. The results are as listed (Table 1).

Measurement no.	Result (2 qubit)	Result (9 qubit)
1	$1|11\rangle$	$1|110011001\rangle$
2	$1|11\rangle$	$1|011111000\rangle$
3	$1|00\rangle$	$1|010111000\rangle$
4	$1|10\rangle$	$1|101111001\rangle$
5	$1|11\rangle$	$1|011011000\rangle$
6	$1|00\rangle$	$1|110011011\rangle$
7	$1|01\rangle$	$1|001101000\rangle$
8	$1|10\rangle$	$1|000100100\rangle$
9	$1|11\rangle$	$1|111100000\rangle$
10	$1|00\rangle$	$1|100110110\rangle$

Table 1 Measurement of superimposed qu-bits

In superposition, the 2-qubit register will be in all 4-possible states simultaneously. However, the act of measurement collapses the state to any one of the computational basis, which for a 2-qubit register is $|00\rangle$, $|01\rangle$, $|10\rangle$ or $|11\rangle$. The notation $1|10\rangle$ means that the qubit has value '10' with probability '1'. Similar inference can be drawn for 9-qubit register. The 9-qubit register is in all 512 states (0–511) simultaneously; however the act of measuring collapses the superimposed state to one of the states in a non-deterministic manner.

4.2 Shor's Quantum Algorithm Simulation

In this sub-section, the results of simulation of Shor's algorithm for quantum factoring are presented (Table 2).

A visual representation of values of computational basis vector (C) and normalized QDFT vector (Z) in form of graphs is presented for further understanding. The graphs for $N = 21$ and $N = 247$ are presented.

Table 2 Result of Shor's quantum algorithm simulation

S. No.	N	x	r	q	Q	Factors
1	21	2	6	9	512	3, 7
2	35	2	12	11	2048	5, 7
3	55	2	20	12	4096	5, 11
4	77	2	30	13	8192	7, 11
5	247	2	36	16	65,536	13, 19

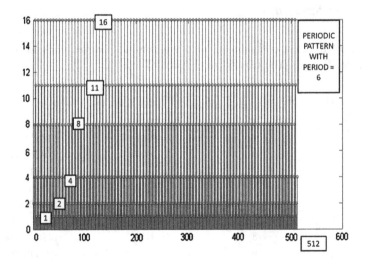

Fig. 1 Plot of values of vector C for $N = 21$. This highlights the order of x modulo N visually

The graph at Fig. 1 visually brings out the pattern 1,2,4,8,11,16, 1,2,4,8,11,16, 1,2,4,8,11,16 …. It can be seen that the period here is 6, which is the order we need to find. The QDFT graph of vector Z for $N = 21$ generated a total of 5 peaks. The plot is given in Fig. 2. Same can be inferred from Figs. 3 and 4 for $N = 247$.

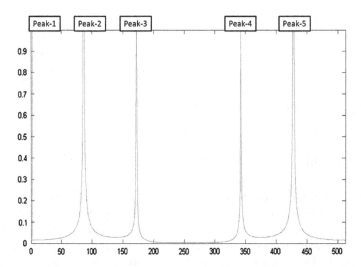

Fig. 2 Plot of values of vector Z for $N = 21$. Plot highlights the places where either 'r' or a factor or 'r' can be found

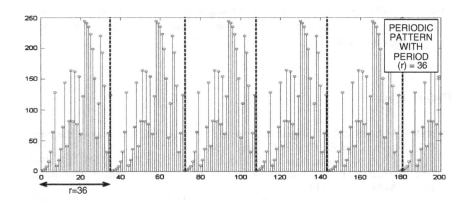

Fig. 3 Plot of values of vector C for $N = 247$. The plot has been clipped for values upto 200 to highlight the period of 36

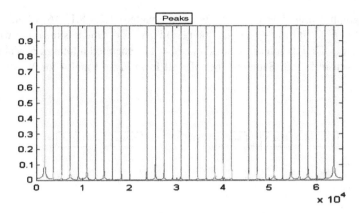

Fig. 4 Plot of values of vector Z for $N = 247$

5 Conclusion

The problem of factoring can be solved and results verified using Shor's algorithm for quantum factoring on a classical computer using primitive simulation. The representation of the results in visual format aids in better understanding of the working of the mathematics behind factorization and helps to arrive at the correct solution faster. The limitations of the present work are that there is no way of making use of superimposed registers which were simulated with the process of factoring. The limitation draws from the fact that the registers on a classical computers cannot be in superposition. The second and a major limitation is that as the value of N increases the order finding, r, becomes exceedingly difficult in the simulation. Although the probability of homing on to correct computational basis is high, in the proposed approach a combination of visual and manual sifting through the data does the trick.

References

1. Nielsen, M.A., Chuang, I.L.: "Quantum Computation and Quantum Information". Cambridge University Press, Part-III, Chap. 8–12, Edition (2002)
2. Feynman, R.P.: Simulating physics with computers. Int. J. Theor. Phys. **21**, 467–488 (1982)
3. Deutsch, D., Jozsa, R.: Rapid solution of problems by quantum Computation. Proc. R. Soc. London **A439**, 553–558 (1992)
4. Simon, D.: "On the power of quantum computation". In: Proceedings of the 35th Annual Symposium on Foundations of Computer Science, 116 pp. (1994), and SIAM J. Comput. **26**, 1474–1483 (1997)
5. Shor, P.: "Algorithms for quantum computation: discrete logarithm and factoring". Proceedings of the 35th Annual Symposium on Foundations of Computer Science, pp. 124–134 (1994), and SIAM J. Comput. **26**, 1484–1509 (1997)

6. Bennett, C.H., Brassard, G.: "Quantum cryptography: public key distribution and coin tossing". In: Proceedings of the IEEE International Conference on Computers, Systems and Signal Processing, Bangalore, pp. 175–179 (1984)
7. Vandersypen, M.K., Steffen, M., Breyta, G., Yannoni, C.S., Sherwood, M.H., Chuang, I.L.: "Experimental realization of Shor's quantum factoring algorithm using nuclear magnetic resonance". Nature **414**, 883 (2001)
8. Nam, Y.S.: "Running Shor's algorithm on a complete, gate-by-gate implementation of a virtual, universal quantum computer". A Dissertation submitted to the faculty of Wesleyan University, Departmental Honors in Physics (2011)
9. Lavpor, C., Manssur, L.R.U, Portugal, R.: "Shor's Algorithm for factoring Large Integers", Lecture notes from graduate courses in Quantum computation given at LNCC, 3 pp. (2008)
10. Politi, A.M., Jonathan, C.F., O'Brien, J.L.: "Shor's quantum factoring algorithm on a photonic chip". J. Sci. **325**(5945), 1221–1221 (2009)
11. Wang, W-Y., Shang, B., Wang, C., Long, G-L.: "Prime factorization in the duality computer". Commun. Theor. Phys. **47**(3), 471 (2007)
12. Browne, D.E.: "Efficient classical simulation of the quantum Fourier transform", New J. Phys. **9**(5), 146 (2007)
13. Lu, C-Y., Browne, D.E., Yang, T., Pan, J-W.: "Demonstration of a compiled version of shor's quantum factoring algorithm using photonic qubits. Phys. Rev. Lett. **99**(25), 250504 (2007)
14. Nagaich, S., Goswami, Y.C. "Shor's algorithm for quantum numbers using MATLAB simulator". In: Fifth International Conference on Advanced Computing and Communication Technologies, pp. 165–168 (2015)

Part II
Communication Technologies

Opportunistic Maximum-Likelihood Channel Assignment in Mesh Based Networks

Amanjot Kaur and Parminder Singh

Abstract Wireless mesh innovation is creating itself as entire world under one rooftop. Mesh Networks are proficiently and viably and besides remotely joining urban areas with economical existing innovation. Couple of years back client joined with one another utilizing some wired access focuses or remote hotspots. Stations automatically choose their safest path to travel from one mesh to another. Yet, in today's reality network innovation is serving many clients at one time crosswise over substantial range. The maximum likelihood has been determining the high priority and free channel to send information to multi-hop mesh based networks. This channel assignment approach is investigating and removing channel fading problem. We have been proposing the notion that recognizes the lapses because of increment the interference in the stations.

Keywords 802.11 · STA · Noise · Channel · Throughput · Unsaturated throughput

1 Introduction

In this paper a novel approach to the channel assignment for mesh based network is discussed. These mesh networks are based on IEEE 802.11 suit of standards. More than two radio interfaces autonomously operate on different channels. Router present in the network can transmit and receive data at the same time. Firstly we initiate the definitions of the basic terms used for wireless mesh networks. Radio spectrum is mainly used to term as the electromagnetic wave frequencies within the scope of near about 3 Hz to 300 GHz. Channels are used to generate a commu-

A. Kaur (✉)
IKGPTU Jalandhar, Jalandhar, India
e-mail: aman.09.jot@gmail.com

P. Singh
CEC Landran, Mohali, India
e-mail: singh.parminder06@gmail.com

© Springer Science+Business Media Singapore 2016   335
R.K. Choudhary et al. (eds.), *Advanced Computing and Communication Technologies*,
Advances in Intelligent Systems and Computing 452,
DOI 10.1007/978-981-10-1023-1_34

nication link between source and destination. Radio band is the spectrum range that can have more than one channel. In wireless communication, wireless links with similar data or transmission strike against each other. If channel assignment algorithm would be better, then it can balance the wireless connectivity and bandwidth.

This paper is systematized as follows: Sect. 1 is giving introduction about the topic. We review the Literature in the Sect. 2. Section 3 gives the proposed model. Section 4 describes the performance evaluation and Sect. 5 discusses the results. Section 6 is conclusion for the method proposed.

2 Related Work

Different types of works has been done with wireless mesh network for channel assignment. Each paper discussed here has a different work done for the channel assignment. Wu and Wang [1] explained Adaptive Distributed Channel Assignment in a wireless mesh network. This protocol used N number of frames and M number of slots for each and every frame. N is fixed for this implementation, and it was taken as nodes in the network. A channel control table and data channel table was maintained by frame and frame was taken care by the node. M value was not fixed, it was set when frame has not sufficient slots for further connections.

In [2], authors proposed a distributed family-based channel assignment (FCA) protocol with multiple channels to distinct varied data rate links. FCA upgraded the node connectivity by assigning channels for rate division of wireless networks with a new metric and heuristic algorithm. In this protocol node connectivity is improved through direct communication between the nodes and also recovered the channel variation by giving multiple paths. Thus, the protocol saved the network capacity. Channel assignment and link scheduling is used for effective deployment of multi-channel multi radio and wireless mesh networks [3]. The traffic load is varied on wireless networks. To achieve the better throughput channel assignment must recheck the network each and every time when traffic demand changes. This paper [3] proposed the model which did the same task to reconsider the communication when there is any change in traffic. In [4] authors presented a performance evaluation of two distributed channel assignment algorithms in a large-scale wireless multi-radio testbed. They compared the performance of a link-based approach, in which channels were assigned to existing links in the network, with an interface-based approach, where channels were assigned to the network interfaces. These algorithms significantly increased the network capacity in small wireless testbeds with less than 15 nodes. Singh [5] has explained WSN in hostile environment.

Ghahfarokhi [6] explained the QoE aware channel assignment method for WLAN. This paper defined the solution for aggregate QoE (Quality of Experience) and user-level fairness and channel assignment was formulated as the optimization problem on maximizing this index. This paper proposed two QoE-aware channel Assignment schemes to improve both fairness and aggregate QoE. Due to widespread use and limited radio spectrum of 802.11 based networks, interference between collocated networks is an important problem [7]. The aim of this system was to dynamically reduce the total interference and increase the aggregate capacity of the network. In comparison to different routing schemes, the distinctive feature of this approach was the channel assignment and the topology control that can be achieved through it. The goal of [8] was to implement an efficient service for maximum users with less interference. The main concentration was on multimedia servers for distributing the streams on a WMN. YUAN et al. [9] presented a channel assignment scheme with the multi-radio link quality source routing (MR-LQSR) protocol, which was called channel assignment with MR-LQSR (CA-LQSR). In this study a physical interference model was established.

Wireless mesh capacity can be increased using Multi-radio multi-channel technology [10]. In this paper, authors proposed a cross-layer architecture that provided an efficient end-to-end communication in multi-radio multi-channel WMNs. First, they proposed to perform channel assignment and routing "on-demand" combined. Singh [11] proposed a new technique for the better performance of vehicular ad hoc network using QOS metrics.

3   Proposed Model

After discussing the literature survey of issues like channel assignment, transmission errors, problem of facing interference on higher layers, the proposed model exuded confident to solve them interference and channel assignment problems. By assignment the channel inside the network has demonstrated in the Fig. 3. Here, we assume the channel c is assigned to the pair of nodes (i, j) and the size of input of each node represented is:

$$\log_2 i + \log_2 j + 2 \tag{1}$$

This binary representation of Eq. (1) should be used in adding more nodes in the network scenario for communication and transmission of information. On the other hand, the size of the network 'n' is also measuring and computing that has shown in the Eq. (2).

$$n = \sum_{(i,j) \in c} [\log_2 i] + [\log_2 j + 2] + [\log_2 k] + 1 (\text{if } i < j) \tag{2}$$

From the Fig. 1, as we assumed the channel assignment time to cover all nodes with channel c to node (i, j), if size |S| of the cover is the number of channel in S, then Fig. 2 is represented as:

Fig. 1 Proposed model mapping with equations

Fig. 2 Topology of
communicating nodes

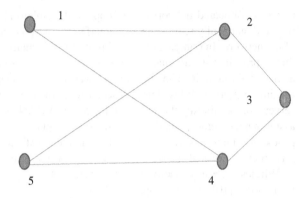

Fig. 3 Topology control of
communicating nodes without
any effect of channel
assignment

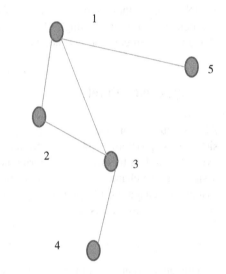

From the Fig. 3, we show the topology control that act like a directed graph where
each node connected to other node and probably the state is changing and these
states change according to control manner and it should be represented in [12]:

$$P_r(X_{n+1} = x | X_n = x_n) \tag{3}$$

The assumption of our model is that we deploying the dynamic network and
there has been chances to change the topology. So we drive the whole scenario
under control manner and behold the whole results calculated with network simulator (NS2) [13]. We have also face some problems like node forbid more than two
nodes to change the x coordinate and y coordinate positions. The probability of
interference in the network model is occur rarely but in some cases, if the interference increases or decreases then we used the proposed Eq. (4) and deploying at
the time of calculating result as of communication of node.

$$F(x) \triangleq \frac{1}{(j-i)} n(t) = 0 \tag{4}$$

Here, $n(t)$ is the noise generated during the transmitted signal by the sender node. If the noise is equals to zero that means the noise is nil; otherwise the destination node did not accept the packets and all such packets drop in the network. The physical channel may consist of paths and these paths connect the number of nodes. The delay path is dependent on the accuracy or bandwidth of the network. More the bandwidth available is lesser the delay between the number of stations. It is also been noted that if the single path channel is available then such type of interference can be synchronized. Moreover, multi-path channel assignment is somewhat difficult and can be solved by time division multiplexing; if this multiplexing is added with Eqs. (2) and (4).

The basic working of this flowchart (Fig. 4) is on the basis of proposed equation as shown in (1), (2), (3) and (4). The deployment of considered nodes formed together with IEEE 802.11 standard and the acknowledgment mechanism is activated when the receiver node receive that data from the sending node. The basic principle of calculating interference is to calculating the noise of the channel. We divide channel time into small slots and employing TDMA approach for assigning each slot, so that minimum interference has been achieving.

4 Performance Evaluation

The proposed model has been simulated using a NS-2 Network simulator and its performance has been compared with earlier work [14], the proposed equation has been written in c language and mapping together with awk scripts and tcl files together with and align with MAC layer. Initially the queue size is empty and the initial value of $Qi = 0$ for all n nodes i.e. 25. The DATA and ACK packets transmitted in opposite directions. The utility function of every round in communication is:

$$R(Fr) = \log(Fr + 0.02) \tag{5}$$

where $R(Fr)$ is the utility function in which the round of communication has been synchronized way. The assumption of all links should be considered in same PHY data rates (Fig. 4).

4.1 Simulation Set-up

We simulated 25 overlapped nodes having radius 250 m, the 50 m as nodes transmission range. The distance between adjacent nodes is 25 m, each of which communicates with its neighboring node and forming routing table. We used 1024

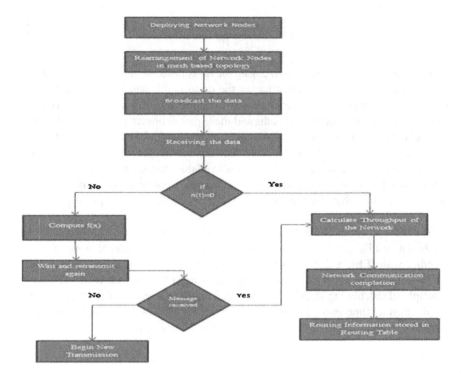

Fig. 4 Proposed flowchart

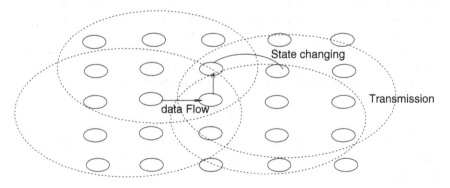

Fig. 5 Proposed scenario

bytes of data with 32 bytes of MAC header where the data rate 11 Mbps is considered for experimental purpose. The simulation scenario is depicted in Fig. 5, which considers the interference among two links in both directions as if the data flow occurs. The proposed scenario has been improving packet flow rate, decentralizing communication and setting up an independent channel. This scenario was taken up for fulfilling the objectives, they needed to be data flow rate and no interference in the multi-hop communication.

5 Results and Discussion

The following results determined from initial layout of the scenario, the scenario can be used to test out. The scenario can be used to see if any interference has been introduced, and to see the interference effects that each station (STA) in the design may be having on its neighboring stations.

5.1 Network Throughput

It has been shown in the Fig. 6, due to heavy traffic, the network throughput will not sharply decline like earlier work. Therefore, we assumed that network resource utilization is significantly improved. In addition, interference is overcoming among this topology network, which is of great practical significance. The number of stations (STA) shows a steady behavior above 5 Mbps and throughput tends to decrease at 150 m where the highest rate 11 Mbps is achievable, whereas at the distance of 200 m. We achieved highest throughput when at the station is 22.

5.2 Unsaturated Throughput

Throughput achieved by our proposed scheme in a topology where 25 STAs are placed at a random distance and each STA generates data at 2 Mbps. The results are sorted according to the throughput achieved per STA; where the X-axis shows the Number of STA. As shown in the Fig. 7 within 100 m distance from the STA can use higher data rates, which eventually increases the per node throughput. Whereas nodes 5, 9, 21 and 22 are within 100–120 m distance and do not always

Fig. 6 Network throughput

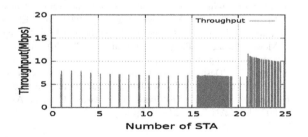

Fig. 7 Network throughput
versus number of errors

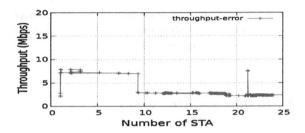

get the highest rate (11 Mbps) for data transmission and nodes that are far away like
1, 20 and 24 achieve the higher data errors as they can transmit only in lower data
rates i.e. 500 kbps.

6 Conclusion

We have proposed a technique that enhances the performance of network
throughput while investigating the interference of the stations. The STC can
broadcast the message at the upper layer and afterwards delivers the casings at the
MAC layer. The error at the MAC and PHY layer is minimum. We have also
examined the fading channel on the basis of interference that is only reason we
disengage the framework by the force of every station. The circulation divert task in
such a route, to the point that the STA has arbitrarily pick the channel and send the
data to recipient side STA. intricacy can be lessened by the task of number of
arbitrary channels. The sign quality of every station is high that of transmission
stream rate is better between co-stations and co-channels so we enhanced the
throughput of the system.

References

1. Wu, C.-M., Wang, Y.-Y.: "Adaptive Distributed Channel Assignment in Wireless Mesh
   Networks", pp. 363–382. Springer (2008)
2. Kim, S.-H., Suh, Y.-J.: "Exploiting Multiple Channels for Rate Separation in IEEE 802.11
   Multi-radio Wireless Mesh Networks", IEEE, pp. 1–9 (2011)
3. Antony Franklin, A et. al.: "Online Reconfiguration of Channel Assignment in Multi-channel
   Multi-radio Wireless Mesh Networks", pp. 2004–2013. Elsevier (2012)
4. Juraschek, F., Seif, S., Günes, M., "Distributed Channel Assignment in Large-Scale Wireless
   Mesh Networks: A Performance Analysis" IEEE ICC 2013—Ad-hoc and Sensor Networking
   Symposium, pp. 1661–1665 (2013)
5. Singh, P.: "Design an Framework of Wireless Sensor Networks by Preventing Malicious
   Nodes Attack", pp. 195–200. International Conference Elsevier, (2014)
6. Ghahfarokhi, B.S.: "Distributed QoE-Aware Channel Assignment Algorithms for IEEE
   802.11 WLANs", pp. 21–34. Springer (2014)

7. Prodan, A., Debenham, J.: "A Measurement Protocol for Channel Assignment Based Topology Control on Multi-radio Wireless Mesh Networks" IEEE, pp. 659–664 (2013)
8. Yang, W.-L., Huang, W.-T.: "The Study of Interference-Free Multicast Using Non-orthogonal Channels for Multi-radio and Multi-channel Wireless Mesh Networks", IEEE, pp. 547–552 (2010)
9. YUAN, F. et al.: "Distributed Channel Assignment Combined with Routing over Multi-radio Multi-channel Wireless Mesh Networks", pp. 6–13. Elsevier (2012)
10. Bononi, L. et al.: "A Cross-Layer Architecture for Efficient Multi-hop Communication in Multi-channel Multi-radio Wireless Mesh Networks", IEEE, pp. 1–6 (2010)
11. Singh, P.: "Comparative Study Between Unicast and Multicast Routing Protocols in Different Data Rates Using VANET", IEEE, pp. 278–284 (2014)
12. https://en.wikipedia.org/wiki/Markov_chain. Accessed 3 Aug 2015
13. www.isi.edu/nsnam/ns/edu. Accessed 3 Aug 2015
14. Fang, M., Malone, D., Duffy, K.R., Leith, D.J. "Decentralised learning MACs for collision-free access in WLANs", Wireless Netw, pp. 83–98 (2013)

A Real-Time (on Premise) Baseline Based DDoS Mitigation Scheme in a Hybrid Cloud

Ankur Rai and Rama Krishna Challa

Abstract Uninterrupted services are the most important factor for building customers trust towards a particular service providers, Distributed denial of service attacks are major threats towards disrupting the customer base for these service providers. Increasing sophistication of these attacks make them stealthier to evade existing perimeter security mechanisms. Hence, there is a need to design a dedicated mechanism to counter these attacks. In this paper we present a real time mitigation approach for DDoS attacks in a hybrid cloud. This approach utilizes a real time hybrid cloud test bed environment implemented with both intrusion detection system (IDS) and intrusion prevention system (IPS) components for result analysis and is utilized to mitigate signature based attacks at layers 3, 4 and 7 of TCP/IP network model. To implement this approach various stages to mitigate these attacks are considered. The results obtained have 100 % detection accuracy in all the scenarios considered.

Keyword DDoS · IDS · IPS · Mitigation · Dedicated mechanism · Hybrid cloud

1 Introduction

Denial of service poses a major threat towards an organization commercial success and also towards receiving unhindered access of services by the legitimate users. Distributed Denial of service (DDoS) attack is one such variation of this threat where large network of compromised systems called as zombies are utilized to perform these attack. These attacks are broadly categorized into three types [1]. First, Volume based attacks where the attackers overwhelm the victim with large

A. Rai (✉) · R.K. Challa
Department of Computer Science & Engineering, NITTTR, Sector 26,
Chandigarh 160019, India
e-mail: anks.rai@gmail.com

R.K. Challa
e-mail: rkc_97@yahoo.com

© Springer Science+Business Media Singapore 2016   345
R.K. Choudhary et al. (eds.), *Advanced Computing and Communication Technologies*,
Advances in Intelligent Systems and Computing 452,
DOI 10.1007/978-981-10-1023-1_35

network traffic such that its resources and services are exhausted and are made unavailable to normal or legitimate users. Typical examples of such type of attacks are Ping flood and TCP-SYN flood attacks which occur at Layer 3 and Layer 4 of TCP/IP network model respectively. Second, Application attacks where various applications can be targeted by the attacker to exhaust the victim's resources. Typical example of such type of attacks is HTTP flood attacks. These attacks occur at Layer 7 of the TCP/IP network model. Third, Low rate attacks where the attacker exhaust the victim's server with minimal bandwidth and connections and hence are hard to detect. Typical example of such type of attack is Slowloris attacks.

2 Related Work

Defeating DDoS attacks [2] Explains the need of complementary solutions in spite of the existence of routers and perimeter security technologies to counter these attacks. They gave their own architecture to provide a complete protection form DDoS attacks. Authors in [3] proposed an approach by merging methods of data mining to detect and prevent DoS attacks, and used multi classification techniques like K-NN, Decision trees to achieve high level of accuracy and also to remove false alert alarms utilizing European Gaza Hospital (EGH) data set to detect the occurrence of DoS attacks.

In [4] the authors proposed an approach in which both the detection and prevention techniques are given. The detection is done utilizing the covariance matrix and TTL counting value and prevention is done using honey pot network. Authors in [5] presented an IDS that utilizes a layered framework integrated with neural network. This proposed system is shown to be more effective than other IDS models. In [6] an analytical approach to address the DDoS attacks with help of binomial distribution is proposed. The results show that the method effectively detects malicious traffic. In [7] a technique using special algorithm, CHARM is presented, which assigns a real–time risk score with the 2 way traffic. When there is an attack then it raises the threshold on the CHARM score and drops the highest risk traffic. Their technique is useful for detecting stealthy attacks. Authors in [8] implemented a virtual experimental setup and focused on three main types of application layer DDoS attacks, low rate, slow send and slow read. Their results show that it is difficult to attacks servers using slow rate and low rate due to their design and implementation. Devi and Yogesh [9] presents a hybrid technique against DDoS attacks based on the client trust value and the entropy. Here the trust value of clients is used to distinguish between the legitimate user and the illegitimate user. Authors in [10] presents an effective method is introduced to differentiate normal web traffic from application DDoS attackers using access matrix and Hidden Markov model. Ajoudanian and Ahmadi [11] explains data availability remains the most problematic area in cloud and gave a security model where security concerns and their solutions are categorized in three layers of security services. Anjali and Padmavathi [12] Explains that IPS devices and firewalls which allow application layer traffic are unable to distinguish between the normal traffic and attack traffic and an effective method is required to mitigate these attacks. In [13] various tools that can be utilized for testing purposes to perform application layer attacks are presented.

3   The Proposed Mitigation Scheme

The main objective of this paper is to develop a mitigation scheme in real time hybrid cloud environment for detecting and preventing signature based DDoS attacks at layers 3, 4 and 7 of TCP/IP model. The proposed scheme works by firstly, determining the attack profiles based upon variations in values of important network and system parameters. Secondly, setting the threshold values for these parameters using the attack profile generated. Finally, these threshold values and attack profiles are used as a baseline to detect and prevent any future DDoS attacks in the network. The proposed scheme works for three common types of attacks like Ping flood attacks, TCP SYN flood attacks, HTTP GET flood attacks. The proposed mitigation scheme follows a mitigation process for its testing and analysis which is shown in Fig. 1 and as explained below:

3.1   DDoS Mitigation Scheme

The mitigation scheme is divided into two phases. Intrusion detection system (IDS) consists of phase 1 and intrusion prevention system (IPS) consists of phase 2 and the entire mitigation process is divided into four steps as shown in Fig. 1. These

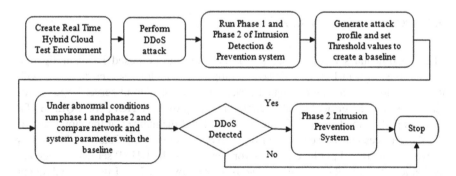

Fig. 1 Process for DDoS mitigation

steps are as explained as follows. In the first step, hybrid cloud environment is setup in the lab. This setup is implemented for testing and analysing the results. This setup provides both Storage as a Service (StaaS) and Software as a Service (SaaS) to the client machines. In the second step, DDoS attacks are performed on the implemented setup using existing DDoS attack tools. We have used three types of DDoS attack tools for performing attacks at the three layers (3, 4 and 7) of TCP/IP network model such as ping flood tool, hping3 tool, HULK (HTTP Unbearable Load King) tool respectively. Thirdly, phase 1 and phase 2 of the proposed mitigation scheme deployed at the server are applied simultaneously to detect the variations in the chosen system and network parameters and to detect number of incoming connections from different IP addresses respectively. The results obtained are used to generate attack profile and threshold values for important network and system parameters which are used to create a baseline for DDoS attack detection. Finally, whenever the system or network parameters show abnormal behavior like (low speed, unavailability of services, etc.) then phase 1 and phase 2 are run again to check the important network and system parameters values and compared with the baseline already created.

If the parameter values comply with the baseline then an attack is considered to be detected and phase 2 of proposed mitigation scheme is run to block the malicious IP addresses.

4 Experimental Results

Proposed IDS and IPS components of the mitigation scheme are implemented using Linux Shell Programming (BASH) and deployed on the system with Red Hat Enterprise Linux 6.4(64 bit) operating system, having Intel Core i7 Processor @ 3.4 GHz, 4 GB of RAM and 500 GB HDD. All the experiments are performed on the system with the mitigation scheme deployed.

4.1 Experimental Setup

The experimental setup for testing and analyzing the proposed DDoS mitigation scheme is shown in Fig. 2. This setup comprises of a cloud server deployed with the mitigation scheme and assigned a static IP address. Cloud server besides providing the cloud services is also configured to provide DHCP, MYSQL, HTTP services and a cloud server website developed and maintained at the cloud server. All the services can be accessed in both private (using Cisco Catalyst 2960 series network switch) and public cloud network except for the DHCP services which are available only for the clients in the private network. The Linux based client

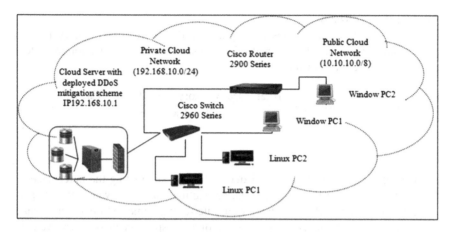

Fig. 2 Hybrid cloud environment setup for DDoS mitigation scheme

machines (Linux PC1 and PC2) with configuration depicted by Resource number 2 in Table 3 are used for performing the DDoS attacks like Ping flood, TCP SYN flood and HTTP GET flood on the cloud server. Windows machine (Windows PC1) with configuration depicted by Resource number 3 in Table 3 is used to determine the cloud server website load time during the DDoS attacks. Windows machine (Windows PC2) with configuration depicted by Resource number 3 in Table 3 and having a static IP address is used to access the cloud services in the public network via an intermediate router (Cisco Catalyst 2900 series). This router is configured with IP addresses of different network addresses (192.168.10.0/24 for private network and 10.10.10.0/8 for public network) on its ports. Next, we installed DDoS attack tools (hping3 and HULK) on the attacking machines (Linux PC1 and PC2) and Extended Status Bar tool for determining cloud server website load time on the Mozilla Firefox browser of Window PC1. After the setup is configured, we begin the process of generating the attack profiles and threshold values for creating a baseline for the DDoS attacks. In order to generate the attack profiles we considered important network and system parameters such as throughput, network latency, memory utilization, cpu utilization and socket utilization and website load time.

(1) Throughput

It is defined as number of bytes of packets transferred per unit time from source to destination. It is represented by Eq. (1) Here, TCP Receive window Size of 87380 bytes (Default value set on the cloud server) and average Round Trip Time are considered for throughput calculation

$$\text{Throughput} \leq \frac{\text{TCP Recieve Window Size}}{\text{Round Trip Time}} \qquad (1)$$

(2) Memory Utilization

Memory utilization is represented by the Eq. (2). Here, Total Memory 4 GB is considered for calculations and other values are determined through system performance tool 'vmstat'.

$$\text{Memory Utilization} = \frac{\text{Total Used Memory} - \text{Buffered} - \text{Cached}}{\text{Total Memory}} \times 100$$

(2)

(3) CPU Utilization

It is represented by Eq. (3) and determined by subtracting the percentage of time CPU or CPU'S were in ideal or waiting state. I/O wait represent percentage of time CPU or CPU'S were idle during some I/O operation to take place, Steal represents the percentage of time spent in involuntary wait by virtual CPU or CPU'S while another virtual processor was being serviced by the hypervisor and Ideal represents the percentage of time CPU or CPU'S were ideal and there was no outstanding I/O disk request for the system and all these values are determined through system statistics tool 'mpstat'.

$$\text{CPU Utilization} = 100\,\% - \text{I/O Wait} - \text{Steal} - \text{Ideal} \tag{3}$$

(4) Socket Utilization

In Linux, every socket when created returns a file descriptor [14]. Hence, socket utilization is determined based upon the percentage of file descriptors used during the attack. Here the default Max Open File Descriptor Limit of 374,382 set at the system is considered for calculations. Other values are determined through linux file descriptor commands such as (cat/proc/sys/fs/file-nr). This command gives both the values of the file descriptors (Max and Used). Socket utilization is represented by the Eq. (4)

$$\text{Socket Utilization} = \frac{\text{Total File Descriptors Used}}{\text{Max Open File Descriptor Limit}} \times 100 \tag{4}$$

(5) Network Latency

It is defined as time taken to send a particular request and receive an acknowledgement of that request. We determined network latency by sending ping packets to one of the client machines before and after the attack. Here only the average Round Trip Time is considered for determining this parameter.

(6) Website Load Time

It is defined as time taken to send a web request to the cloud server and time taken to receive the requested web page. Now, before generation of any type of attack profiles and threshold values, our proposed mitigation scheme is run on the server to determine the initial values of all the parameters considered. This is done to check the variations during and after the attack.

4.2 Experimental Results of Ping Flood

Here, we run total 4 instances of Ping flood tool with a payload of 65,000 bytes on each attacking machine simultaneously. Figures 3, 4 and 5 represents results determined from Phase 1 and Fig. 6 represents results determined from Phase 2 of the mitigation scheme. The attack profile depicted from Figs. 3, 4 and 5 and the threshold values mentioned in Table 1. Represents a baseline for ping flood attack which is used to block the IP addresses (depicted in Fig. 6.) that are contributing maximum to the total received ping request. Figure 3 depicts the values of avg. network latency and avg. throughput before and after the attack. Here, the value of avg. network latency increased to 10719.686 % (62.105 ms) of its initial value (0.574 ms) and the values of Average Throughput, which decreased to 99.07 % (1.406 MB/s) from its initial value (152.229 MB/s) and Fig. 4 depicts the values of system resources (Memory, CPU, Socket) before and after the attack. The results show that these resources were affected mildly. Figure 5 depicts the profile of the website load times during the entire attack duration. The results show that denial of service reached within 50 min ($\sim$1 h) at 3.45 PM and Fig. 6 depicts number of ping requests received from individual IP address and total ping request received during the entire duration of the attack. Based upon the profile depicted in Figs. 3, 4 and 5, threshold values for predicting ping flood attack are shown in Table 1. Both the attack profile and threshold values create a baseline for this attack.

Fig. 3 Average network latency and average throughput

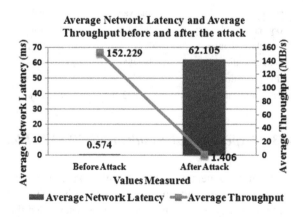

Fig. 4 System resources utilization

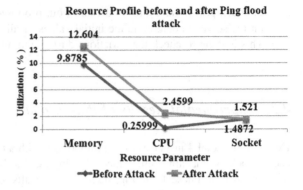

Fig. 5 Website Load time during the attack

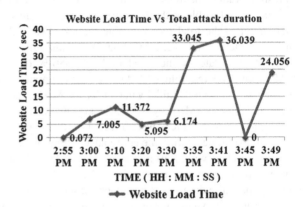

Table 1 Threshold values for ping flood attack

Parameter	Initial value	Final value	Threshold value
Avg. network latency	0.574 ms	62.105 ms	10719.686 % increase
Avg. throughput	152.229 MB/s	1.406 MB/s	99.07 % decrease
Memory utilization	9.8785 %	12.604 %	27.59 % increase
CPU utilization	0.2599 %	2.4599 %	846.15 % increase
Socket utilization	1.4872 %	1.521 %	2.272 % increase

4.3 Experimental Results of TCP SYN Flood

In order to generate profile for this attack, we noted the initial values of the parameters again and run one instance of hping3 tool on each attacking machine simultaneously. During the attack it was determined that this tool brought down the web services in 1 h. Figures 7, 8 and 9 represent the profile generated during the attack and Table 2 depicts the threshold values. Figure 7 represents the change in

Fig. 6 Ping requests from different IP address and total ping request received

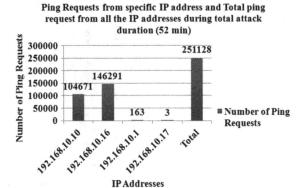

Table 2 Threshold values for TCP SYN flood attack

Parameter	Initial value	Final value	Threshold value
Avg. network latency	0.670 ms	8.352 ms	1146.56 % increase
Avg. throughput	130.4179 MB/s	10.4621 MB/s	91.9 % decrease
Memory utilization	9.8785 %	9.932 %	0.57 % increase
CPU utilization	0.3799 %	0.370 %	2.605 % decrease
Socket utilization	1.4701 %	1.478 %	0.53 % increase

Fig. 7 Average network latency and average throughput

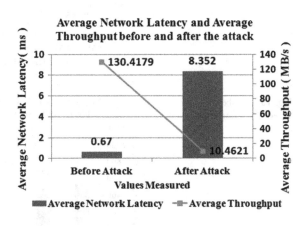

values of average network latency and average throughput before and after the attack. The results depicts that these parameters were affected more than other parameters and Fig. 8 depicts that system resources (memory, CPU and socket) were affected mildly. Figure 9 depicts the website load time during entire attack duration. The results show that denial of service reached in 1 h at 1.38 PM and Fig. 10 depicts number of TCP SYN request received from individual IP address and total TCP SYN request received during the entire attack duration.

Fig. 8 System resources
utilization

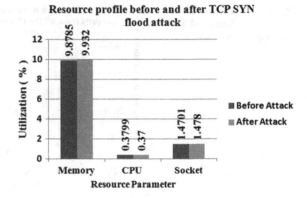

Fig. 9 Website load time

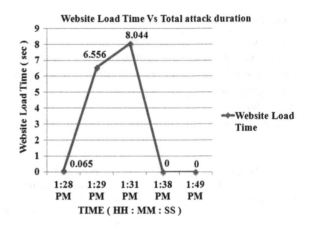

Fig. 10 Number of
TCP SYN requests received

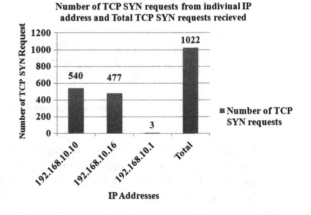

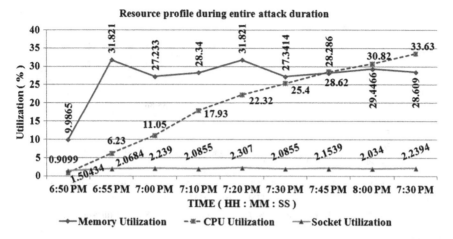

Fig. 11 Resource profile on i7 processor

4.4 Experimental Results of HTTP Flood

HTTP flood attacks traffic being slow-rate and low-rate appears to be as a legitimate traffic and hence are hard to detect. The results from these attacks revealed that stronger servers are affected less than servers with weaker configurations.

The result are determined using server configuration depicted by Resource number 1 in Table 3 with 8 CPU'S used (i7 processor) and with 4 CPU'S used (i4 processor). Figure 11 depicts the resource profile generated with i7 processor. Here the resources are constantly consumed depending upon the connection limit set on the server which can be modified depending upon the load on the server. However, this limit cannot be increased beyond the compile time connection limits. On, the server's where the compile time limits can be increased, these resource utilization can be 100 % and can lead to denial of service of resources and Fig. 12 depicts the resource profile with i4 processor. Here, the maximum memory, CPU, socket consumption reached to 58.704, 45.44, 3.446 % which were 31.821, 33.63, 2.2394 % respectively on i7 processor. Figure 13 depicts the results from phase 2 on i7 processor. Here, number of server sockets kept busy by individual IP address and total number of sockets busy at particular time is shown. At any particular time, IP address contributing maximum to the attack are considered as malicious and are blocked running phase 2 of the mitigation scheme. Similar results are obtained for i4 processor

In order to create a baseline for this attack only the attack profile is considered because Layer 7 attacks such as (HTTP-GET, HTTP-POST flood) are designed to work under the threshold limits. Hence, in order to counter such attacks only attack profile would be an effective solution.

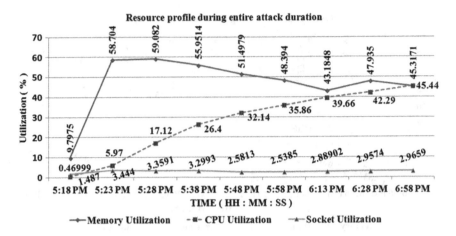

Fig. 12 Resource profile on i4 processor

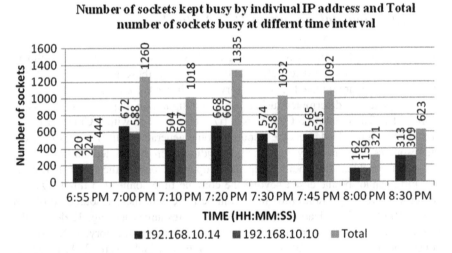

Fig. 13 States of sockets on i7 processor

5 Conclusion

In this paper, we present a real-time approach for DDoS mitigation, with both IDS and IPS components implemented. The proposed scheme besides providing an effective baseline based solution to counter signature based DDoS and DoS attacks can also be utilized for differentiating malicious traffic from legitimate traffic. The results obtained show high accuracy in detecting and preventing malicious IP addresses. The proposed scheme can be deployed in real organizational network. In the future we plan to detect botnet based DDoS attacks using this scheme.

Appendix 1

See Table 3.

Table 3 Resource/Device configuration

Resource number	Resource name	Configuration
1	Server (with File Descriptor Limit (FDL), ServerLimit (SL), and MaxClient limit (MC))	Intel core i7/3.4 GHz, RHEL 6.4(64-bit), kernel 2.6.42–358.el6.x86_64, RAM 4 GB, 500 GB HDD 32 bit or 64-bit, FDL-4096 (default: 1024) SL-200000 (default: 256), MC 200000 (def.256)
2	Linux PC1&PC2	Intel core i7/3.4 GHz, RHEL 6.4(64-bit), kernel 2.6.42–358.el6.x86_64, RAM 4 GB, 500 GB HDD 32 bit or 64-bit
3	Window PC1&PC2	Intel core i3/3.0 GHz, Windows vista (32 bit) RAM 2 GB, 150 GB HDD, 32 bit or 64-bit

References

1. DDoS Defense guide.: http://www.cisco.com. Accessed 6 Oct 2015
2. Defeating DDOS attacks.: http://www.cisco.com/c/en/us/products/collateral/security/traffic-anomaly-detector-xt5600a/prod_white_paper0900aecd8011e927.html. Accessed 6 Oct 2015
3. Tabash, M., Barhoom, T.: An approach for detecting and preventing DOS attacks in LAN. Int. J. Comput. Trends Technol. **18**, 265–271 (2014)
4. Ismail, M.N., Aborujilah, A., Musa, S., Shahzad, A.: New framework to detect and prevent denial of service attack in cloud computing environment. Int. J. Comput. Sci. and Secur. **6**, 226–237 (2012)
5. Srivastav, N., Challa, R.K.: Novel intrusion detection system integrating layered framework with neural network. In: 3rd IEEE International Advance Computing Conference, pp. 682–689 (2013)
6. Shyamala Devi, V., Umarani, R.D.: Analytical approach for mitigation and prevention of DDoS attack using binomial theorem with bloom filter an overlay network traffic. Int. J. Adv. Res. Comput. Commun. Eng. **2**, 3031–3036 (2013)
7. Defending against application-layer DDoS attacks.: http://www.juniper.net/assets/us/en/local/pdf/whitepapers/2000550-en.pdf. Accessed 8 Oct 2015
8. The Impact of Application Layer Denial of Service Attacks.: http://www.cs.unb.ca/~natalia/ApplDDoS.pdf. Accessed 8 Oct 2015
9. Devi, S.R., Yogesh, P.: A hybrid approach to counter application layer DDoS attacks. Int. J. Crypt. Inform. Secur. **2**, 45–52 (2012)

10. Rajesh, S.: Protection from application layer DDoS attacks for popular websites. Int. J. Elect. Eng. **5**, 555–558 (2013)
11. Ajoudanian, S., Ahmadi, M.R.: A novel data security model for cloud computing. Int. J. Eng. Technol. **4**, 326–329 (2012)
12. Anjali, M., Padmavathi, B.: Survey on application layer DDoS attacks. Int. J. Comput. Sci. Inform. Technol. **5**, 6929–6931 (2014)
13. Layer Seven DDoS Attacks.: http://resources.infosecinstitute.com/layer-seven-ddos-attacks/. Accessed 12 Oct 2015
14. Socket Programming. home.iitk.ac.in/~chebrolu/scourse/slides/sockets-tutorial.pdf. Accessed 12 Oct 2015

Clustering Based Single-hop and Multi-hop Message Dissemination Evaluation Under Varying Data Rate in Vehicular Ad-hoc Network

Priyam Singh, Raghavendra Pal and Nishu Gupta

Abstract Enhancement in the existing vehicular communication technologies is important not only for efficient utilization of the available bandwidth but also for event-driven safety message dissemination in Vehicular Ad-hoc Network (VANET). Some of its unique characteristics like geographically constrained topology, unpredictable vehicle density, varying channel capacity etc. constitute VANET as a distinct research field. In this paper we compare and evaluate the performance of clustering based single-hop and multi-hop network for varying data transfer rates in the IEEE 802.11p standard by means of simulations using NS-2. Simulations are performed for highway scenario in vehicle-to-vehicle mode. Results are presented for various Quality of Service parameters such as throughput, packet delivery ratio and end-to-end delay. Significant observations have been made which would serve as an input to enhance further research in this field.

Keywords Clustering · Single-hop · Multi-hop · IEEE 802.11p · V2V · VANET

1 Introduction

A Vehicular Ad-hoc Network (VANET) is a collection of vehicles communicating through wireless technology to improve road safety [1], enhance traffic efficiency and support infotainment applications. In VANET, every vehicle is equipped with

P. Singh (✉)
Electronics and Communication Engineering Department,
Sikkim Manipal Institute of Technology, Majitar, Sikkim, India
e-mail: priyamsingh2006@gmail.com

R. Pal · N. Gupta
Department of Electronics and Communication Engineering,
Motilal Nehru National Institute of Technology Allahabad, Allahabad, India
e-mail: raghavendra.pal3@gmail.com

N. Gupta
e-mail: rel0513@mnnit.ac.in

© Springer Science+Business Media Singapore 2016
R.K. Choudhary et al. (eds.), *Advanced Computing and Communication Technologies*,
Advances in Intelligent Systems and Computing 452,
DOI 10.1007/978-981-10-1023-1_36

wireless communication infrastructure which could collaboratively disseminate data about any road activity (viz. accidents, traffic jams, road construction warning, shadowing effect, bad weather condition etc.) to other vehicles within the broadcasting range [2]. The communication among participant vehicles in VANET is mostly carried out either through single-hop or multi-hop broadcasting technique. Nodes can route their messages towards destination either by using small or large hops. Theoretical knowledge [3, 4] favours using of smaller hops, known as multi-hop, which is considered as more efficient than single-hop. However, the trade-off between reception reliability and transmission count in single-hop or multi-hop technique needs to be carefully considered. Moreover different channel assumptions may have to be made for these schemes.

In single-hop wireless network, the messages are simply broadcasted to all the vehicles in a distance of approximately 100–300 m from each other in urban scenario. This range extends up to around 1000 m in highway scenario. In multi-hop network, communication between two end nodes is carried out through a number of intermediate nodes whose function is to relay information from one point to another. The network turns every participating vehicle into a wireless router or node, allowing vehicles to connect and create a network with a wide range. In this paper, we analyze the capabilities of the legacy standard known as IEEE 802.11p, to give an overview on the limitations of Dedicated Short-Range Communications (DSRC) in VANET technology. In this paper, we propose to evaluate the standard and compare it in terms of different vehicular networking application requirements of travel safety, traffic efficiency and navitainment (an integration of navigation and entertainment) applications. Major contributions of the paper are (i) implementing clustering based single-hop and multi-hop broadcast scenario, (ii) partitioning of the nodes into clusters based upon their mobility range, (iii) evaluating IEEE 802.11p standard on different QoS parameters by varying data transfer rate.

The remainder of this paper is organized as follows. Section 2 presents a brief background of broadcasting techniques in single-hop and multi-hop environment. Section 3 highlights the motivation behind this work. In Sect. 4 we evaluate the performance of single-hop and multi-hop broadcast technique under varying data rate for IEEE 802.11p vehicular ad-hoc network. Section 5 discusses the observations made on the basis of the simulation results. Finally, in Sect. 6 we conclude the paper and present future direction of research.

2 Background

In this section, we highlight the insight of the DSRC operations and the efforts made by the researchers towards protocol enhancement at the Medium Access Control (MAC) layer. IEEE 802.11p was standardised in 2010. To the best of our knowledge, such intensive evaluation and comparison like the one presented in this paper has not been performed since then. The presented work reflects the impact of clustered node mobility on the performance of the IEEE 802.11p MAC in both single-hop and multi-hop scenario, especially for the Vehicle-to-Vehicle (V2V) mode. Moreover, limited literature is available that discusses to optimise the performance of the IEEE 802.11p standard pertaining to mobility aware clustering factors. Through simulation results and analytical means the work in [5] indicates that the IEEE 802.11p standard is susceptible to increased latency and decreased throughput under dense vehicular density. The work presented in [6] studies the saturated performance of the IEEE 802.11 MAC in a single-hop network. The obtained results attest past observations, similar to [5] that the minimal delay requirement is somewhat satisfied to acceptable extent. However, the Packet Delivery Ratio (PDR) decreases drastically as the vehicular density increases. Authors in [7] state that the performance of the legacy standard is severely affected by the inherent nature of Carrier Sense Multiple Access (CSMA) mechanism which is the access scheme employed by IEEE 802.11p for contention-based channels. This is due to the dynamic node mobility and topology of the VANET environment. In [8], the authors state that relaying is also an important paradigm in VANET. Taking into view the dynamic topology and non-uniform vehicular density in VANET, effective message relaying algorithm becomes vital. Further, efficient multi-hop networking can be accomplished by optimizing the channel utilization by spatial reuse of the network resources. The authors in [9] state that multi-hop transmission will always be more energy efficient than single-hop transmission provided the path loss exponent is greater than one.

3 Motivation

The motivation behind this analysis is to address the fundamental requirement of reducing the delivery delay arising due to channel congestion. Secondly, we try to answer what will be the impact on the Quality of Service (QoS) parameters if the messages are relayed over multiple hops instead of being transmitted over single-hop only. By delivering messages through multi-hopping broadcast technique, not only we leverage on the dissemination distance but also alleviate spatial reuse of the network [8]. Unlike in single-hop network, one of the major advantage multi-hop network carries is its capability to avoid hidden terminals. Nodes which cannot transmit messages beyond their transmission range directly are given the possibility to do so via intermediate nodes, making a much larger network feasible.

In Fig. 1, clustered nodes are shown to perform multi-hop broadcast in V2V mode. The nodes in the network are partitioned into clusters based upon their mobility range. Vehicles with high mobility range are grouped into one cluster whereas vehicles with low mobility range are grouped into another cluster. Likewise, different mobility ranges lead to the formation of different clusters. The clustering algorithm does not require any additional messages other than the dissemination of vehicle's status messages (HELLO messages), since the cluster formation is mobility based. Therefore, when the vehicles enter the network region

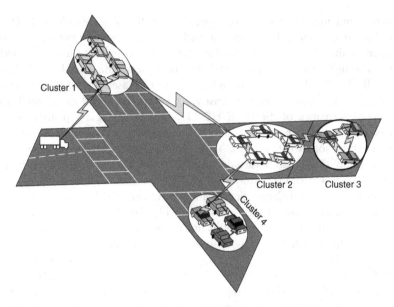

Fig. 1 A scenario depicting clustering based multi-hop broadcast in V2 V mode

for the first time, they send their status messages independently, without any cluster head. Once these messages are received by the nodes in the single-hop distant network range, they form a cluster.

4 Performance Evaluation

In order to signify and highlight the adverse effects of the legacy standard, as they are the ones targeted in this work, three main performance metrics are evaluated in terms of throughput, PDR and end-to-end delay. We implement the clustering mechanism and compare its effectiveness while varying proportionalities in different network conditions. The obtained results raise concerns about Wireless Access in Vehicular Environment (WAVE) capability of providing safety at the road level, and thus justify the need for protocol enhancements that take into account the QoS requirements of vehicular applications.

In order to elaborate the performance and shortcomings of the WAVE, we have implemented the standard in a simulated environment. We evaluate the IEEE 802.11p standard and compare its performance under different parameters by varying their data transfer rate. Table 1 shows the simulation environment and parameters used in the evaluation. Simulations are performed using network simulator (NS-2), version 2.34. The results signify the impact of mobility aware clustering based varying data rate on the performance of the legacy standard for single-hop and multi-hop broadcast scenario. Vehicle speed varies between 15 and

Table 1 Simulation environment and parameters

Parameter	Value
Node speed	15–40 m/s
Simulation time	150 s
Simulation area	3000 m × 4000 m
Transmission rate	3, 6, 12, 24 Mbps
Transmission range of nodes	1000 m
Packet size	512 Bytes

40 m/s. All vehicles follow the WAVE standard MAC parameters for V2V communication in single/multi-hop ad-hoc region. The simulation time is set to 150 s, and the broadcasting range of each vehicle is 1000 m. The message size is taken to be 512 bytes and can be transmitted at varying data rates. The data rate and coverage range are presumed to be as per the IEEE 802.11p standard.

5 Results and Discussion

In Fig. 2a, we compare the throughput of single-hop and multi-hop network for different data rates. It can be seen that for all values of data rate, multi-hop network overshadows single-hop for throughput analysis. Moreover, optimum values are obtained between 3 and 6 Mbps. This result is verified by [10] which states that 6 Mbps is prescribed as the data rate for DSRC safety applications. The reason behind this observation is because of the implementation of the clustering mechanism that allows more data to be delivered per unit time.

Figure 2b compares the PDR for the two broadcast scenario. The results are similar to the ones attained for throughput. Here, for 3 Mbps, multi-hop network shows maximum PDR. As the data rate increases, the PDR drops. PDR of single-hop network is seen to be lower than multi-hop network for all values of the data rate. PDR is a metric that determines the reliability of the network. From the point of safety message dissemination, this parameter is indispensable.

In Fig. 2c, single-hop broadcast wins over its counterpart in terms of end-to-end delay. This is pretty obvious owing to the multi-hop scenario where each intermediate cluster follows a protocol so as to route the message to its one-hop distance cluster. The delay almost remains stable in single-hop scenario for different data rates whereas in multi-hop scenario, it decreases as the data rate increases. However, it can be seen that in contrast to Fig. 2a, b, there is trade-off between delay and the other two parameters in case of multi-hop scenario.

In Fig. 3a–c, we compare the QoS parameters of single-hop and multi-hop network with the cumulative data rates which are in proportion to the range of the cluster mobility. That is, cluster having slowest mobility range is assigned least data rate and cluster having fastest mobility range is assigned maximum data rate in case of increasing data rate. Opposite to this, in case of decreasing data rate, cluster

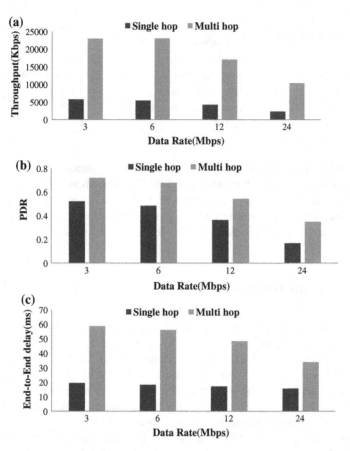

Fig. 2 a Data rate versus Throughput. **b** Data rate versus PDR. **c** Data rate versus End-to-end delay

having slowest mobility range is assigned maximum data rate and cluster having fastest mobility range is assigned least data rate.

Figure 3a compares the throughput of single-hop and multi-hop network for cumulative data rates in the manner discussed above. It is seen that multi-hop network demonstrates better performance than single-hop broadcast for both the cases, that is, when the data rate increases proportionally with increasing cluster mobility and secondly when the data rate decreases with increasing cluster mobility. This goes in line with the theories discussed in [3, 4] which finds multi-hop scenario as more efficient than single-hop.

In Fig. 3b, it is shown that the PDR for multi-hop is better than single-hop network under both increasing and decreasing mobility-data rate comparisons. It accounts to similar reasons as discussed above.

Fig. 3 **a** Cumulative data
rate versus Throughput.
b Cumulative data rate versus
PDR. **c** Cumulative data rate
versus End-to-end delay

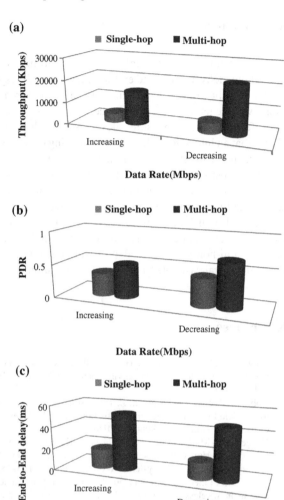

Figure 3c compares the end-to-end delay with the data rates and it is seen that
single-hop broadcasting techniques totally outshines multi-hop technique in both
the cases. The results coincide with the ones presented in [6]. This attributes to the
fact that single-hop network are best suited for delay-intolerant applications but at
the cost of mediocre PDR and throughput values. The decision to choose one
between the two techniques becomes application specific and the extent to which
trade-offs can be accepted.

6 Conclusion and Future Scope

We have compared both single-hop and multi-hop wireless network in terms of throughput, delivery ratio and end-to-end delay with respect to varying data rates and cluster mobility range dependent data rates in increasing and decreasing orders. Moreover, we have addressed the question which network type provides better QoS for different data transmission rates. Based on the obtained results, it can be concluded that an integrated scheme implementing clustering and multi-hop broadcast is a key technique to reliable dissemination of event-driven safety messages in VANET but at the cost of delayed reception. It is yet to be determined whether the delay caused due to multi-hop dissemination under clustering based scheme is within the acceptable limits or not, in order to ensure real-time safety message dissemination. Future work will focus towards development of mathematical model, which will more precisely model power consumption and signal-to-interference-plus-noise ratio in multi-hop network so that the routing strategies can be further investigated especially in safety related message dissemination.

References

1. Schwartz, R.S., Scholten, H., Havinga, P.: A scalable data dissemination protocol for both highway and urban vehicular environments. EURASIP J. Wirel. Commun. Netw. **2013**(1), 1–19 (2013)
2. Abbas, T., Sjöberg, K., Karedal, J., Tufvesson, F.: A measurement based shadow fading model for vehicle-to-vehicle network simulations. Int. J. Antennas Propag. Hindawi Publishing Corporation, Article ID 190607, (2015)
3. Neugebauer, M., Ploennigs, J., Kabitzsch, K.: Evaluation of Energy costs for Single Hop versus Multi Hop with respect to topology parameters. In: IEEE International Workshop on Factory Communication Systems (WFCS-2006), pp. 175–182. Torino (2006)
4. Fedor, S., Collier, M.: On the problem of energy efficiency of multi-hop versus one-hop routing in wireless sensor networks. In: 21st IEEE International Conference on Advanced Information Networking and Applications Workshops (AINAW'07), vol. 2, pp. 380–385. IEEE Computer Society, Ontario (2007)
5. Eichler, S. U.: Performance evaluation of the IEEE 802.11 p WAVE communication standard. In: 66th IEEE Vehicular Technology Conference (VTC-2007) Fall, pp. 2199–203. Baltimore (2007)
6. Chen, X., Refai, H.H., Ma, X.: On the enhancements to IEEE 802.11 MAC and their suitability for safety-critical applications in VANET. Wirel. Commun. Mobile Comput. **10**(9), 1253–1269 (2010)
7. Prakash, A., Tripathi, S., Verma, R., Tyagi, N., Tripathi, R., Naik, K.: Vehicle assisted cross-layer handover scheme in NEMO-based VANETs (VANEMO). Int. J. Internet Protoc. Technol. **6**(1–2), 83–95 (2011)
8. Mittag, J., Thomas, F., Härri, J., Hartenstein, H.: A comparison of single-and multi-hop beaconing in VANETs. In: 6th ACM International Workshop on Vehicul Ar Inter Networking (VANET-2009), pp. 69–78. ACM, New York (2009)

9. Pešović, U.M., Mohorko, J.J., Benkič, K., Čučej, Ž.F.: Single-hop versus Multi-hop–Energy efficiency analysis in wireless sensor networks. In: 18th Telecommunications Forum (TELFOR-2010), pp. 471–474. Belgrade (2010)
10. Chen, Q., Jiang, D., Taliwal, V., Delgrossi, L.: IEEE 802.11 based vehicular communication simulation design for NS-2. In: 3rd ACM International Workshop on Vehicular Ad Hoc Networks (VANET'06), pp. 50–56. ACM, New York (2006)

RF MEMS Based Band-Pass Filter for K-Band Applications

Souradip Hore, Santanu Maity, Jyoti Sarma, Anwesha Choudhury
and Gaurav Yadav

Abstract In today's world RF MEMS based Band-pass Filter is an important field of research and application. In this paper an RF MEMS Band-pass Filter model has been designed and proposed that consists of two metamaterial slots i.e., CSRRs. CSRRs are etched in such a way that effective electrical length is increased leading to increase in inductance and thereby reduction in resonant frequency. The Resonant Frequency has been achieved in K-Band. The main focus has been given on reducing the resonant frequency to less than 20 GHz so that the band-pass filter can be used for K-Band applications.

Keywords Metamaterial · MEMS · Band pass filter · K-band · Radio frequency

1 Introduction

Micro electromechanical systems (MEMS) is a collection of sensors and actuators that uses mechanical movements to achieve an open circuit and short circuit in a transmission line, this mechanical movement can be used for achieving tunability [1–6]. To overcome different problems like power handling capability, actuation voltage, reliability different researchers introduced different RF-MEMS concepts

S. Hore · S. Maity (✉) · J. Sarma · A. Choudhury · G. Yadav
Department of Electronics & Communication Engineering,
National Institute of Technology, Arunachal Pradesh, Yupia, India
e-mail: santanu.ece@nitap.in

S. Hore
e-mail: souradiphore@gmail.com

J. Sarma
e-mail: jyotisarma59@gmail.com

A. Choudhury
e-mail: anwesha.25july@gmail.com

G. Yadav
e-mail: filthypain@gmail.com

© Springer Science+Business Media Singapore 2016
R.K. Choudhary et al. (eds.), *Advanced Computing and Communication Technologies*,
Advances in Intelligent Systems and Computing 452,
DOI 10.1007/978-981-10-1023-1_37

[7–9]. It is a technology that enables batch fabrication of miniature mechanical structures, devices and systems [10]. Bandpass filters can be analogously represented by a simple RLC (resistor inductor capacitor) circuit. It can also be represented as a combination of a high pass filter and a low pass filter. A bandpass filter is a filter which allows to pass frequencies within a specified band only i.e., for which the filter has been designed and it rejects all other frequencies. In this paper we have tried to design a bandpass filter for K-Band applications by bringing down the resonant frequency under 20 GHz. Thus the proposed bandpass filter will increase the data transmission rate in K-Band. For designing the bandpass filters, we have used CSRRs (Complementary split-ring resonator) as metamaterials slots [11]. CSRRs are etched out portions from the signal line. This increases the electrical path length and as the electrical path length increases the inductance also increases. Since resonant frequency is inversely proportional to the inductance therefore the resonant frequency decreases. Metamaterials are also called left handed materials since they have a property of negative refraction. This property is used to increase the inductance and reduce the resonant frequency and thereby boost up the performance and efficiency. Electromagnetic properties for metamaterials depend on the shape of etched out portions i.e., CSRR slots and does not depend on the composition of the materials. In metamaterial, electric, magnetic and wave vector form a left handed triplet and hence they are also known as left handed materials.

In this paper, a CSRRs based bandpass filter performance is shown for K band application. The resonant frequency is tuned by means of varying the lumped parameters with the help of structural changes in CPW as per design consideration. We can implement micromachining technology for obtaining optimum resonant frequency.

2 Design and Formulae

The design parameters of RF MEMS filter are mentioned in Table 1 whereas the dimension of one small unit cell of CSRRs is in Table 2. The bandpass filter model that has been designed using ANSOFT HFSS consists of glass substrate with dielectric constant (ε) 5.5. Gold has been used as the ground and signal line with a thickness of 5 μm. For input and output terminals also gold is the component material. Two CSRRs have been implemented, one at input terminal and the other at output terminal. Top View of RF MEMS based bandpass filter with CSRRs structure and modified ground line of CPW and design of CSRR structure are displayed in Figs. 1 and 2 respectively.

Table 1 Dimensions of the CPW

Parameters	Measurements (μm)
G/W/G of ground	50/100/50
Width of ground at middle, at input & at output	975
Width of ground in front of switches	650
Length of signal line	8200
Length of signal input	3700
Length of the signal output	3700
Length of the substrate	10,000
Width of the substrate	2150
Thickness of the substrate	635
Thickness of the ground	2

Table 2 Dimensions of the metamaterial slot based CSRRs

Parameters	Measurements (μm)
Length of the CSRRs	1000
Width of the CSRRs	80
Thickness of the CSRRs	5
Length of the outer etched portion	1000
Width of the outer etched portion	80
Thickness of the outer etched portion	5
Length of the inner etched portion	960
Width of the inner etched portion	40
Thickness of the inner etched portion	5

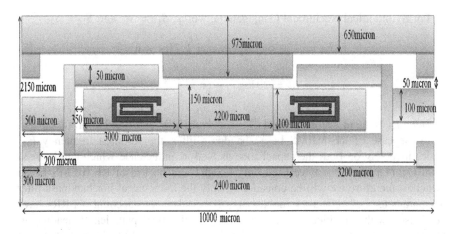

Fig. 1 Top view of RF MEMS based bandpass filter with CSRRs structure and modified ground line of CPW

Fig. 2 Top view of CSRR
(Metamaterial Slot)

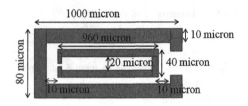

The formulae used for this design are:

$$F_{res} = \frac{1}{2\pi\sqrt{\frac{1}{LC}}} \tag{1}$$

$$S_{11} = \frac{b_1}{a_1}, a_2 = 0 \tag{2}$$

$$S_{12} = \frac{b_1}{a_2}, a_1 = 0 \tag{3}$$

$$S_{21} = \frac{b_2}{a_1}, a_2 = 0 \text{ and} \tag{4}$$

$$S_{22} = \frac{b_2}{a_2}, a_1 = 0 \tag{5}$$

$$\begin{bmatrix} b_1 \\ b_2 \end{bmatrix} = \begin{bmatrix} S_{11} & S_{12} \\ S_{21} & S_{22} \end{bmatrix} \begin{bmatrix} a_1 \\ a_2 \end{bmatrix} \tag{6}$$

For Symmetrical Reciprocal Conjugate networks, $S_{11} = S_{22}$ and $S_{12} = S_{21}$

$$L_A = -20 \log|S_{mn}|; m, n = 1, 2(m \neq n) \tag{7}$$

$$L_R = 20 \log \frac{1}{|S_{nn}|}; n = 1, 2 \tag{8}$$

Here, L_A is the insertion loss and L_R is the return loss, expressed in dB (decibels).

In the case of RF MEMS based bandpass filter, the resonant frequency F_{res} varies inversely on both the capacitance as well as inductance which can be seen from Eq. (1). Since we have used only CSRRs and no use of RF MEMS based switches, therefore we will be considering only the inductance. For bandpass filter S_{11} parameter i.e., reflection coefficient will be taken into account. CSRRs have been used with an objective of increasing the electrical path length leading to increase in inductance so as to bring the S_{11} notch below 20 GHz in K-Band for advanced applications. The entire structure is based on S-Parameters which can be calculated from Eqs. (2), (3), (4), (5) and (6). The insertion and return losses can be calculated from Eqs. (7) and (8) respectively.

3 Simulation Results

Once the CPW structure has been designed, sweep frequency and different excitations can be added using ANSOFT HFSS software with the help of required tools.

Two rectangles are attached at input and output terminals, one at each terminal, which are assigned excitation as waveports. Now, the structure is positioned inside a vacuum box enclosing the entire structure and is assigned excitation with a radiating field in its boundary. The Analysis setup is then done for the structure with a frequency sweep of start frequency 1 GHz and stop frequency of 30 GHz. The solution frequency is given as 10 GHz. The sweep type is set to fast and frequency setup is given as linear count and count is given as 100. From Fig. 3, it is seen that the resonant frequency i.e., S_{11} notch is under 20 GHz. The phase i.e., S_{21} parameter in terms of degrees is shown in Fig. 4. Analyzing the Fig. 3 it is observed that S_{11} notch is obtained at 18.5758 GHz. From this the inference is drawn that the resonant frequency is 18.5758 GHz and the return loss is −37.7752 dB at this frequency. From Fig. 4, it is observed that a clean phase is obtained in terms of S_{21} parameter. This can be implemented for further studies of Phase Shifters.

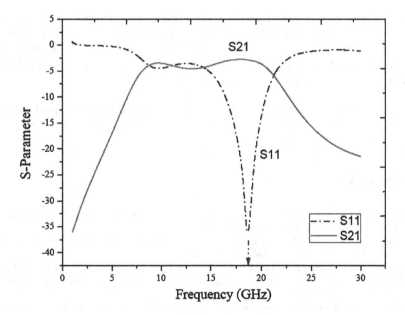

Fig. 3 S-Parameter for band pass filter without switch and with metamaterial

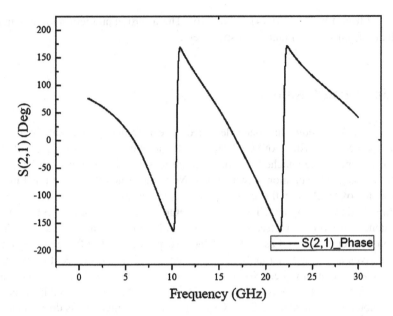

Fig. 4 Phase for bandpass filter without switch and with metamaterial

4   Conclusion

From analysis, it is observed that the resonant frequency is achieved in K-Band which can be implemented for useful K-Band applications. Since the return loss is high, efficiency can be expected to be good. Thus it is concluded that the proposed RF MEMS based bandpass filter model can be used for faster and more efficient data transmission in K-Band with minimum loss.

Acknowledgments We would like to acknowledge support received from Jadavpur University with ANSOFT HFSS software also we are grateful to our friends in preparing the MSS.

References

1. Entesari, K., Rebeiz, G.M.: A 12–18-GHz three-pole RF MEMS tunable filter. IEEE Trans. Microw. Theory Tech. **53**(8), 2566–2571 (2005)
2. Entesari, K., Rebeiz, G.M.: A differential 4 bit 6.5–10 GHz RF MEMS tunable filter. IEEE Trans. Microw. Theory Tech. **53**(3), 1103–1110 (2005)
3. Rebeiz, G.M., Theory, R.F.M.E.M.S.: Design, and Technology. Wiley, New York (2003)
4. Borwick, L.A., et al.: Variable MEMS capacitors implemented into RF filter systems. IEEE Trans. Microw. Theory Tech. **51**(1), 315–319 (2003)
5. Young, R.M.: Low-loss bandpass RF filter using MEMS capacitance switches to achieve a one-octave tuning range and independently variable bandwidth. In: IEEE MTT-S International Microwave Symposium Digest, June 2003, pp. 1781–1784

6. Elfergani, I.T.E., Hussaini, A.S., Rodriguez, J., Marques, P., Abd-Alhameed, R.A.: Tunable RF MEMS bandpass filter with coupled transmission lines. Lecture Notes of the Institute for Computer Sciences, vol. 146, pp. 335–340, 21 May 2015
7. Patel, C.D., Rebeiz, G.M.: A high power () temperature stable RF MEMS metal-contact switch with orthogonal anchors and force-enhancing stoppers. In Proceedings of the IEEE MTT-S International Microwave Symposium Digest (MTT), pp. 1–4 (2011)
8. Rebeiz, G.M., Patel, C., Han, S., Ko, C., Ho, K.M.: The search for a reliable MEMS switch. IEEE Microw. Mag. 14(1), 57–67 (2013)
9. Patel, C.D., Rebeiz, G.M.: A high-reliability high-linearity highpower RFMEMS metal-contact switch for DC–40-GHz applications. IEEE Trans. Microw. Theory Tech. 60 (10), 3096–3112 (2012)
10. Nosrati, M., Vahabisani, N., Daneshmand, M.: Compact MEMS-based ultrawide-band CPW band-pass filters with single/double tunable notch-bands. In: IEEE Transactions on Components, Packaging and Manufacturing Technology, vol. 4, no. 9, Sept 2014
11. Nosrati, M., Vahabisani, N., Daneshmand, M.: A novel ultra wideband (UWB) filter with double tunable notch-bands using MEMS capacitors. In: IEEE Microwave Theory and Techniques Society (MTT-S) International Microwave Symposium Digest (IMS), pp. 1–3, June 2013

Randomized Distance Based Clustering Algorithm for Energy Efficient Wireless Sensor Networks

Rabindra Bista and Anmol Shakya

Abstract Energy efficiency in Wireless Sensor Networks is very important issue now a days because with development of sophisticated hardware and better software, the area of implementation of WSN has widened remarkably. Clustering technique has been widely used in WSN for achieving better management and energy efficiency. LEACH is one of the widely used algorithms in WSN because of its simplicity and easy implementation. However, LEACH is probabilistic based model and it has ignored the effect of different parameters that can affect the energy efficiency of the WSN. In order to overcome this pitfall of LEACH, we introduced new algorithm RDBC, which is both probabilistic as well as distance based approach to achieve better energy efficiency. We simulate our algorithm as well as LEACH in MATLAB and found that RDBC is better than LEACH in different aspects.

Keywords Wireless sensor networks · Clustering · Energy efficiency · Network lifetime · Single-Hop routing · Multi-Hop routing

1 Introduction

The Wireless Sensor Networks (WSN) is a system that comprises of multiple sensor nodes which are linked to one another and Base Station (BS) through the means of wireless channel. Research in this field has been highly motivated since last decade because of its wide range of application in real world. Earlier WSN was mostly used for military application (DARPA projects) but later this has been profoundly used in applications like biomedical health monitoring, industrial process moni-

R. Bista (✉) · A. Shakya
Department of Computer Science and Engineering,
Kathmandu University, Post Box No.: 6250, Dhulikhel-7, Nepal
e-mail: rbista@ku.edu.np

A. Shakya
e-mail: anmol.shakya@student.ku.edu.np

© Springer Science+Business Media Singapore 2016 377
R.K. Choudhary et al. (eds.), *Advanced Computing and Communication Technologies*,
Advances in Intelligent Systems and Computing 452,
DOI 10.1007/978-981-10-1023-1_38

toring, wildlife monitoring and seismic sensing [1–3]. The deployment of sensor nodes are generally performed randomly in outdoor application where large area has to be covered, so it becomes difficult to maintain the network topology in all the applications. In such scenario clustering technique becomes very useful for management of random sensor nodes.

Energy efficiency in WSN is very crucial because sensor nodes have limited amount of energy source (mostly 2 AA sized batteries). Exhaustion of battery is prominent reason for nodes failure in WSN [4]. Once the sensor node fails, it is considered dead. Therefore it is important to implement energy efficient protocols in WSN for reliability and longer lifetime of the WSNs.

Low Energy Adaptive Clustering Hierarchy (LEACH) protocol is most widely used protocol in WSN for energy efficiency. This protocol is almost decade old but still used widely because of its simplicity and easy implementation in most of the WSN scenario. LEACH is a probabilistic approach for selecting the Cluster Head (CH) without consideration of other important parameters like energy or distance. However this approach is better than traditional approach of Direct Transmission and Minimum Transmission Energy [5].

The selection of the CH is one of the crucial operations in clustering algorithm. The process of selecting CH for each cluster should be performed by using minimum communication operation between the sensor nodes and base station. In this paper we proposed a new technique to elect a CH based on the randomized distance parameter, which makes sure that the CH does not have to perform the long distance transmission to BS and same node is not chosen as CH all the time. Here we use both single-hop technique as well as multi-hop technique for transmission of data to the BS. This algorithm uses TDMA scheme same as that of LEACH protocol which allows member nodes to go to sleep state when they have nothing to transmit to the CH.

The remainder of this paper is organized as follow: In Sect. 2 we discuss the related work, later in Sect. 3 the detail of our algorithm is presented followed by Simulation and comparison of our algorithm with LEACH in Sect. 4 and finally we conclude this paper in Sect. 5.

2 Related Works

Many works has been done to improve the various aspects of the WSN using the clustering techniques. The LEACH protocol itself has been modified by many authors in past few years and these modified protocols are generally known as Descendant of LEACH [6]. Apart from LEACH's descendants many other clustering techniques has been developed for better operation of the WSN in different contexts such as lifetime, energy dissipated, packets transmitted, scalability, etc.

LEACH (Low Energy Adaptive Clustering Hierarchy) [5] is a probabilistic approach for selecting the cluster head and forming the cluster. There are two phases in this method: Setup phase and Steady state phase. In Setup phase the election of cluster head is performed using probabilistic approach as shown in Eqs. 1 and 2.

$$T(n) = \frac{p}{1 - p\left(r \bmod \frac{1}{p}\right)} \text{ if } n \in G \tag{1}$$

$$T(n) = 0 \text{ otherwise} \tag{2}$$

where, p = desired percentage of cluster heads, r = current round, n = node ID and G is the set of nodes that have not been cluster head in last 1/p rounds.

A random number between 0 and 1 is chosen, the current node is selected as the CH only if the output of $T(n)$ is less than that random number. After selection of cluster head member nodes are selected according to the strength of signal received from the broadcasting CH.

In steady state phase the member node transmits the data to the cluster head in their assigned time slot in TDMA fashion. This allows the member node to go in sleep mode when they have nothing to transmit to the CH.

The transmission operation performed by both the cluster member and cluster head are single hop. The cluster members have to perform low power transmission but same is not true for the CH. So much energy is used up by the CH while transmitting the data to the BS. CH also has to perform data aggregation which is also an energy intense operation.

LEACH-C (LEACH-Centralized) [7] is similar to the original LEACH but only difference is in Setup phase. LEACH-C uses centralized technique to select the CH. In this approach BS determines which node will become the CH. BS collects the information about the location (through GPS) and remaining energy of all the sensor nodes. The BS calculates the average energy of the overall system. The nodes whose energy level is below this average energy are deprived of being the CH for that round and CH is selected from remaining of the node. The BS uses the Annealing Algorithm to solve the NP-hard problem of finding k optimal clusters. The next stage is the steady state phase which is same as that of original LEACH.

Manzoor et al. [8] introduced the concept of Quadrature-LEACH (Q-LEACH), where the area of deployment is divided into 4 equal parts. The authors stated that by dividing the area, the stability period, throughput and network lifetime improves significantly. They also stated that the Cluster formation is more deterministic in Q-LEACH compared to LEACH. However the cluster formation and CH selection is same as that of LEACH in each quadrature. Since the LEACH is applied in each quadrature the CH selection becomes difficult in later rounds similar to that of the LEACH, which results in low data transmission and low data reliability.

Qiang et al. presented a concept called MS-LEACH [9] which combines the concept of both Multi-hop transmission and Single-hop (Direct) transmission. In this paper the authors have analyzed the problem of energy consumption of the single-hop and multi-hop transmissions in a single cluster. This approach is similar to LEACH as it also has two phases: Setup phase and Data Transmission phase. Setup phase is same as the LEACH and uses the same Eq. 1 to determine the CH. However the transmission phase is different. Here the CH calculates the critical cluster area size $Q_{critical}$ and approximate values of cluster area size Q. If Q is greater than $Q_{critical}$ then CH calculates the routing path tree and broadcast it to its member nodes for multi-hop transmission otherwise the transmission is performed in single-hop by the CH same as in LEACH protocol.

Khediri et al. introduced O-LEACH [10] with some optimization in CH selection. Instead of selecting the CHs based on the probabilistic rotation, the authors are taking the dynamic residual energy for selecting the CHs and concluding that O-LEACH increases the stability of the network and it can be used for dynamic networks. Moreover, it uses the centralized concept of electing the CHs. The BS is responsible for making efficient clusters based on the residual energy. The BS begins the making of clustering process. The CH is elected by considering the energy of the sensor nodes. Nodes which have energy greater than 10 % of minimum residual energy are considered for the selection of CHs. When the energy of nodes is less than 10 % of minimum residual energy the original LEACH method is followed for selecting the CHs.

3 Randomized Distance Based Clustering Algorithm

Randomized Distance Based Clustering (RDBC) Algorithm improves the performance of the LEACH in terms of: (1) the number of working nodes alive in the system at any time, (2) the overall coverage of the WSN area by the sensor nodes and (3) number of useful data transmitted in the system. In this algorithm we include the feature of random selection of the CH along with the effect of the distance on cluster formation. Distance plays an important role while selecting the CH. However, the exact distance of node to the BS is not required; the approximate distance will be enough. The approximate distance of node from the BS can be calculated using the Received Signal Strength (RSS) [11]. Using the basic power-distance model:

$$P_r = P_t \left(\frac{1}{d}\right)^2 \tag{3}$$

where, P_r is the power received at receiver node, P_t is the power transmitted by the sending node, d is approximate distance between the sender and receiver and n is the environment dependent transmission factor (usually 2 for free space). The formula in Eq. 2 is further modified for power in dBm as:

$$P_r(dBm) = P_t(dBm) - 10 \times n \times \lg(d) \tag{4}$$

The calculation of the distance is performed only once in the whole life time of the sensor node. This process is initiated by BS by broadcasting the short message which contains the value of its transmitted power and the header to identify it is from the BS. For generalization the calculated distance is converted into percentage based distance which means that the distance of the furthermost point in the deployed area from the BS is represented by 1and the closest point (i.e. 0 m from BS itself) is represented as 0.

The basic assumptions in RDBC Algorithms are:

- The sensor node in the system always has something to transmit in every round until it is considered dead.
- The deployment of sensor node is completely random and they are deployed in outdoor environment.

RDBC Algorithm has two phases: Setup Phase and Steady State Phase.

3.1 Setup Phase

Setup Phase is initiated each time when the new round is started. At the beginning all the nodes (except dead node) are initialized as non-cluster head node. Cluster Head selection and cluster formation are the two main activities that occurs in setup phase of RDBC algorithm.

At the beginning of each round all the nodes having residual energy greater than the threshold energy (E_{Th}) calculate the parameter called *SCORE* as follow:

$$SCORE_i = Random\left(\frac{1}{D_{i\_to\_BS}(in\%)}\right) \tag{5}$$

where, i $\rightarrow$ Node ID, Random $\rightarrow$ A random number (between 0 and 1) and $D_{i\_to\_BS}$ *(in%)* $\rightarrow$ Distance between current node and BS in percentage.

After calculation of the SCORE, the node with the maximum SCORE broadcast first. Time to wait before broadcasting is calculated as:

$$T_{wait}(i) = \frac{1}{SCORE_i} \tag{6}$$

The first node to broadcast is a winner and declares itself as the CH. The broadcast message is a short message containing the node ID of broadcaster and the transmission power used for broadcasting, $P_{broadcast}$. The non-member nodes which receive the broadcast signal with strength greater than X% of $P_{broadcast}$ become member of that CH. The nodes which receives the broadcast but signal strength less than X% of $P_{broadcast}$ saves the broadcasting node ID as the next hop and waits until it becomes the CH itself or the member of other CH. While broadcasting, the nodes which are near becomes the member of broadcasting node and other saves this node as routing node unless better options are available. However, if the remaining energy of the farther nodes is less than that of threshold energy it becomes the member of the broadcasting CH.

Since this process involves the distance parameter associated with it, there is high possibility that the nodes closer to the BS are chosen to be CHs thus saving the power consumption in long range transmission. The Algorithm 1 shows the steps for Setup Phase.

```
Algorithm 1: Setup Phase
Input: Set of N sensor nodes scatted randomly in MxM area
Output: Selection of CH and Cluster formation
Begin Setup Phase
For each node
  Distance_from_BS=sqrt(Ptx/Prx)
  Distance_in_percent=Distance_from_BS/Max_dis
  Node_type=N //N is for normal node not CH node
  Rem_energy=Rem_energy - Erx X no.of bits received
End For
For each round
  If any node's energy < Eth
    SCORE=0
  Else
    SCORE=random(0,1)X(1/Distance_in_percent)
  End If
  While (any node is still N)
    If(node is N or R and SCORE is maximum)
      Cluster= Cluster + 1
      Node_type=CH
      Broadcast Advertise Message
      For each Broadcast receiving Node
        If(RSS>Threshold Signal Strength or Rem_energy<Eth)
          Node_type=M
          Member_of_node=CH_nodeID
        Else
          Node_type=R
          Next_hop=CH_nodeID
        End If
      End For
    End If
  End While
End For
End Setup Phase
```

3.2 Steady State Phase

The operation performed in the steady state phase is the transmission of the message from the member nodes to the CH nodes and from CH nodes to the BS. In order to avoid the collision of data TDMA scheme is used. Only one unique timeslot is assigned to each node in the system so that they can transmit only on their allocated time and remain in sleep mode for rest of the period. In this algorithm, we have allocated the time slot according to the node ID of the sensor node, which means a timeslot 1 is assigned to the node with node ID 1, time slot 2 is provided to the node with node ID 2 and so on.

The transmission of data is done either in single-hop as in LEACH or in multi-hop where the CH transmit the aggregated data to the next CH and finally to the BS. The CHs whose next hop in the routing table is 0 is the closest CH from the BS. There might be one or more than one CH which transmits directly to the BS.

The transmission of data from the member node to CH is a single-hop and low power consuming operation. The CH receives multiple data from its cluster members and performs the data aggregation operation. The CH then transmits the data either directly to the BS or to other CH which acts as the router to BS. The Algorithm 2 shows the steps for steady state phase.

```
Algorithm 2: Steady State Phase
Input: Resulted Cluster from Setup Phase
Output: TDMA slot, Route to BS and Energy Spent
Begin Steady State Phase
For each node
  TDMA_timeslot←nodeID
End
For each TDMA_timeslot    //TDMA slot == nodeID
  If node_type==M
    Transmit k bits of data to Member_of_node
    Rem_energy←Rem_energy- Etx_k_bits data to CH
    Rem_energy_CH←Rem_energy_CH- Erx_k_bits data + EDA
    /*EDA→ Energy spent for Data Aggregation*/
  Elseif node_type==CH
    If  Next_hop==0
      Transmit k bits of data to BS
      Rem_energy←Rem_energy-Etx_k_bits data to BS
    Else
      Transmit k bits data to Next_hop
      Rem_energy←Rem_energy-Etx_k_bits data to Next_hop
      Rem_energy_CH←Rem_energy_CH-Erx_k bits data + EDA
    End If
  End If
End For
End Steady State Phase
```

3.3 Energy Model

In this algorithm we use the same energy model that was presented by W. Heinzelman in [7] (Fig. 1), which is modeled for transmitting the K-bit packet to distance d and receiver node receives the K-bit packet. In this model the signal has to be amplified before transmitting the data. The amplification factor depends on the distance between transmitter and the receiver. Depending upon the distance, the amplification factor is chosen either as free space amplification or multipath fading amplification. The energy model is shown in Fig. 3.

If the distance, d, between the transmitter and the receiver is greater than threshold distance, d_0 than the multipath (*Emp*) energy model has to be used otherwise we use the free space (*Efs*) energy model. The threshold distance, d_0 is calculated as:

$$d_0 = \sqrt{\frac{Efs}{Emp}} \tag{7}$$

The energy spent to transmit k-bits of data to distance, d is therefore calculated as:

$$E_{Tx} = \begin{cases} E_{elec} \times k + k \times Efs \times d^2, & If\_d < d_0 \\ E_{elec} \times k + k \times Emp \times d^4, & If\_d \geq d_0 \end{cases} \tag{8}$$

and the energy spent to receive the same k-bits of data at receiving end is calculates as:

$$E_{Rx} = k \times E_{elec} \tag{9}$$

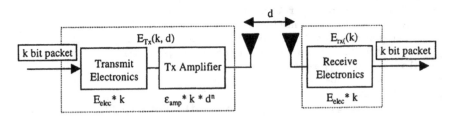

Fig. 1 Radio energy dissipation model

4 Simulation and Results

We used MATLAB version R2014b for simulation of both RDBC and LEACH. We are using the same energy model that was presented by Heinzelman in [7]. The simulation parameters used for the experiment is shown in Table 1.

The BS was fixed at position x-axis = 150 m and y-axis = 50 m. The threshold energy was set to 15 % of the initial energy; however the threshold energy could be modified.

According to [12], the maximum transmission range of MICA2 mote (model no. MPR400CB) in outdoor condition is 150 m (500 ft), but it cannot be achieved completely in practical world due to many environmental factors. The mote's transmission range could be greater than 100 m but it comes with trade-off of maximum packet loss. So in our simulation we set the maximum range for transmission of data 50 m, which according to [13] is achievable with MICA2 mote.

The average remaining energy per round of RDBC algorithm is always greater than that of LEACH for almost all 1200 rounds, except for few rounds at the end of the simulation. The average remaining energy per round of LEACH is greater than that of RDBC at end rounds (Shown in Fig. 2) because in LEACH at later rounds the cluster formation is difficult as the result the nodes becomes in active and does not take part in the communication or any other tasks thus spending no energy at all, but RDBC performs data transmission and other tasks until the last sensor node remains alive.

After 1200 round it was found that LEACH had 1 node still alive where as in case of RDBC all nodes were dead at round 1186. In case of LEACH the first node died on round 547, but in case of RDBC the first node died on round 803 (Fig. 3). The rate of dying of node for RDBC is less at beginning whereas for the LEACH it is opposite.

The packets transmitted per round for both LEACH and RDBC starts with equal packets transmission but in later rounds the data transmission for LEACH drops faster compared to RDBC (Fig. 4). There are numerous occasions (260 times out of

Table 1 Simulation parameters	Parameter	Value
	Simulation area	100 m × 100 m
	Initial energy (Eo)	0.5 J
	Transmitter/receiver electronics (E_{elec})	50 nJ/bit
	Number of nodes	100
	Efs	10 pJ/bit/m^2
	Emp	0.0013 pJ/bit/m^4
	Energy for data aggregation	5 nJ/bit
	Message packet size	4000 bits
	Header packet size	200 bits
	Threshold energy (Eth)	15 % of Eo
	Simulation rounds	1200

Fig. 2 Average remaining
energy graph

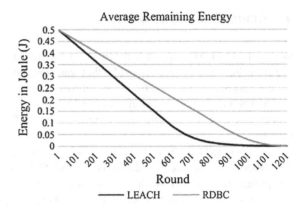

Fig. 3 Dead node per round
graph

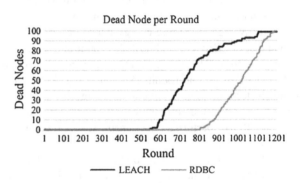

Fig. 4 Data transmitted per
round

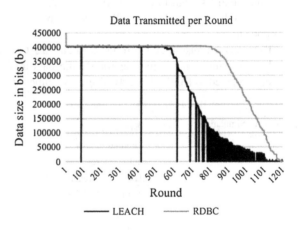

1200 rounds) when LEACH does not transmits any data but RDBC never fails to
transmit the packet, it is because the LEACH fails to form the cluster in those
occasions. The total amount of data transmitted in the network is also greater while
using RDBC as we can see from Fig. 5.

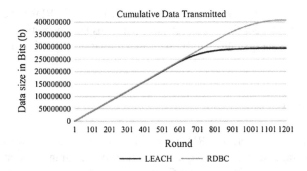

Fig. 5 Cumulative data transmitted in system

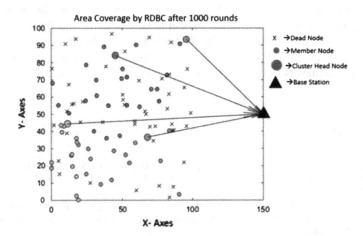

Fig. 6 Area coverage by nodes with RDBC

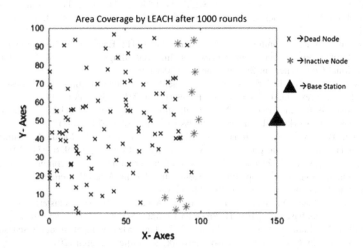

Fig. 7 Area coverage by nodes with LEACH

Unlike LEACH, the coverage area is also better in case of RDBC because more nodes are alive for longer time and each node performs their operation until they are completely out of energy, this can be seen by comparing Figs. 6 and 7.

5 Conclusion

We presented a new algorithm, RDBC, which is a better choice over LEACH for wide area coverage for longer time, larger number of data transmission and longer useful lifetime. Like LEACH, RDBC also implement randomization but with addition of distance parameter. RDBC implements both direct transmission as well as multi-hop transmission. One of the major advantages of this algorithm is its capability to scale up or scale down without having to notify the BS because this algorithm uses distributed technique for CH selection and cluster formation. In later works we can include other parameters such as the number of neighboring nodes and energy parameters to improve the CH selection and efficient cluster formation.

References

1. Lazarescu, M.T., di Elettron, D., di Torino, P., Turin: Design of a WSN platform for long-term environmental monitoring for IoT applications. IEEE J. Emerg. Sel. Top. Circuits Syst. 45–54 (2013)
2. Sukun, K., Pakzad, S., Culler, D., Demmel, J.: Health monitoring of civil infrastructures using wireless sensor networks. In: 6th International Symposium on Information Processing in Sensor Networks, pp. 254–263 (2007)
3. Sendra, S., Lloret, J., García, M., Toledo, J.F.: Power saving and energy optimization techniques for wireless sensor networks. J. Commun. 6, 439–459 (2011)
4. Sepideh, N., Aras, N., Fairouz, A.: SLTP: Scalable Lightweight Time Synchronization Protocol for Wireless Sensor Network, pp. 536–547. Springer-Verlag, Berlin Heidelberg (2007)
5. Heinzelman, W.R., Chandrakasan, A., Balakrishnan, H.: Energy-efficient communication protocol for wireless micro sensor networks. In: Proceedings of the 33rd Annual Hawaii International Conference on System Sciences (2000)
6. Kumar, V., Jain, S., Tiwari, S.: Energy efficient clustering algorithms in wireless sensor networks: a survey. Int. J. Comput. Sci. Issues (IJCSI) 8(5), 259–268 (2011)
7. Heinzelman, W.R., Balakrishnan, H.: An application specific protocol architecture for wireless microsensors networks. IEEE Trans. Wireless Commun. 660–670 (2002)
8. Manzoor, B., Javaid, N., Rehman, O., Akbar, M., Nadeem, Q., Iqbal, A., Ishfaq, M.: Q-LEACH: a new routing protocol for WSNs. Procedia Comput. Sci. 926–931 (2013)
9. Qiang, T., Bingwen, W., Zhicheng, W.C.: MS-Leach: a routing protocol combining multi-hop transmissions and single-hop transmissions. In: Pacific-Asia Conference on Circuits, Communications and Systems, pp. 107–110 (2009)
10. Khediri, S.E., Nasri, N., Wei, A., Kachouri, A.: A new approach for clustering in wireless sensors networks based on LEACH. Procedia Comput. Sci. 32, 1180–1185 (2014)
11. Xu, J., Liu, W., Lang, F., Zhang, Y., Wang, C.: Distance Measurement Model Based on RSSI in WSN, pp. 606–611. Wireless Sensor Network, Scientific Research (2010)

12. Crossbow Technology Inc.: MICA2, Wireless Measurement System. www.xbow.com
13. Anastasi, G., Falchi, A., Passarella, A., Conti, M., Gregori, E.: Performance measurements of motes sensor networks. In: Proceedings of the 7th ACM International Symposium on Modelling, Analysis and Simulation of Wireless and Mobile Systems, pp. 172–181 (2004)

Communication Between Information Systems Using Ordered Weighted Averaging

Waseem Ahmed, Tanvir Ahmad and M.M. Sufyan Beg

Abstract Information system is the prime model in the branch of information technology, the utmost problem encountered here is the communication between information systems. Compatible and consistent homomorphism had played a crucial role in the communication between fuzzy relation information systems. This paper endeavors to establish communication between information systems, even when mapping is not consistent or compatible homomorphism. For this, we propose aggregation operator known as Ordered Weighted Averaging in place of simple relation mapping. Proposed method assures that properties and attribute reduction of an original system and its equivalent image system are identical to one another in the absence of consistent homomorphism. Finally, the utilization of this paper is given with the help of an example.

Keywords Fuzzy relation information system · Ordered weighted averaging · Fuzzy relation mapping · Homomorphism · Attribute reduction

1 Introduction

In this Information age, where information system (IS) is the prime model in the field of information technology, communication between information systems (IS) has remained a big issue [1–3]. Due to varied IS, it is at times important to transfer information within various IS, like decision making, rule extraction and equivalent attribute reduction [4–6]. This entails the central focus of studying the relationship

W. Ahmed (✉) · T. Ahmad
Department of Computer Engineering, Jamia Millia Islamia, New Delhi, India
e-mail: waseem.ahmed86@gmail.com

T. Ahmad
e-mail: tahmad2@jmi.ac.in

M.M. Sufyan Beg
Department of Computer Engineering, Aligarh Muslim University, Aligarh, India
e-mail: mmsbeg@eecs.berkeley.edu

© Springer Science+Business Media Singapore 2016   391
R.K. Choudhary et al. (eds.), *Advanced Computing and Communication Technologies*,
Advances in Intelligent Systems and Computing 452,
DOI 10.1007/978-981-10-1023-1_39

between information systems. Numerous studies on various topics of information systems had been attempted and examined, however, far too little attention has been paid to communication between fuzzy relation information systems(FRISs) [2–4, 7–11]. Communication between FRISs is gripping, intricate and important in granular computing framework. Communication is considered as the issue of transmutations of information systems, along with keeping intact its prime functions and properties. Communication allows us to transfer information from one granular world to another and hence lays a mechanism for interchanging information with other granularity [3]. By mathematical perspective, such type of communication can be explained as analyzing properties of various information systems through mappings. Mapping is effective tool to examine the relationship amongst various IS [2, 3, 6, 8, 10].

Homomorphism mapping had been an important mathematical tool to achieve this transformation [2, 4, 7–10, 12, 13]. The concept of compatible and consistent mapping had played a crucial role in homomorphism between fuzzy information systems. But it may be possible that information system is neither compatible homomorphism nor consistent homomorphism. This paper aims to deal such issues by establishing communication in the absence of these consistent conditions.

This paper extends the above mentioned work by introducing the concept of aggregation operator known as Ordered Weighted Averaging (OWA). Relation mapping make use of supremum operator which takes the maximum value among all. While OWA's remarkable side is its degree of adjustability [14, 15]. Thus OWA operator provides us the property to easily vary the rate of ANDing and ORing accordingly.

By using OWA operator, fuzzy relation mapping can be made flexible and not restricted to the supremum operator. That is by changing the weight vector values, we can change image system according to our need. Also the traditional relation mapping approach can be achieved by using the particular OWA weight vector choice.

The rest of this paper is structured as follows: Sect. 2 summarizes the previous work done on the concept of consistent and compatible mapping. In Sect. 3, we formalize fuzzy relation mapping using OWA. Section 4 shows the result of applying OWA operator to the concept of homomorphism between FRISs. Finally, conclusion and future direction is drawn in Sect. 5.

2 Preliminary Studies

2.1 Fuzzy Relation Mappings

Function mapping is a preliminary mathematical tool to communicate among fuzzy sets. In the similar way, fuzzy relation mappings can be used to actualize this intercommunication between fuzzy information systems. Here, we are defining fuzzy relation mapping using Zadeh's extension principle.

Let U_1 and V_1 are two universes. We will denote all the classes of fuzzy binary relation on U_1 and V_1 by $R(U_1 \times U_1)$ and $R(V_1 \times V_1)$, respectively.

Definition 1 Let $f: U_1 \to V_1$ be a mapping. By Zadeh's extension principle, f can induce a mapping from $R(U_1 \times U_1)$ to $R(V_1 \times V_1)$ and a mapping from $R(V_1 \times V_1)$ to $R(U_1 \times U_1)$, that is

$$f : R(U_1 \times U_1) \to R(V_1 \times V_1), R| \to f(R) \in R(V_1 \times V_1), \forall R \in R(U_1 \times U_1)$$

$$f(R)(x,y) = \begin{cases} sup_{u \in f^{-1}(x)} sup_{v \in f^{-1}(y)} R(u,v), & (x,y) \in f(U_1) \times f(U_1) \\ 0, & (x,y) \notin f(U_1) \times f(U_1) \end{cases}$$

$$f^{-1}: R(V_1 \times V_1) \to R(U_1 \times U_1), f^{-1}(T)(u, v) = T(f(u), f(v)).$$

Here $f(R)$ and $f^{-1}(T)$ are fuzzy relations induced by f on V_1 and U_1, respectively.

2.2 Consistent Functions

This paper aims to prove that properties and attribute reduction of an original system and its equivalent image are identical to one another even when, mapping is not compatible or consistent homomorphism. Most of the work done earlier was proved only under some sufficient conditions [2, 4, 6–9, 12]. It has been found that different authors represent the consistent function in different manner. Definition 2 reviews some of these concepts used in [2].

Definition 2 Let U_1 and V_1 are universe of discourse, $f: U_1 \to V_1$ is their mapping from U_1 to V_1, and $A, A_1, A_2 \in R(U_1 \times U_1)$. Let $[x]_f = \{y \in U: f(y) = f(x)\}$;

For any $u_1, u_2 \in [x]_f$ and $v_1, v_2 \in [y]_f$,

If $A(u_1, v_1) = A(u_2, v_2)$, then the mapping '$f$' is known as compatible with respect to relation A.

For any $x, y \in U_1$, if any of the condition holds:

(1) $A_1(u_1, v_1) \leq A_2(u_1, v_1)$ for any $(u_1, v_1) \in [x]_f \times [y]_f$
(2) $A_1(u_1, v_1) \geq A_2(u_1, v_1)$ for any $(u_1, v_1) \in [x]_f \times [y]_f$, then we say f is consistent with respect to A_1 and A_2.

Now in the next section of our paper, our special emphasis is on the proofs of various theorems using OWA operator.

3 Ordered Weighted Average Aggregation

In this section, we review aggregation operator called as ordered weighted averaging (OWA) operator.

Definition 3 The OWA operator build up an aggregation process in such a way that, a sequence having 'n' values are arranged in non increasing order and then its ordered value is weighted by some weight vector 'W' = w_i, such that

(1) $w_i \in [0, 1]$
(2) $\sum_{i=1}^{n} w_i = 1$ [14].

In particular we say that,

Suppose y_i is the ith largest value in sequence A, $OWA_w(A) = \sum_{i=1}^{n} w_i \cdot y_i$

OWA's provides an aggregation that allow us to lies between two extremes that is 'And' and the 'Or'. So, by changing the weight vector values accordingly the minimum, maximum and other mean values can be achieved. For example

(1) Maximum weight: $w_{max} = w_i$, where $w_1 = 1$ and $w_i = 0$ when $i \neq 1$
(2) Minimum weight: $w_{min} = w_i$, where $w_n = 1$ and $w_i = 0$ when $i \neq n$
(3) Average weight: $w_{avg} = w_i$, where $w_i = 1/n$

3.1 Fuzzy Relation Mappings Using OWA Operator

Here, we formalize fuzzy relation mapping with the help of OWA operator.

Definition 4 Let $f: U_1 \to V_1$ be a mapping. By Zadeh's extension principle, f can induce a mapping from $R(U_1 \times U_1)$ to $R(V_1 \times V_1)$, that is

$$f : R(U_1 \times U_1) \to R(V_1 \times V_1),$$

$$f(R)(x,y) = \begin{cases} owa_{u \in f^{-1}(x)} owa_{v \in f^{-1}(y)} R(u,v), & (x,y) \in f(U_1) \times f(U_1) \\ 0, & (x,y) \notin f(U_1) \times f(U_1) \end{cases}$$

Here, $f(R)$ is fuzzy relation on V_1 generated by f.

Using OWA operator we can make the relation mapping flexible and not restricted to the supremum operator. That is by changing the weight values we can change image system according to the requirement.

As defined above, w_{max} and w_{min} are the special cases of OWA weight vectors.

Where w_{max} gives the maximum value in the sequence i.e. supremum value and w_{min} gives the minimum value in the sequence i.e. infimum value.

Remark By default we are taking the weight value as w_{max}.

Theorem 5 Let $f : U_1 \to V_1$, A, A_1, $A_2 \in R(U_1 \times U_1)$; then, we have the following.

(1) $f(A_1 \cup A_2) = f(A_1) \cup f(A_2)$.
(2) $f(A_1 \cap A_2) \subseteq f(A_1) \cap f(A_2)$; if we take weight vector of OWA as w_{min} then the equality holds.

Proof

(1) For any z_1, $z_2 \in V_1$

$$f(A_1 \cup A_2)(z_1, z_2) = owa_{t_1 \in f^{-1}(z_1)} owa_{t_2 \in f^{-1}(z_2)}(A_1 \cup A_2)(t_1, t_2)$$
$$= owa_{t_1 \in f^{-1}(z_1)} owa_{t_2 \in f^{-1}(z_2)}(A_1(t_1, t_2) \cup A_2(t_1, t_2))$$
$$= (f(A_1) \cup f(A_2))(z_1, z_2)$$

(2)
$$f(A_1 \cap A_2)(z_1, z_2) = owa_{t_1 \in f^{-1}(z_1)} owa_{t_2 \in f^{-1}(z_2)}(A_1 \cap A_2)(t_1, t_2)$$
$$= owa_{t_1 \in f^{-1}(z_1)} owa_{t_2 \in f^{-1}(z_2)}(A_1(t_1, t_2) \cap A_2(t_1, t_2))$$
$$\leq owa_{t_1 \in f^{-1}(z_1)} owa_{t_2 \in f^{-1}(z_2)} A_1(t_1, t_2) \cap owa_{t_1 \in f^{-1}(z_1)} owa_{t_2 \in f^{-1}(z_2)} A_2(t_1, t_2)$$
$$= (f(A_1) \cap f(A_2))(z_1, z_2)$$

Now, by taking the weight vector of owa as w_{min}

$$f(A_1 \cap A_2)(z_1, z_2) = owa_{t_1 \in f^{-1}(z_1)} owa_{t_2 \in f^{-1}(z_2)}(A_1 \cap A_2)(t_1, t_2)$$
$$= owa_{t_1 \in f^{-1}(z_1)} owa_{t_2 \in f^{-1}(z_2)}(A_1(t_1, t_2) \cap A_2(t_1, t_2))$$
$$(\text{i.e. } w_{min} = w_i, \text{ where } w_n = 1, w_i = 0, i \neq n)$$
$$= owa_{t_1 \in f^{-1}(z_1)} owa_{t_2 \in f^{-1}(z_2)} A_1(t_1, t_2) \cap owa_{t_1 \in f^{-1}(z_1)} owa_{t_2 \in f^{-1}(z_2)} A_2(t_1, t_2)$$
$$= (f(A_1) \cap f(A_2))(z_1, z_2)$$

Remark By using owa minimum operator the intersection operation of fuzzy relations can be preserved.

Example 6 Let $U_1 = \{t_1, t_2, t_3\}$, and $V_1 = \{z_1, z_2\}$ and $f: U_1 \to V_1$ is a mapping from U_1 to V_1 and defined as: $f(t_1) = f(t_3) = z_1$; $f(t_2) = z_2$;

A_1 and A_2 are fuzzy relations on U_1, which are given in Table 1. $(A_1 \cap A_2)$ is also given in Table 1. Then, $f(A_1)$ and $f(A_2)$ are computed and presented in Table 2. $f(A_1) \cap f(A_2)$ and $f(A_1 \cap A_2)$ are also given in Table 2. We find that $f(A_1 \cap A_2) \subset f(A_1) \cap f(A_2)$. Here we are using maximum weight owa operator i.e. w_{max}.

Now, by using minimum weight owa operator (i.e. w_{min});
We compute $f(A_1)$ and $f(A_2)$ and presented in Table 3, then $f(A_1) \cap f(A_2)$ and $f(A_1 \cap A_2)$ are also presented in Table 3.
We can verify, $f(A_1 \cap A_2) = f(A_1) \cap f(A_2)$ in this case.

Table 1 Original similarity relations

	Relation A_1				Relation A_2				Relation $(A_1 \cap A_2)$		
	t_1	t_2	t_3		t_1	t_2	t_3		t_1	t_2	t_3
t_1	0.2	0.9	0.5	t_1	0.7	0.8	0.4	t_1	0.2	0.8	0.4
t_2	0.4	0.1	0.9	t_2	0.5	0.2	0.6	t_2	0.4	0.1	0.6
t_3	0.6	0.1	0.5	t_3	0.5	0.4	0.2	t_3	0.5	0.1	0.2

Table 2 Image systems using max operator

	Relation $f(A_1)$		Relation $f(A_2)$			Relation f $(A_1) \cap f(A_2)$			Relation f $(A_1 \cap A_2)$	
	z_1	z_2		z_1	z_2		z_1	z_2		z_1	z_2
z_1	0.6	0.9	z_1	0.7	0.8	z_1	0.6	0.8	z_1	0.5	0.8
z_2	0.9	0.1	z_2	0.6	0.2	z_2	0.6	0.1	z_2	0.6	0.1

Table 3 Image systems using min operator

	Relation $f(A_1)$		Relation $f(A_2)$			Relation f $(A_1) \cap f(A_2)$			Relation f $(A_1 \cap A_2)$	
	z_1	z_2		z_1	z_2		z_1	z_2		z_1	z_2
z_1	0.2	0.1	z_1	0.2	0.4	z_1	0.2	0.1	z_1	0.2	0.1
z_2	0.4	0.1	z_2	0.5	0.2	z_2	0.4	0.1	z_2	0.4	0.1

4 Homomorphism Between Information Systems Using Ordered Weighted Averaging

In this section, we examine the notion of homomorphism, and explore few of the properties of FRISs under homomorphism. First of all, we are introducing the concept of FRISs.

Definition 7 Let U_1 and V_1 be two universe of discourse, $f: U_1 \rightarrow V_1$ be its mapping from U_1 to V_1. Assume $R = \{A_1, A_2, ..., A_n\}$ is collection of fuzzy relations on U_1; then $f(R) = \{f(A_1), f(A_2), ..., f(A_n)\}$. We call the pair (U_1, R) as FRISs, and the corresponding $(V_1, f(R))$ lead to its induced FRISs.

Numerous studies have attempted to explain homomorphism between information systems [2, 4, 7–10, 12, 13]. Following Definition, classify consistent homomorphism and compatible homomorphism as in [2].

Definition 8 Let (U_1, R) be a FRISs. To classify homomorphism satisfying different conditions, the following assignation had been made.

(1) A mapping 'f' is known as consistent homomorphism, if $\forall A_i, A_j \in R$, mapping f is consistent to each A_i and A_j.
(2) A mapping 'f' is known as compatible homomorphism if f is compatible with respect to $\forall A_i \in R$.

Definition 9 Let (U_I, R) be FRISs and if S subset of R follows the following condition:

(1) $\cap S = \cap R$;

(2) $\forall R_i \in S, \cap S \subset \cap (S - \{R_i\})$.

Then S is called as reduct of R.

Theorem 10 verify Theorem 4.9 of [2] without the condition of consistent homomorphism.

Theorem 10 Let (U_I, R) be FRISs and 'f' is mapping from (U_I, R) to $(V_I, f(R))$. Let S be subset of R. Then, S will be reduct of R only when $f(S)$ is reduct of $f(R)$ in the absence of consistent homomorphism.

Proof $\Rightarrow$ Assume $f(S)$ is not a reduct of $f(R)$ and S is reduct of R,

Therefore, $\cap S \neq \cap (S - \{R_i\})$.

Then, there must be $m_1, m_2 \in U_1$,

such that $\cap S(m_1, m_2) < \cap (S - \{R_i\})(m_1, m_2)$, which implies

$$f(\cap (S - \{R_i\}))(f(m_1), f(m_2))$$
$$= owa_{n_1 \in f^{-1}f(m_1)}owa_{n_2 \in f^{-1}f(m_2)} \cap (S - \{R_i\}) (n_1, n_2)$$
$$> owa_{n_1 \in f^{-1}f(m_1)}owa_{n_2 \in f^{-1}f(m_2)} \cap S(n_1, n_2) \qquad (1)$$
$$= f(\cap S)(f(m_1), f(m_2))$$
$$= f(\cap R)(f(m_1), f(m_2))$$

Now, since S is a reduct of R, we have $\cap S = \cap R$. Hence $f(\cap S) = f(\cap R)$

By Theorem 5, $\cap f(S) = \cap f(R)$.

Since $f(S)$ is not reduct of $f(R)$,

Then $\exists R_i \in S$ such that $\cap (f(S) - f(R_i)) = \cap f(S)$.

Since, $f(S) - f(R_i) = f(S - \{R_i\})$. Therefore, $\cap f(S - \{R_i\}) = \cap f(S) = \cap f(R)$

Again, by Theorem 5,

$$f(\cap (S - R_i)) = f(\cap R) \qquad (2)$$

Comparing (1) and (2) we get contradiction to our assumption of $f(S)$ not being reduct of $f(R)$.

$\Leftarrow$ Let S is not a reduct of R and $f(S)$ is reduct of $f(R)$;

then, $\cap f(S) = \cap f(R)$.

Using theorem 5, $f(\cap S) = f(\cap R)$.

Assume that $\cap S \supseteq \cap R$,

Then $\exists m_1, m_2 \in U_1$ such that $\cap R(m_1, m_2) < \cap S(m_1, m_2)$

Now, $f(\cap S)(f(m_1), f(m_2)) = owa_{n_1 \in f^{-1}f(m_1)}owa_{n_2 \in f^{-1}f(m_2)} \cap S(n_1, n_2)$

$\qquad > owa_{n_1 \in f^{-1}f(m_1)}owa_{n_2 \in f^{-1}f(m_2)} \cap R(n_1, n_2) = f(\cap R)(f(m_1), f(m_2))$

This is a contradiction to our assumption. Therefore $\cap S = \cap R$.

Hence S is reduct of R.

Table 4 Original similarity relations

	Relation A_1				Relation A_2			
	m_1	m_2	m_3	m_4		m_1	m_2	m_3	m_4
m_1	0.7	0.9	0.8	1.0	m_1	0.5	0.7	0.6	0.4
m_2	0.5	0.4	0.6	0.7	m_2	0.3	0.7	0.8	0.4
m_3	0.1	0.2	0.5	0.4	m_3	0.9	0.8	0.6	0.5
m_4	0.3	0.8	0.7	0.8	m_4	0.2	0.3	0.2	0.6

Table 5 Original similarity relations

	Relation $A3$				Relation $(A1 \cap A2 \cap A3)$			
	m_1	m_2	m_3	m_4		m_1	m_2	m_3	m_4
m_1	0.6	0.7	0.6	0.5	m_1	0.5	0.7	0.6	0.4
m_2	0.4	0.5	0.6	0.5	m_2	0.3	0.4	0.6	0.4
m_3	0.3	0.7	0.5	0.4	m_3	0.1	0.2	0.5	0.4
m_4	0.2	0.8	0.4	0.7	m_4	0.1	0.3	0.2	0.6

Table 6 Image systems

	Relation $f(A1)$		Relation $f(A2)$		Relation $f(A3)$		Relation $f(A1) \cap f(A2) \cap f(A3)$	
	n_1	n_2		n_1	n_2		n_1	n_2		n_1	n_2
n_1	0.1	0.2	n_1	0.5	0.4	n_1	0.3	0.4	n_1	0.1	0.2
n_2	0.3	0.4	n_2	0.2	0.3	n_2	0.2	0.5	n_2	0.2	0.4

Example 11 Let $U_1 = \{m_1, m_2, m_3, m_4\}$, and $V_1 = \{n_1, n_2\}$ and $f: U_1 \to V_1$ is a mapping from U_1 to V_1 and is given as $f(m_1) = f(m_3) = n_1$; $f(m_2) = f(m_4) = n_2$.

A_1, A_2 and A_3 are fuzzy relations on U_1, which are given in Tables 4 and 5 respectively. $(A_1 \cap A_2 \cap A_3)$ is also given in Table 5.

Then, $f(A_1)$, $f(A_2)$ and $f(A_3)$ are computed and presented in Table 6.

In example 11, we see that $f(A_3)$ is superfluous in $f(R)$ if and only if A_3 is superfluous in R and that $\{f(A_1), f(A_2)\}$ are reduct of $f(R)$ when $\{A_2, A_3\}$ are reduct of R. Therefore, original system can be reduced, by reducing the image system and vice versa.

5 Conclusion and Future Work

We used ordered weighted averaging in place of supremum operator because OWA operator provides the flexibility for being lying between 'And' and 'Or' (that is between maximum and minimum values), by changing the weight vector values accordingly. The concept of consistent and compatible mapping had played a crucial role in homomorphism between fuzzy information systems. We proved that properties and attribute reduction of an Information system and its equivalent image system are identical to one another in terms of its attribute even when mapping is not compatible homomorphism or consistent homomorphism.

This research has relevance applications in many field like, decision making, rule extraction and equivalent attribute reduction, especially when information systems is not consistent or compatible homomorphism.

Future work includes the following:

1. The same OWA approach can be applied in dynamic FRISs.
2. OWA approach can also be used to compress dynamic information systems.

References

1. Pawlak, Z.: Information systems—theoretical foundations. Inform. Syst. **6**(3), 205–218 (1981)
2. Tsang, E.C.C., Wang, C.Z., Chen, D., Wu, C., Hu, Q.: Communication between information systems using fuzzy rough sets. IEEE Trans. Fuzzy Syst. **21**(3), 527–540 (2013)
3. Pedrycz, W., Vukovich, G.: Granular worlds: Representation and communication problems. Int. J. Intell. Syst. **15**, 1015–1026 (2000)
4. Grzymala-Busse, J.W., Sedelow, W.A.: On rough sets, and information system homomorphism. Bull. Pol. Acad. Sci. Tech. Sci. **36**(3–4), 233–239 (1988)
5. Wang, C.Z., Wu, C.X., Chen, D.: A Sytematic study on attribute reduction with rough set based on general binary relation. Inf. Sci. **178**(9), 2237–2261 (2008)
6. Li, D.Y., Ma, Y.C.: Invariant characters of information systems under some homomorphisms. Inform. Sci. **129**, 211–220 (2000)
7. Wang, C.Z., Wu, C., Chen, D., Hu, Q., Wu, C.: Communicating between information systems. Inf. Sci. **178**(16), 3228–3239 (2008)
8. Wang, C., Chen, D., Zhu, L.: Homomorphism between fuzzy information systems. Appl. Math. Letter **22**, 1045–1050 (2009)
9. Zhu, P., Wen, Q.Y.: Homomorphisms between fuzzy information systems revisited. Appl. Math. Lett. **24**(9), 1005–1026 (2011)
10. Wang, C.Z., Chen, D., Hu, Q.: Fuzzy information systems and their homomorphisms. Fuzzy Sets Syst. **249**, 128–138 (2014)
11. Zhu, P., Wen, Q.: Some improved results on communication between information systems. Inform. Sci. **180**(18), 3521–3531 (2010)
12. Lang, G., Li, Q., Guo, L.: Homomorphisms-based attribute reduction of dynamic fuzzy covering information systems. Int. J. Gen Syst **44**(7–8), 791–811 (2015)
13. Wang, C., Wu, C., Chen, D., Du, W.: Some properties of relation information systems under homomorphisms. Appl. Math. Letter. **21**, 940–945 (2008)
14. Yager, R.R.: On ordered weighted averaging aggregation operators in multicriteria decision making. IEEE Trans. Syst. Man, Cybern. **18**, 183–190 (1988)
15. Filev, D., Yager, R.R.: On the issue of obtaining OWA operator weights. Fuzzy Sets Syst. **94**(2), 157–169 (1998)

SDLAR: Self Adaptive Density-Based Location-Aware Routing in Mobile Adhoc Network

Priya Mishra, Saurabh Kumar Raina and Buddha Singh

Abstract Nowadays, location aware routing protocols are accepted potentially, scalable and efficient solution for routing in MANET. The advantages of these routing protocols are that they perform route discovery in a smaller region known as request zone, instead of doing route discovery in the entire network. This shows that the size and shape of the request zone play a major role to enhance the performance of routing procedure. Hence, the paper proposes an efficient scheme which focuses on creation and adjustment of size of request zone to find a stable path with less communication overhead. The protocol confines route discovery within an ellipse shaped request zone to reduce the routing overhead and achieve path stability. Further, the proposed protocol uses a density metric to resize the request zone for successful route discovery. Simulation results show that the proposed protocol can help to improve the path stability with lesser routing overhead than LAR-1.

Keywords MANET · SDLAR · LAR-1 · Request zone · Expected zone

1 Introduction

A mobile ad hoc network is a group of autonomous nodes which can change their topology dynamically. These nodes work as a host as well as routers in the network. Hence, the mobile nodes are the key elements to perform the routing functions. In multi-hop scenarios the nodes have to be dependent on each other and require routing schemes to forward messages. To route the messages in such environments

P. Mishra (✉) · S.K. Raina · B. Singh
Department of Computer Science and Information Technology, JIIT, Noida, India
e-mail: amipriya@gmail.com

S.K. Raina
e-mail: sk.raina@jiit.ac.in

B. Singh
e-mail: b.singh.jnu@gmail.com

© Springer Science+Business Media Singapore 2016   401
R.K. Choudhary et al. (eds.), *Advanced Computing and Communication Technologies*,
Advances in Intelligent Systems and Computing 452,
DOI 10.1007/978-981-10-1023-1_40

(multi-hop), routing protocols initiate the route discovery process by sending the route request packets in the network to make sure that the destination node will receive the route request and then reply. Flooding the request can be costly in terms of wasted bandwidth and can produce huge routing overhead. Therefore, the location based routing protocols are proposed as a novel solution to deal with these issues. The location aware routing protocols enhance the route discovery process by utilizing the location information of nodes and do route discovery in limited request (forwarding) zone to prevent blind flooding of RREQ in the entire network. In addition, by defining the limited request zone they also reduce a number of nodes which reduce the possibility of collisions as well frequency of re-route discovery. These protocols flood request only when they cannot find the path to destination in the request (forwarding) area. As we know the request zone is defined as a region for flooding therefore, the shape and size of the request zone can play a vital role to enhance the performance of routing protocol. If we define smaller request zones, it helps us to reduce the overhead but affect the other routing factors like successful packet delivery and delay. Due to narrow request zone there may be a possibility of holes inside the request (forwarding) region. On the contrary, larger sized request zones are good choice to enhance the possibility for successful route finding but they can increase the routing overhead. Therefore, to avoid these problems the shape and size of request zone should be defined carefully and try to make it optimal to enhance the successful route finding without increasing the routing overhead.

Therefore, the paper aims to propose a self adaptive location aware routing protocol, which focuses on creation and adjustment of request zone to find a stable path with less communication overhead. The contributions of this paper has three folds. First, we define an elliptical shaped request zone region which can be better choice to define a request zone in comparison with larger sized (rectangular) request zone as well as smaller request zone. Secondly, the protocol uses the concept of density to define the threshold value to adjust the request zone. Finally, this threshold value is used to resize the request zone. The simulation results reveal that SDLAR (proposed protocol) can be a better choice to find a stable path with less routing overhead.

The rest of this paper is organized as follows. The Sect. 2 discuss the few existing work which are more related to our proposed work. The SDLAR algorithm is presented in Sect. 3. The simulation details of proposed protocol and results comparisons with LAR-1 are given in Sect. 4. Finally, conclusion and future directions are discussed in Sect. 5.

2 Related Work

In this section, we discuss few existing location aware routing protocols which are more related to our work.

Location aided routing (LAR) protocol [1] had been proposed to reduce the routing overhead by limiting the flooding of routing request packets within a smaller area instead of entire network, known as request zone. The LAR protocol first presented the concept of request and expected zone. The smaller area used for flooding is known as request zone and the circular region around the destination as an expected zone. The authors used two schemes to decide the request zone in LAR named LAR-1 and LAR-2. The LAR-1 scheme defined a rectangular request zone which contains the location of source and the expected zone (circular region) around the destination. The route request flooding was performed within a defined rectangular request zone and the nodes inside the request zone could forward the RREQ packets to other nodes, while nodes reside outside the request zone ignore the requests. In LAR-2 scheme, the nodes forwarded the message to all nodes that are closer to the destination. Tzay and Hsu [2] have proposed a novel method by modifying the request zone area used in LAR-1. The scheme proposed a triangular shaped request zone that was dynamically adopted to reduce the traffic overhead. The protocol increased the request zone by progressive increasing search angle to avoid huge routing overhead and collision, when route discovery failed. Rango et al. [3] have proposed an another definition to defining the request zone in LAR scheme 1. The scheme defined request zone including only expected zone circle. The authors suggested a method to increase the request zone, when route discovery fails. The protocol performed better than rectangular request zone in terms of routing overhead. A concept of adaptive request zone [4] have been proposed when both the source and destination are mobile. The protocol focused to resize the request zone by using the difference of distance between source and destination and expected zone's radius. The protocol also adjusted the expected zone. A density-aware location-aided routing protocol (DALAR) [5] was proposed to improve the performance of the route discovery phase. The protocol forwarded the messages unconditionally but according to the number of its node density. A Distance-Based Location-Aided Routing (DBLAR) [6] for MANET had been proposed to reduce the routing overhead. By tracing the location information of destination nodes and referring to distance change between nodes to adjust route discovery dynamically, the proposed routing algorithm could avoid flooding in the whole networks. A location-based self-adaptive routing algorithm [7] had been proposed, which performed route discovery by flooding route request packets in a cylindrical request zone, to reduce the routing overhead. They suggested a self-adaptive algorithm to adjust the size of request zone by using Bayes' theorem. The protocol also helped to decrease broadcast storm problems associated with the MAC layer.

Modified Location-Aided Routing Protocols for Control Overhead Reduction in Mobile Ad Hoc Networks protocol [8] have been proposed an optimization method to improve the performance of basic LAR scheme. The scheme defined a limited request zone to reduce the route request overhead. The author compared the proposed protocol with basic LAR method especially in dense and highly dynamic ad hoc networks. The authors also discussed enlarging of request zone to reduce the frequency of route failures.

3 Proposed Protocol

As we know the shape and size of request zone plays a major role for efficient routing protocol. If we chose a larger (rectangular) request zone, it produces a huge routing overhead. On the other hand if we define a smaller (triangular shaped) request zone it can help to reduce the routing overhead but does not include enough nodes to find a stable route which leads to the failure of route discovery. Therefore, the proposed protocol focuses on the size of the proper request zone that can find a stable path with less routing overhead. In Fig. 1, node S is considered as a source node, initiates route search to find a path between source S to destination node D. Before starting the route discovery the SDLAR defines an ellipse shaped request zone, shown as the blue dotted line in Fig. 1. The coordinates of source, destination and distance between S and D is denoted by (X_s, Y_s), (X_d, Y_d) and d. The distance (d) is calculated as follows:

$$d = \sqrt{\left((X_s - X_d)^2\right) + ((Y_s - Y_d)^2)} \tag{1}$$

Further, to define an ellipse shaped request zone, we first determine major and minor axis of ellipse because they are the main parameters used for defining a size of routing request zone. We assign the value of major axis (a) as distance from source to destination. However, we should take care of parameter b since it has the most important role to size and resize of request zone. The example shown in Fig. 1

Fig. 1 Defining request zone in SDLAR

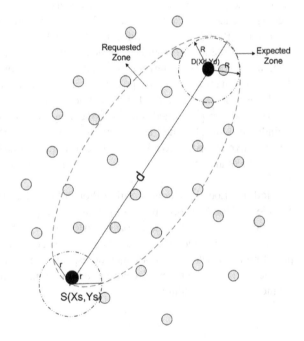

draws the request zone (ellipse) where the vertex of ellipse E is at the centre of source and the other side is covering the expected zone (circular region around destination). The centers of the ellipse are $((X_s + X_d)/2), (Y_s + Y_d)/2)$ and length of major axis is $(d + R)/2$. The request zone is calculated as the area of ellipse as follows:

$$\text{Request Zone (ellipse)} = \pi * \frac{d+R}{2} * r \tag{2}$$

where r is the transmission range of source, the radius of expected zone is R and d denotes the distance between source and destination. Once a Source defines request zone, it sends route request to other nodes. After receiving a request, intermediate nodes determine its membership to become a forwarding node by using own location information. If it is within the request zone, it may be a forwarding node. The node determines its membership by using Eq. (4) (for a predefined 'a' and 'b') as shown in Fig. 1.

$$\frac{(X_i - X_s)^2}{((d+R)/2)^2} + \frac{(Y_i - Y_s)^2}{r^2} \leq 1 \tag{3}$$

where (X_s, Y_s) are the coordinates of the sender node and (X_i, Y_i) are the coordinates of the intermediate nodes. Further, there is a possibility of unsuccessful route discovery in absence of path from source to destination in the defined request zone.

In this scenario, the request zone should be increased to include the sufficient number of nodes to establish the path hence the protocol increases the minor axis (b) to resize the request zone. The second scenario involves collisions which leads to produce a huge routing overhead so we need to decrease the minor axis (b) of the request zone. Hence, the SDLAR proposes self-adaptive method for the adjustment of request zone where the nodes automatically adjust the parameter (minor axis) 'b' of the request zone by node densities and the required number of nodes (N0) inside ellipse (E) are bounded by:

$$N0 = \left(\frac{\pi((d+R)/2)r}{\pi r^2} \right) \tag{4}$$

There are two initially defined thresholds for node density, Th_1 and Th_2, where $Th_1 > Th_2$. When the node density is greater than Th_1, the node senses surplus nodes in the request zone and when the node density is less than Th_2, means node does not have sufficient number of nodes to find a path in request zone. The value of Th_1 is set as N0 and Th_2 is set as 2. Further, we define the adjustment factor with the help of defined threshold values, which represents the small size about the required adjustment in the request zone. To calculate the adjustment step we compare the actual node density (AcNd) with the defined threshold values and if it does not lie between the threshold values. The adjustment step is calculated as follows:

$$\text{Adj\_step} = (\text{AcNd}/\text{N0}) \tag{5}$$

The adjustment step helps to resize the request zone for a successful route discovery with less routing overhead. The size of request zone is increased (or decreased) on increasing (or decreasing) the value of adjustment step. This process is repeated until the source gets route reply from the destination.

4 Performance Evaluation

In this section, we develop a simulation model in MATLAB 7.0 software to evaluate the performance of SDLAR in comparison with LAR-1.

4.1 Performance Metrics

We consider the following network performance metrics to validate our model.

4.1.1 Hop Count

The hop count is the essential metric to evaluate the path length between source and destination and measured as the number of nodes encounter to the path from source to destination.

4.1.2 Route Setup Time

The time taken to construct the path from source to destination is measured as route setup time.

4.1.3 Routing Overhead

The routing overhead is calculated as number of control packets sent by the protocol.

Table 1 Simulation parameters

Simulation parameters	Values
Topology size	1000 * 1000 m^2
Number of nodes	50, 100, 150, 200
Speed	10 m/s
Simulation time	100 s
Bandwidth	2 Mbps
Mobility model	Random way point
MAC layer protocol	IEEE 802.11
Radio propagation model	Two ray ground

4.2 Simulation Model

This section presents the experimental setup used in the simulation scenarios. In experiments, the nodes are uniformly distributed in a 1000 m × 1000 m square region and varied from 50 to 200. Each node has assigned a 200 m transmission range with channel capacity 2 Mbps. The nodes are moving with speed 10 m/s and pause time is taken as 30 s. The "Random Waypoint Mobility Model" is used as the mobility model and Two Ray Ground Model as radio propagation model. The 802.11b is used to simulate the MAC layer. The simulation parameters used in simulation are summarized in Table 1.

4.3 Simulation Results

This section presents the comparative study of results of proposed protocol with LAR-1 protocol. The distribution of routing overhead, route setup time and hop count for all the protocols for varying network size are shown in Figs. 2, 3 and 4. The simulation results in Fig. 2 shows that routing overhead increases linearly for all approaches on increasing the network size and proposed scheme produces less

Fig. 2 Routing overhead at speed 10 m/s

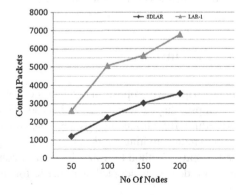

Fig. 3 Route setup time at speed 10 m/s

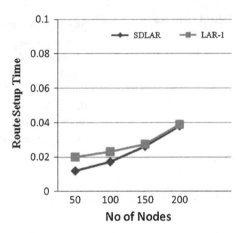

Fig. 4 Average hop count at speed 10 m/s

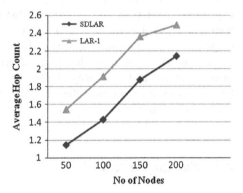

routing overhead in comparison to LAR-1. Figure 3 shows that the route setup time for both the protocols increase as network size increases. The proposed protocol produces lesser routing overhead than LAR-1 in small network size. The hops to reach a path between source and destination is shown in Fig. 4. The result shows that our proposed scheme takes less number of hops to set a shorter path than LAR-1 when speed is set as 10 m/s.

5 Conclusions

In this paper, we propose an efficient location aware routing protocol which defines an ellipse shaped request zone instead of rectangular zone for route discovery procedure. By introducing this new shaped request zone, the proposed protocol can help to reduce routing overhead and enhance path stability. The SDLAR suggests a self-adaptive method to resize the request zone for successful route discovery. The network density is adopted as a metric for resizing the request (forwarding) zone.

The simulation results show that SDLAR performs better in terms of control overhead, route setup time and hop count in comparison with existing LAR-1 routing protocol.

References

1. Ko, Y.B., Vaidya, N.H.: Location-aided routing (LAR) in mobile ad hoc networks. In: Proceedings of the 4th ACM/IEEE International Conference on Mobile Computing and Networking (MobilCom '98) (1998)
2. Shih, T.F., Yen, H.C.: Location-aware routing protocol with dynamic adaptation of request zone for mobile ad hoc networks. Springer Sci. Wireless Netw. (2008)
3. Kirdpipat, P., Thipchaksurat, S.: Impact of mobility on location-based routing with adaptive request zone in mobile adhoc networks. In: 9th International Conference on Electrical Engineering/Electronics, Computer, Telecommunications and Information Technology (ECTI-CON), 2012, pp. 1–4 (2012)
4. De Rango, F., Iera, A., Molinaro, A., Marano, S.: A modified location-aided routing protocol for the reduction of control overhead in ad-hoc wireless networks. In: 10th International Conference on Telecommunications, 2003. ICT 2003, vol. 2, pp. 1033–1037. IEEE (2003)
5. Huang, S.C., Chang, H.Y.: A density-aware location-aided routing protocol for wireless ad-hoc networks. In: Tenth International Conference on Intelligent Information Hiding and Multimedia Signal Processing (IIH-MSP), pp. 670–673. IEEE (2014)
6. Wang, K., Wu, M.: DBLAR: a distance-based location-aided routing for MANET. J. Electron. 26(2), 152–160 (2009)
7. Li, X., Hong, S., Fang, K.: Location-based self-adaptive routing algorithm for wireless sensor networks in home automation. EURASIP J. Embed. Syst. 2011(1), 484690 (2011)
8. Senouci, S.M., Rasheed, T.M.: Modified location-aided routing protocols for control overhead reduction in mobile ad hoc networks. In: Network Control and Engineering for QoS, Security and Mobility, IV, pp. 137–146. Springer US (2007)

Version Control for Authentication in Distributed Reprogramming in Wireless Sensor Networks

Tejal M. Shinde and Reeta Chhatani

Abstract Reprogramming is a technique used in wireless sensor networks (WSN) to update the software part of sensor nodes from remote locations over wireless links during run time of the network. However, reprogramming needs to be integrated with secure and efficient authentication technique to avoid any unauthorized access to sensor nodes. This paper gives an efficient authentication technique for distributed reprogramming using version numbers from version control operation of reprogramming. The Existing version control technique is modified to achieve both authentication and version control in reprogramming. The proposed method is compared with the two other algorithms defined for authentication of users in (Das, Two-factor user authentication in wireless sensor networks. IEEE Trans Wireless Commun 8:1086–1090, 2009) [3] and (He et al, SDRP: a secure and distributed reprogramming protocol for wireless sensor networks. IEEE Trans Ind Electron 59:4155–4163, 2012) [5]. The comparative analysis shows version control method is more efficient and is robust to impersonation attack.

Keywords Reprogramming · Wireless sensor networks · Authentication · Distributed reprogramming · Version control

1 Introduction

Wireless sensor networks are deployed in hazardous environments. Sensor nodes are intended to work for a long time without human intervention which makes manual management and maintenance of nodes difficult. The Reprogramming

T.M. Shinde (✉) · R. Chhatani
Sardar Patel Institute of Technology, Mumbai, Maharashtra, India
e-mail: tejalshinde29@gmail.com

R. Chhatani
e-mail: reeta_gaokar@spit.ac.in

© Springer Science+Business Media Singapore 2016   411
R.K. Choudhary et al. (eds.), *Advanced Computing and Communication Technologies*,
Advances in Intelligent Systems and Computing 452,
DOI 10.1007/978-981-10-1023-1_41

technique given in [1] helps to overcome these problems in many ways by sending software updates to the sensor node over wireless link. This removes the need to approach sensor nodes for their maintenance and provides an easy way to control functions of multiple nodes simultaneously from a remote location. The reprogramming has two approaches called centralized reprogramming and distributed reprogramming. In a centralized approach only the network administrator (NA) has an authority to update software part of nodes while in distributed approach users authorized by network administrator also have an access to update sensor nodes. Distributed reprogramming adds flexibility to network operation as multiple users can access nodes simultaneously, but it also demands proper authentication of users for secure operation. As given in [2] WSN is prone to many security attacks because of its resource constrained nature. Reprogramming can further introduce additional security threats in network operation as any adversary can upload malicious data on sensor node or change functional parameters of sensor nodes by impersonating users. To avoid this sensor node should accept data only from an authorized user. Many well defined authentication algorithms are available, like public key cryptography (PKC), certificate based authentication (CBA), digital signature, Identity based encryption (IBE), Identity based signature (IBS). But many of these methods cannot be used for WSN because they demand high computations and consumes resources. For authentication of users some methods are described in [3–5]. The method given in [3] called two factor authentication performs authentication based on smart card and password. It involves multiple computations and message exchanges. In [4] the two factor method given by Das is improved by adding random secret integers at certain phases. Another method described in [5] uses identity based signature algorithm for distributed reprogramming, but it includes expensive bilinear mapping, merkle hash tree root computations and multiple key computations which makes it less efficient for WSN. Further, both the techniques are vulnerable to impersonation attack.

We propose a technique for authentication of users using version numbers for distributed reprogramming, which requires only one simple comparison for authentication and it does not add extra overheads. Version numbers are used in the version control function of reprogramming given in [1]. Version control tracks software updates with the help of increasing numbers and is well described in [6]. In the proposed scheme version control technique is modified and adopted to achieve authentication in distributed reprogramming. In this paper, we analyze and compare proposed version control technique of authentication with two factor method [3] and IBS method [5] on the basis of security against impersonation attack, complexity, computations, overheads and time consumed. Security analysis shows that version number technique is robust to impersonation attack and is highly efficient for WSN.

2　Related Works

Some techniques have been developed for the authentication of users in WSN. The scheme given in [3] uses the two factor authentication scheme. In this scheme sensor node authenticate user on the basis of a password and a smart card submitted by the user to access data from the node. The gateway node calculates smart card parameters and gives a secret random number to the sensor node for authenticating the user. The secret random number is known to the user and the sensor node only. Later this technique was found to be vulnerable to many security attacks. Authentication methods are discussed in [4, 7] to give security improvement over these vulnerabilities.

In [5] the identity based signature method is used for the authentication of user in distributed reprogramming. The signature and all the keys required for authentication process are calculated by the network administrator and distributed to the user and sensor nodes. To reprogram a sensor node, user calculates and sends a signature to designated node. Sensor node verifies signature using bilinear mapping using a random secret integer as a master secret key which is a known to network administrator, users and sensor nodes. In [8] an inherent security weakness present in [5] is discussed and a mathematical change is suggested to overcome it.

In both the technique given above random secret number is used which is made available at authorized entities at the time of deployment and it remains same throughout the operation period of network or up to next deployment phase. If the adversary predicts this random secret number correctly, it can easily launch an impersonation attack on WSN. Predicting a secret integer is not that difficult as the same integer used throughout the operation of WSN. The same problem is present in the methods given in [4, 7]. This paper describes a technique to use a new random secret number for each reprogramming attempt. The table of random numbers is defined and is loaded at nodes and users at the time of deployment. These numbers are treated as version numbers. Each version number has a sequence number. A counter is used to keep track on sequence of version numbers. Many other techniques given in [9] use well defined authentication methods in WSN but requires more computations and add extra overheads. The proposed version control technique performs authentication without adding any extra overhead with one simple comparison.

3　Version Control

Version control is a technique used in software management. As any software system grows many new objects get developed in it like files, source codes, bitmap graphics, executable codes, etc. It is necessary in software engineering to keep track of changes in these objects and maintain dependencies in old and new versions. In software management maintaining dependencies of system is a challenging job.

As the system grows, it becomes complicated to recognize and mark dependencies. For this version number are used in software engineering to maintain dependencies. Numbers are assigned in increasing order to new developments. Different version number can be assigned to different types of changes or changes in different types of files.

Version control is an important function in network reprogramming [1]. In network reprogramming entity receiving a new code updates accepts it only when its version number is greater than the version number of previously received updates. This automatically maintains dependencies among code images and helps in maintaining freshness of code. But in our proposed method version control is modified to achieve authentication preserving its basic objective of maintaining dependencies.

4 Authentication Using Version Control

In this method the original version control technique is modified to serve the purpose of authentication of user in Distributed Reprogramming. For this the proposed scheme uses version number tables along with counters for authentication.

4.1 Version Number Table

The version number table is created by NA and distributed to users and sensor node at the time of deployment as shown in Fig. 1. The table contains version numbers each associated with a sequence number. Version numbers used in this method are randomly selected and are in random order while in original version control method version number are always in increasing order. In the table, sequence numbers associated with version numbers are in increasing order. The table is a combination of version numbers of variable size and variable types in random order. A single scheme for selecting version numbers is not preferred to make these numbers totally unpredictable. However, in proposed method version numbers sent along with updates are not meant to track on software changes as they are in random orders and serve for the user authentication only. But the sequence number associated with the version number tracks software updates sent by the user.

4.2 Counter

A counter is set at both user and sensor node at the time of deployment. Both the counters are initiated to same value initially. The counter value tracks the sequence number in version number table and also represents the number of the ongoing

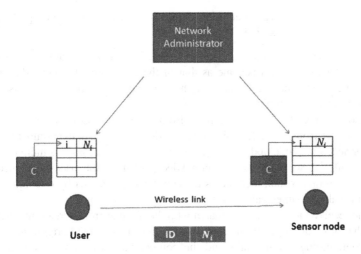

Fig. 1 Authentication using version control

reprogramming attempt. After each successful reprogramming attempt counter value at both the user and the sensor node are incremented by one unit and hence before and after reprogramming counter values at both sensor node and user should be the same.

4.3 Procedures

The proposed method uses version tables and counters to achieve authentication of user in distributed reprogramming. The network scenario for distributed reprogramming involves three entities which are NA, user and sensor node. The user authentication with version control is described according to certain operations performed by each entity during reprogramming. Figure 1 shows authentication operation using version control.

NA completes prerequisites required for authentication as follows

1. NA generates version tables having version number N_i and sequence numbers i as shown in Fig. 1 and distribute same table to sensor node and user.
2. NA initializes counters C at sensor node and user to same value.

When user wants to send updates to sensor node, it sends request to the corresponding sensor node. For that it performs following operations

1. User reads the counter value i.
2. User reads i^{th} version number N_i from the version number table and sends the reprogramming request having version number and identity of user (ID) to the sensor node as shown in Fig. 1.

After receiving the reprogramming request from user sensor node performs the verification operation as given below

1. After receiving the request sensor node first reads counter value. It is assumed that its counter value i is same as that of the user's counter value during each reprogramming attempt. Counters at the user and the sensor node are assumed to be perfectly synchronized.
2. Sensor node reads i^{th} version number from table and accepts updates only when the number from the table of the sensor node matches with the number received. After accepting code updates, the counter value at sensor node is incremented by one unit. Sensor node sends acknowledgement to the user after successful reprogramming which increments the counter value at user by one unit and this again makes counter values at sensor node and user equal.
3. If the received number do not match with the number from version number table at the time of verification, sensor node rejects the reprogramming request, reprogramming attempt fails and the counter values at sensor node and user remains same.

This method gives a very simple approach for authentication of users. In this method during each reprogramming attempt counter value gives the sequence number as shown in Fig. 1 and sequence number in turn represents the number of the reprogramming attempt going on i.e. the sequence number i represents i^{th} reprogramming attempt. It also represents i^{th} reprogramming patch sent by the user. Thereby a sensor node gets the link between software patches sent by user to the sensor node. This helps to achieve version control along with authentication.

The key point of this method is to use a new version number for each reprogramming attempt keeping sequence number secret. If adversary eavesdrop version number, it is of no use for adversary as dropped number is already being used by the user and sensor nodes will not accept repeated number.

5 Security and Performance Analysis

The proposed method to achieve authentication using version numbers is a simple and efficient method for WSN. It gives a simple way to avoid an impersonation attack.

5.1 Resistance to Impersonation Attack

Impersonation attack is always launched after adversary eavesdrops or predicts authentication keys. In the proposed method version numbers are used as authentication keys. But it is not easy for any adversary to predict a version number sent in reprogramming request because of following reasons

- version numbers used as an authentication key are random numbers
- sequence numbers are never sent over wireless links
- type and size of version numbers are variable

This increases the unpredictability of numbers.

In the proposed authentication method even though an adversary eavesdrop version number used for the ongoing reprogramming attempt it is of no use to adversary as it is already being used and the same number will not be accepted later by the sensor node.

Authentication methods used in [3, 4] also uses secret numbers in key calculations. But these numbers remain same throughout the operation of the network. This can make the whole system vulnerable to security attacks if these numbers once get predicted by an adversary. In version control method secret numbers are changed for every new reprogramming attempt and previously used version numbers are not used again. Once all the version numbers are used, the table is dropped and new version number table is deployed by the network administrator at sensor nodes and user.

5.2 Version Control as an Extra Authentication Level

In [5] the protocol designed for distributed reprogramming uses IBS for authentication, but still vulnerable to impersonation attack as given in [8]. If an extra verification level is added using version control in work given in [5] after performing all the IBS verification at sensor node, it can avoid any attempt made for the malicious data upload.

5.3 Complexity and Computations

The two-factor algorithm given in [3] uses hash calculation, bit-wise xor operation and concatenation at each stage. It requires multiple message exchanges among users, sensor node and gateway nodes to accomplish authentication of the user. This increases complexity and adds extra transmission overheads. Similarly the protocol defined for distributed reprogramming in [5] uses complicated bilinear mapping for verification at sensor node. It also multiple includes hash calculations, merkle hash tree, bit-wise concatenation operation and hence demands good computational capabilities. On the other hand the version control when used for authentication requires only one comparison for verification. This reduces computations at sensor node. This is highly desirable for sensor nodes as they are expected to be cheap. The version control authentication technique can be used as an additional verification level in many security protocols. In version control method no extra overheads are added as version number are already present in reprogramming

Table 1 Time consumed by each method for authentication operation

IBS	Two factor	Version control
2213 ms	1489 ms	665 ms

packets. In this way the version number technique reduces complexity, computations and overhead.

5.4 Simulation and Results

The authentication methods in [3, 5] and version control method are implemented in Ns-2 and comparative analysis is done for time consumed by each method to perform authentication. The results are listed in Table 1.

As discussed in 5.3 the version control technique reduces complexity, computation and transmission overheads. This results in reduced time to complete operation as compared to methods given in [3, 5].

6 Conclusion

The proposed method for authentication of users in distributed reprogramming is very simple and gives better solution for impersonation attack. It utilizes one of the function present in reprogramming to achieve authentication and hence do not require any extra message exchanges. This method requires less computation and has least complexity which makes it more efficient and desirable for resource constrained WSN. However it is a naive approach of authentication using new key at each reprogramming access with the help of number tables and counters and can be developed further to make it robust. Counters used to keep track on version numbers are assumed to be perfectly synchronized. Hence there is a scope to add some methods for synchronization of counters. The confidentiality can be further improved by increasing the unpredictability of version numbers used for authentication purpose. As security depends on many parameters this system cannot claimed to be perfectly secure but is simple and efficient for WSN.

References

1. Wang, Q., Zhu, Y., Cheng, L.: Reprogramming wireless sensor networks: challenges and approaches. IEEE Netw. **20**, 49–55 (2006)
2. Perige, A., Stankovic, J., Wagner, D.: Security in wireless sensor networks. Commun. of the ACM **47**, 53–57 (2004)

3. Das, M.: Two-factor User Authentication in Wireless sensor networks. IEEE Trans. Wireless Commun. **8**, 1086–1090 (2009)
4. He, D., Gao, Y., Chan, S., Chen, C., Bu J.: An enhanced two-factor user authentication scheme in wireless sensor networks. Ad Hoc Sensor Wireless Netw. **0**, 1–11 (2010)
5. He, D., Chen, C., Chan, S., Bu, J.: SDRP: a secure and distributed reprogramming protocol for wireless sensor networks. IEEE Trans. Ind. Electron. **59**, 4155–4163 (2012)
6. Agrawal, B.B., Tayal, S.P., Gupta, M.: Software Engineering and Testing. Jones and Bartlett publishers, Massachusette (2010)
7. Chen, L., Wei, F., Mu, C.: A secure user authentication scheme against smart-card loss for wireless sensor networks using symmetric key technique. Int. J. Distrib. Sens. Netw. **2015** (2015)
8. He, D., Chen, C., Chan, S., Bu, J., Yang, L.T.: Security analysis and improvement of a secure and distributed reprogramming protocol for wireless sensor networks. IEEE Trans. Ind. Electron. **60**, 5348–5353 (2013)
9. Patil, S., Kumar, V., Singha, S., Jamil R.: A survey on authentication techniques for wireless sensor networks. Int. J. Appl. Eng. Res. **7** (2012)

Improving the Detection Accuracy of Unknown Malware by Partitioning the Executables in Groups

Ashu Sharma, Sanjay K. Sahay and Abhishek Kumar

Abstract Detection of unknown malware with high accuracy is always a challenging task. Therefore, in this paper we study the classification of unknown malware by two methods. In the first/regular method, similar to other authors (Mehdi et al. Proceedings of the 11th Annual conference on Genetic and evolutionary computation 2009, Moskovitch et al. Intelligence and Security Informatics 2008, Ravi and Manoharan Int J Comput Appl 43(17):12–16 2012) approaches we select the features by taking all dataset in one group and in second method, we select the features by partitioning the dataset in the range of file 5 KB size. We find that the second method detect the malwares with ∼8.7 % more accurate than the first/regular method.

Keywords Malware detection · Computer security · Machine learning · Naive Bayes · Profiling

1 Introduction

Malwares are continuously evolving and are big threats to the leading Windows and Android computing platforms [1]. The attacks/threats are not only limited to individual level, but there are state sponsored highly skilled hackers developing customized malwares [2–4]. These malwares are generally classified as a first and second-generation malwares. In first generation, structure of the malwares remains constant, while in second generation, structure changes in every new variant,

A. Sharma (✉) · S.K. Sahay · A. Kumar
Department of Computer Science and Information System, BITS Pilani,
K. K. Birla Goa Campus, Zuarinagar 403726, Goa, India
e-mail: p2012011@goa.bits-pilani.ac.in

S.K. Sahay
e-mail: ssahay@goa.bits-pilani.ac.in

A. Kumar
e-mail: f2010490@goa.bits-pilani.ac.in

© Springer Science+Business Media Singapore 2016   421
R.K. Choudhary et al. (eds.), *Advanced Computing and Communication Technologies*,
Advances in Intelligent Systems and Computing 452,
DOI 10.1007/978-981-10-1023-1_42

keeping the action same [5]. On the basis of how variants are generated in malware, second generation malwares are further classified into Encrypted, Oligomorphic, Polymorphic and Metamorphic Malwares [6].

Its an indisputable fact that the prolong traditional approach (signature matching) of combating the threats/attack with a technology-centric are ineffective to detect second generation customized malwares. If in time adequate measures has not taken, the consequence of the scale (more 317 million new malwares are reported [7] in the year 2014) at which malware are developed will be very devastating. Nevertheless, the second-generation malware are very effective and not easy to detect. Recently a new malware is reported by McFee which is capable to infect the hard drives and solid state storage device (SSD) firmware and the infection cannot be removed even by formatting the devices or reinstallation of operating systems [8]. Therefore, there is a need that both academia and anti-malwares developers should continually work to combat the threats/attacks from the evolving malwares. The most popular techniques used for the detection of malwares are signature matching, heuristics based detection, malware normalization, machine learning, etc. [6].

In recent years, machine learning techniques are studied by many authors and proposed different approaches [9–11], which can supplement traditional anti-malware system (signature matching).

Hence, in this paper for detection of unknown malware with high accuracy, we present a static malware analysis method which detect the unknown malware with ~8.7 % more accurate then the regular method. The paper is organized as follow, in next section related work is discussed. In Sect. 3 we discuss the data preprocessing and feature selection. Section 4 contains the brief description of Naive Bayes classifier. In Sect. 5 we discuss the method to improve the detection accuracy and results. Finally Sect. 6 contains the conclusion and future direction of the paper.

2 Related Works

To combat the threats/attacks from the second generation malwares, Schultz et al. [12] was the first to introduce the concept of data mining to classify the malwares [12]. In 2005 Karim et al. [13] addressed the tracking of malware evolution based on opcode sequences and permutations. In 2006, Henchiri and Japkowicz [14] reported 37.17 % detection accuracy from a hierarchical feature extraction algorithm by using NB (Naive Bayes) classifier. Bilar [15] uses small dataset to examine the opcode frequency distribution difference between malicious and benign code [15] and observed that some opcodes seen to be a stronger predictor of frequency variation. In 2008, Ye et al. [16] applied association rules on API execution sequences and reported an accuracy of 83.86 % with NB classifier. In 2008, Tian et al. [17] classified the Trojan using function length frequency and shown that the function length along with its frequency is significant in identifying malware family and can be combined with other features for fast and scalable malware classification. Moskovitch et al. [18] studied many classifier viz. NB, BNB, SVM, BDT, DT and ANN by byte-sequence n-grams (3, 4, 5 or 6) and find that NB classifier detect the malwares with 84.53 % accuracy [18]. In 2009 Tabish [19] used 13 different statical features computed on 1, 2, 3 and 4 g by analysing byte-level file content for classification of malwares. In 2009 Syed Bilal Mehdi et al. [20] obtained 64 and 58 % accuracy by NB classifier with good evaluator feature selection scheme on 4 and 6 g features from the executables. Ravi et al. in year 2012 reported 48.69 % accuracy with NB classifier by using 4-grams Windows API calls features [21]. Liangboonprakong et al. [22] proposed a classification of malware families based on N-grams sequential pattern features [22]. They used DT, ANN and SVM classifier and obtained good accuracy. In 2013 Santos et al. [23] used Term Frequency for modelling different classifiers and among the studied classifier, SVM outperform for one opcode and two opcode sequence length respectively. Recently (2014) Salehi et al. construct feature set by extracting API calls used in the executables for the classification of malwares [24].

3 Data Preprocessing and Feature Selection

For the experimental analysis, we downloaded 11088 malwares from malacia-project [25] and collected 4006 benign programs (also verified from virustotal.com [26]) from different systems. In the collected dataset we found that 97.18 % malwares are below 500 KB, (Fig. 1) hence for the classification we took the data set which are below 500 KB.

For classification the features are opcodes of executables obtained by objdump utility available in the linux system and its selection procedure is given in the

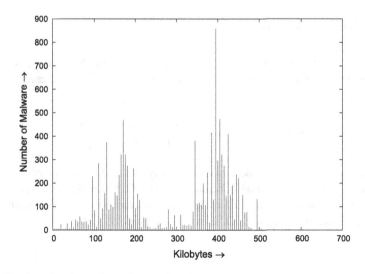

Fig. 1 Number of malwares in the group of 5 KB size

Algorithm 1. To test our methods we select popular Naive Bayes classifier which can handle an arbitrary number of independent variables and is briefly described in next section.

Algorithm 1: Feature Selection

INPUT: Pre-processed data
N_b: Number of benign executables, N_m: Number of malware executables, **n**: Number of features required
OUTPUT: List of features
 BEGIN
 for all benign data **do**
 Add all frequency f_i of each opcode **o** and Normalize them with respect to N_b

$$F_b(o_j) = (\sum f_i(o_j))/N_b$$

 end for
 for all malware data **do**
 Add all frequency f_i of each opcode **o** and Normalize them with respect to N_m

$$F_m(o_j) = (\sum f_i(o_j))/N_m$$

 end for
 for all opcode o_j **do**
 Find the difference of each opcode normalized frequency $D(o_j)$.

$$D(o_j) = |F_b(o_j) - F_m(o_j)|$$

 end for
 return **n** number of opcodes with highest $D(o)$.

---

4 Naive Bayes Classifier

Given a set of features (opcodes), $O = o_1, o_2, o_3, \ldots, o_n$, the Naive Bayes classifier gives the posterior probability for the class C (malware/benign) and can be written as:

$$P(C|o_1, o_2, \ldots, o_n) = \frac{P(C)P(o_1, o_2, \ldots, o_n|C)}{P(o_1, o_2, \ldots, o_n)} \tag{1}$$

where $P(C|o_1, o_2, \ldots, o_n)$ is posterior probability of the class membership, i.e., the probability that a test executable belongs to class C, Since Naive Bayes assumes that the conditional probabilities of independent variables are statistically independent, we can decompose the likelihood to a product of terms:

$$P(o_1, o_2, \ldots, o_n | C) = \prod_{i=1}^{n} P(o_i | C) \tag{2}$$

Here, $P(o_i | C)$ is the probable similarity of occurrence of feature o_i of the class C and can be computed by the equation:

$$P(o_i | C) = \frac{1}{\sqrt{2\pi\sigma_C^2}} e^{-\frac{(o-\mu_C)^2}{2\sigma_C^2}} \tag{3}$$

where o denotes the feature o_i opcode of the test executable and μ_C, σ_C are the mean and variance of the class C.

From the above, the final classification can be done by comparing the posterior probability between both class models, if malware class posterior probability of test executable is high then it is classified to malware else classified as benign.

5 Method to Improve the Detection Accuracy of Unknown Malware

We first study the regular method and then propose our novel partitioned method i.e. grouping the executables, to improve the detection accuracy of unknown malware. Both method are depicted in the Figs. 2 and 3.

5.1 Regular Method

Figure 2 represents the regular method for finding the features for the classification of unknown malwares. In this method, for the classification, we train the classifier by 10558 malwares and 2454 benign executables. For the purpose, the procedure to obtain the features of the dataset is given in Algorithm 1.

We study opcode occurrence of the collected dataset and found (Figs. 4 and 5) that malware opcodes distribution and/or occurrence differ significantly from benign program. Hence, we obtained the promising features by computing the difference of normalized opcodes frequency between benign and malware executables.

We use the machine learning tool WEKA to train and test the chosen (NB) classifier. For the testing 750 malware 610 benign executables are taken from the dataset which are not used for training the classifier. To evaluate classifiers capability, we measure detection accuracy, i.e., the total number of the classifier's hits divided by the number of executables in the whole dataset and is given as:

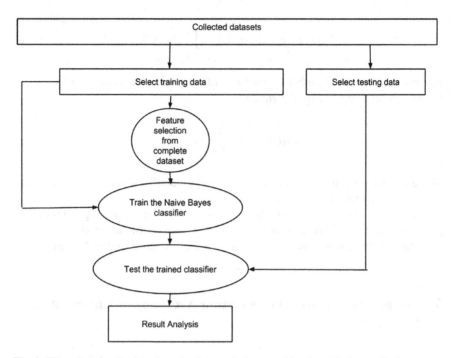

Fig. 2 Flow chart for the detection of unknown malwares without partitioning method

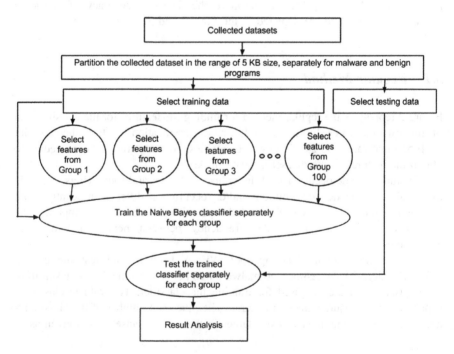

Fig. 3 Flow chart for the detection of unknown malwares by partitioning the executables in groups

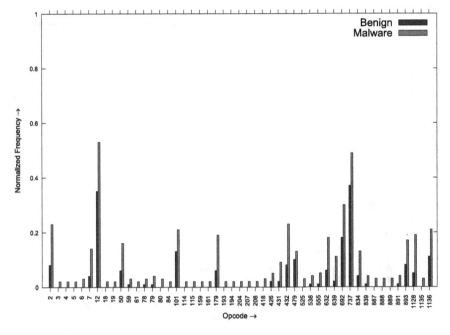

Fig. 4 Opcodes which found more in malwares

$$Accuracy(\%) = \frac{TP + TN}{TM + TB} \times 100 \tag{4}$$

where,

TP   True positive, the number of malwares correctly classified
TN   True negative, the number of benign correctly classified
TM   Total number of malwares
TB   Total number of benign

The accuracy obtained by this method by increasing the features in systematic way are shown in Fig. 6. We find that the accuracy of the classifier is almost flat, if the number of features are greater than 90. The best accuracy obtained by this method is 78.33 %.

5.2   Partitioning Method

In this method as shown in Fig. 3, we partitioned the collected dataset in 5 KB size range. The partition size is based on the study that size of any two malwares

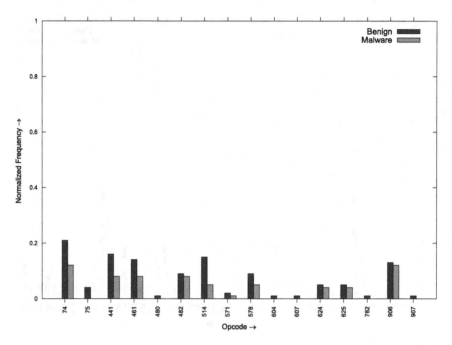

Fig. 5 Opcodes which found more in benign programs

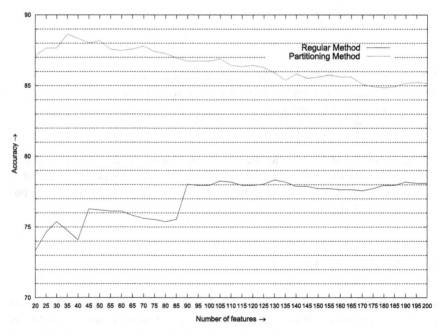

Fig. 6 Detection accuracy obtained by both the methods

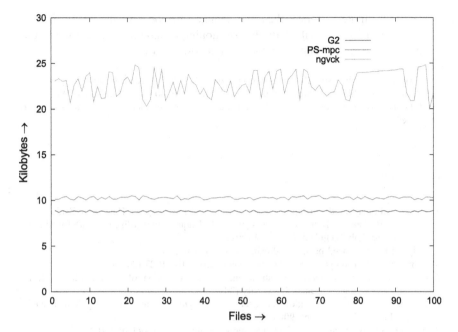

Fig. 7 Fluctuations in the size of malwares generated by G2, PS-MPC and NGVCK

generated by G2 [27], PS-MPC [28] and NGVCK [29] kits does not vary by more than 5 KB size (Fig. 7).

For the comparative experimental analysis we took same classifier, training and testing data which are used in the regular method. In this method procedure of feature selection is same as given in Algorithm 1. However, to improve the accuracy of detection we first listed the features of each group given by Algorithm 1, then serially trained and tested the chosen classifier by systematically increasing the features. We found that the accuracy obtained by this method out-performed the regular method (Fig. 6) and the best accuracy obtained is 87.02 % i.e. the detection accuracy is ~8.7 % more compared to regular method.

6 Conclusion

In this paper we present a novel method to improve the detection accuracy of unknown malwares. We first investigated the variation in the size of malwares generated by G2, PS-MPC and NGVCK and found that the generated malware size vary from 0 to 5 KB. Therefore we partitioned the collected dataset in 100 groups, each in 5 KB range of size and then selected the feature from each group to train classifier for the detection of unknown malwares. We found that, if features are selected by partitioning dataset in the range of 5 KB then the malwares are detected with ∼ 8.7 % more accurate then the regular method. As the result is very promising, in future, we will study the partitioning method in-depth for the further improvement in the detection accuracy of unknown malwares.

Acknowledgment Mr. Ashu Sharma is thankful to BITS, Pilani, K.K. Birla Goa Campus for the support to carry out his work through Ph.D. scholarship No. Ph603226/Jul. 2012/01. We are also thankful to IUCAA, Pune for providing hospitality and computation facility where part of the work was carried out.

References

1. Quick heal quarterly threat report q2 2015. Technical report, Quick Heal (Feb 2015)
2. Bencsath, B., Pek, G., Buttyan, L., Felegyhazi, M.: Duqu: a stuxnet-like malware found in the wild. CrySyS Lab Technical Report 14 (2011)
3. Daly, M.K.: Advanced persistent threat. Usenix, 4 Nov 2009
4. Stone, R.: A call to cyber arms. Science **339**(6123), 1026–1027 (2013)
5. Govindaraju, A.: Exhaustive statistical analysis for detection of metamorphic malware. Master's thesis, San Jose State University (2010)
6. Sharma, A., Sahay, S.: Evolution and detection of polymorphic and metamorphic malwares: a survey. Int. J. Comput. Appl. **90**(2), 7–11 (2014)
7. Internet security threat report. Technical report, Symantec Corporation (2015)
8. Mcafee labs threats report. Technical report, McAfee Labs (May 2015)
9. Allix, K., Bissyande, T.F., Jerome, Q., Klein, J., Le Traon, Y., et al.: Large-scale machine learning-based malware detection: confronting the 10-fold cross validation scheme with reality. In: Proceedings of the 4th ACM Conference on Data and Application Security and Privacy, pp. 163–166, ACM (2014)
10. Gandotra, E., Bansal, D., Sofat, S.: Malware analysis and classification: a survey. J. Inf. Secur. 2014 (2014)
11. Nath, H.V., Mehtre, B.M.: Static malware analysis using machine learning methods. In: Recent Trends in Computer Networks and Distributed Systems Security, pp. 440–450. Springer (2014)
12. Schultz, M.G., Eskin, E., Zadok, E., Stolfo, S.J.: Data mining methods for detection of new malicious executables. In: 2001 IEEE Symposium on Security and Privacy, pp. 38–49, 2001. S&P 2001. Proceedings. IEEE (2001)
13. Karim, M.E., Walenstein, A., Lakhotia, A., Parida, L.: Malware phylogeny generation using permutations of code. J. Comput. Virol. **1**(1–2), 13–23 (2005)
14. Henchiri, O., Japkowicz, N.: A feature selection and evaluation scheme for computer virus detection. In: Sixth International Conference on Data Mining, pp. 891–895, 2006. ICDM'06. IEEE (2006)
15. Bilar, D.: Opcodes as predictor for malware. Int. J. Electron. Secur. Digit. Forensics **1**(2), 156–168 (2007)
16. Ye, Y., Wang, D., Li, T., Ye, D., Jiang, Q.: An intelligent pe-malware detection system based on association mining. J. Comput. Virol. **4**(4), 323–334 (2008)
17. Tian, R., Batten, L.M., Versteeg, S.: Function length as a tool for malware classification. In: 3rd International Conference on Malicious and Unwanted Software, pp. 69–76, 2008. MALWARE 2008. IEEE (2008)
18. Moskovitch, R., Feher, C., Tzachar, N., Berger, E., Gitelman, M., Dolev, S., Elovici, Y.: Unknown malcode detection using opcode representation. In: Intelligence and Security Informatics, pp. 204–215. Springer (2008)

19. Tabish, S.M., Shafiq, M.Z., Farooq, M.: Malware detection using statistical analysis of byte-level file content. In: Proceedings of the ACM SIGKDD Workshop on Cyber Security and Intelligence Informatics, pp. 23–31. ACM (2009)
20. Mehdi, S.B., Tanwani, A.K., Farooq, M.: Imad: in-execution malware analysis and detection. In: Proceedings of the 11th Annual conference on Genetic and evolutionary computation, pp. 1553–1560. ACM (2009)
21. Ravi, C., Manoharan, R.: Malware detection using windows api sequence and machine learning. Int. J. Comput. Appl. **43**(17), 12–16 (2012)
22. Liangboonprakong, C., Sornil, O.: Classification of malware families based on n-grams sequential pattern features. In: 8th IEEE Conference on Industrial Electronics and Applications (ICIEA), pp. 777–782, 2013. IEEE (2013)
23. Santos, I., Brezo, F., Ugarte-Pedrero, X., Bringas, P.G.: Opcode sequences as representation of executables for data-mining-based unknown malware detection. Inf. Sci. **231**, 64–82 (2013)
24. Salehi, Z., Sami, A., Ghiasi, M.: Using feature generation from api calls for malware detection. Comput. Fraud Secur. **2014**(9), 9–18 (2014)
25. Nappa, A., Rafique, M.Z., Caballero, J.: Driving in the cloud: an analysis of drive-by download operations and abuse reporting. In: Detection of Intrusions and Malware, and Vulnerability Assessment. Springer, pp. 1–20 (2013)
26. Canto, J., Dacier, M., Kirda, E., Leita, C.: Large scale malware collection: lessons learned. In: IEEE SRDS Workshop on Sharing Field Data and Experiment Measurements on Resilience of Distributed Computing Systems (2008)
27. Austin, T.H., Filiol, E., Josse, S., Stamp, M.: Exploring hidden markov models for virus analysis: a semantic approach. In: 2013 46th Hawaii International Conference on System Sciences (HICSS), pp. 5039–5048. IEEE (2013)
28. Beaucamps, P.: Advanced metamorphic techniques in computer viruses. In: International Conference on Computer, Electrical, and Systems Science, and Engineering-CESSE'07 (2007)
29. Venkatesan, A.: Code Obfuscation and Virus Detection. Ph.D thesis, San Jose State University (2008)

An Efficient Multi-keyword Synonym-Based Fuzzy Ranked Search Over Outsourced Encrypted Cloud Data

Vandana Saini, Rama Krishna Challa and Neelam S. Khan

Abstract Cloud computing is a paradigm for large-scale distributed computing. The sensitive data should be outsourced to the cloud and stored in an encrypted format to keep it confidential. The existing search schemes do not suggest keywords consequently making the retrieving of documents difficult if user forgets the keyword. In this paper, we propose a multi-keyword synonym based fuzzy ranked search (MSFRS) scheme over outsourced encrypted cloud data which provides efficient search results retaining the security features of the existing schemes. It supports multi-keyword, suggests synonyms if user forgot keyword and returns results after evaluating relevance scores based on frequency of keyword in the documents bearing same rank. The performance analysis of the proposed scheme on the dataset concluded that time taken to update the index file has been reduced up to 45 % over outsourced encrypted cloud data.

Keywords Cloud computing · Search scheme · Index file · Internal ranking · Synonym search

1 Introduction

Cloud Computing is making its way to the advanced stage of entering into the era of centralized service as an outgrowth of third revolution in IT industry. The IT industry draws power from Cloud Computing by reducing its input capital on infrastructure and saving the time witnessing evolution from office access to

V. Saini (✉) · R.K. Challa
Department of Computer Science & Engineering, NITTTR, Chandigarh, India
e-mail: saini.vandana91@gmail.com

R.K. Challa
e-mail: ramakrishna.challa@gmail.com

N.S. Khan
NIT Srinagar, J&K, India
e-mail: nlmkhan01@gmail.com

© Springer Science+Business Media Singapore 2016
R.K. Choudhary et al. (eds.), *Advanced Computing and Communication Technologies*,
Advances in Intelligent Systems and Computing 452,
DOI 10.1007/978-981-10-1023-1_43

anywhere facilitation of pay-per-user which means providing measured services like networks, servers, storage, applications as per their demand. With growing number of cloud users, much of the data including email, file storage and business data are being outsourced on the cloud. Email management software is too on the cloud and managed by the cloud service providers' like Yahoo, Google, and Rediff etc. The users utilize the software's services using internet connection and avails the privileges. However, Cloud service providers' can't be fully trusted which can raise security and privacy concerns of the stored data. When user needs any data, he search using keywords and cloud server returns desired data. Many drawbacks have been reflected in the previous searching schemes like they might fail in absorbing the insight of the user leading to unnecessary network traffic, unwanted decryption and non-retrieval if user forgot the keyword. The aim of this research paper is to propose a multi-keyword synonym based fuzzy ranked search scheme (MSFRS) which reduces the time taken to update the index file and comparing this time with fastest of the available schemes.

The rest of the paper is structured as follows. We first briefly review the related works in Sect. 2 and problem is formulated in Sect. 3. We then propose a search scheme as a solution to increase efficiency in Sect. 4. We discuss the performance analysis of the search scheme and its impact on the parameters in Sect. 5 and finally conclude in Sect. 6.

2 Related Works

The researchers worked on several techniques to take hold of the encrypted data. The initial success was single keyword based searching. As the IT industry grew, some advancement in the search techniques as well as the problems like risk of confidentiality, hacking, phishing came into picture. Several applications need mechanisms for complicated access-control over encrypted data. To overcome this issue, Sahai and Waters [1] introduced Attribute-Based Encryption scheme. It is one-to-many encryption that allows users to encrypt and decrypt data based on some attributes like country, or kind of subscription.

Song et al. [2] introduced the concept of symmetric key cryptography in order to reduce security and privacy risk. But it was not effective when searching is conducted on huge data content. Boneh et al. [3] proposed Public key encryption keyword search. This approach provides data privacy by hiding information about the data from the server. To make indexing secure, Goh [4] proposed a technical report on Secure Indexes to make indexing techniques secure. Secure index data take queries with a trapdoor. The resultant index so generated is secure and no information corresponding to stored data is leaked to the server. However, use of bloom filters resulted in false positives.

Chang et al. [5] proposed Privacy preserving keyword search. This search scheme uses multiple round protocols in which single encrypted hash table index was generated for the entire file collection. When user wants to search file, user compute secret key and send to the server. On the basis of secret key server match the file. But it was rendered inefficient due to large fixed size index. Cao et al. [6] proposed Privacy preserving multi-keyword ranked search (MRSE). In this approach, all the keywords were defined in the dictionary. For data and file indexing two randomly generated invertible matrices were used. Using product of vectors Trapdoor was built but this approach hadn't considered effects of keyword access frequencies and dictionary had to be rebuilt if any new word was added. To solve this problem Xu1 et al. [7] proposed a novel approach called Multi-Keyword ranked query on encrypted cloud data (MKQE) which could greatly reduce the amount of computations when the Keyword dictionary was expanded taking keyword access frequencies into consideration for ranking files.

Li et al. [8] proposed Fuzzy Keyword Search. It enhanced search capability by constructing fuzzy keyword sets so that typing errors and format inconsistencies could be eliminated. Earlier searchable encryption scheme allowed user to search exact keyword such that misspell word cannot generate any result. However the scheme enhanced user searching experience by using edit distance to quantify keyword similarities i.e. wild-card based techniques but it didn't support multi-keyword search. Khan et al. [9] proposed secure ranked fuzzy multi-keyword search over outsourced encrypted cloud data (RFMS). This scheme gave results by matching predefined dictionary keywords or closest possible keywords from dictionary based on similarity semantics when an exact match failed. Moh and Ho [10] developed a system to support semantic search on encrypted data. Three different schemes Synonym based Keyword Search (SBKS), Wikipedia-based Keyword Search (WBKS), and Wikipedia Synonym-based Keyword Search (WBSKS) were developed. But in these schemes time of index construction was very high.

Orencik et al. [11] proposed efficient and secure ranked multi-keyword search. In this, the index is generated using list of keywords with the help of SHA-2 family hash function. and the final output of it is reduced to r-bit binary output. This approach provided keyword privacy, trapdoor security, non-impersonation and data privacy but the number of comparisons required to search for a document was high. Handa et al. [12] proposed a cluster based multi-keyword search on outsourced encrypted cloud data. This scheme reduced the number of comparisons required to search the desired document.

3 Problem Formulation

While studying the concept of searching [6–9] for desired documents over encrypted cloud data, some issues in existing research were identified which needs to be addressed including communication overhead, network traffic, security, efficient searching results and enhanced user searching experiences.

To enable secure and efficient multi-keyword ranked search over outsourced encrypted cloud data, the scheme needs to design a new ranking algorithm which ranks the subset of the top ranked documents having highest relevance scores and a new index generation algorithm which update index file with little overhead. Multi-keyword search query requires to be supported by synonym dictionary in case user forgets particular keyword invites design for new efficient algorithm for suggesting synonyms. When new keyword will be added to keyword dictionary, synonym dictionary gets automatically updated and facilitate the user.

4 The Proposed Search Scheme: An Efficient Multi-keyword Synonym Based Fuzzy Ranked Search Over Outsourced Encrypted Cloud Data (MSFRS)

A framework is proposed for an efficient Multi-keyword Synonym based Fuzzy Ranked Search over Outsourced Encrypted Cloud Data (MSFRS). The cloud data system consists of data owner, cloud server and data user. Working of MSFRS framework is described in the following section.

4.1 Setup Phase

Setup phase begins with generation of *encrypted documents E* from unencrypted documents. Data owner choose keywords which act as trapdoor while searching the encrypted cloud data. Based on these encrypted documents, data owner generates *Encrypted Index I* and *Synonym Dictionary S*. These Encrypted documents along with encrypted index file and synonym dictionary are used by cloud server while performing search operation.

4.2 Search Phase

Search Phase begins when data user wants to retrieve documents; user has to send search query to the cloud server. In return, cloud server will give top ranked document to the data user. After that, user has to decrypt documents using secret key as obtained from data owner. If user forgets the keyword, synonym dictionary facilitates the user by suggesting synonyms.

4.3 Detailed Design of Proposed Scheme

The proposed search scheme is based on the Index file, Ranked Search and Synonym Search.

1. To automatically update Index File, *Index Generation Algorithm* is designed and explained in Algorithm 1.

ALGORITHM 1: Build-index(k_s, E_s, K, I)

1. *Preprocessing :*

   a. Choose unique keywords $K_w = (K_{w1}, K_{w2}, K_{w3} ... K_{wn})$ ∀ documents ∈ the set of sensitive documents D = $(D_1, D_2, D_3...D_m)$.

   b. Build a keyword dictionary $K = (K_1, K_2, K_3, ... K_n)$ ∀ extracted keywords.

2. *Generating Trapdoor Set*

   a. Select document D_i for encryption.

   b. Enter the key E_s for encryption.(E_s is a randomly generated key)

   c. Enter the list of keywords $K_w = (K_{w1}, K_{w2}, K_{w3}... K_{wn})$ from keyword dictionary $K_i \in K$, where K_i is selected for document D_i.

   d. On submitting the above details the scheme works as:

      i. Entire document D_i is encrypted (E_i) .

      ii. Store $C_i(K_i) \rightarrow \{ T_i : i=1,2,3,.............m\}$, where $C_i(K_i)$ = Cipher text of keyword K_i and T_i is the trapdoor set for cipher texts of keywords K_i .

3. *Generating Index File*

   a. Store keywords K_i and their corresponding trapdoor sets $t = (t_1, t_2, t_3...t_m)$ in a table consisting of two columns .i.e. keywords (K_i) and trapdoor set (T_i).

   b. This table is the final index file I.

Output I.

---

2. To rank the subset of the top ranked documents, the detailed design of Ranked *Search Algorithm* is explained in Algorithm 2.

ALGORITHM 2 : Ranked-search (K_i, w_q, E_i)

1. *Search Query (K_i)*

   a. If K_i ← null for any keyword K_i

        Then print "You forgot to enter keyword!"

   Else wq_i ← K_i ∈ index file, where d = min (d).

   b. Multi- keyword query { wq_i : wq_i = wq_1, wq_2,... wq_k) is sent to cloud server C by Data User DU.

   c. Extract trapdoor set T from index file I.

2. *Matching Trapdoors and Encrypted Documents*

   a. If match found for keywords K_i

        i. Calculate frequency of each keyword

        ii. Check subset K_i arrange in ranked order

        iii. Return top k-ranked encrypted documents Ei.

   Else print "Sorry! No documents found"

---

3. To recall actual keyword the proposed search scheme suggest user the synonyms of input word. The detailed design of *Synonym Search Algorithm* is
   explained in Algorithm 3.

ALGORITHM 3 : Synonym-search (S)

1. *Data user guesses some keyword query k = (k₁, k₂, ... kₖ) and send to
   cloud server C.*
2. *Cloud server check kᵢ against synonym dictionary S*
   a. Cloud returns S= S_1, $S_{2...}$ S_k corresponding to k_i.
   b. Else No synonym found.
   c. User chooses keywords from synonym set S and search again.
3. *Returns Eᵢ corresponding to Sᵢ.*
   Output E_i.

After getting encrypted documents from cloud server, data user decrypts these
documents using decryption key D_s. The key exchange must be done in a secure
manner by using a standard key exchange mechanism. The keywords selected for
keyword dictionary must be such that they do not reveal any information about the
content of document.

5 Performance Analysis

The performance analysis of MSFRS scheme was carried out by conducting analytical experiment on real data set of 4000 encrypted documents of size approximately 15 kB and ranging up to 8000 keywords, 03–04 keywords were unique in
many documents. The whole experiment is implemented in PHP language using
Linux server with Eucalyptus cloud installed on it. The experiment was conducted
on Intel core i7 processor running at 3.40 GHz, CentOS 6.3 operating system with
RAM of 06 Gigabytes.

The performance of the scheme regarding the effectiveness of Multi-Keyword
Synonym based Fuzzy Ranked Search (MSFRS) is evaluated for efficient data
retrieval and effective user search experience featuring the time taken for index
generation, search process and ranking.

- *Efficient index construction* An efficient feature of MSFRS over previous
  schemes is that when new keywords were added to keyword dictionary, minimum overhead is incurred in updating the index file. The duration of the index
  construction was compared with the previous schemes MRSE, MKQE and
  RFMS.
  Figure 1 shows that *MSFRS is better than other schemes*. When keyword dictionary is expanded to 8000 keywords, MRSE takes 648.8 s, MKQE takes
  447.4 s and RFMS takes 45.67 s but MSFRS takes 25.06 s only to do so.

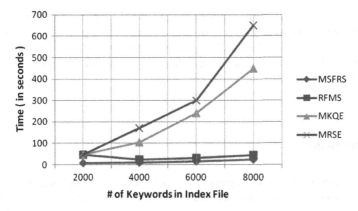

Fig. 1 Index construction time when 2000 files using MSFRS

Table 1 Time taken to rank documents and return results using MSFRS

No. of keywords in search query	100 keywords and 100 documents	1000 keywords and 1000 documents	4000 keywords and 4000 documents	8000 keywords and 4000 documents
1	0.09758866	0.408763495	0.784034955	0.94576552
2	0.06754325	0.598765444	0.688565444	0.86472665
3	0.09876653	0.487313532	0.534313532	0.76911769
4	0.27659443	0.453920866	0.483920866	0.85674543
5	0.19765443	0.568796431	0.843836432	0.97581674
6	0.15433344	0.643267892	0.689031946	0.87546906
7	0.29543347	0.524576742	0.698262524	0.95644324
8	0.27998765	0.698974507	0.798874501	0.91569467
9	0.31876544	0.758788558	0.635432218	0.94545686
10	0.35737032	0.575771254	0.564371254	0.99165398

- *Efficient Search results* The multi keyword synonym based fuzzy ranked search by using internal ranking, rank the subset of top ranked documents so that data user gets most relevant documents. More the number of keywords in search query matches a document, higher is its relevance score. It is clear from Table 1 that time taken is of the order of microseconds only. Time taken to calculate the rank and returning the list of relevant documents is shown in Table 1. It is clear from the Table 1 that time taken is of the order of microseconds only.
- *Secure searching* The MSFRS scheme is secured unless secret key is compromised. The trapdoors created contain special characters and symbols. No guessing attack is possible on the basis of known cipher text as for same keyword different cipher texts are produced.

- *Enhance user searching experience* The scheme enhances user searching experience by providing synonym based search. It help user to search documents if exact keyword is not recalled, by suggesting him the synonyms of the input words. Also users don't have to worry about the typing errors as the proposed scheme also support fuzzy search.

6 Conclusion

In this research paper, multi-keyword synonym based fuzzy ranked search over an encrypted cloud data has been proposed. It allows user to make use of its synonym pool by searching the keywords and index updating at the same time. In Proposed MSFRS scheme, time taken to update index file (25.07 s) is reduced up to 45 % when compared to the fastest of the available search schemes (RFMS—45.67 s). The overhead of generating trapdoor and synonym list for each search query is reduced. In future, it is interesting to study how to increase the synonym base simultaneously reducing the searching time, combining the system with topic based database to get efficient results.

References

1. Sahai, A., Waters, B.: Fuzzy identity-based encryption. In: Proceedings of EUROCRYPT, pp. 457–473 (2005)
2. Song, D., Wagner, D., Perrig, A.: Practical techniques for searches on encrypted data. In: IEEE Symposium on Security and Privacy, pp. 44–55 (2000)
3. Boneh, D., Crescenzo, G.D., Ostrovskyand, R., Persiano, G.: Public key encryption with keyword search. In: International Conference on the Theory and Applications of Cryptographic Techniques, pp. 506–522 (2004)
4. Technical Report: http://eprint.iacr.org/2003/216 (2003)
5. Chang, Y.C., Mitzenmacher, M.: Privacy preserving keyword searches on remote encrypted data. In: 3rd International Conference on Applied Cryptography and Network Security, pp. 442–455. New York (2005)
6. Cao, N., Wang, C., Li, M., Ren, K., Lou, W.: Privacy preserving multi-keyword search over encrypted cloud data. J. IEEE Trans. Parallel Distrib. Syst. **25**, 222–233 (2014)
7. Xu, Z., Kang, W., Li, R., Yow, K., Xu, C.: Efficient multi-keyword rank query on encrypted cloud data. In: 18th IEEE International Conference on Parallel and Distributed Systems, 2012, Singapore, pp. 244–251
8. Li, J., Wang, Q., Wang, C., Cao, N., Ren, K., Lou, W.: Fuzzy keyword search over encrypted data in cloud computing. In: 29th IEEE International Conference on Computer Communications, pp. 1–5, San Diego (2010)
9. Khan Neelam, S., Challa, R.K.: Secure ranked fuzzy multi-keyword search over outsourced encrypted cloud data. In: 5th IEEE International Conference on Computer and Communication Technology (ICCCT), Allahabad (2014)

10. Moh, T.S., Ho, K.H.: Efficient semantic search over encrypted data in cloud computing. In: International Conference on High Performance Computing and Simulation, pp: 382–390. Bologna (2014)
11. Orencik, C., Savas, E.: Efficient and secure ranked multi-keyword search on encrypted cloud data. In: Joint EDBT/ICDT Workshops, pp. 186–195. New York (2012)
12. Handa, R., Challa, R.K.: A cluster based multi-keyword search on outsourced encrypted cloud data. In: International IEEE Conference on Computing for Sustainable Global Development, New Delhi (2015)

Performance Evaluation of Block Ciphers for Wireless Sensor Networks

Kamanashis Biswas, Vallipuram Muthukkumarasamy, Xin-Wen Wu and Kalvinder Singh

Abstract Security is one of the major concerns in many Wireless Sensor Network (WSN) applications. A number of cryptographic algorithms have been developed to provide security services in WSNs. However, selecting an energy-efficient and lightweight cipher is a challenging task due to resource constrained nature of sensor nodes. Systematic evaluation of cryptographic algorithms is, therefore, necessary to provide a good understanding of the trade-off between security performance and operational cost. In this paper, we have examined five block ciphers: Skipjack, Corrected Block Tiny Encryption Algorithm (XXTEA), RC5, Advanced Encryption Standard (AES), and Chaotic-Map and Genetic-Operations based Encryption Algorithm (CGEA). The performance of these ciphers is evaluated on Arduino Pro and Mica2 sensor motes. Then the memory usage, operation time, and computational cost are compared. Finally, some recommendations are provided on evaluated block ciphers and implementation platforms.

Keywords Block ciphers · Memory-and-energy-efficiency · Operational cost · Performance evaluation · Wireless sensor networks

K. Biswas (✉) · V. Muthukkumarasamy · X.-W. Wu
School of ICT, Griffith University, Gold Coast, Australia
e-mail: k.biswas@griffith.edu.au

V. Muthukkumarasamy
e-mail: v.muthu@griffith.edu.au

X.-W. Wu
e-mail: x.wu@griffith.edu.au

K. Singh
Australia Development Lab, IBM, Southport, Australia
e-mail: kalsingh@ibm.edu.au

© Springer Science+Business Media Singapore 2016   443
R.K. Choudhary et al. (eds.), *Advanced Computing and Communication Technologies*,
Advances in Intelligent Systems and Computing 452,
DOI 10.1007/978-981-10-1023-1_44

1 Introduction

Wireless Sensor Networks consist of a large number of low-cost sensor nodes (SNs) which are randomly deployed in hostile environments [7]. These battery-powered SNs have low memory, weak processors, and limited communication capabilities. Therefore, to achieve energy efficiency in WSNs, a number of secure and lightweight block ciphers are proposed in the past few years. However, the experimental results show that many of these ciphers have poor performance compared to conventional ciphers [11]. Thus, performance evaluation is important to provide a benchmark of different cryptographic schemes.

The implementation of security policies has to maintain a trade-off between cost and performance. For example, many WSN applications require complex cryptographic algorithms to provide an enhanced level of security. However, the cost increases as powerful SNs are required to implement the crypto-system. Therefore, it is necessary to clearly understand the relationship between implementation cost and effectiveness. Table 1 presents a comparative view of costs, and hardware specifications of a number of commonly used sensor motes. The table shows that EZ430-RF2500 and Arduino Pro motes are less costly but they also have less memory. Hence, performance evaluation of cryptographic schemes on low-cost SNs is necessary to examine the feasibility of cost-effective platforms.

This paper presents an experimental evaluation of cryptographic algorithms mainly based on actual sensor hardware. A number of block ciphers are implemented in Mica2 and Arduino Pro mote platforms in order to compare the memory efficiency, computational cost, and operation time. Finally, based on the experimental results, some critical insights are provided that will be useful to choose the best cryptographic algorithm and implementation platform.

The rest of the paper is organized as follows: Sect. 2 describes the related works. Section 3 presents an overview of evaluated block ciphers in WSNs. Section 4 details the implementation platforms. Performance evaluation and analysis are presented in Sect. 5. Section 6 discusses and concludes the paper.

2 Related Work

Law et al. present a systematic evaluation framework that considers security properties and memory-and-energy-efficiency of the block ciphers for WSNs [16]. The authors recommend to use Rijndael cipher for security and energy efficiency, whereas MISTY1 is suggested for storage and energy efficiency.

A comparative performance analysis of RC6, AES, and Scalable Encryption Algorithm (SEA) shows that SEA requires less memory compared to AES and RC6 ciphers, whereas AES and RC6 achieve best performance in terms of execution time and bandwidth usage respectively [10].

Table 1 Cost and hardware specification (2014)

Sensor motes	Price (USD)	Package (incl.)	Micro-controller	Bus (bits)	Clock (MHz)	RAM (kB)	Flash (kB)	EEPROM (kB)
SHIMMER	226	2 boards	MSP430F1611	16	4–8	10	48	None
Waspmote	168	4 sensors	ATmega1281	8	14	8	128	4
IRIS	134	No board	ATmega1281	8	8	8	640	4
TelosB	102	3 sensors	MSP430F1611	16	4–8	10	48	1024
Mica2	99	N/Av.	ATmega128L	8	8	4	128	512
EZ430-RF2500	40	Board only	MSP430F2274	16	16	1	32	None
Arduino Pro (328)	25	nrf24L01 radio	ATmega328	8	8-16	2	32	1

Zhang et al. derived and compared computational energy cost of symmetric key ciphers with respect to different block and payload size [21]. Furthermore, the authors recommend the Byte-oriented Substitution-Permutation Network (BSPN) cipher to ensure security and energy efficiency in WSNs.

The effects of symmetric block ciphers on WSN performance and behaviour are analyzed to identify critical network parameters by [1]. AES, RC5, and Skipjack ciphers are implemented on MicaZ and TelosB motes as well as important trade-offs are provided both qualitatively and quantitatively.

Eisenbarth et al. implemented 12 block ciphers on an ATMEL AVR ATtiny45 8-bit microcontroller to achieve comparable results [13]. Lightweight implementation of the ciphers with a footprint of less than 256 bytes RAM and 4 kB code size for encryption and decryption has been performed to provide a good understanding of the cost versus performance trade-off.

Trad et al. measured and compared the memory efficiency, operation time, and energy consumption of AES, RC5, and RC6 algorithms in Mica2 sensor motes [18]. The experimental outcomes show that RC5 is the most suitable block cipher in terms of time and energy-efficiency.

In addition to conventional cryptographic algorithms, several lightweight block ciphers such as HIGHT [15], Simple Lightweight Encryption Scheme [8], and Lightweight Security Protocol [5] are implemented on Mica2 motes. These algorithms are energy-efficient and provide a good level of security in WSNs.

This work implements a number of block ciphers on two different hardware platforms and investigates the security performance and operational cost. The experimental results show the effectiveness of the evaluated block ciphers.

3   Overview of Implemented Block Ciphers

Skipjack, XXTEA, RC5, AES, and CGEA ciphers are described in this section. Table 2 lists the parameters adopted for each block cipher in our experiments.

Skipjack uses an 80-bit key over 64-bit data blocks. It implements an unbalanced Feistel network with 32 rounds. Biiham et al. presented an attack against 31 of 32 rounds using impossible differential cryptanalysis [3]. Moreover, short key length makes Skipjack vulnerable to exhaustive key search attack.

Table 2 Cipher parameters used in experiments

Block ciphers	Key length (bits)	Rounds	Block length (bits)
Skipjack	80	32	64
XXTEA	128	14	64
RC5	128	14	64
AES-128	128	10	128
AES-256	256	14	128
CGEA	256	N/Av.	128

XXTEA has 64-bit block length and 128-bit key length. It implements an unbalanced Fiestel network with variable number of rounds (6–32 full cycles). The last reported attack against the full-round XXTEA presents a chosen plaintext attack based on differential cryptanalysis using 2^{59} queries [20].

RC5 is a flexible block cipher with variable parameters: block size (32, 64, or 128-bits), key size (0–2040-bits), and number of rounds (0–255). It is a widely used block cipher in WSNs. However, 12-round RC5 with 64-bit blocks is vulnerable to a differential attack using 2^{44} chosen plaintext [4].

AES is an iterative block cipher based on a substitution-permutation network and has 128 bits fixed block size. It operates on a 4 × 4 array of bytes. AES running on 10, 12, and 14 rounds for 128, 192, and 256-bits key respectively is vulnerable [9].

CGEA is a lightweight block cipher that uses chaotic map to generate pseudo-random bit sequence [6]. The 256-bit blocks of the sequence is used as key to encrypt or decrypt 128-bit data blocks. The algorithm implements XOR, mutation, and crossover operations on plaintext to generate the ciphertext. Instead of using rounds, it performs crossover operation for each byte of data in plaintext.

4 Implementation Environmnet

4.1 Hardware Specification

Arduino Pro is a microcontroller board based on ATmega168/328 [2]. In experiments, USB powered Arduino Pro (328) motes used with following configurations: Operating voltage—3.3 V, Clock speed—8 MHz, RAM—2 kB, FLASH—32 kB, EEPROM—1 kB, Radio unit—nrf24L01, and Data rate—19.2 kbps.

Mica2 is a low-power sensor mote based on ATmega128L processor [12]. USB powered Mica2 motes used in experiments with following configurations: Operating voltage—3.3 V, Clock speed—8 MHz, RAM—4 kB, Flash—128 kB, EEPROM—512 kB, Radio unit—CC1000, and Data rate—19.2 kbps (Fig. 1).

4.2 Software Specification

The source code of each block cipher is written in Arduino IDE to compile and upload on Arduino Pro motes. In our experiments, two built-in library functions (*microsecondsToClockCycles()*, and *Serial.print()*) are used to obtain and print CPU cycles and encryption time.

A high-level component-based programming language (nesC) [14] is used to implement the ciphers on Mica2 motes. The *LocalTime.get()*, and *prinf()* functions are used to get the execution time, whereas the CPU cycles are obtained by using the ATEMU [17].

(a) (b) (c)

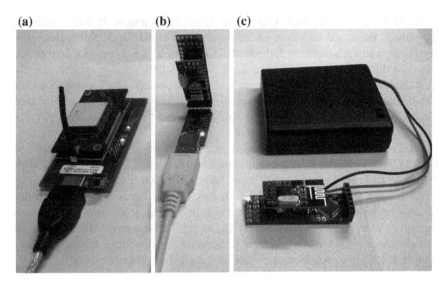

Fig. 1 Experimental setup: **a** USB powered Mica2 mote, **b** USB powered Arduino Pro mote with programmer board, **c** Battery powered Arduino Pro mote

Finally, the avr-size and avr-objdump utilities are used to measure the memory usage on Arduino Pro and Mica2 motes respectively. These two utilities display the header information of object files. The information includes the size of RAM and ROM in terms of text, data, and bss section.

5 Performance Evaluation and Analysis

This section presents a comparative performance analysis of optimized Skipjack, XXTEA, RC5, AES, and CGEA block ciphers implemented on Mica2 and Arduino Pro motes. To make the comparison, three crucial parameters have been selected: memory consumption, computational cost, and operation time.

5.1 Memory Consumption

Memory consumption is a significant performance metric that can be used to select encryption algorithms with less memory overhead. Table 3 shows the amount of memory consumed by each block cipher on Mica2 and Arduino Pro platforms. It can be seen that Skipjack and AES-256 require more memory compared to other algorithms. The memory requirement of AES-128 is slightly lower compared to AES-256 whereas RC5 is the lightest among all algorithms.

Table 3 Memory consumption (bytes)

Block ciphers	Mica2		Arduino Pro	
	RAM	ROM	RAM	ROM
Skipjack	3096	8658	398	4952
XXTEA	542	6312	226	4112
RC5	682	6110	350	3184
AES-128	1074	6296	814	3692
AES-256	1822	7932	1014	4190
CGEA	664	6268	548	3228

Critical observations—Both Skipjack and AES uses a big S-box of 256-bytes, and as a result the algorithms occupy a significant amount of memory. XXTEA, RC5, and CGEA require less memory for execution and is, therefore, suitable for memory constrained SNs like Arduino Pro. One important observation is that the implementation of AES-256 on Arduino Pro mote shows a message regarding low available memory. Therefore, cryptographic algorithms that use excessive memory may experience stability problem on Arduino Pro platform.

5.2 Computational Cost

The energy efficiency of an algorithm can be calculated from its computational complexity. Assuming the energy consumption per CPU cycle is fixed, the amount of consumed energy per byte can be computed by measuring the number of CPU cycles required to process one byte of plaintext. However, Table 4 shows the total number of CPU cycles required by each algorithm to encrypt 32 bytes data. It can be seen that Skipjack is the most energy efficient block cipher, whereas the performance of AES-256 is worst among all algorithms. It is also noted that AES-128 performs two times better than AES-256.

Critical Observations—The key size and number of rounds play a significant role in computational complexity. The implementation of AES-128 block cipher reduces more than half of computational cost required by AES-256 due to small size of key and less number of rounds. It is also noted that RC5 consumes more

Table 4 Computational cost (cycles)

Block ciphers	Mica2	Arduino Pro
Skipjack	9820	12,672
XXTEA	24,064	30,464
RC5	53,014	61,504
AES-128	37,525	43,200
AES-256	80,344	88,896
CGEA	67,786	76,212

CPU cycles compared to AES-128 in spite of having the same size key. The reason is that RC5 executes 14 rounds, whereas AES-128 uses 10 rounds only.

5.3 Operation Cost

Operation speed indicates time efficiency and is defined in terms of encryption time and communication time. Encryption time is the amount of time spent to encrypt the plaintext, whereas the time required to encrypt and successfully send the ciphertext is defined as communication time. Figure 2 shows the execution time required to encrypt 32 bytes data. It can be seen that Skipjack is more than 7 and 6 times faster compared to AES and CGEA ciphers respectively. In addition, AES-128 cipher reduces more than half of AES-256 encryption time. The same results are obtained for communication time experiment as shown in Fig. 3.

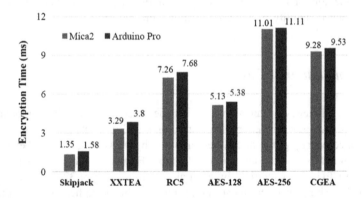

Fig. 2 Encryption time

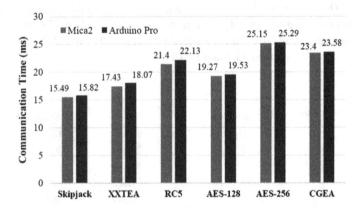

Fig. 3 Communication time

Critical observations—Skipjack algorithm is most efficient since it generates the shortest expanded key among all block ciphers. Similarly, the use of 128-bits key in AES-128 shows better performance compared to AES-256 cipher. XXTEA also requires low encryption time since the cipher is structured with simple XOR and shift operations. The longer word size (32-bit) leads to longer execution times for both key setup and encryption phases in RC5. The CGEA block cipher also takes significant amount of encryption time to perform crossover operations by repeatedly swapping the values at different memory locations.

6 Discussion and Conclusions

According to the experimental results, RC5 is the most memory efficient block cipher. XXTEA and CGEA are also potential candidates for memory constrained SNs like Arduino Pro. On the other hand, Skipjack shows best performance in terms of operation time and computational cost. XXTEA and AES-128 ciphers also consume low energy. However, from a security perspective, Skipjack is a high risk algorithm because of shorter key length. Similarly, XXTEA and RC5 are vulnerable to a number of security attacks such as timing attack, and chosen plaintext attack. Moreover, 128-bits key is not secure against quantum attack. The quantum computing systems are able to break 128-bits key with time 2^{64} [19]. However, AES-256 and CGEA would still be secure against exhaustive search due to 256-bits key length. Therefore, we recommend to use AES-256 or CGEA block ciphers when security is a priority. RC5, XXTEA, and AES-128 ciphers can be used for the applications that require minimum level security.

This paper presents a comparative performance analysis of Skipjack, XXTEA, RC5, AES, and CGEA block ciphers. It is noted that Arduino Pro requires slightly more execution and communication time compared to Mica2 which is negligible. Therefore, it will be cost-effective to use Arduino Pro for common WSN applications such as environmental monitoring instead of time critical and high-memory-demand applications. Our future works will evaluate the performance of stream ciphers and compare the results with block ciphers.

References

1. Antonopoulos, C., Petropoulos, C., Antonopoulos, K., Triantafyllou, V., Voros, N.: The effect of symmetric block ciphers on WSN performance and behavior. IEEE WiMob, pp. 799–806. Barcelona (2012)
2. Arduino Inc.: Arduino Pro (n.d.). https://www.arduino.cc/en/Main/ArduinoBoardPro
3. Biham, E., Biryukov, A., Shamir, A.: Cryptanalysis of Skipjack reduced to 31 rounds using impossible differentials. J. Crypt. **18**(4), 291–311 (2005)
4. Biryukov, A., Kushilevitz, E.: Improved cryptanalysis of RC5. In: Kaisa, N. (ed) EUROCRYPT'98, LNCS. 1403, pp. 85–99. Springer, Heidelberg (1998)

5. Biswas, K.: Lightweight security protocol for WSNs. In: International Symposium on WoWMoM, pp. 1–2. Sydney (2014)
6. Biswas, K., Muthukkumarasamy, V., Singh, K.: An encryption scheme using chaotic map and genetic operations for WSNs. IEEE Sens. J. **15**(11), 2801–2809 (2015)
7. Biswas, K., Muthukkumarasamy, V., Sithirasenan, E.: Maximal clique based clustering scheme for WSNs. IEEE ISSNIP, pp. 237–241. Melbourne (2013)
8. Biswas, K., Muthukkumarasamy, V., Sithirasenan, E., Singh, K.: A simple lightweight encryption scheme for WSNs. ICDCN, LNCS. 8314, pp. 499–504. Coimbatore: Springer, Heidelberg (2014)
9. Bogdanov, A., Khovratovich, D., Rechberger, C.: Biclique cryptanalysis of the full AES. In: Lee, D., Wang, X. (ed) ASIACRYPT. LNCS. 7073, pp. 344–371. Springer, Heidelberg (2011)
10. Çakırolu, M., Bayilmi, C., Özcerit, T., Çetin, Ö.: Performance evaluation of scalable encryption algorithm for WSNs. J. Sci. Res. Essays **5**(9), 856–861 (2010)
11. Cazorla, M., Marquet, K., Minier, M.: Survey and benchmark of lightweight block ciphers for WSNs. SECRYPT (2013)
12. Crossbow Technology.: MICA2, Wireless Measurement System (n.d.). http://www.eol. ucar. edu/isf/facilities/isa/internal/CrossBow/DataSheets/ mica2.pdf
13. Eisenbarth, T., Gong, Z., Güneysu, T., Heyse, S., Indesteege, S., Kerckhof, S., Oldenzeel, L. et al.: Compact implementation and performance evaluation of block ciphers in ATtiny devices. AFRICACRYPT, pp. 172–187. Springer Berlin Heidelberg (2012)
14. Gay, D., Levis, P., Behren, R., Welsh, M., Brewer, E., Culler, D.: The nesC language: a holistic approach to networked embedded systems. ACM SIGPLAN Not. **38**(5), 1–11 (2003)
15. Koo, W., Lee, H., Kim, Y., Lee, D.: Implementation and analysis of new lightweight cryptographic algorithm suitable for WSNs. ISA, pp. 73–76 (2008)
16. Law, Y., Doumen, J., Hartel, P.: Survey and benchmark of block ciphers for WSNs. ACM Trans. Sen. Netw. **2**(1), 65–93 (2006)
17. Polley, J., Blazakis, D., Mcgee, J., Rusk, D., Baras, J.: ATEMU: a fine grained sensor network simulator. IEEE SECON, pp. 145–152 (2004)
18. Trad, A., Bahattab, A.A., Othman, S.B.: Performance trade-offs of encryption algorithms for WSNs. WCCAIS, pp. 1–6. Hammamet (2014)
19. Wu, D.: Introduction to Cryptography, Lecture Notes, Stanford University (n.d.). http://crypto. stanford.edu/ ~ dwu4/notes/CS255LectureNotes.pdf
20. Yarrkov, E.: Cryptanalysis of XXTEA (n.d.). http://eprint.iacr.org/2010/254.pdf
21. Zhang, X., Heys, H., Li, C.: Energy efficiency of symmetric key cryptographic algorithms in WSNs. QBSC Symposium, pp. 168–172. Kingston (2010)

Performance Analysis of Downlink Packet Scheduling Algorithms in LTE Networks

Pardeep Kumar and Sanjeev Kumar

Abstract The Long Term Evolution (LTE) network has been introduced in Third generation partnership project (3GPP) to provide very high data rate for downlink, good services for multimedia, and packet optimized radio access technology. The resource block is the basis entity of a cellular system. So the resource management plays an important role to achieve better performance. For proper utilization of resources there should be some scheduling strategies to allocate the resources among the users in an efficient manner. There exist various scheduling strategies i.e. PF, FLS, MLWDF etc. This article contains the performance analysis of Proportional Fair (PF), Modified Largest Weighted Delay First (M-LWDF) and Exponential-Proportional Fairness (EXP-PF) algorithms, carried out on a single cell affected by interference for different flows such as VoIP, VIDEO and Best Effort. The evaluation is carried out in terms of delay, fairness index, spectral efficiency and throughput.

Keywords LTE · PF · EXP-PF · MLWDF · PDCCH · OFDMA

1 Introduction

The long term evolution abbreviated as LTE is a latest and promising technology which gives 50 times higher performance as compare to matured technology systems. The LTE provides asymmetric data rate for both downlink and uplink data channels. In downlink direction LTE provides date rate up to 300 and 75 Mbps for uplink direction. There are different access techniques used in LTE for channels such as orthogonal frequency division multiple access (OFDMA) in which the

P. Kumar (✉) · S. Kumar
Department of Computer Science & Engineering, Guru Jambheshwar
University of Science & Technology, Hisar, Haryana, India
e-mail: er.pardeepawasthi@gmail.com

S. Kumar
e-mail: sanjukhambra@yahoo.co.in

© Springer Science+Business Media Singapore 2016   453
R.K. Choudhary et al. (eds.), *Advanced Computing and Communication Technologies*,
Advances in Intelligent Systems and Computing 452,
DOI 10.1007/978-981-10-1023-1_45

channel is subdivided into sub-channels to carry data. The OFDMA is used in downlink direction and Single Carrier-Frequency division multiple access (SC-FDMA) is used in uplink direction. The key issues for introducing LTE are high data rate demand of users, backward compatibility with older cellular system, low operational cost and low latency [1, 2].

2 Scheduling

LTE uses dynamic scheduling techniques to make effective use of resources in both time and frequency domain. The scheduler uses Radio Resource Management (RRM) technique for resource allocation. The scheduling algorithms are implemented at the base station. At every Transmission Time Interval (TTI) a resource allocation decision for next incoming TTI is taken by the base station and this judgment information is sent to the user equipment via physical downlink control channel (PDCCH). Thereafter, at each and every TTI, the user equipment computes the channel quality information also known as physical layer information and sends it to the scheduler. On the basis of the received information the base station takes a resource allocation decision, fills up the resource block allocation mask and selects a best modulation and coding scheme (MCS) for data transmission.

This complete information is sent to the user equipment via physical downlink control channel. After reading the information received by physical downlink control channel the UE can determine how it can use physical downlink shared channel (PDSCH) as depicted in Fig. 1 [3–5]. These are the algorithms for downlink packet scheduling.

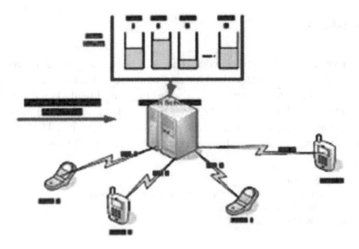

Fig. 1 Packet scheduling architecture

2.1 Proportional Fair (PF) Algorithm

The PF algorithm is not a delay sensitive algorithm due to which it is suitable only for non-real time data flow. The actual goal of this algorithm is to provide a fair share of resources as well as enhance the system throughput. It provides a balance between the system throughput and fairness. In PF, the resource allocation is done on the basis of experienced channel quality and the past user throughput [5, 6].

2.2 Modified-Largest Weighted Delay First (M-LWDF) Algorithm

This algorithm supports real time data flow such video etc. This algorithm is delay sensitive and considers the variation in case of real time transmission service. The users can use many different real time services with different quality of service requirements. The main aim of this algorithm is to balance the weighted packet delay and to measure the efficiency of the channel to make use of it [4–8].

2.3 Exponential-Proportional Fairness (EXP-PF) Algorithm

The EXP-PF algorithm is a combination of Exponential Function and PF algorithm. This algorithm deals with both real time and non real time data flow which means Exponential Function deals with real time data and PF deals with non real time data. This algorithm was designed to support multimedia applications where end to end delay must be bounded for better performance and it also provides better level of fairness [9–11].

3 Experimental Work

This simulation contains performance analysis of three downlink packet scheduling algorithms viz. Proportional Fair (PF), Modified Largest Weighted Delay First (M-LWDF) and Exponential Proportional Fairness (EXP-PF). This study is carried out on a single cell with interference scenario for three types of flows (VIDEO, VoIP and INF_BUF (Best Effort)) and considering measurement parameters i.e. delay, spectral efficiency, fairness index and throughput.

This simulation was carried out on a 32 bit Ubuntu operating system version 14.10, with a core i3 processor. The LTE release −5 was used for simulation and GNU plot was used for graph plotting. Parameters used in simulation are given in Table 1.

Table 1 Simulation parameters

Parameters	Parameters values
No. of cluster	1
Size of cluster N (number of cells in a cluster)	7
Bandwidth	5 MHz
Number of simulations	1
Start users	10
Interval between users	10
Maximum users	40
Radius of cell	1.2 km
VIDEO flow	1
Best effort flow	1
VoIP flow	1
Frame structure type	FDD
User speed	3 km/h
Maximum delay	0.1 ms
Seed	1
VoIP bit rate	8.8
Video bit rate	240 kbps
Scheduler used	PF, M-LWDF and EXP-PF

4 Results and Discussion

4.1 VoIP Flow

The Voice over IP does not use TCP because it is so heavy for real time applications so UDP is used. In UDP, the packets arrival time and order are not considered which means packets can reach to destination at any time and in any order. To solve this problem Real Time Protocol (RTP) is used which permits the receiver to put the packets back into the correct order and restricts long waiting for packets that have either lost their way or are taking too long to arrive (it means there is no need of every voice packet, only continuous flow and packet ordering is required).

The delay for VoIP increases with number of users for MLWDF as depicted in Fig. 2, because the MLWDF is best suited for real time flows but the VoIP is not a real time flow (the RTP is used to make it real time). But the delay for PF and EXP-PF algorithms remains almost unchanged because the PF support non real time data flow and the EXP-PF is a combination of Exponential function and PF algorithm.

The fairness index for VoIP flow is depicted in Fig. 3. Fairness index tells about better utilization of the resources in the system. As the number of users increases the number of resources decreases means both are inversely proportional to each

Fig. 2 VoIP delay

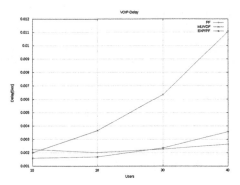

Fig. 3 VoIP fairness index

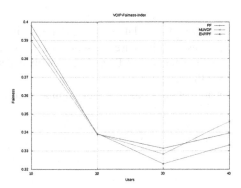

other. So the fairness index decreases with increase in number of users. In the case of fairness index the graph for all scheduler types remain same.

The throughput for VoIP is presented in Fig. 4. From the graph it can be noticed that the overall throughput of all the scheduling algorithms increase with the number of users.

Finally, it can be concluded that the PF and EXP-PF are best suitable for VoIP flow but MLWDF is not suitable for VoIP due to higher delay.

Fig. 4 VoIP throughput

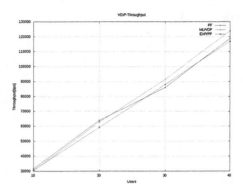

5   VIDEO Flow

The delay for VIDEO flow is shown in Fig. 5. The delay for PF algorithms increases with increase in number of users because PF supports only non real time data flow but the VIDEO is a real time data flow. The delay for MLWDF and EXP-PF almost remains constant because both MLWDF and EXP-PF are delay sensitive algorithms and they support real time data flow.

Fig. 5 VIDEO delay

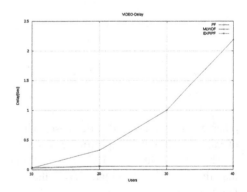

Fig. 6 VIDEO fairness index

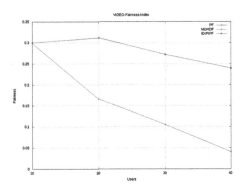

The fairness index for VIDEO flow is depicted in Fig. 6. The number of resources is indirectly proportional to the number of users. From the graph it can be noticed that the fairness index is decreasing rapidly for PF because PF support only non real time data flow but the VIDEO is a real time data flow. But fairness indices for MLWDF and EXP-PF are decreasing very slightly with increase in the number of users because both are real time scheduling algorithms.

The throughput for VIDEO flow is exhibited in Fig. 7. The VIDEO is a real time flow. The overall throughput of MLWDF and EXP-PF is increasing with the number of users because both supports real time data flow and both are delay sensitive algorithms. But the throughput of PF is not good due to higher delay and lower fairness index.

Finally, it can be concluded that the MLWDF and EXP-PF are best suitable algorithms for VIDEO flow but PF is not suitable for VIDEO due to higher delay and lower fairness index.

6   INFINITE_BUFFER (Best Effort) Flow

In best effort flow the trend for delay remains constant for every scheduling algorithm which is 1 ms as depicted in Fig. 8. Because in best effort flow there is no data delivery time boundaries so that's why delay is not a major concern here. So every scheduling algorithm displayed negligible delay.

Fig. 7 VIDEO throughput

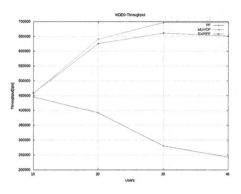

Fig. 8 INF_BUF delay

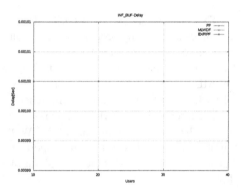

The fairness index for best effort flow increases with number of users but up to a limit because best effort flow provides a best share of resources to each and every user. But the number of resources is indirectly proportional to the number of users. So a best share of resources can be allocated to few users. After that the fairness index decreases. All the schedulers have same fairness index trend for best effort flow as depicted in Fig. 9.

For best effort flow the throughput of all schedulers decreases as long as the number of user increases, as depicted in Fig. 10, due to lower fairness index and higher PLR. Hence, it can be concluded that all the scheduling algorithms are best suited for INF_BUF (Best Effort) flow.

Fig. 9 INF_BUF fairness
index

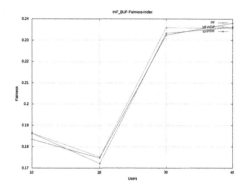

Fig. 10 INF_BUF
throughput

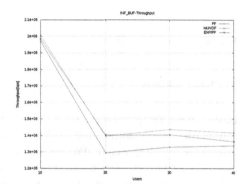

6.1 Spectral Efficiency

The spectral efficiency is shown in Fig. 11. The MLWDF and EXP-PF gave suitable results for every type of flow but PF satisfied the need for some flows because PF supports only non real time data flow and is not delay sensitive. Consequently, MLWDF and EXP-PF support all the flows and assure their QOS needs in terms of PLR, Delay, fairness index and throughput. So they provide better spectral efficiency as compare to PF.

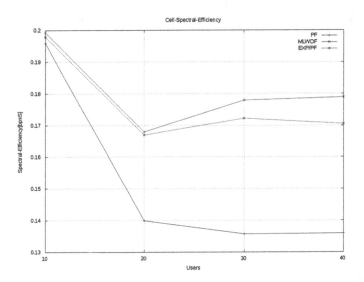

Fig. 11 Spectral efficiency

7 Conclusion

In this paper the performance analysis of different downlink packet scheduling algorithms viz. Proportional Fair (PF), Modified Largest Weighted Delay First (M-LWDF) and Exponential-Proportional Fairness (EXP-PF) algorithms are carried out on a single cell affected by interference for different flows such as VoIP, VIDEO and Best Effort using LTE-Simulator. The performance is measured in terms of delay, fairness index, spectral efficiency and throughput. Finally, it is concluded that the PF and EXP-PF are best suitable for VoIP flow but MLWDF is not suitable for VoIP due to higher delay. On the other hand MLWDF and EXP-PF are best suitable algorithms for VIDEO flow but PF is not suitable for VIDEO due to higher delay and lower fairness index and all the scheduling algorithms are best suited for INF_BUF (Best Effort) flow.

References

1. Kumar, P., Kumar, S.: A survey on inter-cell interference co-ordination techniques of LTE networks. Int. J. Adv. Res. Comput. Eng. Technol. (IJARCET) **04**(06), 2539–2544 (2015)
2. Parkvall, S., Dahlman, E., Furuskar, A., Jading, Y., Olsson, M., Wanstedt, S., Zangi, K.: LTE-advanced—evolving LTE towards IMT-advanced. In: IEEE, pp. 1–5 (2008)
3. Dardouri, S., Bouallegue, R.: Comparative study of scheduling algorithms for LTE networks. Int. J. Comput. Electric. Autom. Control Inform. Eng. **8**(03), 445–450 (2013)

4. Capozzi, F., Piro, G., Gricco, L.A., Boggia, G., Camarda, P.: Downlink packet scheduling in LTE cellular networks: key design issues and a survey. Commun. Surveys Tutorials IEEE, **15**(2) (2013)

5. Kitanov, S., Janevski, T.: Performance evaluation of scheduling strategies for LTE Networks in downlink direction. In: Proceedings of International Conference ETAI IEEE (2010)

6. Sulthana, S.F., Nakkeeran, R.: Study of downlink scheduling algorithms in LTE networks. J. Netw. **9**(12), 3381–3391 (2014)

7. Trivedi, R.D., Patel, M.C.: Comparison of different scheduling algorithms for LTE. Int. J. Emerg. Technol. Adv. Eng. **8**(5), 334–339 (2014)

8. Chandan, S.M., Akki, C.B.: A fair downlink scheduling algorithm for 3GPP LTE networks. Int. J. Comput. Netw. Inf. Security **6**, 34–41(2013)

9. Habaebi, M.H., Chebil, J., AL-Sakkaf, A.G., Dhhawi, T.H.: Comparative study of scheduling algorithms for LTE networks. IIUM Eng. J. **14**(1), 67–76 (2013)

10. Sahoo, B.: Performance comparison of packet scheduling algorithm for video traffic in LTE cellular network. Int. J. Mobile Netw. Commun. Telematics (IJMNCT) **3**(3), 9–18 (2013)

11. Basukala, R., Mohd Ramli, H., Sandrasegaran, K.: Performance analysis of EXP/PF and M-LWDF in downlink 3GPP LTE system. In: F. Asian Himalayas Conference IEEE, pp. 1–5 (2009)

Evaluating Topic Modeling as Pre-processing for Component Prediction in Bug Reports

Monika Jangra and Sandeep K. Singh

Abstract Open bug repositories prove to be very helpful in Software Engineering, since they provide a platform for developers and end-users to report bugs. Along with summary and description, reporters are also expected to assign a component name. Without knowledge of internal structure of a project often a wrong component name is assigned. These incorrect naming may delay process of bug related activities. Pre-requisite to develop any automated component prediction system is to extract relevant features from bug reports. Bug reports are in natural language, therefore, before using them for training process, a vector space of relevant features is built. TFIDF Weighting and Topic Modeling techniques have been examined in this work w.r.t. their ability to choose selective terms from bug reports. This work has done a comparative analysis of two above mentioned preprocessing techniques along with three classifiers—Naive Bayes, SVM and C4.5 in context of correct component prediction.

Keywords BTS · TFIDF · Topic modeling · Bug reports · Pre-processing

1 Introduction

Open Bug Tracking Systems (BTS) are used to report bugs by both developer and user during the development and maintenance of complex software project. According to statistics in [1], number of bugs reported to BTS is about 30 per day. This work has reported a dramatic increase in number of bug reports for complex software. It is important to fix bugs effectively in least possible time frame. Bug triaging as well as component assignment to bug reports is a difficult and time consuming task if done manually. It may also lead to incorrect naming in the

M. Jangra (✉) · S.K. Singh
Jaypee Institute of Information Technology, Noida, India
e-mail: Monikajangra0701@gmail.com

S.K. Singh
e-mail: Sandeepk.singh@jiit.ac.in

© Springer Science+Business Media Singapore 2016
R.K. Choudhary et al. (eds.), *Advanced Computing and Communication Technologies*,
Advances in Intelligent Systems and Computing 452,
DOI 10.1007/978-981-10-1023-1_46

absence of proper and complete knowledge of project's structure. Most of the reported bugs have correct summary and description, but an incorrect component name.

The incorrect bug fixing at development phase may prove costly at deployment phase. A wrong component name may leave the bugs unresolved or a delayed bug fix. Average time to fix a bug with correct component label is 136 days and incorrectly a labeled bug is 190 days as reported in [1]. So, it is necessary to automate the process of bug component prediction to help users of BTS. Any system that can automate such procedures needs to be trained first with pre-processed data. This pre-processed data is in the form of vectors called training vectors. These training vectors are obtained from the raw bug reports in bug repository.

Wang et al. [1] presents an automated technique for component prediction in bug reports. They used TFIDF technique in the pre-processing phase to generate feature vector from bug reports, where vector elements are unique terms with their TFIDF values in bug reports. Manual classification of Bug Reports by Herzig [2] classifies bug reports as bugs and other categories. Also, work in [3] concludes that Topic modeling approach gives better classification results than TFIDF Weighting approach. Classifying Bug Reports to Bugs and Other Requests Using Topic Modeling [4] compared Topic based model with word based model. They showed that when number of topics is taken 50, Topic model proves to be better than Word based model.

TFIDF Weighting and Topic Modeling techniques have been examined in this work w.r.t. their ability to choose selective terms from bug reports. This work has done a comparative analysis of two above mentioned preprocessing techniques along with three classifiers—Naive Bayes, SVM and C4.5 in the context of correct component prediction. We have chosen JIRA as bug repository.

Our research work is centered on investigating following questions:

RQ1: Which technique is better for feature selection: TFIDF weighting or Topic modeling?
RQ2: Which is best classifier out of Naive Bayes, SVM and C4.5 for component prediction?
RQ3: Does the performance of classifiers vary with the number of topics in case of topic modeling?

Rest of the paper is organized is as follows: Sect. 2 describes the Related Work. Section 3 explains the two approaches of pre-processing taken up for comparative study; Sect. 4 details the Experimental Setup and the results of all three research questions. Section 5 summarizes the Conclusion and Future work. Finally, paper ends with the Threats to Validity in Sect. 6.

2 Related Work

2.1 States of a Bug Report in a BTS

The bug report is defined using fields like BugID, status, resolution, summary, description, component etc. From ASSIGNED to FIXED status, a bug report traverses through many stages. Initially, report's status is NEW, when a developer is given the responsibility to fix it, its status changes to ASSIGNED. A RESLOVED bug may further be labeled as VERIFIED or CLOSED. The resolution may be WON'T FIX, INVALID, DUPLICATE, FIXED and WORKSFROME also.

2.2 Word Stemming

Porter [5], invented a language called SNOWBALL, which defines the rule of word stemming. SNOWBALL stemming is improvement over PORTER algorithm. Stemming maps all grammatical variations of a word to same token. Earlier work [6] shows that word stemming has a little impact on performance improvement.

2.3 Prediction Techniques in Component Assignment

Three algorithms Naive Bayes, Logistic regression and decision tree are used for bug fixing in [4] and Naive Bayes proved superior among three. Work in [1] applied Naive Bayes, LibSVM, LibLINEAR for bug triaging and showed that Naive Bayes reaches up to 59.31 % accuracy, LibSVM reaches up to 84.12 % and LibLINEAR up to 80.81 %. Anvik et al. [7] used Naive Bayes, LibSVM and C4.5 for developer assignment. Sureka [8] compared TFIDF scheme and DLM implemented in LingPipe for assigning components to bugs and component reassignment prediction.

2.4 Topic Modeling

Topic Modeling is a natural language processing scheme which identifies the top k topics from the corpus of the documents. The value of k may be set heuristically, based on some empirical study or often a random choice by developer. Latent Dirichlet Allocation (LDA) [9] explains the topic modeling as a generative probabilistic model for extracting discrete topics from text corpora.

3 Approaches for Pre-processing

3.1 TFIDF Weighting

Step1   Summary and description of bug reports are collected. During data cleaning, html tags, punctuations, numeric characters and stop words are removed from the corpus.

Step2   Snowball is applied to each bug report. Unique words from this new corpus are taken to construct the new feature space. Chi-squared feature selection is used to rank the unique terms with ranker threshold 0.

Step3   TFIDF is calculated for each chi-squared selected unique word.

$$\text{TFIDF} = \text{TF} * \log(N/DF) \tag{1}$$

TF   Term frequency defined as number of times a term appears in a document

N   Total number of documents

DF   Document frequency defined as the total number of documents in which term is present

Step4   Naive Bayes, LibSVM and C4.5 models are trained with the unique word vectors built on the basis of TFIDF scores. An open source Data Mining Tool Weka [10] is used for classification task.

3.2 Topic Modeling

Step1 and Step2   Both are same for Topic Modeling as done in TFIDF Approach except that no feature selection is employed in step 2 as topic modeling itself gives top k results.

Step3   We extracted topics from the bug reports using open source LDA Tool [11]. The number of topics in our experiment is heuristically decided viz. 10, 20, 30 and 50.

Step4   With topic feature vector space from step3, classifiers are trained and tested (Fig. 1).

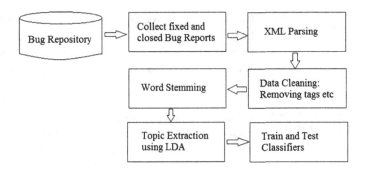

Fig. 1 Bug component prediction using topic modeling approach

4 Experimental Setup and Results

4.1 Experimental Setup

The closed fixed bug reports for experimental work are taken from the JIRA repository [12]. The dataset details are shown in Table 1.

During chi-squared feature selection, the Ranker search method is used with 0 ranking threshold. In our experimental work, classifiers are applied with default parameters. The three classifiers whose accuracy is evaluated are Naïve Bayes, SVM and C4.5. Naïve Bayes is a Probabilistic classifier used in previous works [1, 6, 7] and proved to be a better choice for bug classification problems. SVM is a non linear classification algorithm that proved to be effective for text categorization problem [12]. The manual method used for bug component classification is somewhat similar to traversing a decision tree. Thus, we have also evaluated C4.5 algorithm. Topic modeling is carried out using LDA tool and classifiers are implemented using weka tool. LibSVM library is used for implementation of SVM.

4.2 Results

Results are presented in terms of the prediction accuracy of classifiers achieved after using TFIDF Weighting in one case and Topic Modeling in another case to create feature vectors. The results are divided into following Result Sets.

Table 1 Bug counts

Project	BugId	#Bug reports	#Components
Lucene	Lucene-4444 to 6403	343	24
HTTPClient	HTTPClient-587 to 1619	434	12
JCR	JCR-2431 to 3861	659	31

Table 2 Prediction accuracy for TFIDF approach

Project	Naive Bayes	SVM	C4.5
Lucene	58.35	50.70	52.40
HttpClient	35.33	62.81	68.36
JCR	53.86	49.77	56.29

Table 3 Prediction accuracy for topic modeling approach

Project	#Topics	Naive Bayes	SVM	C4.5
Lucene	10	40.22	53.54	47.87
	20	42.20	44.19	45.60
	30	49.29	57.22	51.55
	50	49.57	50.42	53.54
HttpClient	10	40.87	56.81	49.65
	20	32.71	57.14	47.92
	30	39.86	57.14	49.53
	50	49.77	56.22	50
JCR	10	17.14	49.77	40.81
	20	20.33	49.77	39.75
	30	17.90	49.77	42.33
	50	17.90	49.77	45.67

RS1: Prediction accuracy with TFIDF feature vector space is shown in Table 2.
RS2: Prediction accuracy with Topic feature vector space is shown in Table 3.

4.3 Discussion

RQ1: Which technique is better TFIDF weighting or Topic Modeling?

According to conclusions in Table 4, derived from Tables 2 and 3, it seems that selecting a technique for pre processing is project and classifier specific. So, making a comparative study between two techniques before actually designing an automated system may prove effective. Since, according to Tables 2 and 3 the prediction accuracy difference between two techniques varies from 0 to 33.53 %.*RQ2: Among three classifiers, which is best in both approaches?*

Table 4 Summary from Tables 2 and 3

Project	Naive Bayes	SVM	C4.5
Lucene	TFIDF	Topic modeling	Topic modeling
Httpclient	Topic modeling	TFIDF	TFIDF
JCR	TFIDF	TFIDF/topic modeling	TFIDF

Table 5 Best performing classifier among Naïve Bayes, C4.5 and SVM

Project	TFIDF	10 topics	20 topics	30 topics	50 topics
Lucene	Naive Bayes	SVM	Naive Bayes	C4.5	C4.5
HttpClient	C4.5	SVM	SVM	SVM	SVM
JCR	C4.5	SVM	SVM	SVM	SVM

Table 6 Average performance change for varying number of topics

Model	10 topics	20 topics	30 topics	50 topics
Naive Bayes	32.74	31.75	35.68	39.08
SVM	53.37	50.37	54.71	52.14
C4.5	46.11	44.42	47.80	49.74

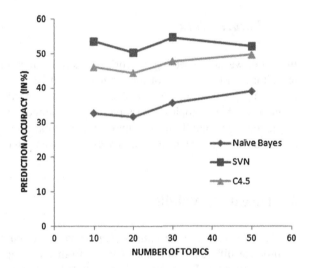

Fig. 2 Average prediction results for topic modeling approach for all three classifiers

From Table 5, conclusions from Tables 2 and 3 in terms of best performing classifier, it is clear that Topic Modeling works better with SVM and for TFIDF approach, C4.5 is better choice. *RQ3: What can be stated about the performance of classifiers for varying number of topics?*

From Table 2, it is evident that C4.5 performs better with 50 topics and SVM with 30 topics. Naive Bayes also works better with 30 except for JCR project. The average results of all the projects from Table 2 are given in Table 6 and same finding is represented in Fig. 2, SVM is the best performing classifier for Topic Modeling.

5 Conclusion and Future Work

5.1 Conclusion

Pre-processing structured bug reports gathered from bug repositories is an important initial step to train models for component prediction. We have evaluated TFIDF and Topic modeling as two pre-processing techniques. Among Naïve Bayes, SVM and C4.5, except for SVM, which performs better with topic feature space, both Naive Bayes and C4.5 perform better with TFIDF feature space. The number of topics is set heuristically and we are not able to find any regular pattern to decide what should be the optimal number of topics, since it depends on the training set. From our dataset, we conclude for SVM and C4.5, the best number of topics is 30 and 50 respectively.

5.2 Future Work

In future, we are expecting to work on more challenging data sets with different repositories. More values of topic numbers can be experimented, that can help in generating good results. Some objective measure needs to be formulated to find optimal number of topics. The topic length (number of words in topic) and topic extraction technique if changed may have different accuracy results. Moreover, we aim to develop a system for automated component prediction in Bug Reports.

6 Threats to Validity

- *Data set may not be challenging*. Data set when taken in larger size may produce different result and if taken from other repository, having additional categorizing fields, which when included in training process, may affect the prediction results also.
- The percentage split of bug reports to individual component class is not taken into account. The classes with low number of instances may get dominated by the majority classes, which are generally known as class imbalance effect.
- The comparative model is not generalized. The results shown by us are project specific. To develop a generalized model we need to train classifier with multiple projects taken at same time as single training set.

References

1. Wang, D., Zhang, H., Liu, R., Lin, M., Wu, W.: Predicting bugs' components via mining bug reports. J. Softw. **7**(5), 1149–1154 (2012)
2. Herzig, K., Just, S., Zeller, A.: It's not a bug, it's a feature: how misclassification impacts bug prediction. In: Proceedings of the 2013 International Conference on Software Engineering, pp. 392–401. IEEE Press (2013, May)
3. Ramage, D., Dumais, S.T., Liebling, D.J.: Characterizing microblogs with topic models. ICWSM **5**(4), 130–137 (2010)
4. Pingclasai, N., Hata, H., Matsumoto, K.I.: Classifying bug reports to bugs and other requests using topic modeling. In: Software Engineering Conference (APSEC, 2013 20th Asia-Pacific), pp. 13–18. IEEE (2013, December)
5. Porter, M.F.: Snowball: a language for stemming algorithms. http://snowball.tartarus.org/texts/introduction.html (2008). Accessed 11-03 2008
6. ʹCubraniʹc, D., Murphy, G.C.: Automatic bug triage using text classification. In: Proceedings of Software Engineering and Knowledge Engineering, pp. 92–97 (2004)
7. Anvik, J., Hiew, L., Murphy, G.C.: Who should fix this bug? In: Proceedings of the 28th international conference on software engineering, pp. 361–370. ACM (2006, May)
8. Sureka, A.: Learning to classify bug reports into components. In: Objects, Models, Components, Patterns, pp. 288–303. Springer, Berlin Heidelberg (2012)
9. Blei, D.M., Ng, A.Y., Jordan, M.I.: Latent Dirichlet allocation. J. Mach. Learn. Res. **3**, 993–1022 (2003)
10. Weka 3—Data mining with open source machine learning software in java. http://www.cs.waikato.ac.nz/ml/weka
11. Topic-modeling-tool—A graphical user interface tool for topic modeling-google project hosting. https://code.google.com/p/topic-modeling-tool
12. Joachims, T.: Text categorization with support vector machines: learning with many relevant features, pp. 137–142. Springer, Berlin Heidelberg (1998)
13. Yang, Y., Pedersen, J.O.: A comparative study on feature selection in text categorization. In: ICML, vol. 97, pp. 412–420 (1997, July)
14. Issue Navigator-ASF JIRA. http://issues.apache.org/jira

A Hybrid Approach for Simultaneous Compression and Encryption of an Image in Wireless Media Sensor Networks

A. Karthikeyan, V. Srividhya, Pranjay Gupta and Naveen Rai

Abstract In this paper, the advantages of discrete cosine transform and discrete wavelet transform have been harnessed by considering them together for compression, followed by an energy efficient Secured Force encryption technique. This was done in order to simultaneously compress and encrypt the image that has to be transmitted through Wireless Media Sensor Network (WMSN). The proposed technique provides better and faster compression as compared to the existing techniques. Results show that, there is an average increase of 4 % in compression with respect to DCT based compression and DWT based compression. Also, in the proposed approach, Chroma cr. matrix is used for quantization, which has reduced the total processing time by a huge amount as compared to the other quantization matrices.

Keywords WMSN (wireless media sensor network) · DCT (discrete cosine transform) · DWT (discrete wavelet transform) · SF (secure force) · Compression · Encryption

1 Introduction

With the evolution of wireless sensor network (WSN) technology, the applications for which it was used initially, also diversified. In addition to transmit scalar data [1] such as temperature, humidity, rainfall, information etc. wireless sensor

A. Karthikeyan (✉) · V. Srividhya · P. Gupta · N. Rai
SENSE,VIT University, Vellore 632014, Tamil Nadu, India
e-mail: karthikeyan.anbu@vit.ac.in

V. Srividhya
e-mail: v.srividhya2015@vit.ac.in

P. Gupta
e-mail: Pranjay.gupta2012@vit.ac.in

N. Rai
e-mail: Naveen.rai2012@vit.ac.in

© Springer Science+Business Media Singapore 2016   475
R.K. Choudhary et al. (eds.), *Advanced Computing and Communication Technologies*,
Advances in Intelligent Systems and Computing 452,
DOI 10.1007/978-981-10-1023-1_47

networks are nowadays used for transmitting and receiving multimedia messages such as images, videos etc. The multimedia applications have led to the introduction of wireless multimedia sensor networks (WMSN) [1] which differs from WSN in terms of memory, processing power, and architecture [2]. Although, the concept of wireless technology remains the same but, the algorithms used for processing information are different from wireless sensor network.

Though, WMSN has a huge advantage over WSN in terms of amount of data that can be transferred but, there are certain limitations that hinder the performance of WMSN's such as, large memory consumption, high processing time and unreliable transmission. Since, wireless media sensor node deals with the images, most of the memory consumed is because of allocation of space to each and every pixel. As the image is composed of many pixels, there is a possibility that at some places, there is a little variation in the definition of pixels which, indicates a lot of redundant information. There are generally three types of redundancies (1) Spatial Redundancy: Where there is less variation in adjacent pixels, (2) Temporal Redundancy: Here we are concerned with number of bits required to represent given image and, (3) Spectral Redundancy: Where there is a correlation between different spectral domains. In this paper, we are more concerned about reducing temporal and spatial redundancies so that, we can compress the amount of data that has to be transferred across Wireless Media Sensor Network (WMSN). In order to remove these redundancies we have used a Hybrid approach [3] which incorporates DWT, followed by DCT algorithm. The main reason behind using this approach is that, we can reduce number of significant elements to a larger extent as compared to DCT or DWT compression done individually [3]. In DWT [4] four wavelets are created out of one image, amongst which, three are highly sparsified i.e. they contain a lot of zeros. While, in DCT [5, 6], the coefficients generated have most of the information i.e. the low frequency content, concentrated in the upper left corner of the compressed matrix. Even though, a lot of research has been done for compressing [2], or encryption the input image [7, 8], a scant amount of work considers simultaneous compression and encryption of media files that are to be transmitted [9], through wireless media sensor network. Various cryptographic algorithms can be used to encrypt an image [10] but, for WMSN we require an algorithm with less computational complexity and high reliability. For this purpose, we have chosen Secured Force encryption technique [11]. This technique performs well on the grounds of code complexity and complex key expansion where, it takes only five rounds for the encryption and performs the expansion process at data sink which, in turn, saves a lot of energy at the sensor nodes.

In the proposed algorithm, to maintain a constant PSNR and simultaneously increase the compression rate, different types of quantization matrixes are used for sparsifying the components after DWT and DCT coefficients are obtained. The considered matrices are JPEG, LUMA, Chroma, Chroma_Cr and Chroma_Cb [12, 13]. The matrices were evaluated on the basis of MSE, PSNR, Compression percentage and Time required for processing.

2 Methodology

2.1 Hybrid DWT-DCT Compression

Firstly, the two Dimensional DWT [14] technique is performed on image to obtain one approximate and three detailed coefficient of the image where, detailed coefficient matrix is highly sparsified containing small amount of significant elements to represent the image. It is done by performing 1D wavelet transform on row and columns respectively for a given image (Fig. 1), to obtain two columns representing detailed and approximate coefficient. After obtaining the approximate component, again the same procedure is followed to obtain three detailed and one approximate coefficient. Main advantage of using wavelet transform is that, representing image in form of wavelets avoids blocking which, gives a better results when the image is reconstructed.

Even though, DWT transforms the image using wavelets, compression is comparatively less as compared to DCT. DCT is done by grouping complete image into set of high frequency and low frequency image blocks. For compression, the high frequency image blocks can be neglected as, they don't have much contribution in defining the image at the output. After performing DCT on each of the wavelet, we get highly sparsified Approximate and Detailed Coefficients, which are then encrypted using SF encryption Technique.

2.2 Secure Force Encryption Technique

SF encryption Technique is a low complexity encryption algorithm that is used in WSN for fast and reliable encryption. Following are the steps involved in encrypting an image using SF encryption:

(a) Take the image whose number of columns and rows are multiple of 8.
(b) Take a manual key of 64 bits from the user.
(c) Using this key, 4 keys of size 16 bits are formed. This is also called Key Expansion.
(d) Key expansion is followed by Key management where using the LEAP [15] protocol, the 4 keys are updated in each wireless sensor node and a node Id is formed using these keys. Later, this algorithm helps in finding, if any node is imitating the base station.
(e) Using the 4 sub keys and one key that is derived from XOR of 4 sub keys, we encrypt the image and a highly confused and diffused image is formed. It is done by swapping and shifting the elements of image on the basis of 5 rounds which involve each of the generated keys.

2.3 Quantization Matrix

We have compared performance of 5 matrices for finding the optimized one for the proposed algorithm. They are, JPEG, LUMA, Chroma, Chroma_Cr and Chroma_Cb. The inability of JPEG to create a highly sparsified matrix encourages us to go for Chroma, Chroma_Cr and the Chroma_Cb matrix which are based on crude approximation of Gaussian Blur [13].

3 Results and Discussion

First, simultaneous compression and encryption using DCT and DWT were analyzed separately and were compared with the hybrid compression-encryption approach. They were evaluated on the basis of various performance metrics as shown in Table 1.

Here, time indicates the total time required by the MATLAB to execute the process.

It is evident that, the Hybrid DWT-DCT compression (Fig. 2) followed by, SF encryption (Fig. 3) technique has outperformed the other techniques on the grounds of compression rate, which is our primary concern. The main reason behind this is, the reduction in number of significant elements required for reconstructing the image (Fig. 4). Since, we are sparsifying twice using DWT and DCT, there is a significant rise in compression percentage but, at the cost of PSNR (Fig. 5).

Table 2 represents use of different quantization matrices to sparsify the DCT coefficient matrix and, it can be observed that, the matrix named Chroma_cr has the best compression percentage. It is because, the elements involved in this matrix have significantly high value. So, when we divide the coefficient matrix with

Table 1 Performance analysis of hybrid compression approach

Approach	PSNR (dB)	MSE	Compression (%)	Time (s)
DWT	20.149	628.1949	37.304	949
DCT	16.4094	1.49E+03	80.85	224.458646
DWT-DCT	16.5372	1.44E+03	87.9	116.7

Table 2 Comparison of different quantization matrix on hybrid compression

MATRIX	PSNR (dB)	MSE	Compression (%)	Time (seconds)
Chroma_Cr	16.4971	1.46E+03	94.33	65.700684
Chroma_Cb	16.507	1.45E+03	93.066	71.4406
Chroma	16.5061	1.45E+03	91.69	82.977216
LUMA	16.5364	1.44E+03	87.9	113.207213
Normal JPEG	16.5372	1.44E+03	87.9	116.7

Chroma_Cr matrix and approximate the elements to nearest integer, we find that a lot of elements become zero. Thus, the image gets highly compressed. The Chroma_Cr matrix and the hybrid compression approach gives the maximum compression. This is further used with the SF encryption technique. The input images along with the intermediate results of each stage are shown below.

3.1 Formulas

(a) DWT

For N * N dimensioned image which is divided into blocks of n * n dimension on which DWT is performed. Following are the equations used for finding the coefficient after performing DWT and finding inverse DWT from sparsified matrix.

$$W_\varphi(j_o, k_1, k_2) = \left(\frac{1}{\sqrt{n^2}}\right) \sum_{i_1=0} \sum_{i_2=0} f(i_1, i_2) \varphi_{j_0,k_1,k_2}(i_1, i_2) \tag{1}$$

$$W_\psi^q(j_1, k_1, k_2) = \left(\frac{1}{\sqrt{n^2}}\right) \sum_{i_1=0}^{n-1} \sum_{i_2=0}^{n-1} f(i_1, i_2) \psi_{j_1,k_1,k_2}^q(i_1, i_2) \tag{2}$$

$$f(i_1, i_2) = \left\{ \left(\frac{1}{\sqrt{n^2}}\right) \sum_{i_1} \sum_{i_2} W_\varphi(j_o, k_1, k_2) \varphi_{j_0,k_1,k_2}(i_1, i_2) \right.$$
$$\left. + \left(\frac{1}{\sqrt{n^2}}\right) \sum_{q=H_1,V_1,D_1} \sum_{j_1=0}^{\infty} \sum_{K_1} \sum_{K_2} W_\psi^q(j_1, k_1, k_2) \psi_{j_1,k_1,k_2}^q(i_1, i_2) \right\} \tag{3}$$

where, the following represents:
$W_\varphi(j_o, k_1, k_2)$ Approximate coefficient
$W_\psi^q(j_1, k_1, k_2)$ Detailed coefficients
$\varphi_{j_0,k_1,k_2}(i_1, i_2)$ Scaling function
$\psi_{j_1,k_1,k_2}^q(i_1, i_2)$ Wavelet function

$q = \{H_1, V_1, D_1\}$ gives index of LL, HH, HL Bands where LL has approximate and others have detailed coefficients.

(b) **DCT**

The DCT operation is governed by the following equation

$$F(u,v) = \frac{2}{n}C_uC_v \sum_{i_1=0}^{n-1}\sum_{i_2=0}^{n-1} f(i_1,i_2) \cos\left[\frac{\pi(2i_1+1)u}{2n}\right] \cos\left[\frac{\pi(2i_2+1)v}{2n}\right] \quad (4)$$

Similarly the inverse DCT is performed on the compressed block of image by using the following equation

$$f(i_1,i_2) = \frac{2}{n}C_uC_v \sum_{i_1=0}^{n-1}\sum_{i_2=0}^{n-1} F(u,v) \cos\left[\frac{\pi(2i_1+1)u}{2n}\right] \cos\left[\frac{\pi(2i_2+1)v}{2n}\right] \quad (5)$$

where u, v = 0, 1...n − 1 and

$$C_u, C_v = \begin{cases} \frac{1}{\sqrt{2}} & u,v = 0 \\ 1 & u,v \neq 0 \end{cases}$$

where $f(i_1,i_2)$, is the image block of size 256 * 256.

Fig. 1 Input image

Fig. 2 Compressed wavelets

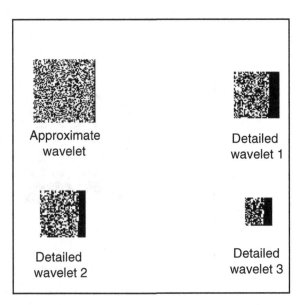

Fig. 3 Wavelets after encryption

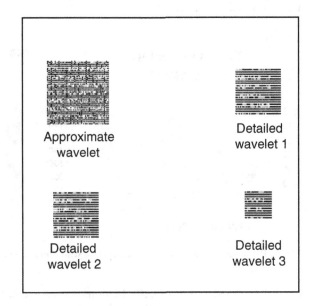

3.2 Analysis

In order to optimize the results of our hybrid approach we considered different quantization matrices and found that Chroma_cr has the highest compression percentage of 94.33 with a PSNR of 16.4971 which is approximately equal to other

Fig. 4 Reconstructed image

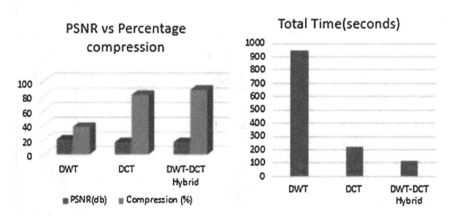

Fig. 5 Performance evaluation of SF encryption with different compression techniques

matrices, as evident from Fig. 6. Main reason behind such a high compression in Chroma_cr as compared to JPEG is, the elements that are used in Chroma are based on approximation of Gaussian blur in spectral domain giving rise to more weighted elements at lower right corner of Chroma matrix. This distribution helps in elimination of more high frequency components which carry little information, leading to a high compression, without any significant change in PSNR. The number of elements to be encrypted are highly reduced leading to great reduction in processing time. Time reduces by an average of 33 s in Chroma_Cr as compared to LUMA

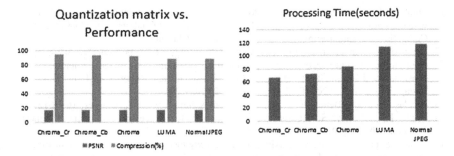

Fig. 6 Performance evaluation of hybrid compression and encryption with different quantization matrix

and JPEG. Thus, Hybrid approach with quantization matrix Chroma_cr, gives the most optimized results.

References

1. Akyildiz, I.F., Melodia, T., Chowdhury, K.R.: A survey on wireless multimedia sensor networks, computer networks, vol. 51, pp. 921–960. Elsevier (2007)
2. ZainEldin, H., Elhosseini, M.A., Ali, H.A.: Image compression algorithms in wireless multimedia sensor networks: a survey, Ain Shams engineering journal, vol. 6, pp. 481–490. Elsevier (2015)
3. Nandhini, A.S., Sankararajan, R., Rajendiran, K.: Video compressed sensing framework for wireless multimedia sensor networks using a combination of multiple matrices, computers and electrical engineering, vol. 44, pp. 51–66. Elsevier (2015)
4. Chowdhury, M.M.H., Khatun, A.: Image compression using discrete wavelet transform. IJCSI Int. J. Comput. Sci. Issues 9(4), No. 1 (2012)
5. Jaffar Iqbal Barbhuiya, A.H.M, Laskar, T.A, Hemachandran, K.: An approach for color image compression of JPEG and PNG images using DCT and DWT. In: International Conference on Computational Intelligence and Communication Networks (CICN), pp. 129–133 (2014)
6. Mohta, J, Pathak K.K.: Image compression and gamma correction using DCT. In: International Conference on Futuristic Trends on Computational Analysis and Knowledge Management (ABLAZE), pp. 322–326 (2015)
7. Alfalou, A., Brosseau, C.: Optical image compression and encryption methods. Adv. Opt. Photon 1, 589–636 (2009)
8. Alfalou, A., Mansour, A., Elbouz, M., Brosseau C.: Optical compression scheme to multiplex and simultaneously encode images. In: Optical and Digital Image Processing Fundamentals and Applications, pp. 463–483. Wiley, NewYork (2011)
9. Sawlikar1, A.P., Khan, Z.J., Akojwar, S.G.: Power optimization of wireless sensor networks using encryption and compression techniques. In: 2014 International Conference on Electronic Systems, Signal Processing and Computing Technologies, pp. 222–227 (2014)
10. Rajput, A.S., Mishra, N., Sharma, S.: Towards the growth of image encryption and authentication schemes. In: 2013 International Conference on Advances in Computing, Communications and Informatics (ICACCI), pp. 135–147 (2013)

11. Ebrahim, M., Chong, C.W.: Secure force: a low-complexity cryptographic algorithm for wireless sensor network (WSN). In: 2013 IEEE International Conference on Control System, Computing and Engineering, pp. 557–562. Penang, Malaysia (2013)
12. Zimbico, A., Schneider, F., Maia, J.: Comparative study of the performance of the JPEG algorithm using optimized quantization matrices for ultrasound image compression. In: Biosignals and Biorobotics Conference (2014): Biosignals and Robotics for Better and Safer Living (BRC), 5th ISSNIP-IEEE, pp. 1–6 (26–28 May 2014)
13. Nicolas Robidoux: Better JPEG quantization tables. http://www.imagemagick.org (2012)
14. Gupta, D., Choubey, S.: Discrete wavelet transform for image processing. Int. J. Emerg. Technol. Adv. Eng. 4(3), 598–602 (2015)
15. Zhu, S., Setia, S., Jajodia, S.: LEAP: Efficient security mechanisms for large scale distributed sensor networks. In: Proceedings of the 10th ACM Conference on Computer and Communications Security, pp. 62–72. Washington DC USA (2003)

A Novel Approach of Aspect Ranking Based on Intrinsic and Extrinsic Domain Relevance

Sheikh Amanur Rahman and M.M. Sufyan Beg

Abstract In this paper, we present a novel approach for ranking product aspects from customer feedbacks. Traditionally, the extraction of the aspects is based on the dependency analysis and sentiment classification of customer feedbacks. But there are two main limitations of above mentioned approaches. Firstly, it gives the large number of aspects which increases the complexity of ranking algorithm. Secondly, sometime it gives non-aspects as aspects of the product. Our proposed approach overcomes the above mentioned problem. First, it extracts the candidate feature using dependency analysis then it filters the genuine aspects with the help of intrinsic and extrinsic domain relevance score and finally aspect ranking algorithm is applied of the product aspects. The importance of this work is that it reduces the time complexity and space complexity of ranking algorithm and it gives only important and relevant aspects of any product in a ranked manner.

Keywords Opinion mining · Opinion features · Customer review · Dispersion · Deviation

1 Introduction

As the e-commerce industry is expanding, people are buying products online. They use to share their experiences about the product/service on social site which is termed as feedback comments or review. The identification of the aspects/features/attributes along with their rating or feedback from the review is termed as Opinion Mining. Formally, the objective of Opinion Mining (also known as sentiment analysis) is to analyse people's opinions, sentiments and attitudes

S.A. Rahman (✉) · M.M.S. Beg
Department of Computer Engineering, Aligarh Muslim University,
Aligarh, UP, India
e-mail: sheikhamanur@gmail.com

M.M.S. Beg
e-mail: mmsbeg@hotmail.com

© Springer Science+Business Media Singapore 2016   485
R.K. Choudhary et al. (eds.), *Advanced Computing and Communication Technologies*,
Advances in Intelligent Systems and Computing 452,
DOI 10.1007/978-981-10-1023-1_48

toward entities such as products, services and their aspects [1]. In recent years, most of aspect identification work is based on sentiment classification which depends on the concept of natural language processing. The computation and processing of natural language is not an easy task. So to avoid the complexity of sentiment classification, pruning, etc., it can be replaced with the concept of Intrinsic and Extrinsic Domain Relevance Method (IEDR) [2].

2 Related Work

There are several studies on analysing feedback comments in online shopping applications [3–6]. Supervised and unsupervised classes are the two categories of existing product aspect identification techniques. Some of the widely used supervised learning approaches are Hidden Markov Models, Conditional Random Fields, Maximum Entropy, Class Association Rules and Naive Bayes Classifier. Some of the unsupervised approaches are K-means, Mixture Models, Hierarchical Clustering, etc. In [7] frequent nouns and noun phrases are considered aspects for product reviews. In [8] it is further proposed to apply lexical knowledge patterns to improve the aspect extraction accuracy. Some work group aspects into clusters, assuming aspect opinion expressions are given [9]. Recently a semi-supervised algorithm [10] was proposed to extract aspects and group them into meaningful clusters as supervised by user input seed words. Mei et al. [11] utilized a probabilistic topic model to capture the mixture of aspects and sentiments simultaneously. For aspect sentiment classification, lexicon-based and supervised learning are the two major approaches. The bootstrapping strategy is usually employed to generate a high-quality lexicon. For example, Hu and Liu [7] started with a set of adjective seed words for each opinion class (i.e., positive or negative). Ding et al. [12] presented a holistic lexicon-based method to improve Hu's method [7]. In [7] frequent nouns and noun phrases are considered aspects for product reviews, and an opinion lexicon is developed to identify opinion polarities. A product may have hundreds of aspects and it is necessary to identify the important ones [13]. Wang et al. [14] developed a latent aspect rating analysis model, which aims to infer reviewer's latent opinions on each aspect and the relative emphasis on different aspects. Snyder and Barzilay [15] formulated a multiple aspect ranking problem. Most of the existing work is based on the pruning and sentiment classification which depends on processing of natural language and computation of natural language is a tedious task. Another problem with existing aspect ranking methods is that sometimes they identify irrelevant aspects as relevant aspects, hence increasing the complexity of ranking algorithm. According to [7], 15–20 % aspect belongs to such class.

3 Proposed Work

Our proposed method is aspect ranking based on intrinsic and extrinsic domain relevance (ARIEDR) and it consists of three parts: (i) Aspect Identification, (ii) Intrinsic and Extrinsic Domain Relevance [2] and (iii) Aspect Ranking. Figure 1 gives an overview of proposed work. Firstly, the potential candidate aspects are extracted from the customer feedback. Then potential candidate is filtered with the help of IEDR thresholding which gives the aspects (only relevant) of the given domain. Finally, any of the efficient ranking algorithms can be applied to get the list of aspects along with their ranking. The details of the proposed work are given below:

3.1 Potential Candidate Extraction

According to previous studies [16], candidates are usually noun and noun phrases. For identifying candidates in the text reviews, we can employ any existing approach. In particular, we first split the text reviews into sentences and then applied Stanford Parser [17] on each sentences. Stanford parser is used for part of speech (POS) tagging. It tags each word of sentences with their part of speech. The words with POS tag of noun and noun phrases are saved along with their frequencies and only the frequent nous are considered potential candidate.

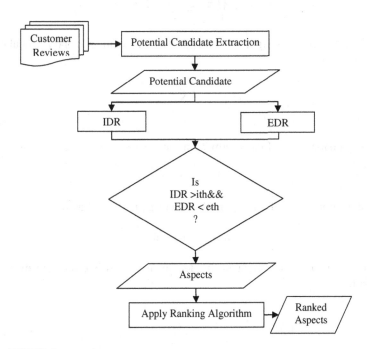

Fig. 1 ARIEDR framework

3.2 Intrinsic and Extrinsic Domain Relevance

The Intrinsic and Extrinsic Domain Relevance [IEDR] is based on [2]. The overview of the method is discussed below. Domain relevance shows that how much a term is related to a particular domain. Intrinsic domain relevance (IDR) represents the statistical association of the candidate to the given domain corpus whereas Extrinsic domain relevance (EDR) represents the statistical relevance of the candidate to the domain independent corpus. Intrinsic and Extrinsic Domain Relevance (IEDR) is the combination of IDR and EDR. The domain relevance is based on two kinds of statistics, i.e., dispersion and deviation. The computation of dispersion and deviation depends on *term frequency-inverse document frequency (TF-IDF)* term weights. The weight w_{ij} of term T_i in document D_j is calculated as follows:

$$w_{ij} = \begin{cases} \left(1 + \log TF_{ij}\right) \times \log \frac{N}{DF_i} & \text{if } TF_{ij} > 0, \\ 0 & \text{otherwise.} \end{cases} \tag{1}$$

where TF_{ij} is the term frequency of each term T_i in document D_j, DF_i is the global document frequency of term T_i, $i = 1, \ldots, M$ for a total number of M terms, and $j = 1, \ldots, N$ for a total number of N documents in the corpus. The standard variance s_i for term T_i is defined as follows:

$$s_i = \sqrt{\frac{\sum_{j=1}^{N} \left(w_{ij} - \bar{w}_i\right)^2}{N}}. \tag{2}$$

where $\bar{w}_i$ is the average weight of term T_i through all documents and is calculated as follows:

$$\bar{w}_i = \frac{1}{N} \sum_{j=1}^{N} w_{ij}. \tag{3}$$

The dispersion $disp_i$ of each term T_i in the corpus is calculated as follows:

$$disp_i = \bar{w}_i / s_i. \tag{4}$$

The deviation $devi_{ij}$ of term T_i in document D_j is defined as follows:

$$devi_{ij} = w_{ij} - \bar{w}_j. \tag{5}$$

where $\bar{w}_j$ is the average weight in the document D_j and is calculated over all M terms as follows:

$$\overline{w_j} = \frac{1}{M} \sum_{i=1}^{M} w_{ij}. \tag{6}$$

The domain relevance dr_i for term T_i in the given corpus can be calculated as follows:

$$dr_i = disp_i \times \sum_{j=1}^{N} devi_{ij}. \tag{7}$$

The calculation of IDR and EDR score solely depends on domain relevance dr_i score. When the domain relevance dr_i is calculated on domain specific review corpus, it is called IDR score and when domain relevance dr_i is calculated on domain independent corpus, it is called as EDR score. A candidate feature is termed as aspect if it's IDR score would be greater than minimum IDR threshold ith and it's EDR score would be lesser than maximum EDR threshold eth.

3.3 Aspect Ranking

There are lot of aspect ranking algorithm which works quite well. For ranking, we can employ any of the existing technique, but to make the system simple we chosen *term frequency* method. The reason behind choosing term frequency method is two: (i) to keep the system and computation simple and easy and (ii) the aspects identified in previous step is not only relevant to the domain also the important ones.

The *term frequency* is total number of occurrences of specific term in the given domain dependent review corpus. Depending upon the frequencies, the aspects are arranged in the descending order and can be ranked accordingly.

4 Result and Discussion

We have crawled 2,666 reviews on Apple iphone6 from *amazon.in* and *amazon.com*. For aspect identification, the evaluation is based on precision, recall and F-measure and compared with IEDR [2]. For aspect ranking, the evaluation is based on NDCG@5 and NDCG@10 and compared with Yu et al. [18]. NDCG@k [19] is widely used *Normalized Discounted Cumulative Gain at top k* evaluation metric for ranked aspects. It is calculated as follows:

$$NDCG@k = \frac{1}{Z} \sum_{i=1}^{k} \frac{2^{t(i)} - 1}{\log(1+i)}. \tag{8}$$

where Z is a normalization term derived from the top-k aspects of a perfect ranking and t(i) is the importance degree of the aspect at position i.

4.1 Evaluation of Aspect Identification

Table 1 shows the comparison of results between the ARIEDR and IEDR [2]. The results are not satisfactory because of the two reasons: (i) our method is based on English language dataset (we are first to use English dataset) whereas IEDR uses Chinese language dataset and (ii) review used by IEDR is mostly short in nature whereas review used by ARIEDR is mostly long in nature.

4.2 Evaluation of Aspect Ranking

Table 2 shows the comparison of results between the ARIEDR and Yu et al. [18]. The result shows that ARIEDR is not much better than Yu et al. but very close to that work. Considering that ARIEDR uses very simple ranking algorithm, the result can be improved by using some efficient algorithm.

4.3 Advantages of Proposed Work

There are mainly three advantages of ARIEDR. First, it helps in reducing the cost factor such as space complexity and time complexity of the system. By varying the value of ith and eth, desire number of aspects can be extracted and the space and time complexity of ranking algorithm depends on number of aspects. Second, it reduces the burden of several steps such as pruning and sentiment classification. Third, since ARIEDR gives only relevant and desired number of aspects, common people would not get confuse while making the decision regarding product.

Table 1 Comparison of results of aspect identification

Methods	Precision (%)	Recall (%)	F-measure (%)
IEDR	65.60	61.71	63.60
ARIEDR	43.63	69.89	53.72

Table 2 Comparison of results of aspect ranking

Methods	NDCG@5	NDCG@10
Yu et al.	0.948	0.902
ARIEDR	0.916	0.910

5 Conclusion

In this paper, we propose the novel approach of aspect ranking based on Intrinsic and Extrinsic domain relevance. This approach is quite effective and very useful in reducing cost factor of the system. It also helps in decision making of the common person about any product. The result of the proposed work is not very satisfactory but we will try to improve it in the future as we are in our initial phase of research. In future, we can make our system more efficient by employing more efficient parsing technique as well as aspect ranking algorithm.

References

1. Liu, B.: Sentiment analysis and opinion mining, In: Synthesis Lectures on Human Language Technologies, vol. 5, no. 1, pp. 1–167 (2012)
2. Hai, Z., Chang, K., Kim, J.J., Yang, C.C.: Identifying features in opinion mining via intrinsic and extrinsic domain relevance. In: IEEE Transactions On Knowledge and Data Engineering, vol. 26, no. 3 (2014)
3. O'Donovan, J., Smyth, B., Evrim, V., McLeod, D.: Extracting and visualizing trust relationships from online auction feedback comments. In: Proceedings of IJCAI, pp. 2826–2831. San Francisco, CA, USA, (2007)
4. Lu, Y., Zhai, C., Sundaresan, N.: Rated aspect summarization of short comments. In: Proceedings of 18th International Conference World Wide Web, pp. 131–140. New York, USA (2009)
5. Gamon, M.: Sentiment classification on customer feedback data: noisy data, large feature vectors, and the role of linguistic analysis. In: Proceedings of 20th International Conference on COLING (2004)
6. Hijikata, Y., Ohno, H., Kusumura, Y., Nishida, S.: Social summarization of text feedback for online auctions and interactive presentation of the summary. Knowl. Based Syst. 20(6), 527–541 (2007)
7. Hu, M., Liu, B.: Mining and summarizing customer reviews. In: Proceedings of 4th International Conference KDD, pp. 168–177 (2004)
8. Qiu, G., Liu, B., Bu, J., Chen, C.: Opinion word expansion and target extraction through double propagation. Comput. Linguist. 37(1), 9–27 (2011)
9. Zhai, Z., Liu, B., Xu, H., Jia, P.: Constrained LDA for grouping product features in opinion mining. In: Proceedings of 15th PAKDD, pp. 448–459 (2011)
10. A. Mukherjee and B. Liu: Aspect extraction through Semi-supervised modeling. In: Proceedings of 50th ACL, vol. 1, pp. 339–348 (2012)
11. Mei, Q., Ling, X., Wondra, M., Su, H., Zhai, C.X.: Topic sentiment mixture: modeling facets and opinions in weblogs. In: Proceedings of 16th International Conference of World Wide Web, pp. 171–180. Banff, AB, Canada, (2007)
12. Ding, X., Liu, B., Yu, P.S.: A holistic lexicon-based approach to opinion mining. In: Proceedings of WSDM, pp. 231–240, New York, NY, USA (2008)
13. Zha, Z.J., Yu, J., Tang, J., Wang, M., Chua, T.S.: Product aspect ranking and its applications. IEEE Trans. Knowl. Data Eng. 26(5) (2014)
14. Wang, H., Lu, Y., Zhai, C.X.: Latent aspect rating analysis on review text data: a rating regression approach. In: Proceedings of 16th ACM SIGKDD, pp. 168–176. San Diego, CA, USA (2010)

15. Snyder, B., Barzilay, R.: Multiple aspect ranking using the good grief algorithm. In: Proceedings of HLT-NAACL, pp. 300–307. New York, NY, USA (2007)
16. Liu, B.: Sentiment analysis and subjectivity. Handbook of Natural Language Processing. Marcel Dekker, Inc., New York, NY, USA (2009)
17. The stanford natural language processing group. http://nlp.stanford.edu/software/lex-parser.shtml
18. Yu, J., Zha, Z.J., Wang, M., Chuam, T.S.: Aspect ranking: identifying important product aspects from online consumer reviews. In: Proceedings of the 49th Annual Meeting of the Association for Computational Linguistics, pp. 1496–1505
19. Jarvelin, K., Kekalainen, J.: Cumulated gain-based evaluation of IR techniques. ACM Trans. Inform. Syst. 20(4), 422–446 (2002)

A Compact H Shaped MIMO Antenna with an Improved Isolation for WLAN Applications

Nirmal Pratima and Nandgaonkar Anil

Abstract A simple H shaped MIMO antenna with side to side spacing of 6 mm with three pair of unequal length of slits in the ground plane is designed at 5.8 GHz for WLAN application. A pair of unequal slit length is etched in the ground plane to reduce the mutual coupling between the MIMO antenna element is studied and proposed in this paper. High isolation of more than 45 dB is achieved using this proposed design. The MIMO antenna is spaced $0.311\lambda o$ apart from the center sharing the same ground plane. The isolation is increased from 24 to 46 dB with the help of unequal length of slit in the ground plane. The structure is fabricated on the 40 mm × 24 mm finite ground plane leading to small size and coaxial feeding is used to feed the patch. The proposed design is simple to construct on FR4 substrate leading to low cost with an advantage of high isolation. The proposed H shaped MIMO structure is fabricated and tested. The effect of various parameters such as spacing between slits, the width of the slit, number of slits of equal length and unequal length of slits on mutual coupling is studied.

Keywords Multiple input multiple output (MIMO) antenna · Isolation · Mutual coupling reduction · Defected ground structures

1 Introduction

MIMO antenna makes use of many numbers of antennas for transmitting and receiving the signals. The capacity of system can be increased by increasing no of antennas. The use of MIMO antenna has following benefits: higher capacity, increased data rates, reduced co-channel interference, high speed and a wide

N. Pratima (✉) · N. Anil
Department of Electronics and Telecommunication Engineering,
Dr. BATU Technological University, Lonere, Raigad, India
e-mail: pratima.nirmal@gmail.com

N. Anil
e-mail: abnandgaonkar@yahoo.com

© Springer Science+Business Media Singapore 2016   493
R.K. Choudhary et al. (eds.), *Advanced Computing and Communication Technologies*,
Advances in Intelligent Systems and Computing 452,
DOI 10.1007/978-981-10-1023-1_49

coverage area [1]. The difficulty in design of MIMO antenna is to place the multiple antennas closer to each other with a reduced mutual coupling and improved isolation characteristics. Among varieties of antennas, microstrip antenna is widely used as it has planar structure, low profile, low cost, easy to fabricate and integrate [2, 3]. So design of MIMO antenna using microstrip patch antenna will avail the benefits of low profile and planar structure.

Various way of reducing the mutual coupling is reported with its advantages and disadvantage. EBG structures, basically are periodic structure is placed between antenna to lowers the mutual coupling [4, 5]. EBG structure suppresses the propagation of the surface wave and it also improves the gain of antenna and reduces the back radiation. The drawback of EBG is it occupies a large area and is difficult to design. An incorporating parasitic element between the antenna element is another way of reducing mutual coupling and improving isolation is being reported [6].

The parasitic element helps to introduce the double coupling path by creating a reverse coupling to lower the mutual coupling. The amount of reverse coupling produced depends upon the position between the antenna element and the parasitic element, the number of parasitic elements and spacing between parasitic elements.

By placing U shaped parasitic element between the two antenna, mutual coupling can be lowered [7]. A U shaped section is placed in such a way that it creates an indirect signal that cancel outs the signal coming from one element and hence reduce the coupling from one element to other elements. The advantage of using this method is improved isolation and ease of integration of U shaped microstrip section with printed antennas. Isolation of more than −45 dB is reported by using metamaterial having negative permittivity and permeability [8]. The advantage of using metamaterial is its compact size. Inductance and capacitance of designed structure determines the resonant frequency of these antennas.

Defected Ground Structure (DGS) is used to minimize the mutual coupling by introducing a defect in the ground plane which disturbed the surface current distribution [9]. The isolation of 40 dB is achieved between the microstrip antennas with the help of slot structure in the ground [10]. The various parameter such as dimension and the spacing of the slot is optimized to obtain maximum isolation. PIFA Antenna with slits in the ground plane is studied to obtain the isolation of more than 20 dB.

In this paper, an easy way to improve isolation between MIMO antenna using a pair of unequal length of slits is proposed. The unequal length of slit is introduced in the ground plane which disturbs the surface current distribution and improves the isolation as compared to an equal length of slit in the ground plane. This technique helps to minimize the mutual coupling between antenna elements. H shaped MIMO antenna with the centre to centre distance of $0.311\lambda o$ is designed on low cost FR-4 substrate. The effect of various parameters such as length, position, width and spacing between slit is studied to lower the mutual coupling and improve the isolation.

2 Proposed Design

An H shaped MIMO antenna with the dual element operating at 5.8 GHz is as shown in Fig. 1. The length and width of H shaped MIMO antenna are 10.1 mm × 15 mm. The MIMO antenna is placed $0.311\lambda o$ apart from the center sharing the same ground plane of $0.774\lambda o \times 0.464\lambda o$. When two antenna elements are placed on the common ground plane, current is induced in the ground plane led to the increasing coupling between the antennas. The suppression of coupling current in the ground is adequate to improve isolation in the far field coupling. But in near field coupling, direct radiation from antenna element as well current induced in the ground plane leads to mutual coupling among the antennas. Hence by placing a pair of slit of unequal length in the ground plane, coupling can reduce by a good amount. The three unequal pair of slit of length Ls0, Ls1 and Ls2 is introduced at the middle of the ground to increase isolation among antenna elements. The length, position, width and spacing of slit is optimized to have rejection band at 5.8 GHz. The proposed antenna is constructed on the FR-4 substrate of dielectric permittivity 4.4, a thickness of 0.159 mm and loss tangent of 0.002. The antenna is fed by a coaxial probe of 50 Ω. The structure is optimized to operate over 5.725–5.875 GHz, ISM band. The antenna is fabricated on 40 mm × 24 mm ground plane.

The simulated result of the H shaped MIMO antenna without DGS and with DGS is as shown in the Figs. 2 and 3 respectively. The H shaped MIMO antenna without DGS have mutual coupling of more than −20 dB across the desired band. When a pair of unequal slit length is introduced in the ground plane, a large amount of current is being trapped by the slit in the ground plane next to the radiating patch which leads to less current propagating across the other slits which leads to

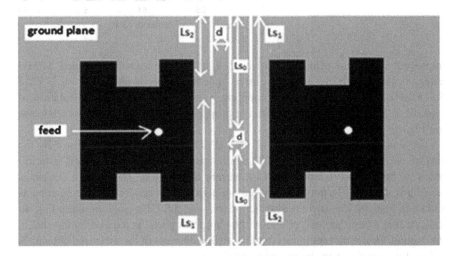

Fig. 1 Proposed structure

Fig. 2 Simulated S
parameter without DGS

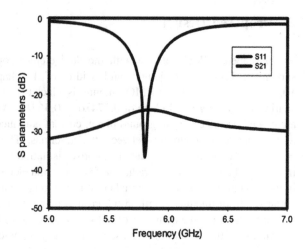

Fig. 3 Simulated S
parameter with DGS

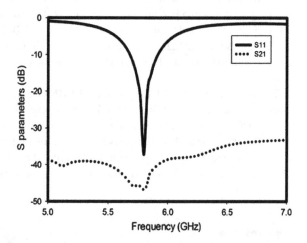

improved mutual coupling as compared to the conventional H shaped MIMO antenna without DGS. The measured impedance bandwidth is from 5.725–5.875 GHz with reduced mutual coupling over the band. The high isolation of −46 dB at 5.8 GHz is obtained with the proposed antenna with DGS as shown in the Fig. 3.

The mutual coupling behavior of the MIMO antenna is well explained by the surface current distribution graph. The port 1 is excited and second port is terminated by 50 Ω load to obtain surface current distribution. The surface current is flowing from port 1 to port 2 which lead to high mutual coupling in the antenna as shown in Fig. 4a. The current distribution graph at 5.8 GHz is as shown in Fig. 4b indicate that a large amount of current flowing from port 1 is trapped by the slits in the ground. Hence current induced in the second element of the proposed antenna is reduced as compared to the antenna without slits.

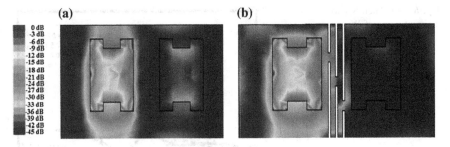

Fig. 4 Surface current distribution at 5.8 GHz. **a** Without slits and **b** with the slits

3 Parametric Study

The mutual coupling of about −24 dB is reported without DGS at 5.8 GHz, but when an unequal pair of length is introduced in the ground plane about 22 dB reduction in mutual coupling is achieved. In order to optimize the design structure to have a good amount of isolation and reduced mutual coupling, the effect of slitted ground parameter is studied. The following parameter affect the mutual coupling and isolation are: (a) Spacing between the slits (b) Width of the slit (c) Number of slits of equal length and (d) Unequal Length of slits.

The Spacing between slit in the ground plane (d) is varied from 0.5 to 2 mm as shown in the Fig. 5. It has been observed that the optimum spacing is required between the slits to have a good amount of isolation. So optimized Spacing of 0.5 mm is chosen in the proposed design. The other parameter is Width of the slit (W). When the width of the slit is change from 0.25 to 1 mm, the largest isolation

Fig. 5 Simulated S_{21} for different value of d

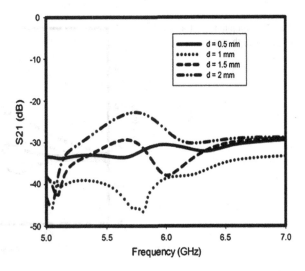

Fig. 6 Simulated S_{21} for
different value of W

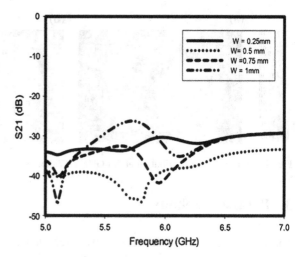

appears at various frequency as shown in Fig. 6. The maximum isolation of -46 dB
is obtained at a resonant frequency 5.8 GHz for width w = 0.5 mm.

The next parameter is the number of slits of equal length of 11 mm is varied to
optimize the structure as shown in Fig. 7. As the numbers of slits in the ground
plane increases from 1 to 3, there is a reduction in mutual coupling of about dB. But
as we increase the slit to 4 or 5, there is degradation in mutual coupling because of
more interaction between the antenna element and no of slot. Hence the no of slit is
chosen to be 3 for our proposed design. The other parameter is the slits of unequal
length of Length Ls0, Ls1 and Ls2 is varied as shown in Fig. 8. When the slits of
unequal length Ls0, Ls1 and Ls2 is added to ground plane, more current is trapped
by slit adjacent to the patch. Due to the unequal length of the slits introduced in the

Fig. 7 Simulated S_{21} for no.
of slits length (N)

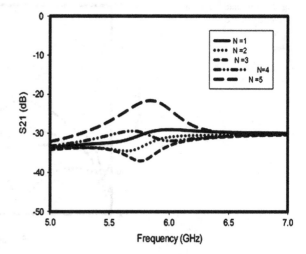

Fig. 8 Simulated S21 parameter of equal for different length for 3 slit

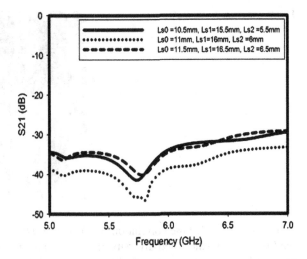

ground plane, a large amount of current is trapped leading to reduced mutual coupling among the antennas. The Length Ls0, Ls1 and Ls2 is optimized to have high isolation between the antenna. The optimized dimensions are Ls0 = 0.5 mm, Ls1 = 16 mm and Ls2 = 6 mm

4 Fabricated Antenna and Measured Results

An H shaped MIMO antenna is fabricated and tested using vector network analyzer. The Fabricated antenna is as shown in the Figs. 9 and 11a, b show the simulated and measured S_{11} and S_{21} of the proposed antenna structure. The measured result follows the simulated result. It can be seen that mutual coupling S_{21} is less than -35 dB over 5.725–5.875 GHz frequency band. The correlation coefficient is another parameter which signifies the diversity achieve in MIMO antennas. The correlation coefficient should be less than 0.5 dB so that the good amount of diversity is achieved. The correlation coefficient is being calculated from scattering parameter using the following formula:

$$\rho = \frac{|S11 * S12 + S21 * S22|}{|(1 - |S11^2| - |S21^2|)(1 - |S22^2| - |S12^2|)|} \quad (1)$$

and the diversity gain is calculated as follows:

$$G_{app} = 10 \times \sqrt{1 - |\rho|} \quad (2)$$

Fig. 9 Fabricated antenna

It can be seen that the correlation coefficient of 0.001 dB is achieved as shown in the Fig. 10. The calculated diversity gain is more than 9 dB is achieved through the proposed antenna over 5.725–5.875 GHz band (Figs. 11).

Fig. 10 Correlation coefficient

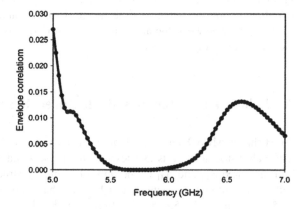

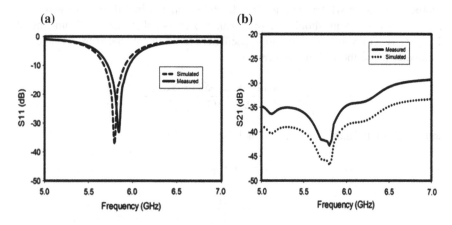

Fig. 11 Measured and simulated. **a** S_{11} and **b** S_{21} of the proposed structure

5 Conclusion

This paper proposes a simplest method to lower the mutual coupling between two closely spaced MIMO antenna elements. The structure is easy to fabricate, low cost, small size and is suitable for WLAN application. A good amount of isolation is achieved through this proposed structure. The effect of various parameters such as length, number of slits, width and spacing between slits is studied to reduce the mutual coupling. The correlation coefficient of less than 0.001 dB is achieved throughout the required band. These techniques can be applied for more than two element of MIMO antenna.

References

1. Jamil, A., Yusoff, M.Z., Yahya, N., Zakariya, M.A.: Design and parametric study of multiple element MIMO antennas for WLAN. In: Intelligent and Advanced Systems (ICIAS), 2012 4th International Conference, vol. 1, pp. 194–199, Kuala Lumpur Malasiya (2012)
2. Jamil, A., Yusoff, M.Z., Yahya, N., Zakariya, M.A.: Current issues and challenges of MIMO antenna designs. In: Intelligent and Advanced Systems (ICIAS), 2010 International Conference, vol. 1, pp. 1–5, Kuala Lumpur Malasiya (2010)
3. Kumar, G., Ray, P.K.: Broadband microstrip antennas. Artech House, Norwood (2003)
4. Manh, C.T., Han, X., Ouslimani, H.H., Priou, A., Marteau, A., Collignon, G.: Reduction of the coupling between microstrip arrays antennas using high impedance surface (HIS) structure. In: European Microwave Conference (EuMC), pp. 1453–1456, Paris (2010)
5. Ma, N., Zhao, H.: Reduction of the mutual coupling between aperture coupled microstrip patch antennas using EBG structure. In: Wireless Symposium (IWS), 2014 IEEE International, pp. 1–4, X'ian (2014)
6. Zhengyi, L., Zhengwei, D., Takahashi, M., Saito, K., Ito, K.: Reducing mutual coupling of MIMO antennas with parasitic elements for mobile terminals. In: IEEE Transactions on Antennas and Propagation, vol. 60, pp. 473–481. IEEE Press (2012)
7. Farsi, S., Aliakbarian, H., Schreurs, D., Nauwelaers, B., Vandenbosch, G.A.E.: Mutual coupling reduction between planar antennas by using a simple microstrip U-section. In: IEEE Antennas and Wireless Propagation Letters, vol. 11, pp. 1501–1503. IEEE Press (2012)
8. Abdalla, M.A., Ibrahim, A.A.: Compact and closely spaced metamaterial MIMO antenna with high isolation for wireless applications. In: IEEE Antennas and Wireless Propagation Letters, vol. 12, pp. 1452–1455. IEEE Press (2013)
9. Habashi, A., Nourinia, J., Ghobadi, C.: A rectangular defected ground structure (DGS) for reduction of mutual coupling between closely-spaced microstrip antennas. In: 20th Iranian Conference on Electrical Engineering, (ICEE2012), pp. 1347–1350. IEEE Press (2012)
10. Ouyang, J., Yang, F., Wang, Z.M.: Reducing mutual coupling of closely spaced microstrip MIMO antennas for WLAN application. In: IEEE Antennas and Wireless Propagation Letters, vol. 10, pp. 310–313. IEEE Press (2011)

Design and Testing of Low Power Reliable NoCs for Wireless Applications

R. Ganesan, G. Seetharaman, Tughrul Arslan, C. Vennila
and T.N. Prabakar

Abstract Reducing the power consumption of Network on Chip is a careful trade-off between performance parameters such as speed, power, area and reliability in general. In this paper, various power reduction techniques are proposed, that can significantly reduce power consumption in NoCs. First, node traffic is studied between different cores in NoCs and high traffic nodes are placed near each other. Secondly power on unused paths is switched off. Also power is monitored on each path and if it exceeds a predetermined level, then that path is blocked off for some time. Thirdly voltage scaling technique is employed between different paths in NoCs which leads to significant power reduction. Fourth, structure based approach is adopted for design of NoCs which implies placement of nodes at predetermined places in NoCs. For this, Xilinx and Altera tools are employed to assign position for the design using macros and User Constraints File (UCF) and Logic-Lock features. Finally, FSM controllers are replaced with R3TOS (Reliable Reconfigurable Real time Operating System) for the design of NoCs which significantly reduce power consumption. R3TOS also increases the reliability by relocating the critical tasks in runtime, if the underlying silicon is damaged or faulty. From the proposed methods, it is observed that the design of NoCs must result in low power consumption

R. Ganesan (✉) · G. Seetharaman · T.N. Prabakar
Oxford Engineering College, Trichy 620009, India
e-mail: rganesaa@gmail.com

G. Seetharaman
e-mail: jgsraman@gmail.com

T.N. Prabakar
e-mail: tnprabakar@gmail.com

T. Arslan
System Level Integration Group, University of Edinburgh, Edinburgh EH9 3JL, UK
e-mail: t.arslan@ed.ac.uk

C. Vennila
Saranathan College of Engineering, Trichy 620012, India
e-mail: venilla-ece@saranathan.ac.in

© Springer Science+Business Media Singapore 2016   503
R.K. Choudhary et al. (eds.), *Advanced Computing and Communication Technologies*,
Advances in Intelligent Systems and Computing 452,
DOI 10.1007/978-981-10-1023-1_50

without sacrificing performance metrics. Also the above said techniques can very well be adopted for Application Specific NoCs based on wireless applications such as Software Defined Radios (SDR) and Cognitive Networks.

Keywords ASNoCs · R3TOS · Custom topology · Macro · UCF · Low power · Reliability

1   Introduction

Networks-on-Chip (NoC) has emerged as a feasible solution to overcome communication problem in a System on Chip (SoC). NoC brings the concept of packet switched network as compared with circuit switched network on to the chip. In NoC, I/O blocks are connected through routers. Network topology is the way by which the interconnection link and the IP cores are connected to form a network. Standard topologies like Mesh, Ring, Star and Binary tree are mainly used to interconnect routers and IP blocks. Standard topologies are suitable for NoCs that are reusable in many applications. In addition to the standard regular topologies, custom made irregular topology is also available for Application Specific NoC (ASNoC). However, for ASNoC, such standard topologies would lead to poor performance such as increased area, increased power consumption and latency thereby limiting the use of standard topologies for ASNoC. Hence, for ASNoCs tailor made custom topology has to be designed to increase their performance metrics. The custom topology utilizes fewer resources like routers and interconnection links that lead to less area and power consumption [1, 2].

Power consumption forms a major concern in modern VLSI systems. While leakage and short circuit currents contribute to static power consumption, bit transition contributes the major part of dynamic power consumption. Interconnects are the major power source of power consumption in NoCs. While aggressive inter wire spacing and shielding presents reduction of power at physical level, elimination of redundant bit transmission presents a solution at system layer of NoC [3]. In this paper, a large number of techniques are analyzed to reduce power consumption of ASNoC based wireless applications.

NoC structures are being increasingly used in diverse application specific networks. The NoCs used for applications such as desktops require low power consumption due to related thermal issues whereas mobile application require low power NoC architecture for long lasting battery requirements. The motivation for this paper stems from the need for exploring various techniques in NoCs for reducing power without increasing the clock speed.

The organization of the rest of the paper is as follows: In Sect. 2, an overview of Network on Chip is described. In Sect. 3, Software tools for NoCs are presented. Section 4 summarizes existing power reduction methods and Sect. 5 presents the proposed research work. Section 6 summarizes the conclusions and future works.

2 Network on Chip Architecture

Miniaturization of transistors and improved manufacturing methods reduce the size of IC chips. As the size of IC chips grow smaller, number of IP blocks increase on a single chip. SoCs are too complex structure to use traditional bus or crossbar communication links. NoCs are communication block of traditional SoCs. Higher operating frequencies, reduced wire congestion and reuse of IPs are the principal advantages where NoCs score over traditional routing methods shown in Fig. 1. Conventional NoC topologies are shown in Fig. 2. There are two switching schemes circuit based and packet based. Packet switching refers to breaking down the entire data into small number of packets called flits. Wormhole switching is the most general form of packet switching. Wormhole switching is susceptible to deadlock in which a set of packets are blocked forever in its wait for network access. Packet switching reduces the total power consumption as compared with circuit switching.

In addition, application specific topologies are suggested after studying the traffic pattern for particular NoC application. In application specific routing, algorithm, area and power optimization is achieved for a particular application. Routing algorithms form an integral part of NoC structure. Routing algorithms are divided into two sub groups oblivious and adaptive. Oblivious is further subdivided into deterministic and stochastic. Deterministic routing refers to sending the packet between any two nodes only on a predetermined path every time whereas stochastic routing depends on randomness. Adaptive routing takes into account congestion on the routing path to establish connection between cores.

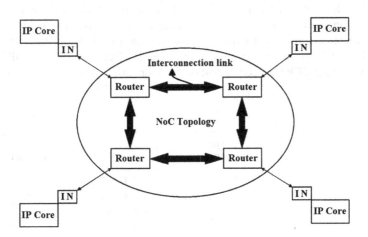

Fig. 1 Traditional NoC structure

Fig. 2 Standard topologies for NoCs

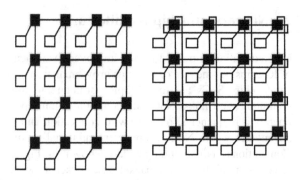

3   Software Tools for NoCs

Interconnects are the major power source of power consumption in NoCs. While aggressive inter wire spacing and shielding reduces power at physical level, elimination of redundant bit transmission presents a solution at system layer of NoC [3]. Various techniques are proposed in this paper to reduce power consumption of ASNoC based applications. The power analysis in our proposed paper is determined after placement of cores using Xilinx Macro, UCF and Altera LogicLock which are described below [4, 5].

3.1   User Constraints File (UCF)

The UCF file is an American Standard Code for Information Interchange (ASCII) file specifying constraints on the logical design. These constraints affect how the logical design is implemented in the target device. Figure 3 illustrates the UCF flow. The UCF file is an input to Native Generic Database (NGD) build. The constraints in the UCF file become part of the information in the NGD file produced

Fig. 3 UCF flow

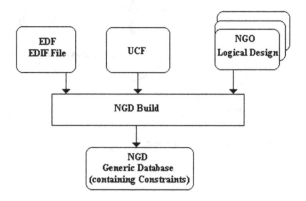

by NGD Build. The syntax for various assignments through UCF is given in literature. In this syntax, NET is used for the individual pin assignments and INST is used for assigning the position for the sub modules.

UCF file is automatically created in the Xilinx ISE 8.1 software after opening the new project. In Xilinx ISE 8.1 software, the Constraints Editor may be used to create constraints within a UCF file. The constraints may be entered with any text editor. After specifying the constraints in the UCF, the file is saved with different name to avoid confusion. Otherwise, the version and revision name for the particular design have to be remembered because for each implementation, the tool overwrites UCF file.

3.2 Macros

Macro is a pre-defined primitive that has a known implementation in the target technology. For instance, many FPGAs provide specialized logic for addition. Macros are a form of design using intellectual property core. A macro for addition provides a simple way for the designer to perform the addition. The CAD tools identify the macro and map it into the specialized addition logic. If the logic is later targeted to a different device that does not provide this specialized logic, it is still possible to rewrite the macro in terms of generic logic. Macros provide a fairly painless way for designers to make use of specialized features.

3.3 LogicLock

The LogicLock block-based design flow enables to design, optimize, and lock down a design one section at a time. With the LogicLock methodology, designer can independently create and implement each logic module into a hierarchical or team-based design. With this method, designer can preserve the performance of each module during system integration. Additionally, designer can reuse logic modules in other designs, further leveraging resources and shortening design cycles. The LogicLock design flow allows to "lock" the placement and routing of nodes in a region of a device so that the relative placement of logic within the LogicLock region or regions remains unaltered. Quartus II software then places the LogicLock region into the top-level design with these constraints.

3.3.1 LogicLock Region Properties

A LogicLock region is defined by its size (height and width) and location (where the region is located on the device). Designer can specify the size and/or location of a region, or Quartus II software can generate them automatically. The Quartus II software bases the size and location of the region on its contents and the module's timing requirements. Quartus II software cannot automatically define a region's size if the location is locked. Therefore, the size of the partition must be specified, if the region is to be placed in a particular location.

3.3.2 Hierarchical (Parent/Child) LogicLock Regions

With the LogicLock design flow, designer can define a hierarchy for a group of regions by declaring parent/child regions. Quartus II software places a child region completely within the boundaries of its parent region and allows us to further constrain the module locations. Additionally, parent/child regions allow us to further improve a module's performance by constraining the nodes in the module's critical path. The advantage of LogicLock region compared to UCF is that not only the interconnect length is reduced but also the position of the complete design is locked. After assigning the position using LogicLock region, the design is back-annotated and exported. The advantage of UCF compared to LogicLock region is that it is not only used for assigning the position but also used for connecting the I/O's to the light emitting diodes (LEDs), liquid crystal display (LCD) and port pins. Using these assignments, the designers can verify the circuit after downloading the design to the FPGA. The advantage of LogicLock compared to MACRO is that no manual routing is required for the integration of top level design with clock and skew blocks.

4 Existing Power Reduction Methods

Reduction of inter wire spacing and shielding has been proposed at physical level for power reduction [6]. Power synthesis algorithms are proposed to be introduced at system level if the traffic between cores is known as a priori. Interconnects consume a large share of power in NoCs. Hamming distance forms the basis for data encoding technique. If the change in bits over successive data transmission over links changes by half the number of bits, then different encoding techniques were adopted. Some of the methods are Bus Invert (BI), adaptive coding, gray coding and transition methods among others. In addition adaptive data compression techniques have been employed for power reduction. In our proposed research work, power reduction for an application specific NoC is studied by placement and clustering of cores using software tools.

5 Power Analysis in NoCs

Existing method of interconnection of cores with planar metal interconnect results in significant power consumption. Power saving can be achieved if these metal interconnects are replaced with wireless links [7].

Power efficiency is one of the most important concerns in NoC design. Consider a 10×10 tile-based NoC, assuming a regular mesh topology and 32 bit link width in 0.18 μm technology and minimal spacing, under 100 Mbit/s pair-wise communication demands, interconnects will dissipate 290 W of power [8]. Thus, reducing the power consumption on global interconnects is a defect to the success of NoC design. Processing Element (PE) and router are the main components in the NoC. The PE is a module that generates/receives packets based on a traffic model like Local, uniform and hotspot. Routers receive packet on their input channels and after routing a packet based on the routing algorithm and destination address, the packet is sent to the selected channel.

NoCs are the communication link of SoCs. Interconnects define communication between different nodes of NoCs. In our work, MPEG decoder is taken as an example of ASNoC. Communication volumes between different cores of the NoCs are studied first. High volume nodes are placed near each other using Xilinx UCF. Macros define the communication link between the different cores. In addition different cores are clustered together to form a cluster approach using Xilinx UCF and Macros. Power dissipation is studied and found to be lower if high communication nodes are placed near each other. [9, 10] Placement of clusters and nodes are attempted for the objective of equal power dissipation between different nodes and different clusters.

Altera LogicLock defines a functional block in the physical layout. Using Altera LogicLock physical placement of cores in NoCs can be defined. Using Altera LogicLock tool, physical placement of cores and clusters for MPEG ASNoC is being attempted and power consumption is considerably reduced if high volume nodes are placed near each other.

In NoC communication between different cores are packet based. The flits contain header information which is decoded using Finite State Machine (FSM). When different nodes request the same communication channel, an arbiter scheme generally employing Round Robin algorithm is used to resolve the conflict. A FSM is employed in arbitration [11]. In our proposed research work, for both of these FSM structure, a R3TOS structure is proposed to be employed as elaborated in Fig. 4 [12, 13]. In R3TOS system hardware tasks can be allocated to non-damaged FPGA resources thereby enhancing reliability. The reconfigurable area is defined using Xilinx UCF and Macros and in case of Altera Tools using LogicLock. As reconfiguration results in effective use of physical resources, considerable power savings is envisaged. In addition reliability is enhanced as faulty FPGA computation links are isolated in real time.

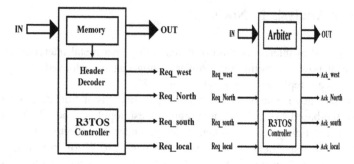

Fig. 4 R3TOS structure to be used in place of FSM controller in ASNoC

6 Conclusions

Interconnects form the main power consuming block of typical NoCs. As no software simulation tool accounts for interconnect delay, power reduction by specifying placement constraint is studied in this paper for reducing power. In this paper, an example ASNoC MPEG decoder is first implemented as NoC. Power reduction in MPEG ASNoC is studied by physical placement of nodes and physical definition of communication links using Xilinx UCF and Macrotools and LogicLock tool in case of Altera Quartus software. It is concluded that placement of high volume nodes near each other results in significant power reduction and the ASNoC structure is most suited for the heterogeneous structure. Further reduction in power consumption will result if reconfigurable structure is implemented in place of FSM for header decoding and communication channel arbitration. Because in R3TOS hardware tasks are scheduled to non-faulty FPGA resources, reliability is considerably increased. In general, NoC design is a tradeoff between speed, area, reliability and power. Our proposed work increases reliability while reducing power consumption at the same time. Detailed implementation results will be presented in subsequent work. In the next phase of the proposed work, multi hop wires in the router interconnects will be replaced with single hop wireless links.

References

1. Maheswari, M., Seetharaman, G.: Design and implementation of low complexity router for 2D mesh Network on Chip using FPGA. In: The Proceedings of International Conference on Embedded System Application. Los Vegas, USA, 18–21 July 2011
2. Maheswari, M., Seetharaman, G.: Implementation of application specific Network on chip architectures on reconfigurable devices using topology generation algorithm with genetic algorithm based optimization technique. In: Venugopal, K.R., Patnaik, L.M. (eds.) ICIP 2012, CCIS 292, pp. 436–445. Springer-Verlag, Berlin Heidelberg
3. Dally, S.J., Towels, B.: Route packets, not wires: on chip inter connection networks. In: Proceedings of Design Automation Conference, pp. 684–689

4. Xilinx Tutorials. www.xilinx.com
5. Quartus Tutorial. www.altera.com
6. Deb. S., et.al.: Wireless NoC as interconnection backbone for multicore chips: Promises and Challenge. IEEE J. Circuits Syst. **2** (2012)
7. Hu, Y., Chen, H., Zhu, Y., Chien, A.A., Cheng, C.: Physical synthesis of energy-efficient network-on-chip through topology exploration and wire style optimizations. Design (ICCD), pp. 111–118 (2005)
8. Ganguly, A., Pande, P.P., Belzer, B., Grecu, C.: Design of low power & reliable networks on chip through joint crosstalk avoidance and multiple error correction coding. J. Electron. Test. Theor. Appl. (JETTA), Special Issue on Defect and Fault Tolerance 67–81 (2008)
9. Maheswari, M., Seetharaman, G.: Implementation of application specific network on chip architectures on reconfigurable devices using topology generation algorithm with genetic algorithm based optimization technique. In: Venugopal, K.R., Patnaik, L.M. (eds.) ICIP 2012, CCIS 292, pp. 436–445. Springer-Verlag, Berlin Heidelberg (2012)
10. Ganguly, A., Feero, B., Belzer, B.: Design of low power & reliable networks on chip through joint crosstalk avoidance and forward error correction coding. In: Proceedings of DFT, pp. 466–476 (2006)
11. Balasubramanian, S., Prasun, B., Jawad, K., Ranga, V.: LiPaR: A light weight parallel router for FPGA-based Networks-on-chip. In: Proceedings of GLSVLSI 2005, pp. 452–457 (2005)
12. Iturbe, X., Benkrid, K., Erdogan, A.T., Arslan, T., Azkarate, M., Martinez, I., Perez, A.: 'R3TOS: a reliable reconfigurable real-time operating system. In: NASA/ESA Conference on Adaptive Hardware and Systems (2010)
13. Iturbe, X., Benkrid, K., Senior Member, IEEE, Hong, C., Ebrahim, A., Torrego, R., Martinez, I., Arslan, T.: R3TOS: a novel reliable reconfigurable real-time operating system for highly adaptive, efficient, and dependable computing on FPGAs. IEEE Trans. Comput. **62**(8) (2013)

Hardware and Software Architecture for Embedded Distributed Control System Using Adaptive Hybrid Communication Channel

Pramod U. Chavan, M. Murugan and Pratibha P. Chavan

Abstract The paper addresses the design and development of an embedded distributed control system using heterogeneous communication channel. This design provides the fail proof communication using two completely different communication channels, viz; wired communication using RS 485 and wireless communication using Xbee. The system consists of a master unit and multiple slave units. The control of system lies with the master, which decides the type and time duration of the protocol to be used. Appropriate switching algorithm has been applied for auto-switching between communication protocols to achieve continuous transmission of data between master and slaves. Hardware and software architecture for the proposed system is developed using the ARM processors and the Visual Basic (VB) software for proper query response and controlling actions respectively.

Keywords Distributed system architecture · Control system · Master-slave · Event triggered · Heterogeneous

1 Introduction

The Distributed Control System (DCS) is a broad concept in which controllers are not central but distributed over a network. These controllers, in turn, control the sub-system components of the network, thus providing a distributed control. The controllers are interconnected to each other using wired and/or wireless

P.U. Chavan (✉)
Department of ETCE, Sathyabama University, Chennai, TN, India
e-mail: pramodp.chavan@gmail.com

M. Murugan
Department of E&CE, SRM's Valliammai Engineering College, Chennai, TN, India
e-mail: dr.murugan.m@gmail.com

P.U. Chavan · P.P. Chavan
Department of E&TC, K J College of Engineering & Management Research, Pune, India
e-mail: pratibhap.chavan1981@gmail.com

© Springer Science+Business Media Singapore 2016
R.K. Choudhary et al. (eds.), *Advanced Computing and Communication Technologies*,
Advances in Intelligent Systems and Computing 452,
DOI 10.1007/978-981-10-1023-1_51

communication channel [1, 2] for the purpose of monitoring and control. These communication protocols define various computer or electrical buses to connect the processors to the input/output modules and to the system. These buses also act as a bridge between the distributed controllers and the central controller. Depending on the type of DCS, wired or wireless buses can be used to implement the system. The processors act as a middleware between the modules by receiving the information from input modules and sending it to the output modules. With the increasing market needs, a number of DCS have been implemented using new technologies.

This work aims at the development of a layered software architecture [3, 4] for the distributed control system (DCS). It consists of single master and multiple slaves. The computer system is used as a master unit or server unit and multiple embedded controllers as slave units or client units. The master unit is equipped with server software and connectivity to the multiple slaves. It is used for information processing, sensing and controlling of slave units. The Slave/Client units are equipped with hardware interfaces called modules. These modules are different input and output components of robotic system. Each module is providing certain services for its functionality and diagnostic features to the unit. These services provide hardware abstraction to the requester and also isolate diagnostic features from the module's functionality.

Each unit provides services to the other units or a master unit to execute certain task on a distributed system. Different units or a master unit can access individual module services via a communication network. As robotic component based DCS are complex in nature, software architecture [5] is necessary to tackle complexity issue which provides the foundation to analyze the behavior of the software system before building the system. This helps early identification of pitfalls in design of complex robotic systems. Architecture also suggests modularity [6] in the design which helps us to find similar functionalities by reducing the design time, cost and risk of mistakes.

2 Related Work

The evolution of the distributed control system took place in early 1980s with the centralized control as base. Then number of DCS with various wired and wireless model came into existence. The traditional DCS was only used for process control and could not provide flexible multi-operation techniques. The shortcomings of traditional DCS were eliminated using the programmable PLC interrelated DCS. Later the hybrid architecture DCS provides parallel computing and sequencing of commands. The open architecture DCS had the advantages of much less shut down as compared to the other DCS. The problems of collision of packets persisted consistently. The Predictive Distributed Control model which operated on Distributed Control Function (DCF) succeeded enough in avoiding this problem.

Joint layer design based DCS then became prominent providing more stable parameters for distributed control [7]. The competition for providing multi vendor capability to DCS increased substantially in the years ahead.

In the presented work, authors have implemented a master slave protocol where the controlling actions are performed by the master with the help of the user. Thus the slaves are informed about the events to be performed. The system is based on the timing controls and an event based message, where specific actions are performed using either timing based actions or event driven actions. In timing based action e.g. timeout is given for a reply from the slaves after which the master takes proper action. In an event driven system, in case of failure of any communication channel, switching algorithm will automatically switch the system to alternate channel being used. This provides fail proof communication for the system to be more reliable [7]. Each slave is having its own wireless and wired communication medium in the form of Xbee [8, 9] as wireless medium and RS485 [10] as wired. Synchronization between all modules of the system exists which makes the system more intelligent and adaptive as per the runtime requirement.

3 Hardware Architecture

It consists of the master unit as Laptop/PC. The master is connected to RS 485 bus using USB to RS 485 convertor and the Xbee module using USB to serial converter. RS-485 [10] is an electrical-only standard defines the electrical characteristics of drivers and receivers. It is used to implement a balanced multipoint transmission line. The communication between units is established using one or two pairs of wires. The Xbee [9] is used for wireless communication in star topology to fulfill the needs of low power consumption.

Figure 1 shows the hardware architecture of DCS. It consists of three slave units as Robotic Arm, IR Counter and Load cell designed using ARM processor. The Robotic Arm is constructed using three DC motors for arm like movement and pick and place functions. The IR counter counts number of objects moving on a conveyor belt. The Load cell continuously gives weight readings of object using analog to digital converter and signal conditioning circuit.

4 Layered Software Architecture

In the proposed system, layered functionality is implemented in the software architecture [1]. Figure 2 shows the layered software architecture.

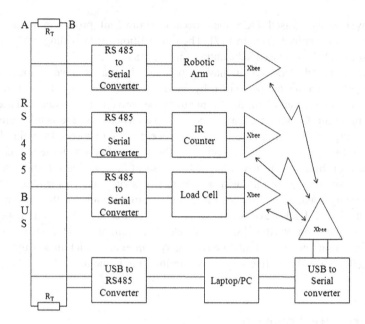

Fig. 1 Hardware architecture of DCS

Fig. 2 Layered software
architecture

Application Layer
Presentation Layer
Inter Unit Communication Layer
Unit Logic Abstraction Layer
Inter Module Communication layer
Module Logic Abstraction Layer
Hardware Abstraction Layer

4.1 Hardware Abstraction Layer

Each hardware interface such as DC motor, IR sensor and Load cell connected to a system is treated as a module. Hardware abstraction layer implements necessary software routines to interact with these hardware modules.

Fig. 3 Module
implementation

Module Identification
Device Identification
Module Data
User Function
Diagnostic Function

4.2 Module Logic Abstraction Layer

This layer provides the diagnostic features as well as services provided by each
module. It also provides the encryption of these features when they are executed in
the user mode. Figure 3 shows the module implementation.

4.3 Inter Module Communication Layer

The communication between the two modules is achieved using this layer. In this
layer, the request from each module is placed in the form of memory queues so that
every module can avail of the services offered by other modules.

4.4 Unit Logic Abstraction Layer

It consists of service tables which show the services provided by the various
modules inside a unit. It also provides the unit level diagnostic features for the
security purpose. Figure 4 shows the unit implementation.

4.5 Inter Unit Communication Layer

This layer provides communication between the different units of the system by
means of wired as well as wireless communication paths or channels. As the system

Fig. 4 Unit implementation

Unit Type
Unit Address Number
Unit Data
User Services
Diagnostic Services

consists of master slave protocol, any kind of service is initiated by the master. Thus, slaves are required to request the needed service to the master. As a result of such communication, system benefits in terms of: Multiple addressing, Communication acknowledgement, Data collision avoidance, Error detection and Media/Topology conversion.

4.6 Presentation Layer

The presentation layer provides API's to use the services provided by different units. This layer simplifies the complex communication interfaces to the end user. Thus when end user calls a particular API, presentation layer will send the required communication packets for designated unit and acknowledge its response. This layer maintains a record of all services provided by the different units.

4.7 Application Layer

In application layer, a user can execute different algorithms or a task by utilizing different API's provided by the presentation layer. The application can be run in diagnostic mode or user mode by providing proper validation.

5 Experimentation and Result

The system consists of three slave units implemented using ARM7 (LPC 2148) processor as Robotic Arm, IR sensor and Load cell. To operate these units independently and/or simultaneously, slave ID is given to each unit. For each unit, two serial interrupts are sequentially scanned to match the slave ID. Once the slave ID is matched, corresponding slave unit receives the commands from the master unit and operate the robotic modules. On matching the slave ID of Robotic Arm, user can operate different motors connected as per the requirement. The IR counter and/or Load cell sends the corresponding count and/or Load cell data to master unit to display. The communication channel between master and slaves is initiated and established as per the requirement as wired or wireless. If one channel fails, the system automatically switches over to another channel without any interruption.

In earlier implementation [1], DCS has been built using wireless communication channel. The limitation of system or slave fail is eliminated using two completely different communication channels. The channel auto switching characteristics are improved as compared to the implementation using AVR controller [2].

Fig. 5 DCS hardware implementation

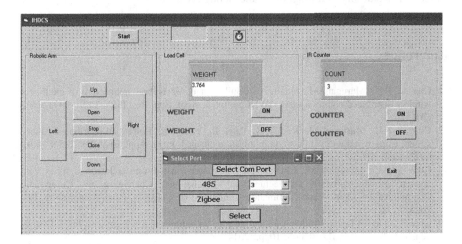

Fig. 6 DCS software implementation

Figures 5 and 6 shows the hardware and software implementation of DCS. Software is implemented using VB. Three slave systems are shown, in which Robotic Arm controlling is achieved by rotating motors. The Load cell is showing the weight of the object and IR counter is counting the number of objects moved on the conveyor belt. For communication between master and slave units, initially, user needs to select particular channel. If the selected channel fails, the system will automatically switch over to the alternate channel without any interruption.

6 Conclusion

The authors have presented the design and development of the hardware and software architecture for the distributed control of a modular control system. The system consists of master and slave units communicated via wired and/or wireless media. The master is equipped with the server software developed in VB and is used to send query messages to locate and identify slaves. Upon getting response from the slaves, the master is ready to take controlling action if any. The results show that the slaves are actively synchronized with the master for response and control. The results also show that the developed system is intelligent enough to switch over between two communication channels. Thus, providing fail proof communication over shorter as well as longer distance i.e. within the operating regions of communication channel. The authors suggest that the developed system can be improved and enhanced in terms of key issues like platform independency, hardware modularity, code maintainability, testability and diagnostic and debugs features.

References

1. Chavan, P.U., Murugan, M., Singh, A., Unnikkannan, V.: Hybrid architecture of intelligent wireless distributed control system (IWDCS). Int. J. Appl. Eng. Res. ISSN 10(9), 0973–4562 (2015)
2. Chavan, P.U., Murugan, M., Kumar, S., Patel, P., Yadav, P.: Embedded distributed control system using wired and wireless communication channel. Int. J. Appl. Eng. Res. ISSN 10(9), 0973–4562 (2015)
3. Chavan, P.U., Murugan, M., Chavan, P.P.: A review on software architecture styles with layered robotic software architecture. In: International Conference on Computing, Communication, Control and Automation Conference (ICCUBEA), Pune, Maharashtra, Feb 2015
4. Choi, J., et al.: A layered middleware architecture for automated robot services. Int. J. Distrib. Sensor Netw. Article ID 201063, 10pp., 2014
5. Iñigo-Blasco, P., et al.: Robotics software frameworks for multi-agent robotic systems development. Int. J. Robot. Auton. Syst. 60: 803–821 (2012)
6. Moubarak, P., Ben-Tzvi, P.: Modular and reconfigurable mobile robotics. Int. J. Robot. Auton. Syst. 60, 1648–1663 (2012)
7. Matthias, R., Bihlmaier, A., Worn, H.: Robustness, scalability and flexibility: key-features in modular self-reconfigurable mobile robotics. In: 2012 IEEE international conference on multisensor fusion and integration for intelligent systems (MFI) Hamburg, Germany, pp. 457–463, 13–15 Sept 2012
8. Fuzi, M.F.M., Ibrahim, A.F., Ismail, M.H., Halim, N.S.A.: HOME FADS: a dedicated fire alert detection system using Zig Bee wireless network. In: IEEE 5th Control and System Graduate Research Colloquium (ICSGRC), pp. 53–58 (2014)
9. Huang, Y., Pang, A., Hsiu, P., Zhuang, W., Liu, P.: Distributed throughput optimization for ZigBee cluster-tree networks. In: IEEE Transactions on Parallel and Distributed Systems, vol. 23, no. 3, March 2012
10. Wang, Y., Mao, W., Li, J., Zhang, P., Wang, X.: A distributed rectifier testing system based on RS-485. In: 5th IEEE Conference on Industrial Electronics and Applications (ICIEA), pp. 779–781, 15–17 June 2010

Analysis of Performance Limitations in Optical Wireless Communication System Due to Laser Source Impairment (RIN) and Background Noise

Mohammed Abdulhafedh Mohsen, Mazen Radhe Hassan and Shankar Duraikannan

Abstract Optical Wireless Communication (OWC) is a promising technology in several science and industry sectors, owing to its unique features: easy to install, no licenses needed, high speed communication, low power consumption and secure. However, OWC terrestrial applications suffers from limitations due to noise caused by natural phenomena such as fog, haze and rain and the Relative Intensity Noise (RIN) of the laser. In this paper, the performance evaluation, in terms of Signal to Noise Ratio (SNR) and the corresponding Bit Error Rate (BER) of terrestrial OWC is analysed in presence of laser RIN and background noise under different weather conditions and link distances. Simulation results of the mathematical model presented in this paper shows that, according to the typical values of parameters used here, RIN has, in general, negligible effect on the system performance at relatively high laser output power (which corresponds to laser injection current) for all weather conditions. In addition, background noise power (solar radiation) is higher in clear weather conditions and this may affect the OWC system even when the attenuation of channel is low at clear weather condition. Moreover, in general, SNR is relatively high (BER is low (improved)) for clear weather condition. However, when the weather becomes worse due to haze, fog or rain, attenuation of channel increases due to decreasing of visibility and this leads to decrease SNR and degrades BER (system performance).

Keywords Optical wireless communication (OWC) · Relative intensity noise (RIN) · Background noise (P_{BG}) · Signal to noise ratio (SNR) · Bit error rate (BER)

M.A. Mohsen (✉) · M.R. Hassan · S. Duraikannan
School of Engineering, Asia Pacific University of Technology and Innovation,
Kuala Lumpur 57000, Malaysia
e-mail: me200580@ymail.com

M.R. Hassan
e-mail: mazen@apu.edu.my

S. Duraikannan
e-mail: shankar@apu.edu.my

© Springer Science+Business Media Singapore 2016   521
R.K. Choudhary et al. (eds.), *Advanced Computing and Communication Technologies*,
Advances in Intelligent Systems and Computing 452,
DOI 10.1007/978-981-10-1023-1_52

1 Introduction

Recently, optical communication has been developed and become significant system in communication field since it replaces copper wires and it has very high data speed compared to other communication systems. There are two types of optical communication channel mediums that include fiber optic communication system and optical wireless communication system (OWC) or sometimes is called free space optical (FSO). Terrestrial Laser OWC system is developing quickly in different aspects of Telecommunication science and industry and it becomes important and exciting possibilities for signaling and modem design. This technology is useful where physical wires are impractical because of cost. Although the laser optical wireless communication OWC system has some advantages such as easy to install, no licenses needed, high speed communication and more secure [1], there are some disadvantages of OWC due to noises that caused by different weather conditions such as haze and fog and also the noise that caused by the system itself due to laser system components produce noise from the laser source [1].

2 Optical Wireless Communication System

2.1 Constructional Details

Generally, terrestrial OWC system consists of transmitter, atmospheric channel and receiver as shown in Fig. 1. The optical transmitter converts the electrical signal to an optical signal which propagates thorough the channel into the receiver.

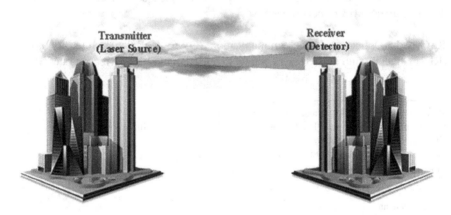

Fig. 1 Schematic of OWC system

The transmitter includes modulator which converts analogue information to digital signal. Also, it includes the telescope which aligns the LED or laser sources to collimate beam to the receiver side. In the channel, the signal is attenuated and blurred due to noise and absorption. The receiver includes telescope for collecting the incoming signal and drives it to filter, filters remove the radiation background, photo detector which converts the optical signal to electrical signal and decision unit which can determine nature of information bits based on amplitude of signal pulse [1].

2.2 Relative Intensity Noise

Relative Intensity Noise (RIN) describes the instability in the power signal level of the laser source and may affect the signal to noise ratio (SNR) and BER by enhancing the noise value. The laser noise occurs mainly due to spontaneous emission and other reasons such as cavity vibration and fluctuations in the laser gain medium. it is important to characterize it to achieve the corresponding improvement. RIN depends mainly on laser cavity parameters and it varies with the modulation frequency. In addition, RIN increase around specified value of modulation frequency that is: resonance frequency [2, 3].

2.3 Background Noise

This type of noise arises to the detection of inserted photons generated by the environment which contributes to background light such as sun. Background noise limits the performance of the system in terms of degradation of BER. In addition, the background light may cause, in certain situations, a considerable decrease in the SNR in free space communication channel and even totally disrupt its operation for some period of time [4].

3 Mathematical Analysis

Analyzing the RIN noise and background noise theoretically, mathematical models of optical signal and noise are formulated to determine the signal to noise ratio (SNR) of laser optical wireless communication system.

3.1 Signal Model

The signal power at the detector by the Friis transmission formula [1] is,

$$P_R = P_T G_T G_R T_F T_A L_{FS} \tag{1}$$

where the P_T is transmitter power, G_T is transmitter telescope gain, G_R is receiver telescope gain, T_F is optical wireless transmissivity, T_A is atmosphere transmission coefficient and L_{FS} is free space loss. If the source is contained within the receiver's field of view, backround power noise is [4],

$$P_{BG} = A_r H_\lambda(\Delta\lambda) T_F T_A \tag{2}$$

where H_λ is peak spectral irradiance. A_r is receiver effective primary area, $\Delta\lambda$ is optical filter bandwidth.

3.2 Noise Model

Mathematical models for optical noises should be determined for obtaining signal to noise ratio SNR for laser optical wireless communication system. The variance in detector for RIN and background noise are [1],

$$\sigma_{RIN}^2 = (RIN)(\Re P_R)^2 B \tag{3}$$

where RIN is relative intensity noise detector's current caused by laser. R is detector responsively and B is electronic bandwidth.

$$\sigma_{BG}^2 = 2q\Re P_{BG}B \tag{4}$$

where q is electron charge. So the SNR can be determined by including the signal power over total noises [1].

3.3 Bit Error Rate (BER)

To determine the BER of the proposed system, by using erfc which is the error function probability Gaussian Distribution. BER is expressed by [5],

$$BER = \frac{1}{2}erfc\left(\frac{1}{2\sqrt{2}}\sqrt{SNR}\right) \tag{5}$$

4 Simulation Result and Discussion

The design specification of the proposed system is based on Fabry-Perot laser diode for minimum physical assumptions. In addition, parameters of laser are also valid for single mode distributed feedback (DFB) laser diodes assuming that the mirror loss is dominated the photon lifetime [1, 6]. Five different laser power levels are selected in order to analyze the performance of OWC system under different weather conditions with respect to different links.

4.1 Relative Intensity Noise (RIN)

RIN can be changed if the injection current is changed because of photon density and resonance frequency. Therefore, when the photon density of the injection current increases, the resonance frequency is decreased correspondingly.

Figure 2 illustrates the effect of optical received power including RIN under different weather condition with respect to link distances. It is observed that when laser power increases, the effect of RIN decreases and the values of RIN are very small. This is because of the loss inside laser cavity and increase the photon density dazed by the extra injected carriers. In addition, the associated increase of resonance frequency with injection current leads to damp and shifts away the resonance peak, hence RIN decreases accordingly. For instance, when power laser value is 30.191 dBm, RIN variance is 10^{-14} for 2 km link distance. As result, RIN variance has, in general, negligible effect on the system performance at relatively high laser output power (which corresponds to laser injection current) for all weather conditions.

Fig. 2 RIN under different weather conditions

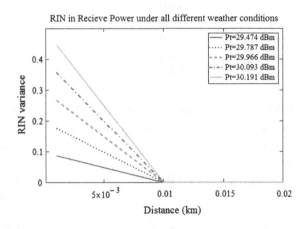

4.2 Background Noise (P_BG)

The mathematical evaluation of the relation between Background Noise Power (Sun Light) and different atmospheric attenuations of proposed system are tabulated as follows.

Tables 1, 2 and 3 shows background radiation power (sunlight) for 1, 2 and 3 km heights, respectively. Background solar radiation impacts the photodetector of proposed system under different weather conditions. The table shows the spectral irradiance at the sea level for solar radiation at zenith sun angle of 60° and typical 300 K earth object blackbody temperature was assumed [4, 7]. In addition, the analysis shows that the background power is taking place where there are higher

Table 1 Background noise under different weather condition for 1 km

Weather status	Atmospheric coefficient T_A	Background power P_{BG} (A)
Very clear	0.967	8.389×10^{-3}
Dense fog	9.158×10^{-33}	9.476×10^{-33}
Light fog	0.022	1.868×10^{-6}
Haze	0.539	4.682×10^{-3}
Light haze	0.839	7.284×10^{-3}
Heavy rain	0.118	1.02×10^{-3}
Light rain	0.633	5.491×10^{-3}

Table 2 Background noise under different weather condition for 2 km

Weather status	Atmospheric coefficient T_A	Background power P_{BG} (A)
Very clear	0.934	8.11×10^{-3}
Dense fog	0	0
Light fog	4.631×10^{-4}	4.019×10^{-6}
Haze	0.291	2.525×10^{-3}
Light haze	0.704	6.113×10^{-3}
Heavy rain	0.014	1.199×10^{-4}
Light rain	0.4	3.474×10^{-3}

Table 3 Background noise under different weather condition for 3 km

Weather status	Atmospheric coefficient T_A	Background power P_{BG} (A)
Very clear	0.903	7.84×10^{-3}
Dense fog	0	0
Light fog	9.966×10^{-6}	8.649×10^{-3}
Haze	0.157	1.362×10^{-3}
Light haze	0.591	5.131×10^{-3}
Heavy rain	1.623×10^{-3}	1.409×10^{-3}
Light rain	0.253	2.198×10^{-3}

atmospheric attenuations due to high visibility. For instance, the effect of solar radiation in fog weather condition is very low while the effect of solar radiation in clear weather is very high because of high visibility. The calculations were performed for different standard internationality code weather conditions [8, 9, 10] (Table 4).

Table 4 BER with laser values under different weather conditions

Status	Transmitted power P_T (dBm)	Distance (km) at BER (10^{-9})
Very clear weather condition	29.474	0.53
	29.787	0.76
	29.966	0.93
	30.093	1.08
	30.191	1.2
Dense fog weather condition	29.474	0.0576
	29.787	0.0646
	29.966	0.0684
	30.093	0.0708
	30.191	0.0738
Light fog weather condition	29.474	0.33
	29.787	0.37767
	29.966	0.4174
	30.093	0.4665
	30.191	0.4852
Haze weather condition	29.474	0.4646
	29.787	0.6332
	29.966	0.7315
	30.093	0.8299
	30.191	0.9282
Light haze weather condition	29.474	0.8208
	29.787	0.7175
	29.966	0.8861
	30.093	0.9984
	30.191	1.1249
Heavy rain weather condition	29.474	0.3661
	29.787	0.4644
	29.966	0.5393
	30.093	0.5815
	30.191	0.6283
Light rain weather condition	29.474	0.4786
	29.787	0.6332
	29.966	0.7737
	30.093	0.858
	30.191	0.9984

4.3 Bit Error Rate (BER)

Figures 3, 4, 5, 6, 7, 8 and 9 show the relationship between the BER including optical received power, background noise, RIN and other parameters with the link distance (km) under different weather condition with different visibilities and different values of laser power as shown in this figures. Generally, it is observed that the SNR decreases when the link distance increases due to increase in propagation. In addition when laser power increases, the received power increases and more link distance can be covered with significantly high power. The SNR decreases when attenuation coefficient increases due to lower visibilities.

Under weather conditions, the BER dependence on the link distances were studied through mathematical simulation. Refereeing to the standard performance

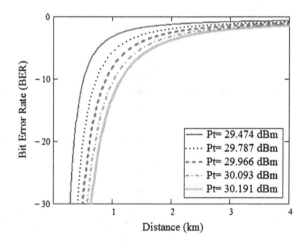

Fig. 3 BER under clear weather condition with respect to link distance

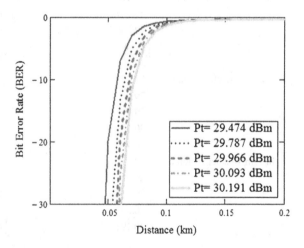

Fig. 4 BER under dense fog weather condition with respect to link distance

Fig. 5 BER under light fog weather condition with respect to link distance

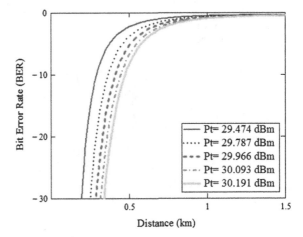

Fig. 6 BER under light haze weather condition with respect to link distance

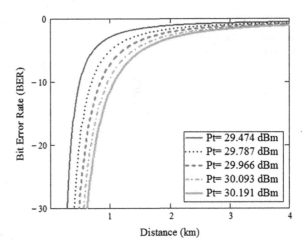

Fig. 7 BER under haze weather condition with respect to link distance

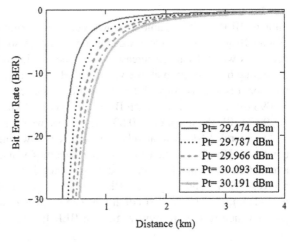

Fig. 8 BER under light rain
weather condition with
respect to link distance

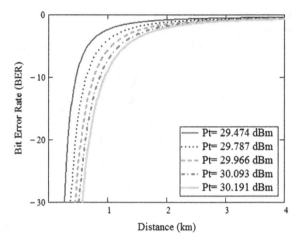

Fig. 9 BER under heavy rain
weather condition with
respect to link distance

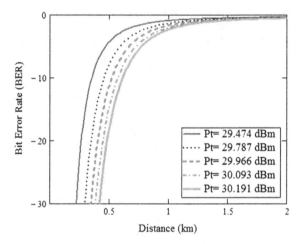

limit of BER = 10^{-9}, Table 2 shows that the maximum link distance of 1.2 km
around BER = 10^{-9} in clear weather condition. Also, it can be shown that BER
decreases when distance increases due to decreasing of SNR which degrades due
increasing of attenuation at low visibility. Also, the more power laser is transmitted,
the more distance is got with less BER. For instance, when the power laser value is
30.093 dBm, the distance when BER (10^{-9}) is 1.08 km while the power laser value
is 29.474 dB, the distance is 0.53 km in clear weather.

In general, SNR is relatively high (BER is low (improved)) for clear weather
condition. However, when the weather becomes worse due to haze, fog or rain,
attenuation of channel increases due to decreasing of visibility and this leads to
decrease SNR and degrades BER (system performance). The corresponding
degradations due to weather conditions can be overcome by increasing laser output
power which results in improvement in BER for relatively longer link distance.

5 Conclusion

This paper has analyzed, and simulated mathematically the performance of terrestrial wireless optical communication in presence of RIN noise and background radiation noise and under different weather conditions. It has been shown that, for all weather conditions, the RIN has, in general, negligible effect on the system performance due to relatively high laser output power that leads to damp RIN significantly and shifts the resonance peak away towards very high frequency region. In addition, background noise power (solar radiation) is higher in clear weather conditions due to higher solar radiance received by the photodetector, hence, this may affect the OWC system even when the attenuation of channel is relatively low, i.e. clear weather condition (high visibility). Moreover, in general, SNR is relatively high (BER is low (improved)) for clear weather condition due to low attenuation. However, when the weather becomes worse due to haze, fog or rain, attenuation of channel increases due to decreasing of visibility and this leads to decrease SNR and degrades BER (system performance). The corresponding degradations due to weather conditions can be overcome by increasing laser output power which results in improvement in BER for relatively longer link distances.

References

1. Manor, H., Arnon, S.: Performance of an optical wireless communication system as a function of wavelength. Appl. Opt. **42**(21), 4285 (2003)
2. Kallimani, K., O'Mahony, M.: Relative intensity noise for laser diodes with arbitrary amounts of optical feedback. IEEE J. Quantum Electron. **34**(8), 1438–1446 (1998)
3. Ghassemlooy, Z., Popoola, W., Rajbhandari, S.: Optical wireless communications. CRC Press, Boca Raton (2013)
4. Kopeika, N., Bordogna, J.: In background noise in optical communication system, pp. 1571–1577 (1970)
5. YoussefElganimi, T.: Performance comparison between OOK, PPM and PAM modulation schemes for free space optical (FSO) communication systems: analytical study. Int. J. Comput. Appl. **79**(11), 22–27 (2013)
6. Ab-Rahman, M., Hassan, M.: Theory of elimination of emitted optical pulse/s remotely. IEEE J. Quantum Electron. **48**(1), 27–35 (2012)
7. Wolfe, W., Zissis, G.: The Infrared Handbook. The Office, Washington (1978)
8. Ali, M., Mohammed, M.: Effect of atmospheric attenuation on laser communications for visible and infrared wavelengths. J. Al-Nahrain Univ. **16**(3), 133–140 (2013)
9. Rashed, A.: High efficiency wireless optical links in high transmission speed wireless optical communication networks. Int. J. Commun. Syst., p. n/a-n/a (2013)
10. Sharma, V., Kaur, G.: Degradation measures in free space optical communication (FSO) and its mitigation techniques—a review. Int. J. Comput. Appl. **55**(1), 23–27 (2012)

Part III
Antenna Design and Power Systems

PV-Based Distributed Generation Location Using Mixed Integer Non-linear Programming in Deregulated Electricity Market

Manish Kumar, Ashwani Kumar and K.S. Sandhu

Abstract In this paper, a mixed integer nonlinear programming (MINLP) approach for determining optimal location and number of PV-based distributed generators in pool based electricity market is presented. For obtaining the PV output power, solar modeling has been carried out for a given data. The solar PV has been integrated into the system finding its optimal location and numbers based on the transmission loss minimization. The analysis has been carried out for IEEE 24 bus test system in pool electricity market model. The realistic load model as ZIP load has been considered and the results are compared for constant P, Q load model and ZIP and model. The non-linear mixed integer programming problem has been solved using MATLAB and GAMS interfacing.

Keywords Distributed generation · Mixed integer nonlinear programming · Optimal location

1 Introduction

The proper and strategically placement of DG at specific site is required in order to obtain the maximum benefits, reliability of the system improved, power losses reduction, improving the voltage profile, distribution and transmission system congestion management, operating cost and for better quality of power [1]. The optimal sizing and location of DG reduces the transmission loss of the line and also reduces the overall cost of the power network. Many studies performed for optimization to minimize transmission loss, for optimal location and sizing of DGs in power network [2–6]. In [7], GA-based Tabu Search method for the optimal location of DG units with multiple objectives is proposed in hybrid/pool market. In [8],

M. Kumar (✉) · A. Kumar · K.S. Sandhu
National Institute of Technology, Kurukshetra 136119, India
e-mail: khanagwal.manish@gmail.com

© Springer Science+Business Media Singapore 2016   535
R.K. Choudhary et al. (eds.), *Advanced Computing and Communication Technologies*,
Advances in Intelligent Systems and Computing 452,
DOI 10.1007/978-981-10-1023-1_53

a scheme is proposed for identifying optimal location of DGs in distribution system, improved voltage stability and minimize network power losses The minimization of the power loss for multiple-DG placement and sizing is considered using neural network because of the nonexistence of an analytical equation [9]. In [10] was proposed three alternative analytical expressions to determine the optimum size and operating strategy of distributed (DG) units for minimization of power loss.

In this paper, an approach is proposed to determine optimal location and number of distributed generator by using a mixed integer nonlinear programming (MINLP) approach for considering minimization of transmission loss with the presence of PV-based DG. The total real and reactive power loss, percentage reduction in active power loss, optimal DG location has been obtained. The results have also been obtained for minimization of transmission loss with zip load also. The results have been obtained for IEEE 24 bus reliability test system.

2 Solar PV Modeling

The solar irradiance for each for hourly of the day is modeled by the Beta probability density function (PDF) based on historical data which have been collected for 3 years [11]. The solar irradiance samples are produced using MCS and mean solar irradiance of hourly shown in Fig. 1.

In this paper, we have utilized four type of PV modules and capacity factor (CF) of each module show in Figs. 2 and 3 respectively. The four PV modules have been named as module A, module B, module C, and module D. Active power of each module in MW is shown in Fig. 2. Capacity factor for each module is shown in Fig. 3. It is observed that capacity factor is higher for module D compared to other modules. A ratio of the average output power to the real power is called capacity factor [12].

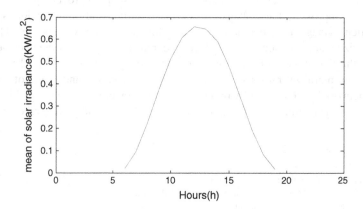

Fig. 1 Mean of per hour for 24 h solar irradiance (kW/m^2)

Fig. 2 Active power of
module A, module B, module
C and module D in MW

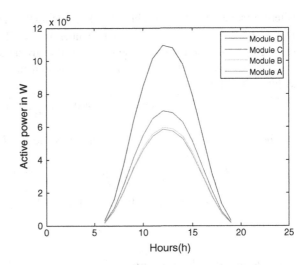

Fig. 3 Capacity factor
(CF) of different PV module

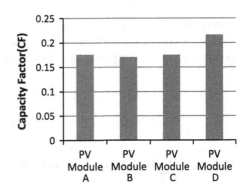

3 General OPF Formulations of Distributed Generation Considering Minimization of Total Transmission Loss with Constant Load

General objective of MINLP approach is:

$$\text{Min } F\left(h, g, \zeta^{\text{int}}\right) \tag{1}$$

Subject to equality and inequality constraints defined as

$$x\left(h, g, \zeta^{\text{int}}\right) = 0 \tag{2}$$

$$u\left(h, g, \zeta^{\text{int}}\right) \leq 0 \tag{3}$$

where h (state vector of variables V, δ), g (the control parameters, P_{gk}, Q_{gk}, P_{PVk}, Q_{PVk}) and ζ^{int} (integer variable with values 0, 1). The integer variable zero it represents no DG and one represent present of distributed generator in the network.

Objective function F is minimization of the transmission loss and is represented as given in Eq. (4).

$$\text{Min } F\left(h, g, \zeta^{int}\right) = P_{kjl} + P_{jkl} \qquad (4)$$

The line flows from bus-k to bus-j and bus-j to bus-k are given as:

$$P_{kjl} = V_k^2 G_{kj} - V_k V_j \left(G_{kj} \cos\left(\delta_k - \delta_j\right) + B_{kj} \sin\left(\delta_k - \delta_j\right)\right) \qquad (5)$$

$$P_{jkl} = V_j^2 G_{kj} - V_k V_j \left(G_{kj} \cos\left(\delta_k - \delta_j\right) - B_{kj} \sin\left(\delta_k - \delta_j\right)\right) \qquad (6)$$

3.1 Equality Constraints

With the presence of distributed generation, the power balance equations for all the buses are modified for both the real and reactive power adding the DG power. The modified equations for power flow can be represented as:

$$P_k = P_{gk} + \zeta_k^{int} * P_{PVk} - P_{dk} \quad \forall k = 1, 2, \ldots N_b \qquad (7)$$

$$Q_k = Q_{gk} + \zeta_k^{int} * Q_{PVk} - Q_{dk} \quad \text{for all } k = 1, 2, \ldots N_b \qquad (8)$$

$$P_k = \sum_{j=1}^{N_b} V_k V_j \left[G_{kj} \cos\left(\delta_k - \delta_j\right) + B_{kj} \sin\left(\delta_k - \delta_j\right)\right] \quad \forall k = 1, 2, \ldots N_b \qquad (9)$$

$$Q_k = \sum_{j=1}^{N_b} V_k V_j \left[G_{kj} \sin\left(\delta_k - \delta_j\right) - B_{kj} \cos\left(\delta_k - \delta_j\right)\right] \quad \forall k = 1, 2, \ldots N_b \qquad (10)$$

3.2 Inequality Constraints

The inequality constraints for power genertaion, voltage limits, and phase angle limits can be represented as:

(a) Real power generation limit of generators at bus-k

$$P_{gk}^{min} \leq P_{gk} \leq P_{gk}^{max}, \quad k = 1, 2, \ldots N_g \qquad (11)$$

(b) Reactive Real power generation limit generators and other reactive sources at bus-k

$$Q_{gk}^{\min} \leq Q_{gk} \leq Q_{gk}^{\max}, \quad k = 1, 2, \ldots N_q \tag{12}$$

(c) Voltage limit of $V_k^{\min}, V_k^{\max}$ at bus-k

$$V_k^{\min} \leq V_k \leq V_k^{\max}, \quad k = 1, 2, \ldots N_b \tag{13}$$

(d) Phase angle limit of $\delta_k^{\min}, \delta_k^{\max}$ at bus-k

$$\delta_k^{\min} \leq \delta_k \leq \delta_k^{\max}, \quad k = 1, 2, \ldots N_b \tag{14}$$

(e) Line flow limit based on thermal and stability considerations.

$$\left| S_{kj} \right| \leq S_{kj}^{\max}$$

3.3 Power Generation Limit of Generators at Bus-k with Constraint on Optimal Number of Generators

(a) Real power generation limit

$$P_{PVk}^{\min} \leq P_{PVk} \leq P_{PVk}^{\max}, \quad k = 1, 2, \ldots N_{PV} \tag{15}$$

(b) Reactive power generation limit

$$Q_{PVk}^{\min} \leq Q_{PVk} \leq Q_{PVk}^{\max}, \quad k = 1, 2, \ldots N_{PV} \tag{16}$$

(c) Optimal number of distributed generators

$$N_{PV} = \sum_{k=1}^{Npv} \zeta_k^{int} \leq N_{PV}^{\max} \tag{17}$$

4 ZIP Load Model

The load is modeled as polynomial load and can be represented as:

$$P_{dz} = P_o\left(A_p V^2 + B_p V + C_p\right) \tag{18}$$

$$Q_{dz} = Q_o\left(A_q V^2 + B_q V + C_q\right) \tag{19}$$

$$\left(A_p + B_p + C_p\right) = \left(A_q + B_q + C_q\right) = 1 \tag{20}$$

where V node voltage in p.u., P_o, Q_o the real power and reactive power consumed at the specific node under the reference voltage. A_p, A_q are the parameters for constant impedance (constant Z) load component. B_p, B_q are the parameters for constant current (constant I) load component; C_p, C_q are the parameters for constant power (constant P and Q) load component. The values of A_p, A_q, B_p, B_q and C_p, C_q are determined for different load types in distribution systems. Usually experimental or experience values could be used.

4.1 Without PV-Based DG

The real and reactive power injection equations can be modified in the presence of zip load as:

$$P_k = P_{gk} - P_{dzk} \quad \forall k = 1, 2, \ldots N_b \tag{21}$$

$$Q_k = Q_{gk} - Q_{dzk} \quad \forall k = 1, 2, \ldots N_b \tag{22}$$

4.2 With PV-Based DG

With distributed generation from Eqs. (7) and (8) the real and reactive power constraints are modified in the presence of zip load as:

$$P_k = P_{gk} + \xi_k^{int} * P_{PVk} - P_{dzk} \quad \forall k = 1, 2, \ldots N_b \tag{23}$$

$$Q_k = Q_{gk} + \xi_k^{int} * Q_{PVk} - Q_{dzk} \quad \forall k = 1, 2, \ldots N_b \tag{24}$$

5 Results and Discussion

The proposed approach is applied to IEEE 24-bus reliability test system [13] for finding optimal distribution generation location. The results have been obtained for voltage profile, total real and reactive power loss, PV-based DG output and %age reduction in the transmission loss with PV-based DGs. The results are also obtained with zip load variation at each bus for comparison with constant P, Q load model. The results are given in tabular form in Tables 1 and 2 respectively.

Optimal location and number of DGs have been obtained solving the mixed integer programming problem and MATLAB and GAMS interfacing has been utilized for obtaining the solution. The results of different cases with different number of PV-based distributed generators have been obtained and it categorized as:

Case1 (Without PV-based distributed generator), Case2 (With 1 PV-based distributed generator), Case3 (With 2 PV-based distributed generators), Case4 (With 3 PV-based distributed generators) and Case5 (With 4 PV-based distributed generators).

5.1 Results for Minimization of Transmission Loss with Constant P,Q Load

The simulation of transmission loss minimization has been determined by solving nonlinear optimization problem as explained in Sect. 3.

In Table 1, the results for total active and reactive loss named as PLT and QLT, minimum voltage, number of PV based DGs with their optimal location at a bus is presented. It also represents the percentage reduction in total active power loss which is calculated by the formula given in 25.

$$\% \text{ reduction in loss} = \frac{PLT_{without\,dg} - PLT_{with\,dg}}{PLT_{without\,dg}} \times 100\% \qquad (25)$$

In case1, without PV-based DG the total reactive and active power loss in the system is -2.9975 p.u. MVar and 0.2877 p.u. MW respectively. The total active and reactive load is 28.5(p.u. MW) and 5.8 p.u. MVar respectively which remain constant in each case. In case2, the total active and reactive power losses are 0.2731 p.u. MW and -3.1605 p.u. MVar with the optimal bus location of PV-based DG at bus 3. In case3, the total active and reactive power loss are 0.2641 p.u. MW and -3.3453 p.u. MVar with the optimal bus location of DG at 3 and 4. In case4, the total active and reactive power loss are 0.2520 p.u. MW and -3.3630 p.u. MVar with the optimal bus location of PV-based DG at 3,4,10. In case5, the total active and reactive power loss are 0.2428 p.u. MW and -3.3962 p. u. MVar with the optimal bus location of PV-based DG at 3, 4, 6, and 10.

Table 1 Results for minimization of total transmission loss with constant load

	Case1: without DG (PV)	Case2: with 1DG (PV)	Case3: with 2DG (PV)	Case4: with 3DG (PV)	Case5: with 4DG (PV)
PLT (p.u. MW)	0.2877	0.2731	0.2641	0.2520	0.2428
QLT (p.u. MVar)	−2.9975	−3.1605	−3.2453	−3.3630	−3.3962
Minimum voltage	0.9661 (at bus 4)	0.9667 p.u. (at bus 4)	0.9654 p.u. (at bus 4)	0.9704 p.u. (at bus 4)	0.9648 p.u. (at bus 3)
Total active load (p.u. MW)	28.5	28.5	28.5	28.5	28.5
Total reactive load (p.u. MVar)	5.8	5.8	5.8	5.8	5.8
%age reduction in loss		5.07 %	9.41 %	12.40 %	15.60 %
Optimal bus location of DG (PV)		3	3, 4	3, 4, 10	3, 4, 6, 10
Total DG (PV) size (p.u. MW)		0.33	0.51	0.84	1.05
Total DG (PV) size (p.u. MVar)		0.04	0.06	0.07	0.08

In each case when number of PV-based DG increases, it is observed that the percentage reduction in loss increases. This is attributed to the fact that generation is available locally and the power flow patterns changes in such a way that there is loss reduction in each line contributing to overall loss reduction in the system. The best case is found to be case5. The voltage, PLT and PV-based DG size and location are shown in Figs. 4, 5 and 6 respectively.

Fig. 4 Voltage profile with and without PV-based DG with constant load

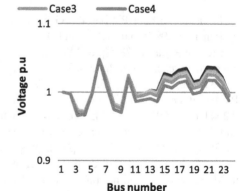

Fig. 5 Total active power
loss with and without
PV-based DG with constant
load

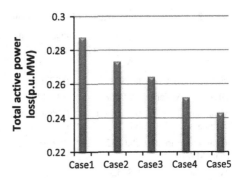

Fig. 6 PV-based DG size in
p.u. with constant load

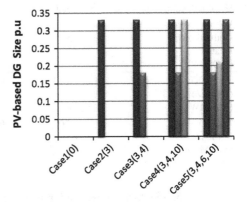

5.2   Results with ZIP Load

The results obtained with zip load are obtained and are given in Table 2. In Table 2,
the results for all the cases for total real power and reactive power loss, optimal
output obtained from PV-based DGs and the location of PV-based DGs at the
respective buses are given. The %age reduction in the total loss is also provided.
The Zip load coefficients taken are: Ap = 0.2, Bp = 0.2, Cp = 0.6, Aq = 0.2,
Bq = 0.2, Cq = 0.6.

In case1, the total active and reactive power loss in the system are 0.2880 p.
u. MW and −2.9868 p.u. MVar respectively. The total active and reactive load is
28.6242 p.u. MW and 5.8252 p.u. MVar respectively and the slight change in each
case is observed due to the voltage dependency of the load. In case2, the total active
and reactive power loss are 0.2736 p.u. MW and −3.1294 p.u. MVar with the
optimal bus location of PV-based DG at bus 3. In case3, the total active and reactive
power loss are 0.2601 p.u. MW and −3.2746 p.u. MVar with the optimal bus
location of PV-based DG at buses 3, 10. In case4, the total active and reactive
power loss are 0.2484 p.u. MW and −3.3301 p.u. MVar with the optimal bus
location of PV-based DG at buses 3, 6, 10.

Table 2 Result for minimization of total transmission loss with zip load

	Case1: without DG (PV)	Case2: with 1DG (PV)	Case3: with 2DG (PV)	Case4: with 3DG (PV)	Case5: with 4DG (PV)
PLT (p.u. MW)	0.2880	0.2736	0.2601	0.2484	0.2401
QLT (p.u. MVar)	−2.9868	−3.1429	−3.2746	−3.3301	−3.4083
Minimum voltage	0.9655 (at bus 3)	0.9686 p.u. (at bus 4)	0.9674 p.u. (at bus 4)	0.9627 p.u. (at bus 3)	0.9624 p.u. (at bus 3)
Total active load (p.u. MW)	28.6242	28.6050	28.5564	28.4505	28.4437
Total reactive load (p.u. MVar)	5.8252	5.8213	5.8115	5.7899	5.7886
%age reduction in loss		5.00 %	9.68 %	13.75 %	16.63 %
Optimal bus location of DG (PV)		3	3, 10	3, 6, 10	3, 4, 6, 10
Total DG (PV) size (p.u. MW)		0.33	0.66	0.87	1.05
Total DG (PV) size (p.u. MVar)		0.04	0.05	0.06	0.08

In case5, the total active and reactive power loss are 0.2401 p.u. MW and −3.4083 p.u. MVar with the optimal bus location of PV-based DG at buses 3, 4, 6 and 10. In each case with increases in number of PV-based DGs, it is observed that the percentage reduction in loss increases. The best results are found for case5 with a maximum reduction in the transmission losses. The voltage, PLT and PV-based DG size and location are shown in Figs. 7, 8 and 9 respectively.

Fig. 7 Voltage profile with and without PV-based DG with zip load

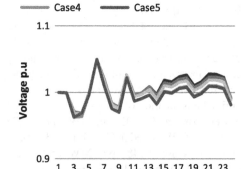

Fig. 8 Total active power
loss with and without
PV-based DG with zip load

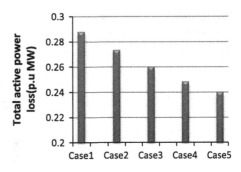

Fig. 9 PV-based DG size in
p.u. with zip load

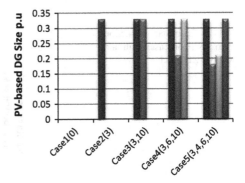

5.3 Comparison of Total Real Power Loss and Percentage Reduction in All Cases

From Fig. 10 it is observed that in case1 the total losses are maximum with ZIP load (0.2880 p.u. MW) and minimum with constant load (0.2877 p.u. MW). In case2 the losses are minimum with constant load (0.2731 p.u. MW) and maximum with ZIP load (0.2736 p.u. MW). In case3 the losses are minimum with ZIP load (0.2601 p.u. MW) and maximum with constant load (0.2641 p.u. MW). In case4 the losses are minimum with ZIP load (0.2484 p.u. MW) and maximum with constant load (0.2520 p.u. MW). In case5 the losses are minimum with ZIP load (0.2401 p.u. MW) and maximum with constant load (0.2428 p.u. MW).

From Fig. 11 it is observed that in case2 the percentage reduction in loss is minimum with ZIP load (5 %) and maximum with constant load (5.07 %). In case3 the percentage reduction in loss is minimum with constant load (9.41 %) and maximum with ZIP load (9.68 %). In case4 the percentage reduction in loss is minimum with constant load (12.40 %) maximum with ZIP load (13.75 %). In case5 the percentage reduction in loss is minimum with constant (15.6 %) and maximum with variable ZIP load (16.63 %). Based on the loss reduction comparison, the percentrage reduction in loss is more for ZIP load model compared to constant P, Q load except for case 2 where loss reduction is comparable.

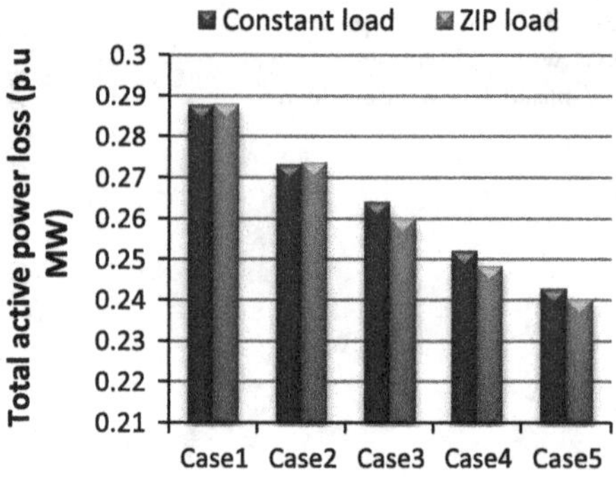

Fig. 10 Total real power loss
in all cases with and without
ZIP load

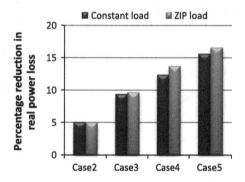

Fig. 11 Percentage reduction
in losses in all cases with and
without ZIP load

6 Conclusions

In this work the PV-based DG system is presented, the beta PDF model is used to
describe the probabilistic nature of solar irradiance. The solar irradiance samples are
produced using MCS. The four different type of the PV plant are used to design the
PV-based DG system, it also measured the capacity factor of each PV module. It
presented the comparison of optimal size and location of distributed generation for
minimization of total transmission loss with constant and zip load model. In every
case of load (zip as well as constant load) the total reactive and active power losses
are calculated by the optimization problem. The optimization problem is formulated
as MINLP, using a GAMS environment. It is observed that in case of zip load the
losses are more in the system without PV-based DG as compared to the constant
load model. But when PV-based DGs are added in the system, there is more reduction in active power loss in case of zip load as compared to the constant P, Q
load. With increase in the number of PV-based DGs, the power loss reduces
considerably. The comparison in total real power loss and percentage reduction in
all Cases, it obtained case1 (without PV-based DG) and case2 (one PV-based DG)
the total real power loss are maximum in ZIP load and minimum in constant load
model but the number of PV-based DGs connected to system (case3 (two PV-based
DG), case4 (three PV-based DG) and case5 (four PV-based DG)) the total real

power loss are maximum in constant load and minimum in ZIP load model. The percentage reduction in loss is maximum with constant model for the case2 (two PV-based DG) only. All other cases the maximum percentage reduction in real power loss with ZIP load model and minimum with constant load model is observed.

References

1. Chiradeja, P.: Benefit of distributed generation: a line loss reduction analysis. In: IEEE/PES Transmission and Distribution Conference and Exhibition: Asia and Pacific, China Dalina, pp. 1–5 (2005)
2. Wang, C., Nshrir, M.H.: Analysis approach for optimal placement of distributed generation source in power system. IEEE Trans. Power Syst. **19**(4), 2068–2076 (2004)
3. Moradi, M.H., Abedini, M.: A combination of genetic algorithm and particle swarm optimization for optimal DG location and sizing in distribution systems. Int. J. Electr. Power Energ. Syst. **34**(1), 66–74 (2012)
4. Kim, K., Song, K., Joo, S., Lee, Y., Kim, J.: Multi-objective distributed generation placement using fuzzy goal programming with genetic algorithm. Euro. Trans. Electric Power. **8**, 217–230 (2008)
5. Ghosh, S., Ghoshal, S.P., Ghosh, S.: Optimal sizing and placement of distributed generation in a network system. Int. J. Electr. Power Syst. **32**, 849–856 (2010)
6. Porkar, S., Poure, P., Abbaspur-Tehrani-fard, A., Saadate, S.: A novel optimal distribution system planning framework implementing distributed generation in a deregulated electrical market. Int. J. Electr. Power Syst. **60**, 828–837 (2010)
7. Mohammadi, M., Nafar, M.: Optimal placement of multitype DG as independent private sector under pool/hybrid power market using GA-based Tabu search method. Electr. Power Energy Syst. **51**, 43–53 (2013)
8. Kumar, S., Kumar Prema, N.: A novel approach to identify optimal access point and capacity of multiple DGs in a small, medium and large scale radial distribution systems. Electr. Power Energy Syst. **45**, 142–151 (2013)
9. Ugranli, F., Karatepe, E.: Multiple-distributed generation planning under load and different penetration levels. Int. J. Electr. Power Energy Syst. **46**, 132–144 (2013)
10. Hung, D.Q., Mithulananthan, N., Bansal, R.C.: Analytical strategies for renewable distributed generation integration considering energy loss minimization. Appl. Energy **105**, 75–85 (2013)
11. Hung, D.Q., Mithulananthan, N., Lee, K.Y.: Determining PV penetration for distribution systems with time-varying load model. IEEE Trans. Power Syst. **29**(6), 3048–3057 (2014)
12. Atwa, Y.M., El-Saadany, E.F., Salama, M.M.A., Seethapathy, R.: Optimal renewable resources mix for distribution system energy loss minimization. IEEE Trans. Power Syst. **25**, 360–370 (2010)
13. IEEE Reliability Test System: A report prepared by the reliability test system task force of the applications of probability methods subcommittee. IEEE Trans. Power Apparatus Syst **98**:2047–2054 (1979)

Investigation of Deterministic and Random Cyclic Patterns in a Conventional Diesel Engine Using Symbol Sequence Analysis

Rakesh Kumar Maurya

Abstract Current study focuses on cyclic variability analysis of indicated mean effective pressure (IMEP) and total heat release (THR) in a conventional single cylinder diesel engine using symbol sequence statistics. The objective is to characterize the cyclic variations and estimate the deterministic pattern in experimental data series. Results presented in this study can be beneficial for the design of effective control strategy. Experiments are performed at constant speed for five different engine operating conditions. Engine cylinder pressure data is acquired for 2500 consecutive cycles. Engine combustion parameters (IMEP and THR) are calculated from measured cylinder pressure on cycle to cycle basis and resulting data series is analyzed by symbol sequence method. Results revealed that selection of symbol sequence length and number of partitions is dependent on engine operating conditions and chosen combustion parameter.

Keywords Diesel engine · Symbol sequence · Cyclic variability · IMEP · Control

1 Introduction

Most significant and widely used prime mover in automotive industry is internal combustion (IC) engines. Mainly two types of IC engines (spark ignition and compression ignition) are used for automotive applications. Combustion process in IC engines is affected by several input parameters (air-fuel ratio, intake air temperature and pressure, humidity etc.) resulting into complex dynamics. Combustion processes in IC engines have cycle to cycle variations even though duration of engine cycle is same at constant speed. These cyclic variations are dependent on engine operating conditions and can be very large in some of the operating

R.K. Maurya (✉)
Department of Mechanical Engineering, Indian Institute of Technology Ropar,
Rupnagar 140001, Punjab, India
e-mail: rakesh.maurya@iitrpr.ac.in

© Springer Science+Business Media Singapore 2016
R.K. Choudhary et al. (eds.), *Advanced Computing and Communication Technologies*,
Advances in Intelligent Systems and Computing 452,
DOI 10.1007/978-981-10-1023-1_54

conditions [1, 2]. Cyclic variations can be significant for even well controlled average ignition timings [3]. Cyclic variations in combustion lead to lower power and higher emissions for the same fuel consumptions. Therefore understanding cyclic variability patterns can be beneficial to eliminate the variability and improve fuel conversion efficiency of engine. The IC engine combustion process can be controlled by effective use of cylinder pressure based combustion parameters using various actuators such as fuel injector/EGR valve etc. [4]. Due to comparatively larger cyclic variations in spark ignition engines, they are often main focus of investigation [5, 6]. Stringent emission legislations and fuel economy requirements leads to investigation of cyclic variability in conventional compression ignition engines also [7–9] and as well as advanced premixed compression ignition engines [10]. Mostly experimental studies conducted have used statistical methods for cyclic variability analysis. Effective controllers can be developed by understanding the dynamics of various combustion parameters on cycle to cycle basis. Few studies are conducted to understand the dynamics to variations using nonlinear dynamic analysis and chaotic approach in diesel engines [7–9]. Wavelets are also used for characterization of cyclic variability in diesel engines [8, 11]. Symbol sequence statistics can provide the insight into deterministic pattern in experimental data series. Symbol sequence method is applied in few studies to characterize the cyclic variability in spark ignition engines [6, 12]. Characterization of cyclic variability using symbol sequence statistics method is completely missing in published literature for diesel engines combustion parameters.

In this study symbol sequence statistics method is used to characterize the variations in diesel engine. This method provides important insight into the behavior observed for different combustion parameters in an engine operation [6, 12]. Symbolization of experimental data can very useful and effective for distinguishing the pattern in noisy data [13]. Symbolization can estimate deterministic effect of previous cycle correctly for appropriately chosen partition number. Symbol sequence technique can also be beneficial in onboard diagnostics and real time control of engine due to their capability of data compression. In this study, cyclic variations of indicated mean effective pressure (IMEP) and total heat release (THR) are investigated to determine their behavior for stochastic and deterministic ranges.

2 Experimental Methodology

For the current study, a direct injection conventional diesel engine having swept volume 0.66 L was used. Engine has rated power of 3.5 kW at rated speed (1500 rpm) for compression ration 17.5. Experiments were conducted at constant engine speed (1500 rpm) and compression ratio of 17.5. The in-cylinder pressure was acquired using piezoelectric transducer at resolution of 1 crank angle degree (CAD) for 2500 consecutive cycles for all the test points. Acquired in-cylinder pressure data was processed for estimation of IMEP and THR on cycle to cycle basis. Heat release rate is calculated from in-cylinder pressure data using the following heat release equation

$$\frac{dQ}{d\theta} = \frac{\gamma}{\gamma - 1} P \frac{dV}{d\theta} + \frac{1}{\gamma - 1} V \frac{dP}{d\theta} + \frac{dQ_{ht}}{d\theta} \tag{1}$$

THR was computed by integration of calculated heat release rate from cylinder pressure data for each engine cycle during combustion duration.

IMEP is defined as the ratio of calculated indicated work and displacement volume. IMEP was calculated using following equation

$$IMEP = \frac{W_{ind}}{V_d} \tag{2}$$

Indicated work is defined as work transferred from combustion chamber gases to piston during compression and expansion stroke of piston movement. Indicated work is calculated as:

$$W_{ind} = \frac{2\pi}{360} \int_{-180}^{180} \left(P(\theta) \frac{dV}{d\theta} \right) d\theta \tag{3}$$

Using these equation, time series of IMEP and THR was calculated and analyzed.

3 Results and Discussion

In present section, analysis of cyclic variability using symbol sequence statistics are presented and discussed for indicated mean effective pressure and total heat release. Experimental data series of 2500 engine cycles for IMEP and THR are computed by equations presented in Sect. 2. Figure 1 shows the coefficient of variation (COV) in THR and IMEP data series at several engine load conditions. It can be noticed from the figure that COV has higher values at low load conditions for both test parameters. This observation suggests that low load conditions have higher cyclic variability due to lower combustion chamber temperature. Lower combustion chamber temperature is attained at lower load test condition because relatively less quantity of fuel burnt in the cylinder.

Symbolization can be a useful tool to evaluate the dynamic patterns in combustion data series. Symbolization converts the data into discrete bins, which are represented by a specific symbol. Figure 2a illustrates the conversion of IMEP data series in binary series of symbol '0' and '1'. IMEP data series is divided into two equiprobable partition resulting into equal number of data points in each partition. This is simplest partition to convert IMEP series into a binary series of 2500 length. Number of partitions to be used for any particular time series needs to be optimized

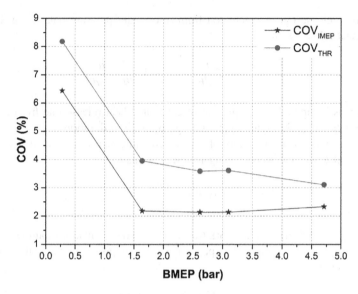

Fig. 1 Variations of COV values with engine load for IMEP and THR data series

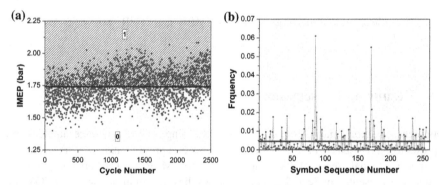

Fig. 2 Illustration of **a** binary partition symbolization and **b** symbol sequence histogram for sequence length '8' at lowest engine load condition using IMEP

to avoid the effect of noise. Sequences of symbolized time series are considered for calculating their relative frequency.

Figure 2b shows the relative frequency of occurrence of symbols for sequence length 8. Total number of possible sequence for binary partition for 8 sequence length is 2^8 or 256. All possible binary numbers for sequence length 8 is shown in decimal format and represents x axis of Fig. 2b. In general for the 'X' number of partition and 'Y' sequence length, total number of possible sequence is $N = X^Y$. For each possible number in any sequence length, frequency of occurrence is computed from symbol series obtained after symbolization. Relative frequency is the ratio of number of occurrences to the total number of possible occurrence for that sequence length. This relative frequency is plotted on Y axis (Fig. 2b) for each 256 numbers.

Relative frequency of purely random data sequence will be equal due to equi-probable partitioning. Random frequency (F_r) for a particular sequence length (Y) and number of partition (X) will be $F_r = (1/X)^Y$. This frequency obtained for truly random data is plotted with thick blue line in Fig. 2. Significant deviation above this line indicates deterministic structure and cyclic correlation in experimental time series [6]. Peaks above baseline frequency (Fig. 2b) represent the repeating deterministic events in time series. It can be noticed from figure that numbers have higher frequency than background frequency but some of them is noteworthy. For example numbers 17, 34, 85, 170, 234 etc. (their corresponding numbers in binary 00010001, 00100010, 01010101, 10101010, 11101010 respectively) have relatively higher frequency. Sequence number 85 and 170 has highest frequency, which means alternate high and low IMEP sequence is most common. Choosing optimal sequence length is need for better characterization of deterministic patterns.

A modified form of Shannon entropy is used to quantify the deviation of symbol statistics from randomness [12]. Shannon entropy value '1' suggests purely random series of data and for values less than '1' suggests correlation between sequential points. Shannon entropy can be used for estimation of optimum sequence length for presenting symbol sequence histogram. The modified Shannon entropy H_s is defined by

$$H_s = \frac{1}{\log n_{seq}} \sum_k p_k \log p_k$$

where p_k is the probability with which sequence 'k' occurs and n_{seq} is total number of sequence with non-zero probability.

Figure 3 shows the variation of modified shannon entropy for IMEP and THR for different engine operating conditions using binary partition. Minimum value of

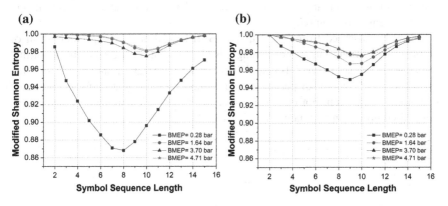

Fig. 3 Modified Shannon entropy variations using binary partition for **a** IMEP and **b** THR

shannon entropy at particular sequence length indicates, highest presence of determinism in data series. It can be noticed from Fig. 3a that shannon entropy depends on engine operating conditions and combustion parameters. Entropy values are lowest for lowest load conditions for both IMEP and THR. This suggest that IMEP and THR values have higher correlation and determinism at lower loads. At the same lowest load, IMEP has lower entropy values compared to THR. As load increases, minimum entropy value is higher for IMEP, which mean THR data values are more correlated and higher determinism compared to IMEP at higher loads. Therefore THR is more recommended control parameter. It is also observed from Fig. 3 that minimum entropy values is at sequence length '10' for IMEP (except lowest load) and around '9' sequence length for THR. This minimum shannon entropy sequence length is the optimal sequence length for binary partition. It can be summarised that effective control strategy must take '9' previous cycles into consideration using binary partition.

Appropriate number of partitions are also essential for evaluation of deterministic components in experimental data and accordingly determination of effective control strategy. To estimate the optimal combination of number of partition and sequence length, a matrix of values ranging 2–10 is evaluated for both combustion parameters.

Figure 4 shows the variations of modified Shannon entropy for IMEP at different engine load conditions. Similar trend is also observed for THR data series but only IMEP variation is plotted (Fig. 4). It can be observed from Fig. 4 that entropy values are minimum around '9' sequence length for binary partition for all the test conditions. At lowest load conditions optimum partition is binary partition. At higher loads minimum entropy values are also present for other partitions such as '4' '6' and '10'. It can be noticed from figure that optimum sequence length (lower entropy values) for higher number of partition is lower as compared to binary partition. This observation suggests that by using higher number of partition controller needs to remember lower number of engine cycles as compared to binary partition. Lower sequence length will save memory and enables better real time control and onboard diagnostics of engine. Higher number of partition might not suitable for noisy engine data. Therefore a balance of number of partition and sequence length is required.

It can be observed from Fig. 4 that at BMEP 0.28 bar optimal combustion of number of partitions and sequence length is '2-8' and at other engine loads optimal possible combinations are '4-5' and '6-4'. This observation shows that optimal combination of number of partitions and sequence length is dependent on engine operating constitution. For effective use of deterministic component in data series, optimal combination is beneficial.

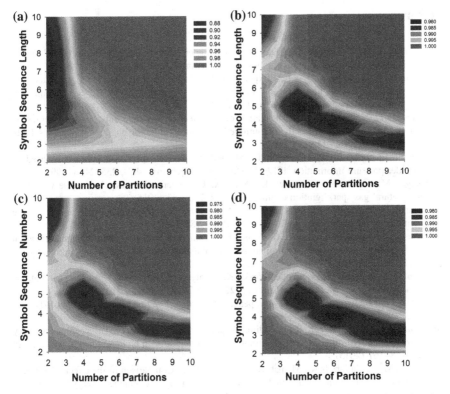

Fig. 4 Variations of modified Shannon entropy with sequence length and number of partition for IMEP at BMEP. **a** 0.28 bar. **b** 1.64 bar. **c** 3.70 bar. **d** 4.71 bar

4 Conclusions

Cyclic variations in a conventional diesel engine are analyzed at 1500 rpm at different load conditions for estimation of deterministic pattern in data series using symbol sequence method. It is found that cyclic variations of THR and IMEP experiences clear shift from stochastic to deterministic behavior at different load conditions. Optimal combination of number of partition and sequence length is dependent on engine operating conditions and selected combustion parameters. It is found that at lowest engine load optimal combination of number of partitions and sequence length is '2-8' and higher engine loads optimal possible combinations are '4-5' and '6-4'. Hence for effective engine control strategy, informations of more number of cycles is beneficial than just the immediate previous cycle. In summary, symbol sequence analysis can be very useful tools for understanding nonlinear cyclic combustion dynamics and developing effective engine control strategy.

References

1. Wagner, R.M., Drallmeier, J.A., Daw, C.S.: Characterization of lean combustion instability in premixed charge spark ignition engines. Int. J. Engine Res. **1**, 301–320 (2000)
2. Maurya, R.K., Agarwal, A.K.: Experimental investigation on the effect of intake air temperature and air fuel ratio on cycle-by-cycle variations of HCCI combustion and performance parameters. Appl. Energ. **88**, 1153–1163 (2011)
3. Klos, D.T., Kokjohn, S. L.: Investigation of the effect of injection and control strategies on combustion instability in reactivity-controlled compression ignition engines. J. Eng. Gas Turb. Power 138, 011502-011502-9 (2015)
4. Leonhardt, S., Muller, N., Isermann, R.: Methods for engine supervision and control based on cylinder pressure information. IEEE/ASME Trans. Mechatron. **4**(3), 235–245 (1999)
5. Reyes, M., Tinaut, F.V., Giménez, B., Pérez, A.: Characterization of cycle-to-cycle variations in a natural gas spark ignition engine. Fuel **140**, 752–761 (2015)
6. Finney, C.E.A., Green, Jr. J.B., Daw, C.S.: Symbolic time-series analysis of engine combustion measurements. SAE Technical Paper 980624 (1998)
7. Yang, L.P., Ding, S.L., Litak, G., Song, E.Z., Ma, X.Z.: Identification and quantification analysis of nonlinear dynamics properties of combustion instability in a diesel engine. Chaos **25**(1), 013105 (2015)
8. Sen, A.K., Longwic, R., Litak, G., Górski, K.: Analysis of cycle-to-cycle pressure oscillations in a diesel engine. Mech. Syst. Signal Pr. **22**(2), 362–373 (2008)
9. Litak, G., Longwic, R.: Analysis of repeatability of diesel engine acceleration. Appl. Therm. Eng. **29**, 3574–3578 (2009)
10. Gharehghani, A., Hosseini, R., Mirsalim, M., Jazayeri, S.A., Yusaf, T.: An experimental study on reactivity controlled compression ignition engine fueled with biodiesel/natural gas. Energy **89**, 558–567 (2015)
11. Ali, O.M., Abdullah, N.R., Mamat, R., Abdullah, A.A.: Comparison of the effect of different alcohol additives with blended fuel on cyclic variation in diesel engine. Energy Procedia **75**, 2357–2362 (2015)
12. Kaul, B.C., Vance, J.B., Drallmeier, J.A., Sarangapani, J.: A method for predicting performance improvements with effective cycle-to-cycle control of highly dilute spark ignition engine combustion. P. I. Mech. Eng. D-J. Aut. **223**(3), 423–438 (2009)
13. Crutchfield, J.P., Packard, N.H.: Symbolic dynamics of noisy chaos. Physica D **7**, 201–223 (1993)

Design of a Slotted Square Patch Antenna for 3G and 4G Wireless Communication

Purnima Sharma, Nidhi Pandit, S.K. Jha and P.P. Bhattacharya

Abstract In this paper, a reconfigurable square patch microstrip antenna has been designed. The proposed antenna consists of a slot and two metal strips for switching operation. In today's scenario where technology is merging, reconfigurable antennas for 3G and 4G technologies are needed. Particularly the proposed design is unique and attractive since it supports the third generation and fourth generation technologies for mobile phones where multiband operation is essentially required. The presented antenna operates in the ranges 2.25–3 GHz without metal strips (OFF-OFF state) and 1.96–2.76 GHz with metal strips (ON-ON state). Characteristics parameter such as return loss-S11 and radiation pattern are analysed using HFSS 13.0.

Keywords Third generation · Fourth generation · Microstrip patch antenna · MIMO · Reconfigurable

1 Introduction

Microstrip patch antennas are widely used because of their properties such as low profile, light weight and ease to manufacture [1]. The applications of Microstrip patch antennas are in various fields such as in mobile communication, medical applications, satellites and in the area of military as rockets, aircrafts and missiles etc. [2]. The demand of mobile communication is to design reconfigurable antenna

P. Sharma (✉) · N. Pandit · S.K. Jha · P.P. Bhattacharya
Department of Electronics and Communication Engineering,
Mody University of Science and Technology, Lakshmangarh, India
e-mail: purnimasharma.1412@gmail.com

N. Pandit
e-mail: npandit7@gmail.com

S.K. Jha
e-mail: ersantoshjha@yahoo.co.in

P.P. Bhattacharya
e-mail: hereispartha@gmail.com

© Springer Science+Business Media Singapore 2016   557
R.K. Choudhary et al. (eds.), *Advanced Computing and Communication Technologies*,
Advances in Intelligent Systems and Computing 452,
DOI 10.1007/978-981-10-1023-1_55

suitable for 3G and 4G technologies. A reconfigurable antenna is an alternative to achieve a wide impedance bandwidth by switching ON and OFF some parts of the antenna [3]. Switches, such as radio-frequency micro-electro-mechanical systems (RFMEMS), varactor diodes and pin diodes, can be used in structures to form frequency reconfigurable antenna [4, 5]. The reconfigurable antennas have been used for applications such as cognitive radio, MIMO channels, and space applications [5]. An asymmetric M-shaped patch antenna was designed for dual/triple band operation [6]. In [7], a triple band circularly polarized microstrip antenna is presented, which operates at 2.4, 3.5 and 5.7 GHz bands. The characteristics of mobile wireless devices is to provide faster access, improved processors, additional connectivity with Wi-Fi, GPS, third generation (3G), and fourth generation (4G) world access [8]. The third generation system operates in 2.1 GHz (1880–2170 MHz) and fourth generation in the 2.5 GHz (2570–2690 MHz). Therefore, antenna system must be capable of simultaneously switching between these two separate bands [9]. Bekali et al. have designed a reconfigurable dual frequency microstrip patch antenna [10]. In which a circular patch with slot is created on a FR4 epoxy substrate with a thickness of 1.5 mm and a relative permittivity of 4.4. To make the antenna reconfigurable PIN diodes were incorporated.

2 Proposed Design

The design proposed by Bekali et al. [10] is modified by incorporating a square patch. The proposed design consists of a square patch on a FR4 substrate with truncated ground as shown in Fig. 1a.

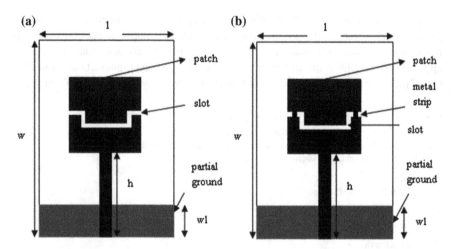

Fig. 1 Geometry of proposed square patch slotted antenna **a** without metal strip (OFF-OFF state) **b** with metal strip (ON-ON state)

There are some important parameters which are to be considered for the designing a square microstrip patch with slot antenna suitable for mobile communication. The antenna has been designed for the ranges 2.25–3 GHz (OFF-OFF state) and 1.96–2.76 GHz (ON-ON state); which covers both 3G and 4G bands at 2.1 and 2.5 GHz respectively. The frequency of resonance is chosen 2 GHz. The dielectric material used for this design is FR4 epoxy substrate with dielectric constant of 4.4 and height of 1.5 mm.

In this paper a square patch symmetric slotted microstrip antenna has been proposed for analysis. The partial ground plane is used in this antenna design. It is observed that shaping the ground plane significantly affects the refection coefficients, input impedances and current distributions [11]. Figure 1b shows the geometry of proposed antenna with metal strips to achieve reconfigurable performance. The antenna parameters are summarized in Table 1.

A slot is created in the square patch of the proposed antenna because slot antennas are popular design style for multiband applications. Figure 2 shows the structure of slot which was created in the square patch of antenna for wide bandwidth operation.

The dimensions of the slot are as follows y1 = 13 mm, y2 = 12 mm, y3 = 3 mm, y4 = 0.5 mm and y5 = 1.5 m. To design reconfigurable antenna various techniques have been proposed. Presently we have used narrow metal strips at two ends of slot to provide step impedance for achieving diversity in frequency. For

Table 1 Antenna parameters

Dimension	Value (in mm)
FR4 substrate	38 × 45 × 1.5
Square patch	16 × 16
Width of microstrip feed line	2.8
Length of feed line	16
Ground plane	38 × 10.6

Fig. 2 Geometry of slotted patch

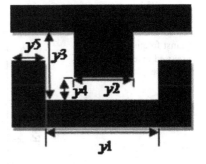

best results the locations and width of the metal strips were optimized. The position of metal strip 1 is (12, 21, 1.55) and metal strip 2 is (26, 21, 1.55). The widths of metal strips are 0.11 mm and height = 1 mm.

3 Simulation Results and Discussion

The proposed antenna design was simulated using HFSS 13.0. The performance parameters as return loss, VSWR and radiation pattern were obtained from simulations. The proposed design was compared with the conventional partial ground circular patch.

3.1 Square Patch Slotted Microstrip Antenna Without Metal Strip

The reference antenna [10] has been modified by replacing circular patch with square patch. This design was simulated and results in terms of return loss and radiation pattern are shown below.

Return Loss Figure 3 presents the return loss with respect to frequencies. The minimum return loss obtained was −49.08 dB at 2.73 GHz. The minimum return loss of reference antenna in the band of (2.33–3.02) GHz was –18 dB [10].

In comparison to the reference antenna the return loss is reduced by a significant amount. The bandwidth of proposed antenna is 2.29–3.38 GHz, which covers the frequency band suitable for 4G mobile communication. The bandwidth is 1.09 GHz which is more than that obtained by reference antenna. Thereby in terms of return loss and bandwidth proposed antenna is better than reference antenna. The VSWR value is found to be 1.03 at the resonating frequency.

Fig. 3 Simulated return loss against frequency for proposed antenna in OFF–OFF state

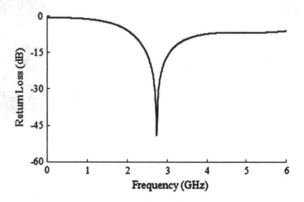

3.2 Square Patch Slotted Microstrip Antenna with Metal Strip

Stepped impedance resonator (SIR) technique was introduced in the proposed design to make this antenna reconfigurable by adding metal strips. A SIR compromises of transmission lines (TLs) with two different impedance sections. The impedance and length ratio of these two sections can be adjusted to tune the resonance frequency of a SIR [12]. While fabricating, the metal strips are to be replaced by PIN diodes.

Return Loss To further examine the effect of incorporating metal strip on the proposed antenna design, the frequency response of return loss is analyzed and shown in Fig. 4. According to simulation result it is found that the proposed antenna achieves two resonating modes over the frequency ranges of 1.96–2.76 and 3.42–4 GHz for $|S11| < -10$ dB, which covers operational bands—2.1 GHz for third generation and 3.5 GHz for Wi-MAX. The minimum value of return loss for first band is −49 dB and for second band is −22 dB which are desirable. For reference antenna the return loss was around −12 dB in the band of 2–2.26 GHz [10]. So in comparison with the reference antenna improvement is clearly visible in terms of return loss and bandwidth. Also for the proposed antenna design, VSWRs at resonating frequencies are less than 1.5 which is desirable.

Radiation Pattern The radiation pattern of proposed antenna in OFF–OFF state is shown in Fig. 5a. Its shape is like dipole. The radiation pattern for ON-ON state is omnidirectional as shown in Fig. 5b, with almost equal radiation in the X–Z plane, which is suitable for mobile applications.

Fig. 4 Simulated return loss of proposed antenna in ON–ON state

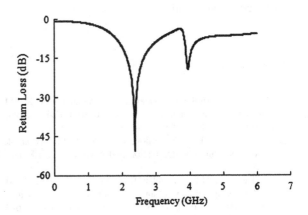

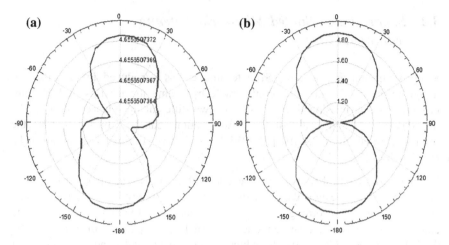

Fig. 5 **a** Radiation pattern of proposed antenna in OFF–OFF state (without metal strips) at 2.5 GHz. **b** Radiation pattern of proposed antenna in ON–ON state (with metal strips) at 2.1 GHz

4 Conclusion

In this work, a square patch slotted antenna has been designed and analyzed. By incorporating square patch better results in terms of return loss and bandwidth were found in comparison with reference antenna. Around −31 dB improvement in return loss was observed. Frequency reconfigurable behavior was observed with the application of metal strips. The designed antenna is suitable for 3G, 4G and Wi-MAX applications. In future, this work may be extended to improve the directivity and reduce the back lobe radiation. The present design may be fabricated and tested for verification.

References

1. Mathew, S., Anitha, R., Deepak, U., Aanandan, C.K., Mohanan, P., Vasudevan, K.: A compact tri-band dual-polarized corner-truncated sectoral patch antenna. IEEE Trans. Antennas Propag. **63**, 5842–5845 (2015)
2. Kaivanto, E.K., Berg, M., Salonen, E., Maagt, P.D.: Wearable circularly polarized antenna for personal satellite communication and navigation. IEEE Trans. Antennas Propag. **59**, 4490–4496 (2011)
3. AbuTarboush, H.F., Nilavalan, R., Cheung, S.W., Peter, K.M.T., Budimir, D., Al-Raweshidy, H.: A reconfigurable wideband and multiband antenna using dual-patch elements for compact wireless devices. IEEE Trans. Antennas Propag. **60**(1), 36–43 (2012)
4. Yang, S., Zhang, C., Pan, H.K., Fathy, A.E., Nair, V.K.: Frequency reconfigurable antennas for multiradio wireless platforms. IEEE Microwave Mag. **10**, 66–83 (2009)

5. Mirzamohammadi, F., Nourinia, J., Ghobadi, C.: A novel dual-wideband monopole-like microstrip antenna with controllable frequency response. IEEE Antennas Wirel. Propag. Lett. **11**, 289–292 (2012)
6. Peng, L., Ruan, C.-L., Wu, X.H.: Design and operation of dual/triple-band asymmetric M-shaped microstrip patch antennas. IEEE Antennas Wirel. Propag. Lett. **9**, 1069–1072 (2010)
7. Bao, X.L., Ammann, M.J.: Printed triple-band circularly polarised antenna for wireless systems. IET Electronics Letters **50**, 1664–1665 (2014)
8. F.M. Caimi: Antenna design challenges for 4G. IEEE Wireless Commun. **18**, 4–5 (2011)
9. Cui, Y.H., Li, R.L., Wang, P.: Novel dual-broadband planar antenna and its array for 2G/3G/LTE base stations. IEEE Trans. Antennas Propag. **61**, 1132–1139 (2013)
10. Bekali, Y.K., Essaaidi, M.: Compact reconfigurable dual frequency microstrip patch antenna for 3G and 4G mobile communication technologies. Microwave Opt. Technol. Lett. **55**, 1622–1626 (2013)
11. Ooi, P.C., Selvan, K.T.: The effect of ground plane on the performance of a square loop CPW-FED printed antenna. Prog. Electromagnet. Res. Lett. **19**, 103–111 (2010)
12. Makimoto, M., Yamashita, S.: Bandpass filters using parallel coupled stripline stepped impedance resonators. IEEE Trans. Microw. Theory Tech. **28**, 1413–1417 (1980)

Multiple DG Synchronization and De-synchronization in a Microgrid Using PLC

O.V. Gnana Swathika, K. Karthikeyan and S. Hemamalini

Abstract Microgrid is an aggregate of generating units and loads at the distribution level. It operates in two modes: Grid connected mode and Islanded mode. Power continuity is very essential for critical loads in microgrid and maybe achieved vide 100 % standby diesel generator sets in a microgrid network. This paper considers a real time 24/7 Information Technology (IT) campus where even a very short power break will lead to huge commercial loss. Hence it is very important to bring the standby power during the utility power failure. This is achieved through multiple diesel generator sets synchronized in a single panel through a proper sequential logic vide programmable logic controller (PLC). In the event of fault occurrence, based on the receipt of signals from the respective relays, the PLC identifies the fault and issues appropriate tripping commands to the microgrid to safely isolate the fault. The significance of neutral isolating contactor (NIC) and neutral grounding resistor (NGR) in microgrid protection is also analyzed.

Keywords Microgrid protection · PLC · Grid connected mode · Islanded mode · Neutral grounding resistor protection

1 Introduction

A microgrid is an aggregate of many DG micro-sources and loads connected to the distribution system [1] with system capacity between several kW and several MW. Prominent contribution of microgrid to the main-grid are its ability to reduce

O.V. Gnana Swathika (✉) · S. Hemamalini
School of Electrical Engineering, VIT University, Chennai, India
e-mail: gnanaswathika.ov@vit.ac.in

S. Hemamalini
e-mail: hemamalini.s@vit.ac.in

K. Karthikeyan
Engineering Design and Research Center, Larsen and Toubro, Chennai, India
e-mail: kskk@lntecc.com

© Springer Science+Business Media Singapore 2016   565
R.K. Choudhary et al. (eds.), *Advanced Computing and Communication Technologies*,
Advances in Intelligent Systems and Computing 452,
DOI 10.1007/978-981-10-1023-1_56

congestion, cater to the instant need for additional generation, improve stability of system, react to variation in loads and provide continuity of supply to the customers [2, 3]. Distributed generator penetration is challenging as it [1] may lead to grid instability or failure, if not properly controlled. The control is achieved using synchronization algorithm and current controller. Various grid synchronization algorithms with their role and influence in the control of Distributed Generator penetration on normal and faulty grid condition is elaborated in [4, 5]. Programmable Logic Controller (PLC) based Diesel Generator synchronization is a popular industry practice.

This paper considers a real-time IT campus microgrid network with critical loads. This test system is monitored, controlled and protected using a PLC. The PLC is responsible in providing uninterrupted supply to the critical loads in the microgrid network. The purpose of neutral isolating contactor (NIC) and neutral grounding resistor (NGR) in PLC based microgrid protection is also elaborated.

2   Programmable Logic Controller Based DG Synchronization and De-synchronization

Figure 1 explains the Grid connected mode of a PLC monitored microgrid network, where the utility power meets the load requirements and this is highlighted in red color. Figure 2 indicates the islanded mode of operation of the microgrid network.

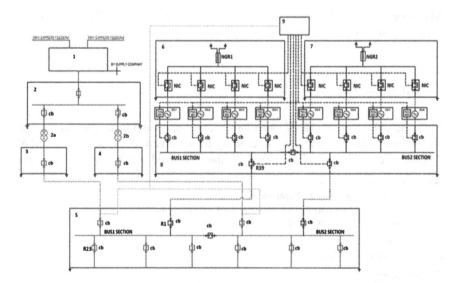

Fig. 1 Grid connected mode of PLC controlled microgrid (main supply meets load requirements)

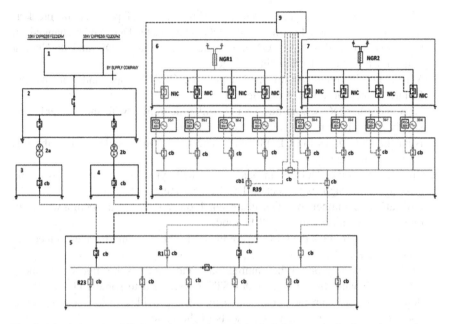

Fig. 2 Islanded mode of PLC monitored microgrid (only DG meets the load requirements)

Legend:

1   33KV HT panel with Utility Metering
2   33KV HT panel with Consumer metering
2a 33/11KV Power transformer 1
2b 33/11KV Power transformer 2
3   11KV HT Panel 1
4   11KV HT Panel 2
5   11KV Main HT Panel
6   Neutral Isolating Contactor (NIC) cum Neutral Grounding Resistor (NGR) Panel 1
7   Neutral Isolating Contactor (NIC) cum Neutral Grounding Resistor (NGR) Panel 2
8   11 KV DG synchronization panel
9   Programmable logic controller (PLC)
cb Vacuum circuit breakers

2.1 PLC Operation in Islanded Mode

Initially the standby diesel generators and its associated panels and components in the microgrid are kept in isolation mode. During the utility power failure condition, the following sequential logic is performed to achieve proper DG synchronization:

1. PLC takes the power failure signal from Main 11 kV HT panel circuit breaker.
2. Let DG1 be considered as the Master DG. The Master DG maybe changed on sequential basis. PLC gives ON command to the Master DG and its Neutral contactor and Neutral grounding resistor.
3. Master DG incoming circuit breaker gets the CLOSE command from its DG Control Panel on the dead bus Sect. 1.
4. Now the other DG sets in bus Sect. 1 will synchronize with the available bus supply through its respective DG Control Panel.
5. Next the PLC gives the CLOSE command to the bus coupler so that the bus Sect. 2 is also charged.
6. Similarly the other DG sets in bus Sect. 2 will synchronize with the available bus supply through its respective DG Control Panel.
7. Now all the 8 numbers of DG sets are synchronized in one bus section with only one neutral contactor in CLOSED condition.
8. Finally PLC provides the CLOSE command to the outgoing breaker to feed the downstream loads.
9. PLC continues to monitor the running loads and if it finds the supply to be more than the demand, the PLC provides STOP command to some DG sets. If the demand is more than the supply, then the PLC provides START command to some DG sets to meet the requirements of the system.

2.2 PLC Operation in Grid Connected Mode

When the utility power resumes, the following sequential logic is performed to have a proper DG de-synchronization:

1. PLC keeps monitoring the healthiness of the utility power. The moment when the utility power resumes, PLC gives a TRIP command to the outgoing circuit breakers. The incomer circuit breaker of the utility power gets a CLOSE command from its inbuilt interlock so that the loads are fed from the utility power.
2. The PLC now removes the START command from all DG sets' DG Control Panels which in turn trips all DG sets' incomer circuit breakers and the Neutral isolating contactor.
3. The DG sets will STOP after the cool down time and is kept ready for the next operation.

From the above sequence of operation it is evident that a proper DG synchronization or de-synchronization is achieved using PLC based monitoring, protection and control system to provide uninterrupted supply to the critical loads.

3 Protection Requirements of PLC Controlled Microgrid System

The following are the minimum protection that must be provided to the microgrid system shown in Fig. 1.

Protection for external faults (Backup protection):
Overvoltage, Overcurrent, Thermal overload and Negative phase sequence protection during unbalanced loading.

Stator winding Protection for internal faults (Primary protection):
Earth fault, Differential, Interturn and Short circuit.

Field excitation Protection (Field failure):
Field earth fault and Loss of excitation.

Mechanical protection:
Overspeed, Vibration, Bearing and Reverse power.

Figures 3 and 4 illustrate the typical relay coordination curve in islanded mode during phase fault and ground fault respectively for a fault near relay R23 i.e. the downstream relay. The primary protection is provided by R23. The backup protection is given by R1 followed by R39.

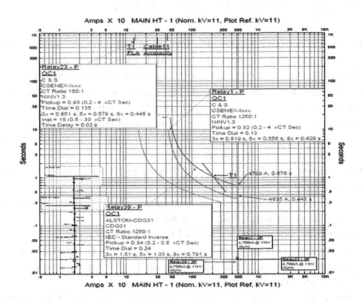

Fig. 3 Relay coordination curve during phase fault

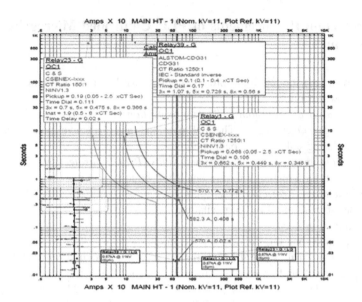

Fig. 4 Relay coordination curve during ground fault

3.1 Neutral Grounding Resistor Protection

In addition to the above protection, it is very essential to provide Neutral grounding resistor protection to the microgrid test system. If the star point of a star connected three phase system is grounded, then it is said to have solid earthing. In this system, the net resistance is contributed by soil. This causes heavy current in the order of hundreds of amperes to flow in windings during faults. This high rating current may cause damage to the windings. To save the windings from such damage, a series resistance called neutral grounding resistor (NGR) is added at the star point. This resistance limits the current during earth faults. In the proposed PLC based microgrid monitoring system, the current flowing in NGR is monitored to activate the earth fault relay as and when required.

Typical NGR sizing calculation for alternator full load current of one number 2MVA, 11 kV diesel generator alternator set is:

Alternator full load current of 2MVA DG is 105 Amperes.

Voltage available at neutral point = Line Voltage/$\sqrt{3}$ = 11,000/1.732 = 6351 V.

Therefore using ohms law, the resistance required to limit the current is:

R = V/I = 6351/105 = 60.48 Ω.

Thus the NGR is selected for a rated current of 105 amps, 60.48 Ω resistance for duration of 10 s. Using the same method, NGR sizing calculations can be done for remaining DG sets.

3.2 Neutral Isolating Contactor

Neutral isolating contactor is an important factor while synchronizing one or more DG sets. Ideally no unbalanced current flows in the neutral during balanced load condition. But during earth fault or overload fault on any phase, unbalanced loading occurs and an unbalanced current flows through the neutral. To facilitate convenient discharge of this fault current to ground, the neutral is grounded.

An electrical system with parallel operation of two or more low voltage diesel generator sets must meet the following conditions:

(i) Neutral of only one generator needs to be earthed to avoid the flow of zero sequence current.

(ii) During independent operation, neutrals of both generators are required in low voltage switchboard to obtain three phase, 4 wire system including phase to neutral voltage.

(iii) Restricted earth fault protection (REF) for both the generators whilst in operation.

Since the earthing neutral of only one generator is required, a contactor called neutral contactor of suitable rating is fixed in neutral to earth circuit of each generator. Neutral contactors are interlocked in such a way that only one contactor remains closed during parallel operation of generators. During independent operation of any generator its neutral contactor is always closed. Breaker auxiliary contacts automate these neutral contactors.

4 Conclusion

This paper proposes a PLC based monitoring, protection and control system for a real time microgrid network of a 24/7 IT campus with critical loads. During the utility power failure, continuity of supply to these critical loads is achieved through multiple diesel generator sets synchronized in a single panel controlled through PLC. The PLC is also responsible in clearing the fault by tripping the appropriate breakers vide receiving signals from the respective relays. The significance of neutral isolating contactor (NIC) and neutral grounding resistor (NGR) in this microgrid network is also analyzed to ensure that the system is protected in totality during DG synchronization.

References

1. Sachit Gopalan, A., Victor Sreeram, Herbert Iu, H. C.: A review of coordination strategies and protection schemes for microgrids. Renew. Sustain. Energy Rev. **32**, 222–228 (2014)
2. Blaabjerg, F., Chen, Z., Kjaer, S.: Power electronics as efficient interface in dispersed power generation systems. IEEE Trans. Power Electron. **19**, 1184–1194 (2004)
3. Blaabjerg, Frede, Teodorescu, Remus, Liserre, Marco, Adrian Timbus, V.: Overview of control and grid synchronization for distributed power generation systems. IEEE Trans. Industr. Electron. **53**, 1398–1409 (2006)
4. Balaguer, I.J., Lei, Q., Yang, S., Supatti, U., Peng, F.Z.: Control for grid-connected and intentional islanding operations of distributed power generation. IEEE Trans. Industr. Electron. **58**, 147–157 (2011)
5. Gnana Swathika, O.V., Hemamalini, S.: Kruskal aided Floyd-Warshall algorithm for shortest path identifcation in microgrid. ARPN J. Eng. Appl. Sci. **10**, 6614–6618 (2015)

Demand Side Response Modeling with Controller Design Using Aggregate Air Conditioning Loads and Particle Swarm Optimization

S. Sofana Reka and V. Ramesh

Abstract In this paper, the authors present a current issue of balancing the task between the demand and the generated power with a new controller design. A controller is modelled for reducing the peak demand by an intelligent design scheduler using Particle swarm optimization technique. The power requirement when the peak load exists tends to create a more non eco friendly transformation in power generation. In this research work, Particle swarm optimization creates a minimal load schedule in order to minimize the peak demand to comfort the users. Simulation is carried out using MATLAB with the performance of intelligent demand response modeling using direct load control technique. The highlight of this research work is to develop a stable controller for the realistic demand response model for thermostatic air-conditioning loads.

Keywords PSO · Demand · Direct load control · Controller design · Demand response

1 Introduction

In today's world of electrical technology, it is very clear that new innovations are needed to reduce carbon footprint. Increased power stability can be achieved through control design algorithms on the load side for efficient management of load and demand. The challenging needs of modernized electric grid can be responded to some new assessment techniques. One major technology that responds to modern

S. Sofana Reka (✉) · V. Ramesh
School of Electrical Engineering, VIT University, Vellore 632014,
Tamil Nadu, India
e-mail: chocos.sofana@gmail.com

V. Ramesh
e-mail: vramesh@vit.ac.in

© Springer Science+Business Media Singapore 2016   573
R.K. Choudhary et al. (eds.), *Advanced Computing and Communication Technologies*,
Advances in Intelligent Systems and Computing 452,
DOI 10.1007/978-981-10-1023-1_57

smart grid is demand side response [1, 2]. The controlling of demand of end consumers in accordance with the variable pricing scheme is called Demand response (DR) [3]. The DR exhibits to reduce the power imbalance in the new smart grid vision in the future era [3]. Many objectives have been considered such as the peak sharing and valley filling. Residential homes and buildings are the primary end users of the grid enhance the vital role of demand response modeling in smart grid. The existing real time Demand Response (DR) management can be embedded with smart meters with automatic execution of optimal operation of appliances in residential consumers [4].

The operational period of residential appliances have been classified into deferrable/non-deferrable loads [4] and interruptible/non-interruptible ones [4] based on the preference of DR strategy. Household heating and cooling handles the maximum share of peak demand in most parts of energy usage. As per survey the energy consumption consists of 21 % of all the energy usage for all the major appliances [5]. In this regard effective operation control with the smart meter interface is needed in order to improve the efficiency of the grid in near future. Based on different control operations the loads are classified as non-shiftable, time-shiftable and power shiftable [6]. There is a rapid increase in contribution of air-conditioning (AC) loads in the household appliances as they are user preference loads in a day. They perform as ideal elements for Demand Side Management (DSM) projects because of the cold water storage. Conventional air-conditioning systems in which the air gets cooled and dehumidified in which the return of the cold water temperatures are instantly fixed with 7–12 °C. Naturally in summer climates, the AC loads can contribute 20 % of the houses with a normal middle class family ratio. There are different techniques which can be involved in smart grid DSM programmes like Direct load control [7]. Using this approach there is possibility to shift the loads of the AC to the approachable valleys of daily load profile in residential consumers which results a flatten load in the day. It is first important to develop a controller design to carefully understand cooling and the Aggregated load control and modeling structure have been intensively studied in the literature survey about HVAC [8, 9]. Developing the need of efficient DR model of electrical appliances and optimization algorithms with their coordinated operations are the key problems existing in energy management system in a smart home, has been a coordinating attention in recent research [10, 11]. This paper proposes a control design with the idea of using optimization algorithm to schedule the domestic air conditioning loads in a complete demand response modeling and achieve practical results. The paper is organized as follows: In Sect. 2 problem formulation of the optimization control technique and the model is analyzed with necessary expressions. In Sect. 3 the proposed control model is developed with analysis of peak load reduction and explains the optimization algorithm and how it is applied for the current control design. Simulation results with sample data are provided and discussed in Sect. 4. Conclusions and future work is drawn in Sect. 5.

2 Methodology

In the paper, the control model uses the air conditioning systems and PSO creating a design with an energy scheduler to shift the loads from the peak demand period in a day.

2.1 Control Model

On the crisis of effective control, the domestic AC loads with less impact on consumers, the two main parameters involved should be predicted. The optimal control of AC systems have typical load control in variable air volume systems, temperature control unit (TCU), air handling unit (AHU) and pressure control. AHU exhibits the indoor temperature control and also outdoor ventilation flow control. Figure 1 shows the basic block diagram of typical air-conditioning system.

In the future smart grid environment, the appliance scheduling is very important with the use of smart meters installed in the residential house. The linear dynamic control model is developed with the analysis of smart meters fixed at the household premises as a case work. The module of controller is split into two steps. At the first step, the data from the appliances considered have been processed in a pattern recognition system with the known value of AC loads [12–15] in the case considered 1500 W. As the second step the temperature level is analyzed from the system, this is analyzed from the previously well renowned model considered [12]. The indoor temperature linear dynamic model with accordance with energy consumption on per day basis is analyzed mathematically in (1) as objective control function.

$$T^{in}(t) = T^{in}(t-1) + a\left[T^{out}(t) - T^{in}(t-1)\right] + be_{\alpha,t} \tag{1}$$

With the constraint of energy consumption range

$$E_{n\alpha}^{min} \leq e_{\alpha,t} \leq E_{n\alpha}^{max} \tag{2}$$

$$T^{mdes} \leq T^{in}(t) \leq T^{mxdes} \tag{3}$$

$$R_a(e_{\alpha,t}) = k_1\left(T^{in}(t) - T^{need}\right)^2 - k_2 \tag{4}$$

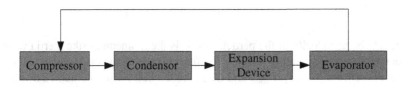

Fig. 1 Typical block diagram of AC system

where the indoor temperature by the air conditioner is $T^{in}(t)$, $e_{\alpha,t}$ is the energy consumed and relation made between the temperatures using the constraints depending on the model parameters, a, b are the thermal situations around the ac system, $T^{iout}(t)$ is the outdoor temperature with $T^{in}(0)$ considered as the initial room temperature. The additional constraints are considered on the energy consumed in the peak sharing time and operated in a balanced temperature range on shifting the loads using a controller design. They are used as constraints in Eqs. (2) and (3), where T^{mdes}, T^{mxdes} are the maximum and minimum required temperatures [13–16]. The other parameters involved in control design for the optimization of values are k_1 and k_2 constants, T^{need} is the comfortable temperature decided by the consumer is explained in Eq. (4). In the summary of the methodology the model is achieved from the initial conditions with constraints for peak sharing [17–19] results and the values are much predicted using the above objective function which is developed as a controller and scheduler in the smart meters in the consumer side. The analysis is done with the deduced objective function and the solution is optimized without losing the comfort level.

2.2 Particle Swarm Optimization

PSO is an optimization technique based on the analysis of combined population introduced by Kennedy and Eberhart [20]. The algorithm is developed from the behavior of birds. The total population is understood by representing randomly initializing parameters as swarms. PSO is a swarm intelligent based technique widely used by many researchers due to the numerous advantages such as derivative free algorithm, can be used for stochastic optimization technique, simple in design and easy to implement. The swarm particles move around in a search space with two parameters of velocity and position. The concept behind PSO is rather not established from evolutionary natural selection but exhibited in a much intelligent behavior of swarms of birds, fishes etc. [20]. In this technique the choosing of the swarm particles are done randomly with the updation of velocity and position. This process is continued until the best solution is obtained among the population.

The velocity and position of the particle is updated by using the following equation:

$$P_{v_i}^{t+1} = RP_{v_i}^t + k_1 r_{k_1}(P_{rBest_i}^t - z_i^t) + k_2 r_{k_2}(G_{rBest_i}^t - z_i^t) \tag{5}$$

$$z_i^{t+1} = z_i^t + P_{v_i}^{t+1} \tag{6}$$

where, P_{v_i} is the velocity of the particle and R is the momentum (the inertia weight factor), k_1, k_2 are accelerating learning coefficients, z_i is the position carried off from the set of particles $\{0.....i\}$, P_{rBest} and G_{rBest} are the best position obtained by all the particles in the variations with other particles. Using PSO, the representation of particles considered for study are obtained by updating the parameters from Eqs. (5) and (6). After the n number of iterations the global best gives the required solution of the problem initiated. In this analysis of parameter control, for discrete optimization problems as per [20] the PSO algorithm is more approachable in this proposed model for the involvement of finding the optimal load schedule. This schedule is obtained using a set point of optimizer block in the controller design for minimizing of peak demand with effect of maximize of user's comfort.

3 Proposed Control Design

The total generation capacity, peak load demand will constitute a vital portion of demand. This demand gradually moves on to the stress of power systems by building new power plants which indeed increases carbon emission worldwide. Thus by flattening the load profile gradually increases the efficiency and indeed the stability of the system. Hitherto, various desired load profiles are been used by the utility market providers to carry on different objectives by DSM. Direct load control [7] in general is the analysis of controlling appliances and scheduling from both ways of residential consumer and utility. It provides a raw balance between them. The scheduling of appliances and considering the major parameters are important by flattening the peak. The PSO [20] based controller for DLC [7] is given in Fig. 2, in order to minimize the peak demand, costs and increase comfort level. The mathematical model is explained in Sect. 2 where the inputs given to the distribution level is the acquired demand from the system and consumer demand exhibited from the smart meter level. The analysis of control design depends on the control signal.

From the control set up, the analysis is done setting a temperature point with the cooling effect of the room and the desired value is provided in accordance with the distribution side. With PSO controller there are particles used to revolve around the space explained in Eqs. (5) and (6). The condition of the acquired particle movement and reaching the fitness function is obtained by in Eq. (7).

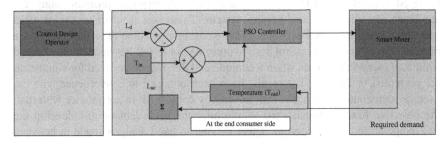

Fig. 2 Control schematic for the proposed method

Fig. 3 Simulation of appliance for uncontrolled system

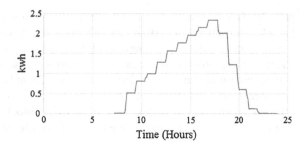

$$P_f = \sum_{j=0}^{N} (C_{\mathrm{maxd}} - C_{des})R_1 + (L_d - L_{ac})R_2 \tag{7}$$

The fitness function determines the final function obtained that is involved to minimize the peak load demand. The parameters are $C_{\mathrm{max}\,d}$ is maximum desired set point of cooling on user comfort with C_{des} as the desired level. The desired load levels are given as L_d. A interpose exists between the demand parameter and the load by reducing the power from the AC system to be decreased. In Fig. 3 explains the load simulation of the appliance for an uncontrolled system. With final analysis, the consideration of the user's comfort is a main parameter and intolerance occurs when the indoor temperature is higher than the required value, and when the temperature reaches the maximum value, AC will be operated. The energy consumption of the typical AC load is in Fig. 3. The cooling down level will exist to move on until the temperature attains a specified set value. Since the AC task is a non-deferrable load and vice versa the cooling procedure should be done within the predefined comfort horizon of the users $T_{\min}$ and $T_{\max}$ value. The control model is evaluated with a proper relationship between the indoor, outdoor and energy consumed by the consumer.

4 Results

To validate the performance of the proposed system simulations are carried out using MATLAB/SIMULINK for the new Demand response program using the appliance of typical air conditioning system. There is a need of communication between the utility market provider and the consumers where the smart meters are installed. The direct load control system presented in Fig. 2 has been optimized and simulated using PSO optimization technique and the parameters used for simulation are presented in Table 1. The results have been obtained in convergence with the average temperature adjustable with the energy consumed. In accordance with that, the objective function is analyzed to minimize the peak demand and develop the AC temperature in a day period. There is hand in hand analysis made in between

Table 1 Parameters used in simulation

Simulation parameter	Obtained value	Simulation parameter	Obtained value
Maximum No. of swarm	100	K_2	1.0
R	2.0	R_1	11.0
Computations	50	R_2	1.0
Max. voltage	1.5	R_d	$R_{d(t)}/12(2)$
K_1	1.0	T_{kd}	T_{kmax}

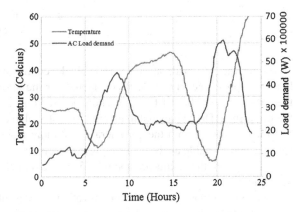

Fig. 4 Temperature and AC load for the uncontrolled system

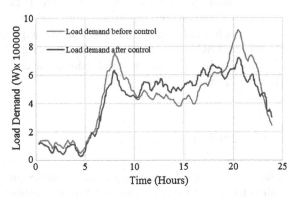

Fig. 5 Load demand response system without and with PSO algorithm

the demand with the cooling analysis as in by reduction of power made. The results are obtained with the sample data obtained from the smart data collected [6]. In Fig. 4 with the proposed controller in smart meter design, from controlling of load it is demonstrated that the peak clipping of the loads and valley filling.

The analysis from the results presented in Figs. 4 and 5 is that the first peak load on peak demand during the day according to the data is reduced to 120 kW and further at the immediate second peak of the day is reduced among them as 170 kW. According to the simulation obtained there exhibits approximately the peak load time for about 600–850 kW.

Fig. 6 Aggregated load
demand and temperature
response using PSO algorithm

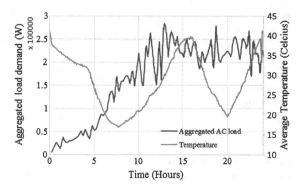

Time (Hours)

As per the AC loads at about 1600 W depending on the sizes installed at the home are presented in Fig. 6. If considered for the above rating in the simulation there it means for about 1 to 5 household ac used houses can be controlled easily in the peak time. The values obtained are more of scalable and can be deduced to the distribution system with the consideration of more residential loads in a home AC system. The probable analysis of saving cost from the energy providers and with the consumers from the results are about $1.5 and be reduced to much savings on a day schedule.

The values obtained after the optimization level results in w_1 and w_2. Hence it can be modified by two varied conditions when the peak demand increases and the user comfort level are very much compromised. The limitable temperature control with PSO optimization technique is kept in the range of necessary limits. The total or aggregated AC loads are maintained in a flat range during the day, as the need of using AC loads are in summer and maintained constant throughout the day are presented in Fig. 6.

5 Conclusion

In this paper a particle swarm optimization technique with an effective optimized method for demand side response intelligent modeling using air conditioning systems in residential consumers is presented. The main advantage of using this technique is the ability to obtain effective and accurate results from the designed control model. The approach implemented in this paper exhibits the objective function of temperature control of indoor system and the cooling system with thermostatic loads in order to schedule the loads with smart meters installation in the household premises. The proposed system in this work enhances a control structure using PSO controller on DLC. PSO method was found to good tracking ability even under varying temperature conditions and results have been obtained using MATLAB which shows an effective energy management DSM schemes with regard to minimize the cost and power quality.

References

1. Balijepalli, V.S.K.M., et al.: Review of demand response under smart grid paradigm. Innovative Smart Grid Technologies-India (ISGT India), IEEE PES (2011)
2. Palensky, P., Dietrich, D.: Demand side management: demand response, intelligent energy systems, and smart loads. IEEE Trans. Ind. Inform. 7.3, 381–388 (2011)
3. Siano, P.: Demand response and smart grids—a survey. Renew. Sustain. Energy Rev. **30**, 461–478 (2014)
4. Chen, Zhi, Lei, Wu, Yong, Fu: Real-time price-based demand response management for residential appliances via stochastic optimization and robust optimization. IEEE Trans. Smart Grid 3(4), 1822–1831 (2012)
5. DoE. U. S. Energy efficiency trends in residential and commercial buildings. US Department of Energy, Washington, DC (2008)
6. Zhu, Z., Tang, J., Lambotharan, S., Chin, W.H., Fan, Z.: An integer linear programming based optimization for home demand-side management in smart grid. In: Proceedings of 2012 IEEE PES, ISGT, Washington, DC, Jan 2012
7. Ruiz, N., Cobelo, I., Oyarzabal, J.: A direct load control model for virtual power plant management. IEEE Trans. Power Syst. 24.2, 959–966 (2009)
8. Banash, S., Fathy, H.: Modeling and control insights into demand-side energy management through set point control of thermostatic loads. In: Proceedings of American Control Conference, June (2011)
9. Zhang, W., Kalsi, K., Fuller, J., Elizondo, M., Chassin, D.: Aggregate model for heterogeneous thermostatically controlled loads with demand response. In: Proceedings of IEEE PES General Meeting, San USA, July 2012
10. Mohsenian-Rad, A.H., Leon-Garcia, A.: Optimal residential load control with price prediction in real-time electricity pricing environments. IEEE Trans. Smart Grid 1.2, 120–133 (2010)
11. Ugrinovskii, V.A., Petersen, I.R.: Finite horizon minimax optimal control of stochastic partially observed time varying uncertain systems. Math. Control Signals Syst. 12.1, 1–23 (1999)
12. Zhang, Y., Lu, N.: Demand-side management of air conditioning cooling loads for intra-hour load balancing. In: Innovative Smart Grid Technologies (ISGT), pp. 1–6 (2013)
13. Zhang, Wei, Jianming, Lian, Chang, C.Y., Kalsi, K.: Aggregated modeling and control of air conditioning loads for demand response. IEEE Trans. Power Syst. **28**(4), 4655–4664 (2013)
14. Rezeka, S.F, Attia, A.H., Saleh, A.M.: Management of air-conditioning systems in residential buildings by using fuzzy logic. Alexandria Eng. J. 54.2, 91–98 (2015)
15. Bashash, Saeid, Fathy, Hosam K.: Modeling and control of aggregate air conditioning loads for robust renewable power management. IEEE Trans. Control Syst. Technol. **21**(4), 1318–1327 (2013)
16. Beil, I., Hiskens, I.A., Backhaus, S.: Round-trip efficiency of fast demand response in a large commercial air conditioner. Energy Build. **97**, 47–55 (2015)
17. Tsui, K.M., Chan, S.C.: Demand response optimization for smart home scheduling under real-time pricing. IEEE Trans. on Smart Grid 3.4, 1812–1821 (2012)
18. Rahman, M.M., Hettiwatte, S., Gyamfi, S.: An intelligent approach of achieving demand response by fuzzy logic based domestic load management. In: IEEE Power Engineering Conference (AUPEC), pp. 1–6 (2014)
19. Thomas, M.S., Bansal, P., Taneja, P.: Smart home energy management by Demand Response controller design. In: 6th India International Conference on Power Electronics (IICPE), pp. 1–6 (2014)
20. Kennedy, J., Eberhart, R.: Particle Swarm Optimization. In: Proceedings of IEEE International Conference on Neural Networks, pp. 1942–1948 (1995)

Particle Swarm Optimization and Schelkunoff Unit Circle to Minimize the Interference Near to the User

Smita Banerjee and Ved Vyas Dwivedi

Abstract This paper analyses the mathematical design of a linear antenna array in overcoming the problem of interfering signal near to the user signal. A mathematical modelling of modified version of Schelkunoff polynomial method with Particle Swarm Optimization has been presented. The radial displacement and phase on Schelkunoff unit circle are fixed for maintaining the direction of user and interferer. Reduction of sidelobe level constraint is done by searching the best location of phases. Parameters like sidelobe level and directivity have been considered in showing the usefulness of this technique. Effectiveness and limitations of placing nulls near to the main beam have been shown by relevant examples through variation of interferer positioning.

Keywords Linear antenna array · Schelkunoff unit circle · Particle swarm optimization · Sidelobe level · Directivity · First null beam width

1 Introduction

The interference can be rejected or suppressed by either putting low or minimum gain towards that direction or reducing the sidelobes. Sidelobes causes degradation of the actual signal and hence reduce the efficiency of the antennas. Placing null in the sidelobe region can reduce the effect of interference. However, this will lead to an increase of the sidelobe level (SLL). This increment of SLL will depend upon the number of interferers and their locations. If the interferers are near to the main beam, then it will directly affect the beam width: this condition needs to be further studied [1–4].

S. Banerjee (✉)
School of Engineering, RK University, Rajkot, India
e-mail: smita161@gmail.com

V.V. Dwivedi
C. U. Shah University, Wadhwan, India
e-mail: vedvyasdwivediphd@gmail.com

© Springer Science+Business Media Singapore 2016   583
R.K. Choudhary et al. (eds.), *Advanced Computing and Communication Technologies*,
Advances in Intelligent Systems and Computing 452,
DOI 10.1007/978-981-10-1023-1_58

Different pattern synthesis techniques such as adjusting the element position, inter-element spacing, amplitude excitation, phase excitation and complex excitation are available in the literature are useful to cancel the undesired interferers [5, 6]. Selection of the parameters alone or combination is cumbersome to achieve desirable properties. Hence, the synthesis problems are highly non-linear in nature and need to be solved using non-linear optimization algorithm [7–11].

The analytical method of synthesis using Schelkunoff Polynomial Method (SPM) was developed to put nulls in the desired direction. The main beam location and the main beam width (also known as first null beam width-FNBW) will depend upon the number of nulls and their location [12]. Several studies of SPM for the synthesis of antenna array have been reviewed [13–19]. The characteristics of the main beam have been synthesized using the largest degree of the sub-polynomial in [13]. From the specific nulls along with target points, the radiation pattern is synthesized to minimize the error between the desired and optimized radiation pattern [14]. The beamforming approach using the null points and target points is proposed in [15, 16]. Design of conformal antenna array and the effect of various parameters on the radiation pattern is discussed in [17]. By dividing the polynomial into different sets and controlling one of the sub-polynomials, the number of optimized parameters required were lesser than that of the classical complex synthesis method [18]. From a given number of antenna element and number of nulls, the radiation pattern is synthesized for reduced SLL [19].

In this study, modified SPM (MSPM) is used to put main beam and null along the specific direction by fixing the radial position and phase on the Schelkunoff unit circle (SUC). Other left over roots are taken as the optimization parameters in Particle Swarm Optimization (PSO) with cost function to reduce the SLL near to the main beam. This paper offers a detailed comparative study of the available polynomial techniques used to put the null in the desired direction. The parameters for comparison are maximum SLL, mean SLL, directivity, number and position of the interferer keeping direction of main beam and FNBW constant.

The rest of the paper is arranged as follows: Sect. 2 describes the mathematical analyses of SPM for linear antenna array, Sect. 3 describes the design procedure using PSO, Sect. 4 compares the results and Sect. 5 concludes the whole study.

2 Mathematical Analysis of Schelkunoff's Polynomial Method for Linear Antenna Array

Consider a linear array of N radiating antenna elements which are equally separated by a distance d arranged along a line. Mathematically, the AF is given by [1],

$$AF(\theta) = \sum_{n=1}^{N} w_n e^{j(n-1)\Psi} \tag{1}$$

where $w_n = a_n * \exp(jb_n)$ = complex array weights at each antenna element, a_n = amplitude weight and b_n = phase weight, $\Psi = kd\sin\theta + \beta$ = phase variation due to time delay between the elements, $k = 2\pi/\lambda$ = wave vector which specifies the variation of the phase as a function of position, θ = incidence angle w.r.t. array normal and β = progressive phase.

SPM is the classical approach for the synthesis of antenna array based on the number and location of interferers. The advantage of this method lies in its placement of minimum gain or nullification of the interference from undesirable directions [12]. Apply Euler's relation $z = x + jy = e^{j\Psi} = e^{j(kd\sin\theta + \beta)}$ and rewrite Eq. (1) in terms of z as:

$$AF(z) = \sum_{n=1}^{N} w_n z^{(n-1)} \tag{2}$$

From the Fundamental Theorem of Algebra, for an N element array, the array factor can be viewed as polynomial of degree $(N-1)$ and can be expressed as a product of $(N-1)$ linear terms which represents $(N-1)$ roots

$$AF(z) = W_N(z - z_1)(z - z_2)(z - z_3)\ldots(z - z_{N-1}) = w_N \prod_{n=1}^{N-1}(z - z_n) \tag{3}$$

where $z_1, z_2, z_3 - - - z_{N-1}$ are the roots of the polynomial. $z_n = \exp(j\Psi_n)$; $\Psi_n = kd\sin\theta_n$ is the phase of the nth root. The complex variable z can be rewritten as: $z = |z|e^{j\Psi} = |z|\angle\psi = 1\angle\psi$.

Instead of varying the weights w_n, appropriate placement of all the roots on the SUC is carried out for certain array pattern.

In Modified form of SPM (MSPM), the direction and beam width can be achieved by fixing 2 roots z_{M1} and z_{M2} for main beam out of $(N-1)$ roots. I interferers from undesired direction can be rejected by placing null in the required pattern and fixing the roots z_{In} on the SUC. Rest of the $(N - I - 3)$ roots can be used to control the other constraint of the array pattern. The value of $z_{Rn} = \exp(j\Psi_{Rn})$; Ψ_{Rn} are the phase of the nth root. The equation can be written as [18]:

$$AF(z) = w_N \prod_{n=1}^{2}(z - z_{Mn}) \prod_{n=3}^{I+2}(z - z_{In}) \prod_{n=I+3}^{N-1}(z - z_{Rn}) \tag{4}$$

3   Design Procedure

Figure 1 shows the position of the user and the interferer which are placed near to the main beam at the peak of the first and second sidelobes. Following are the steps in the design procedure:

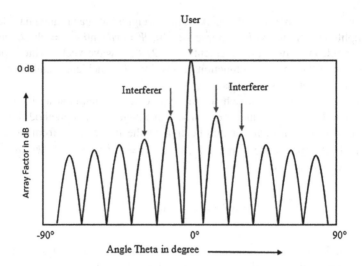

Fig. 1 User and interferer position

Step (1) Specify the size of the array as the number of interferers that can be introduced in the design will be one less than the size of the array.

Step (2) Specify the position of the main beam and calculate the phase of the 2 roots responsible to form first nulls. MSPM fixes the main beam position accurately by placing the phases Ψ_{M1} and Ψ_{M2} on the SUC.

Step (3) Specify the points $\theta_{In}(n = 3, 4, - - I)$ where I is the number of interferers. Calculate the phases of the interferers $\Psi_{In}(n = 3, 4, -I)$. MSPM fixes the interferer position accurately by placing the phases on the SUC.

Step (4) Remaining $(N - 3 - I)$ roots are optimized to search the location of Ψ_{Rn} roots on the SUC. In this work PSO is used to achieve the best array pattern to minimize the effect of interferer near to the main beam.

The following steps shows how PSO is used to search the best Ψ_{Rn} roots [11]

Step (i) Initially a population (*npop*) of 100 particles are taken at random and the number of iterations (*imax*), tuning parameters ($\phi 1 \ and \ \phi 2$) and weights (w) are set. The $(N - 1 - I)$ roots other than the main lobe and interferer are chosen as the variable $\Psi_{Rn}(i)$ in the optimization problem. Initially the lower $\Psi_{Rn}(i, \min)$ and upper $\Psi_{Rn}(i, \max)$ limit of phase are chosen for the design variable.

Step (ii) Initialize the position for the kth variable in the population by $\Psi_{Rn}(i, k) = \Psi_{Rn}(i, \min) + (\Psi_{Rn}(i, \max) - \Psi_{Rn}(i, \min))u(i)$ where $k = 1, 2, ...npop$ and $u(i)$ is the random number generated between 0 and 1. Initialize the velocities of the kth variable as $v(i, k) = 0$.

Step (iii) The appropriate fitness function for suppressing the interferer is the reduction of sidelobe level. For each set of possible phase angles, the

Step (iv)

SLL is evaluated. And objective function is to minimize the SLL. $FF(\Psi_{Rn}(i,k)) = [SLL \,|\, AF(z, \Psi_{Rn}(i,k))|]$.

Compute best fitness of the particle $pbest(i,k) = Fitness$ $function\,(\Psi_{Rn}(i,k))$ and global best $gbest(i,k) = \min\,(pbest(i,k))$. The location of $pbest(k)$ and $gbest$ are given by $p(\Psi_{Rn}ik)$ and $g(i\Psi_{Rn})$.

Step (v)

Update the velocity $v(i+1,k) = w * v(i,k) + \phi1(p(\Psi_{Rn}ik) - \Psi_{Rn}(i,k))$ $u(i) + \phi2(g(i\Psi_{Rn}) - \Psi_{Rn}(i,k))u(i)$ and position $\Psi_{Rn}(i+1,k) = \Psi_{Rn}(i,k) + v(i+1,k)$ for each particle.

Step (vi)

From the new position and velocity, update the fitness $FF(\Psi_{Rn}(i+1,k)) = [SLL \,|\, AF(z, \Psi_{Rn}(i+1,k))|]$.

Step (vii)

If FF $(\Psi_{Rn}(i+1,k)) < pbest(i,k)$ then, $pbest(i+1,k) = FF$ $(\Psi_{Rn}(i+1,k))$. Update $gbest\,(i+1,k) = \min\,(pbest(i+1,k))$.

Step (viii)

The selection continues until maximum number of iterations is reached. If $i < imax$, then increment i and go to step (5) or else the solution $gbest\,(i+1,k)$ is the location of the phase angle where minimum SLL is obtained.

4 Numerical Simulation Results

In order to show the effectiveness of this method, a 16 element linear antenna array with $\lambda/2$ interelement spacing is taken. Synthesis using SPM is applied: this is considered as the reference. Therefore the number of nulls that can be placed using SPM is 15. Two cases are studied for different interferer position. The simulation is done using MATLAB.

In case 1, main beam is at angle $0°$ and 2 interferers are assumed at the peaks of first sidelobe near to the main beam. The two phase angle for the main beam roots z_{M1} and z_{M2} are chosen as $\Psi_{M1} = -21.93°$ and $\Psi_{M2} = 21.93°$ to form a main beam along $0°$ with first null beam width (FNBW) of $14°$. The phase angle for the 2 interferers are $\Psi_{I3} = -34.34°$ and $\Psi_{I4} = 31.25°$ to suppress the gain at the peak of the first sidelobes $\theta_{I3} = -11k\,and\,\theta_{I4} = 10°$. It has been observed that the maximum SLL deteriorates from -11.40 to -4.20 dB. Hence rest of the 11 roots have been considered for optimization to reduce the maximum SLL In case 2, main beam is at angle $0°$ and 4 interferers are assumed at the peaks of first and second sidelobe. The main direction and main beam width (MBW) are considered same as that of case 1. The phase angle for the 4 interferers required to suppress the interference are $\Psi_{I3} = -55.62°$, $\Psi_{I4} = -34.34°$, $\Psi_{I5} = 31.25°$ and $\Psi_{I6} = 55.62°$. The maximum SLL increased from -10.36 to -1.93 dB. Only 9 roots are put in the optimization to reduce the maximum SLL. Figures 2 and 3 shows the synthesized radiation pattern after SPM, MSPM and PSO. Table 1 shows the location of the roots on the SUC and Table 2 shows the computed element complex excitation for the optimized radiation pattern of Figs. 2 and 3.

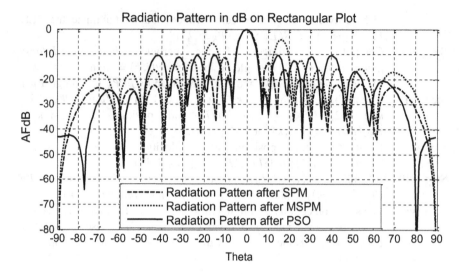

Fig. 2 Best radiation pattern found by PSO for 16 element antenna array with interferer at the peak of first sidelobe

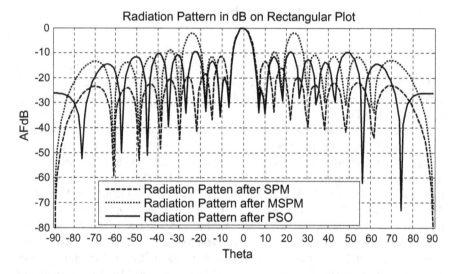

Fig. 3 Best radiation pattern found by PSO for 16 element antenna array with interferer at the peak of first and second sidelobe

Table 3 shows a comparison of SPM, MSPM and PSO. It is observed that as the number of interferers increased from two to four, the maximum SLL deteriorates more after MSPM. An improvement of 6–7 dB in maximum SLL is achieved from the optimized array pattern as compared to the MSPM. This however comes at a

Table 1 Location of roots on the SUC

Root	Radial position	Figure 2		Figure 3	
		Phase θ	Phase Ψ_n	Phase θ	Phase Ψ_n
1	1	−7.00	−21.93	−7.00	−21.93
2	1	7.00	21.93	7.00	21.93
3	1	−11.00	−34.34	−18.00	−55.62
4	1	10.00	31.25	−11.00	−34.34
5	1	−77.17	−175.51	10.00	31.25
6	1	−58.01	−152.67	18.00	55.62
7	1	−49.58	−137.05	−75.11	−173.96
8	1	−36.36	−106.73	−57.78	−152.29
9	1	−27.56	−83.28	−44.68	−126.57
10	1	−19.27	−59.42	−35.23	−103.85
11	1	18.83	58.12	−26.43	−80.13
12	1	26.24	79.59	26.53	80.42
13	1	35.16	103.66	34.48	101.91
14	1	46.62	130.83	43.86	124.72
15	1	57.95	152.57	57.78	152.29

Table 2 Relative complex excitation of each antenna element

Element	Figure 2		Figure 3	
	Amplitude	Phase (°)	Amplitude	Phase (°)
1	1.0000	0.0000	1.0000	0.0000
2	0.8122	0.7724	0.3684	2.5817
3	0.1381	−115.1845	0.0852	−175.9010
4	0.3330	−2.5476	0.4123	5.4397
5	0.5207	−2.5370	0.2662	−3.9045
6	0.4134	−11.4519	0.5865	−10.6377
7	0.6620	−6.5656	0.5541	−3.3381
8	0.8710	−12.2772	0.4985	−3.8959
9	0.6387	−7.7301	0.5874	−3.3358
10	0.8710	−3.1829	0.4985	−2.7757
11	0.6620	−8.8946	0.5541	−3.3335
12	0.4134	−4.0082	0.5865	3.9661
13	0.5207	−12.9231	0.2662	−2.7671
14	0.3330	−12.9125	0.4123	−12.1113
15	0.1381	99.7244	0.0852	169.2294
16	0.8122	−16.2325	0.3684	−9.2533

cost of mean SLL. Compared to SPM, though optimization could not improve the mean SLL, it has been shown to improve the directivity by 3 dB as compared to the initial SPM pattern.

Table 3 Maximum SLL, Mean SLL, FNBW and directivity

Parameters	Value				
	After SPM	Figure 2		Figure 3	
		After MSPM	After PSO	After MSPM	After PSO
FNBW (°)	14.04	14.04	14.04	14.04	14.04
Max. SLL (dB)	−11.40	−4.20	−10.36	−1.93	−9.37
Mean SLL (dB)	−18.61	−13.79	−16.21	−10.82	−13.34
Directivity (dB)	6.97	8.52	9.93	9.08	10.01

5 Conclusion

This paper describes the mathematical design of a linear antenna array using a modified version of Schelkunoff polynomial method with Particle Swarm Optimization. The roots of the main beam are fixed to maintain the position and beamwidth of the user. The roots of the interferer are also kept constant to provide lower value of gain in the undesired direction. The remaining roots are varied to reduce the SLL. It has been observed that as some of the roots are fixed, the complexity of the optimization algorithm reduces as lesser number of variables are used in the optimization. This study shows the performance of null placement and its dependence on maximum and mean SLL, FNBW, directivity, number and position of nulls. The numerical results shows a good performance in terms of directivity and SLL. Simulated result shows successful placement of −30 dB gain towards the interferers at the peak of first and second sidelobes as well as considerable reduction in SLL keeping the beam width constant. Although it is implemented for linear antenna array, but it can be further studied for planar and conformal antenna array. The proposed approach can also be helpful in designing and developing microstrip patch antenna array to change the radiation pattern by changing the amplitude and phase of each of the array element.

References

1. Balanis, C.A.: Antenna Theory: Analysis and Design, 3rd edn. Willy, New York (2005)
2. Haupt, R.L.: Synthesizing low sidelobe quantized amplitude and phase tapers for linear arrays using genetic algorithm. In: International Conference on Electromagnetics in Advanced Application, Italy, pp. 221—224 (1995)
3. Che, X.Q., Bian, L.: Low-side-lobe pattern synthesis of array antennas by genetic algorithm. In: 4th International Conference on Wireless Communications, Networking and Mobile Computing WiCOM, pp. 1–4 (2008)
4. Jeyali Laseetha, T.S., Sukanesh. R.: Synthesis of linear antenna array using genetic algorithm with cost based roulette to maximize side lobe level reduction. WSEAS Trans. Commun. **10**, 385—394 (2011)
5. Steyskal, H., Shore, R., Haupt, R.: Methods for null control and their effects on the radiation pattern. IEEE Trans. Antennas Propag. **34**(3), 404–409 (1986)

6. Ismail, T.H., Dawoud, M.M.: Null steering in phased arrays by controlling the element positions. IEEE Trans. AP **39**, 156–1566 (1991)
7. Guney, K., Akdagli, A.: Null steering of linear antenna arrays using a modified Tabu search algorithm. Progr. Electromagnet. Res. PIER **33**, 167–182 (2001)
8. Khodier, M.M., Christodoulou, C.G.: Linear array geometry synthesis with minimum sidelobe level and null control using particle swarm optimization. IEEE Trans. Antennas Propag. **53**(8), 267–2679 (2005)
9. Goswami, B., Mandal, D.: A genetic algorithm for the level control of nulls and side lobes in linear antenna arrays. J. King Saud University, Comput. Inform. Sci. 117–126 (2013)
10. Banerjee, S., Dwivedi, V.V.: Review of adaptive linear antenna array pattern optimization. Int. J. Electron. Commun. Eng. (IJECE) **2**(1), 25–42 (2013)
11. Arora, R.K.: Optimization: Algorithms and Applications, 1st edn. CRC Press, New York (2015)
12. Schelkunoff, S.A.: A mathematical theory of linear arrays. Bell Syst. Techn. J. **22**, 80—107 (1943)
13. Mismar, M.J., Abu-Al-Nadi, D.I.: Analytical array polynomial method for linear antenna arrays with phase-only control. Int. J. Electron. Commun. AEU **61**, 485—492 (2007)
14. Marcano, D., Duram, F.: Synthesis of antenna arrays using genetic algorithms. IEEE Antennas Propag. Mag. **42**(3), 12–20 (2000)
15. Recioui, A., Azrar, A., Bentarzi. H., Dehmas. M., Chalal. M.: Synthesis of linear arrays with sidelobe level reduction constraint using genetic algorithms. Int. J. Microw Opt Technol. **3**(5) (2008)
16. Canabal, A., Jedlicka. R.P., Pino. A.G.: Multifunctional phased array antenna design for satellite tracking. Elsevier J. **57**(12), 887–00 (2005)
17. Surendra, N., Subhashini, K.R., Manohar, G.L.: Cylindrical antenna array synthesis with minimum side lobe level using PSO technique. In: Conference on Engineering and Systems SCES, pp. 1–6 (2012)
18. Abu-Al-Nadi, D.I., Ismail, T.H., Al-Tous, H., Mismar, M.J.: Design of linear phased array for interference suppression using array polynomial method and particle swarm optimization. Wirel. Pers. Commun. **63**, 501–513 (2012)
19. Banerjee, S., Dwivedi, V.V.: Linear array synthesis using Schelkunoff polynomial method and particle swarm optimization. In: IEEE International Conference on Advances in Computer Engineering and Applications ICACEA, pp. 727–730 (2015)

Author Index

© Springer Science+Business Media Singapore 2016
R.K. Choudhary et al. (eds.), *Advanced Computing and Communication Technologies*,
Advances in Intelligent Systems and Computing 452,
DOI 10.1007/978-981-10-1023-1

Printed in the United States
By Bookmasters